Coal Preparation Technology

Volume 1

Coal Preparation Technology

Volume 1

D. G. Osborne

Manager, Coal Technical Services
PT Kaltim Prima Coal
Jakarta, Indonesia

formerly of

Kilborn Engineering (B.C.) Ltd.
Vancouver

BP Coal Ltd.
London

Graham & Trotman Limited

A member of the Kluwer Academic Publishers Group
LONDON/DORDRECHT/BOSTON

First published in 1988 by

Graham & Trotman Limited
Sterling House
66 Wilton Road
London SW1V 1DE
UK

Graham & Trotman Inc.
101 Philip Drive
Assinippi Park
Norwell, MA 02061
USA

British Library Cataloguing in Publication Data

Osborne, D. G.
  Coal preparation technology.
  1. Coal. Processing
  I. Title
  662.6′2

ISBN 185333 0922 (Vol. 1 and 2 set)
ISBN 086010 995 X (Vol. 1)
ISBN 086010 996 8 (Vol. 2)
LCCCN 88-1314

© D. G. Osborne 1988

This publication is protected by international copyright law. All rights reserved. No part of this publication may be reproduced, stored in a retrieval system, or transmitted in any form or by any means, electronic, mechanical, photocopying, recording or otherwise, without the prior permission of the publishers.

Typeset by Santype International Limited, Salisbury, Wiltshire
Printed in Great Britain at the Alden Press, Oxford

FOREWORD

This book fills a substantial void in the coal-preparation engineering literature by being a text for students, a source book for other engineers wishing to learn coal preparation in depth, a research source book, and a reference book for operating personnel, design engineers, supervisors and administrators. The standard books of the past quarter-century have typically had a much narrower audience in mind, and the available books fill only some of the needs. This new book has the distinct advantage over most of its predecessors of being written by only one author, and has a cohesiveness not generally found in a book written by several authors.

The book is efficiently organized, and the introduction to each chapter clearly delineates the importance of the topic being discussed, the problems to be solved and how the subject at hand forms a part of the whole. The first 12 chapters (volume 1) cover the elements of coal uses, properties, size reduction, sizing, concentration, dewatering and process-evaluation techniques. This part will be invaluable as a basic text for students, as a tutorial for non-specialists, or as a refresher course, as well as being an excellent reference source. Volume 2 introduces a variety of specialized topics which will be particularly useful to advanced students, design engineers and, most especially, to operating personnel.

Dr David Osborne is ideally suited to produce this excellent blend of theory and application, by nature of his extensive, world-wide experience in both industry and the academy, covering every conceivable aspect of coal preparation — teaching, research, development, property and process evaluation, design, start-up, operations, supervision, administration and consulting. He is also well versed in ore-dressing techniques, and skilfully augments the conventional wisdom about them. His blend of knowledge from both coal and ore beneficiation is reflected in both the breadth and depth of the book.

This welcome addition to the coal-preparation literature has been a labour of love by Dr Osborne who has generously shared with us his considerable experience in the art, science and engineering of coal preparation. It should be in the library of every coal-preparation engineer.

F. F. Aplan
Professor of Mineral Processing
The Pennsylvania State University, USA

PREFACE

The personal desire that an up-to-date book be written on the subject of coal-preparation technology probably goes back to my days as a university student, or perhaps even before that to the time when I served my apprenticeship as a draughtsman with a company which specialized in this field.

The personal desire to produce such a book is much more recent and would probably not have progressed to this satisfying stage without the initial encouragement and assistance of my wife and a few close friends and colleagues, in particular Tony Walters who was originally to have been my co-author.

The potential value of a monograph such as this is obvious—one author, one style and in my case, a fairly simple approach in plain English wherever possible. The main disadvantage is that one author cannot possibly know everything about everything and is therefore dependent upon the resources of technical libraries, personal notes and records, and the contributions by others more expert in the numerous specialized fields covered by this book. Responsibility for errors and omissions, however, rests with this author. It is therefore my hope that readers will decide that in this case the advantages of having a single author outweigh the disadvantages.

Volume 1 deals with what might be regarded as the basics of coal preparation. On the other hand, Volume 2 covers most of the more specialized aspects and the so-called new technologies. Notably significant is the rise in importance of the role of chemicals and of computers, and the appreciation of the environmental impact of pollutants in modern coal-preparation technology. This is acknowledged by specific chapters which describe the current status of each of these important topics. Both volumes, are, it is hoped, fully complementary and will be of value as a general text for use by engineers in the field, college and university students, and anyone else in the coal industry primarily interested in coal treatment and utilization.

In writing a book of this type one realizes that whilst the task itself is a formidable one, the task of acknowledging all the people who have in some way or another contributed to its content is truly impossible. To those whose names have not appeared in the Acknowledgements but who know that, at one of the various stages of my career, knowledge has passed from them to me, I take this opportunity to express my sincere gratitude.

Finally, to those who made the most direct contribution, my wife Hazel, and Marie Parker and Daniel Smith of the publishing company, I extend my warmest appreciation.

D. G. Osborne
Jakarta, March 1988

Contents

FOREWORD v
PREFACE vi
ACKNOWLEDGEMENTS xiii
DEDICATION xiv

VOLUME 1

CHAPTER 1 INTRODUCTION 1

1.1 Coal—a fossil fuel 1
1.2 Future role 3
1.3 Rank and quality 4
1.4 Utilization 5
1.5 Coal preparation 12
1.6 Steam-coal cleaning 20
1.7 Coking coal cleaning 22
1.8 Other uses of coal 24

CHAPTER 2 PROPERTIES OF COAL 34

2.1 Introduction 34
2.2 Coal petrology 35
2.3 Coal rank and classification 40
2.4 Molecular structure of coal 49
2.5 Mineral matter in coal 50
2.6 Coal behaviour on heating 54
2.7 Coal utilization and the specific property parameters involved 55

CHAPTER 3 SIZE REDUCTION 58

3.1 Introduction 58
3.2 Mechanism of size reduction 60
3.3 Crushing and comminution testing 62
3.4 Size-reduction equipment 69
3.5 Selection of size reduction equipment 105

CHAPTER 4 SCREENING AND CLASSIFICATION 114

4.1 Introduction 114
4.2 Screening 116

4.3	Screening applications in the preparation plant	127
4.4	Types of screens	132
4.5	Screening efficiency	139
4.6	Classification	140

CHAPTER 5 SIZE ANALYSIS AND FLOAT–SINK TESTING 153

5.1	Introduction	153
5.2	Obtaining the samples required	154
5.3	Size analysis and presentation of data	156
5.4	Float–sink analysis	172
5.5	Equipment and procedures used in float–sink testing	174
5.6	Construction and use of washability curves	179
5.7	The M-curve	186
5.8	Important dos and don'ts	197
5.9	Special acknowledgement	197

CHAPTER 6 DENSE-MEDIUM SEPARATION 199

6.1	Introduction	199
6.2	Basic principles of dense-medium separation	202
6.3	Forms of media commonly used	206
6.4	Properties and methods for testing magnetite	210
6.5	Magnetite preparation and recovery	223
6.6	Types of dense-medium separators	250
6.7	Dense-medium cyclone cleaning of fine coal	276

CHAPTER 7 HYDRAULIC AND PNEUMATIC SEPARATION 288

7.1	Introduction	288
7.2	The movement of solids in fluids	290
7.3	Trough and upwards-current separators	299
7.4	Jigging separators	303
7.5	The Baum jig	313
7.6	Laterally pulsed air jig	334
7.7	Flowing-film separators	347
7.8	Reichert cone separator	365
7.9	Water-only cyclones	367
7.10	Slurry distributors	371
7.11	Pneumatic (or dry) separation	373

CHAPTER 8 EFFICIENCY TESTING OF GRAVITY CONCENTRATORS 387

8.1	Efficiency definitions	387
8.2	Theory of partition (distribution) curves	390
8.3	The partition curve	390
8.4	Other performance test data	395
8.5	Conducting an efficiency test	400
8.6	Relationship between écart probable moyen and grain size	402
8.7	Relationship between écart probable moyen and density of separation	408
8.8	Use of density tracers	410

CONTENTS

CHAPTER 9 FLOTATION, AGGLOMERATION AND SELECTIVE FLOCCULATION — 415

- 9.1 Introduction — 415
- 9.2 Physicochemical process fundamentals — 416
- 9.3 Flotation fundamentals — 427
- 9.4 Testing procedures and factors affecting flotation — 443
- 9.5 Weathering and oxidation effects — 452
- 9.6 Pyrite reduction — 454
- 9.7 Flotation circuits and practice — 457
- 9.8 Agglomeration fundamentals — 460
- 9.9 Testing procedures and factors affecting agglomeration — 466
- 9.10 Selective flocculation fundamentals — 470
- 9.11 Concluding comments — 473

CHAPTER 10 SOLID–LIQUID SEPARATION — 478

- 10.1 Introduction — 478
- 10.2 Principles of sedimentation — 479
- 10.3 Thickening and clarification — 482
- 10.4 Thickening and clarification equipment — 488
- 10.5 Principles of filtration — 502
- 10.6 Vacuum filtration — 504
- 10.7 Pressure filtration — 515
- 10.8 Principles of dewatering screening — 522
- 10.9 Dewatering screens — 524
- 10.10 Principles of centrifugation — 526
- 10.11 Centrifuges — 528

CHAPTER 11 THERMAL DRYING — 543

- 11.1 Introduction — 543
- 11.2 Principles of thermal drying — 545
- 11.3 Thermal dryer types — 547
- 11.4 Convection dryers — 549
- 11.5 Fluidized-bed dryers — 553
- 11.6 Other forms of direct dryer — 558
- 11.7 Conduction or radiant-heat dryers — 560
- 11.8 Novel drying systems — 565
- 11.9 Dryer selection — 567
- 11.10 Environmental factors — 570

CHAPTER 12 CHEMICALS IN COAL PREPARATION — 576

- 12.1 Introduction — 576
- 12.2 Materials handling — 579
- 12.3 Reagents in coal beneficiation — 583
- 12.4 Solid–liquid separation — 592
- 12.5 Environmental control — 595
- 12.6 Case study — 596
- 12.7 The value of chemicals test work — 599

VOLUME 2

CHAPTER 13 INSTRUMENTATION AND CONTROL — 601

- 13.1 Introduction — 601
- 13.2 Basic principles of control — 602
- 13.3 Process control instrumentation — 612
- 13.4 Process control loops — 617
- 13.5 Information for process control — 619
- 13.6 Control valves — 624
- 13.7 Computer control — 626

CHAPTER 14 MATERIALS HANDLING — 637

- 14.1 Introduction — 637
- 14.2 Belt conveyors — 640
- 14.3 Scraper conveyors, bucket elevators and other forms of mechanical conveyors — 658
- 14.4 Feeders — 660
- 14.5 Skip hoisting and coal tipplers — 663
- 14.6 Pipelines and pumps — 664
- 14.7 Pipeline transportation systems — 688
- 14.8 Railways, roads and water — 694
- 14.9 Material storage and reclamation systems — 703

CHAPTER 15 BLENDING AND HOMOGENIZATION — 711

- 15.1 Introduction — 711
- 15.2 Blending and homogenization methodology — 711
- 15.3 Principles of homogenization — 715
- 15.4 Homogenization systems — 720
- 15.5 Blending systems — 732
- 15.6 Environmental considerations — 735
- 15.7 Justifying the blending system — 736

CHAPTER 16 PLANT DESIGN — 740

- 16.1 Introduction — 740
- 16.2 Design criteria and data requirements — 746
- 16.3 Detailed plant design — 771
- 16.4 The role of the computer in plant design — 790
- 16.5 Important dos and don'ts in plant design — 790

CHAPTER 17 TESTING OF COAL FOR THE MARKET — 793

- 17.1 Introduction — 793
- 17.2 National and international standards for categorizing and testing coals — 794
- 17.3 Standard laboratory test methods for coal — 800
- 17.4 Standard laboratory analytical methods for coal — 806
- 17.5 Standard laboratory analytical methods for coal ash — 827
- 17.6 Metallurgical and thermal coal-specific analytical testing — 831
- 17.7 Sales contracts and usual forms of guarantees — 862

CHAPTER 18 ECONOMICS OF COAL PREPARATION — 869

- 18.1 Introduction — 869
- 18.2 Capital costs — 872
- 18.3 Operating costs — 880
- 18.4 Capital budgeting — 884
- 18.5 Optimization of proceeds — 893
- 18.6 Cost trends in coal exportation — 897

CHAPTER 19 WASTE DISPOSAL AND ENVIRONMENTAL FACTORS — 900

- 19.1 Introduction — 900
- 19.2 Waste handling and disposal — 901
- 19.3 Waste disposal site evaluation and development — 909
- 19.4 Chemical and acid generation — 924
- 19.5 Dust generation and control — 927
- 19.6 Acid rain, caused by coal combustion — 929
- 19.7 Spontaneous combustion of coal — 932
- 19.8 Noise emissions and controls — 934
- 19.9 Coal-waste re-treatment and utilization — 945

CHAPTER 20 COMPUTER APPLICATIONS — 958

- 20.1 Introduction — 958
- 20.2 The role of computers in plant design — 961
- 20.3 The role of computers in plant operational control — 978
- 20.4 The role of computers in plant accounting — 997

CHAPTER 21 PLANT PERFORMANCE TESTING — 1000

- 21.1 Introduction — 1000
- 21.2 Plant types — 1001
- 21.3 Methods for efficiency testing — 1012
- 21.4 Comprehensive tests — 1012
- 21.5 Acceptance and performance tests and guarantees — 1015
- 21.6 Performance tests — 1021

CHAPTER 22 COAL SAMPLING — 1026

- 22.1 Introduction — 1026
- 22.2 Fundamentals of sampling — 1027
- 22.3 Sampling procedure — 1039
- 22.4 Mechanized sampling systems — 1066
- 22.5 Sampling slurries and slurry samples — 1090

CHAPTER 23 SPECIALIZED FINE-COAL BENEFICIATION METHODS — 1095

- 23.1 Introduction — 1095
- 23.2 High-gradient magnetic separation — 1096
- 23.3 Electrostatic separation — 1103
- 23.4 True heavy-liquids separation — 1107
- 23.5 Air separation — 1112
- 23.6 Biological leaching — 1112
- 23.7 Chemical comminution and cleaning — 1112

APPENDIX 1	Units, equivalents and conversion factors	1119
APPENDIX 2	Graphical symbols for coal-preparation plant flowsheets	1125
APPENDIX 3	Coal-moisture terms and definitions	1134
GLOSSARY		1139
ACKNOWLEDGEMENTS (continued)		1153
INDEX		1157

ACKNOWLEDGEMENTS

The author wishes to acknowledge the following for their assistance in providing either information, or illustrations, or both for use in this book. In many cases, much more was provided than was requested, and these contributions have extended greatly the book's practical appeal and its usefulness as a general text on the subject. This includes most of the subsection on Effective Screening Area (pp. 120–125), most of Section 4.5, Screening Efficiency (pp. 139–140) and Figs 4.7 to 4.10 (pp. 121–124) in Chapter 4. All this material is reproduced (with slight modifications) from Chapter 8, Screening (pp. 213–241) of the second edition of L. Svarovsky (Ed.), *Solid–Liquid Separation*, 1981, by kind permission of the publishers, Butterworths and Company (Publishers) Limited ©. Acknowledgement is also given to the Canada Centre for Mineral and Energy Technology (CANMET) for figures and textual material which are reproduced with permission of the Minister of Supply and Services, Canada. Much of the textual material in Section 22.2 and many of the figures in Chapter 22 are reproduced by permission of the Standards Council of Canada. It is the author's hope that all such material has been fairly presented and described. However, the reader is recommended always to contact the source for the most up-to-date technical information and for specific aspects of consultancy and advice. A complete list of acknowledgements appears on p. 1153.

The contribution provided by one company, Kilborn Engineering (BC) Limited requires special acknowledgement. The author produced the majority of this book during the course of four years' employment by this company. During this period, very valuable support in the form of typing the manuscript, and correspondence with the publisher and the many potential contributors, was generously provided. Special thanks are expressed to Mr Doug Beaumont, Vice-President and General Manager who steadfastly supported the project, and many other members of his staff in Vancouver, notably the following: Romain Lathioor, Tony Walters, Gerard Laman, Mike Streeter, Rosemary Taylor, Marie Rende, Margot Vandaelle, Jocelyne Torok, Theresa Dawes, Barbara Johnson and Linda Gibb.

Dedication

This book is dedicated to the memory of the following three gentlemen who together provided the foundation stone of my knowledge of coal-preparation technology.

Ernie Sanderson—former Office Manager and Chief Draughtsman, Head Wrightson Colliery Engineering Limited, Manchester

Len Needham—former Managing Director, Head Wrightson Colliery Engineering Limited, Sheffield

Herbie Robinson—former Senior Lecturer in Coal Preparation and Ore Dressing, University of Newcastle upon Tyne

There are many like me who owe them a lot.

Chapter 1

INTRODUCTION

1.1 COAL—A FOSSIL FUEL

Coal is the world's most abundant fossil fuel, as shown in Table 1.1.

World total resources of coal are currently estimated at about 11.75×10^6 tonnes and, in terms of energy content, are over 20 times the total resources of crude oil and over 1.5 times the total resources of crude oil and all other fossil fuels combined. Of more practical significance, however, are the resources of each fuel that are considered to be economically extractable under present technology. These are known as economically recoverable reserves, and Table 1.2 shows that

TABLE 1.1   World Total Resources *in situ* of Fossil Fuels

Fuel	Quantity ($t \times 10^6$, except for natural gas)	Energy content ($J \times 10^{18}$)	(% of world total)
Crude oil	257 500	11 134	2.8
Natural gas	$281\,743 \times 10^9$ m^3	10 509	2.6
Oil shale	2 511 600 (as oil)	108 602	26.9
Tar sand	364 155 (as bitumen)	15 746	3.9
Peat	222 378	1 861	0.5
Coal	11 769 779	255 246	63.3
Total		403 098	100.0

TABLE 1.2   World Economically Recoverable Reserves of Fossil Fuels

Fuel	Quantity ($t \times 10^6$, except for natural gas)	Energy content ($J \times 10^{18}$)	(% of world total)
Crude oil	85 000	3 675	15.2
Natural gas	$81\,147 \times 10^9$ m^3	3 027	12.5
Oil shale	29 540 (as oil)	1 227	5.1
Tar sand	27 312 (as bitumen)	1 181	4.9
Peat	23 417	196	0.8
Coal	690 672	14 823	61.4
Total		24 129	100.0

Owing to rounding, percentages do not add up to exactly 100.

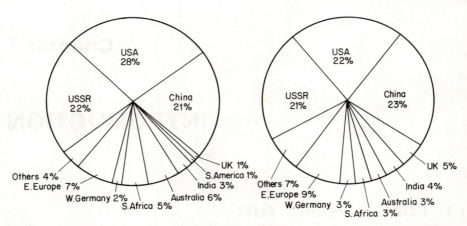

Fig. 1.1 (a) World hard-coal reserves (economically extractable). (b) World hard-coal production.[2]

for coal they amount to about 700×10^9 tonnes, a figure which can obviously be improved upon with further technological developments in extraction methods and utilization. Figure 1.1 shows how world fossil-fuels reserves and production compared in 1980, and Fig. 1.2 shows the location of world coal resources based on current exploration data.

It is now widely agreed that the availability of oil in international trade is likely to diminish over the next two decades. This is to some extent partly reflected by the steady growth in coal exportation which has occurred in the past 20 years. Most forecasts agree that, in spite of recession periods, this will continue to be an increasing trend accompanied by technological development of coal con-

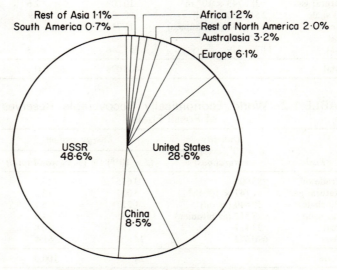

Fig. 1.2 Location of world coal resources.

INTRODUCTION

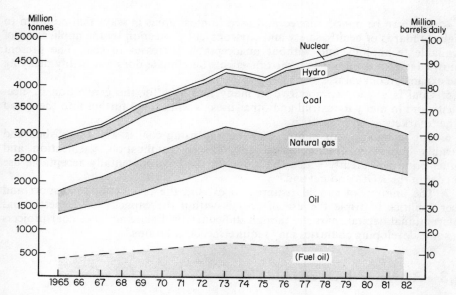

Fig. 1.3 Non-communist world primary energy consumption 1965–82 (courtesy BP Limited).[3]

version into liquid and gaseous fuel products. Figure 1.3 shows the gradual expansion which is occurring in coal consumption in terms of oil equivalent.

Vigorous conservation, the development and rapid implementation of programmes for nuclear power, natural gas, unconventional sources of oil and gas, solar energy and new technologies will not be sufficient to meet the growing energy needs of the world. Coal will, therefore, become increasingly utilized to meet this shortfall.

1.2 FUTURE ROLE

The conclusions of the last major world coal study, known as the WOCOL Study,[1] completed in 1980, provide an important summary to the foregoing:

'1. Coal is capable of supplying a high proportion of future energy needs. It now supplies more than 25% of the world's energy. Economically recoverable reserves are very large—many times those of oil and gas—and capable of meeting increasing demands into the future.

2. Coal will have to supply between one-half and two-thirds of the additional energy needs of the world during the next twenty years. To achieve this goal, world coal production will have to increase 2.5–3 times, and the world trade in thermal coal will have to grow 10–15 times above the 1979 levels.

3. Many individual decisions must be made along the chain from coal producer to consumer to ensure that the required amounts are available when needed. Delays at any point affect the entire chain. This emphasizes the need for prompt and related actions by consumers, producers, governments and other public authorities.

4. Coal can be mined, moved and used in most areas in ways that conform to high standards of health, safety and environment protection by the application of available technology and without unacceptable increases in cost. The present knowledge of possible carbon dioxide effects on climate does not justify delaying the expansion of coal use.

5. Coal is already competitive in many locations for the generation of electricity and in many industrial and other uses. It will extend further into these and other markets as oil prices rise.

6. The technology for mining, moving and using coal is well established and steadily improving. Technological advances in combustion, gasification and liquefaction will greatly widen the scope for the environmentally acceptable use of coal in the 1990s and beyond.

7. The amount of capital required to expand the production, transport and user facilities to triple the use of coal is within the capacity of domestic and international capital markets, though difficulties in financing large coal projects in some developing countries may require special solutions.'

1.3 RANK AND QUALITY

Understanding the differences between various kinds of coal is important in determining their suitability for different uses, and to start at the very beginning with a description of how coal was formed provides the proper basis for such an understanding.

All coals are rocks formed from the altered remains of what was originally luxuriant vegetation. The variety of coals encountered relates to the amount of alteration that the original material has undergone and the conditions that have influenced its formation into rock. The concept of rank is used by coal technologists to denote the degree of alteration and coals of the highest rank are those

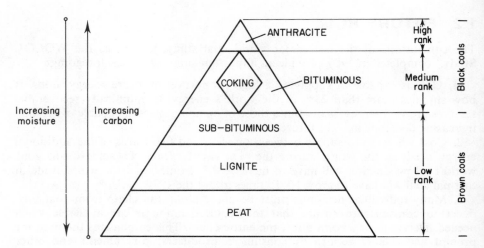

Fig. 1.4 Broad classification of coals by rank.

INTRODUCTION

TABLE 1.3 Trend of Some Coal Properties with Rank

Coal rank	Low rank Peat Lignite	Medium rank Sub-bituminous Bituminous	High rank Anthracite
Moisture content	⇦ HIGH	Decreases with increasing rank	LOW
Volatile content	⇦ HIGH	Decreases with increasing rank	LOW
Carbon content	LOW	Increases with increasing rank	HIGH ⇨
Calorific value	LOW	Increases with increasing rank	HIGH ⇨
Complexity of geological setting	LOW	Increases with increasing rank	HIGH ⇨
Proportion of reserves minable by open cast	⇦ HIGH	Decreases with increasing rank	LOW
Cost of extraction	LOW	Increases with increasing rank	HIGH ⇨
Unit realization price	LOW	Increases with increasing rank	HIGH ⇨

coals which have undergone the greatest change—metamorphosis, as it is termed. Figure 1.4 shows a simplified classification of coals by rank. The terms 'black' and 'brown' coals are commonly used to distinguish high and low ranks.

In a later chapter, a more detailed classification is presented and described, but by arranging the various types of coal in an ascending order of rank, we achieve what is known as the coalification series. In proceeding up this series from low to high rank, the character and composition of the material changes by the expulsion of water and volatiles and consequent increase in carbon content. For combustion, to provide thermal energy, coals of higher rank will have higher calorific values, but coal of the lowest rank still burns. Unfortunately, because their origin usually depends upon them having been deeply buried and geologically disturbed, high-rank coals are usually more difficult to mine and more costly to clean than the lower-rank brown coals. These occur much as they were deposited, in thick water-laden seams close to surface from where they can be easily won by open-cast mining. Table 1.3 describes some of the major variations that occur with changes in rank.

1.4 UTILIZATION

Rank more than anything else determines the category into which a particular coal can be placed with regard to its utilization. Next comes economics of mining, cleaning and handling, and finally other factors such as availability, specific quality, etc. As far as utilization is concerned, coals can be divided into three broad categories:

(a) *Carbonization of coking coals* for metallurgical purposes, i.e. coals that possess the special properties which produce a strong, porous, solid residue when heated in an oven or retort, and are used principally in iron and steel making and for certain non-ferrous reduction applications. See Fig. 1.5.

(b) *Thermal, energy or steam coals*, i.e. coals which are burnt either for direct heating or for the production of hot water and steam for heating and power

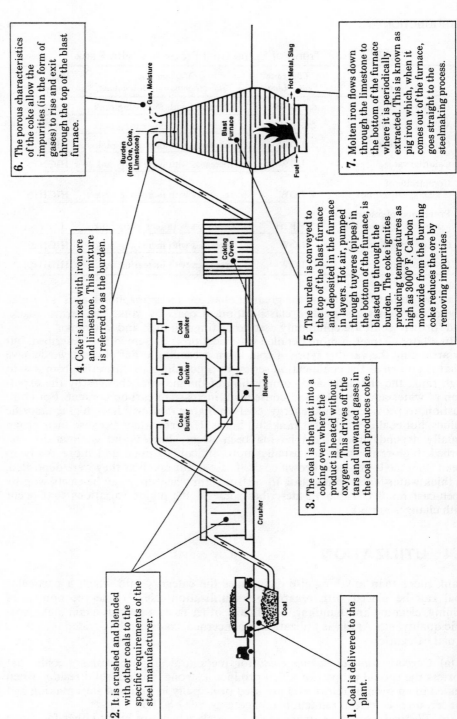

Fig. 1.5 Steelmaking.

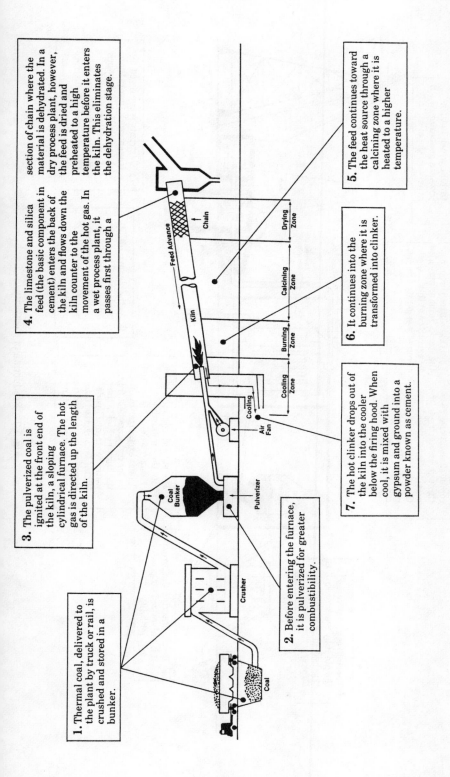

Fig. 1.6 Cement production.

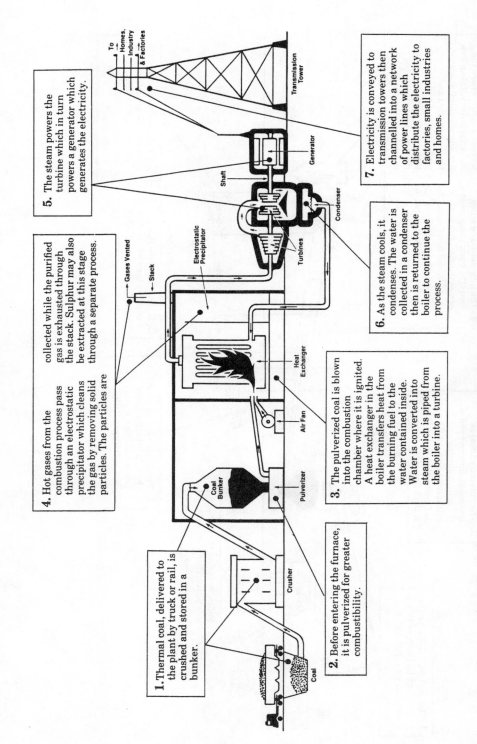

Fig. 1.7 Production of electricity.

generation. One example of direct heating is cement manufacture as shown in Fig. 1.6. The use of coal for electrical power generation is shown in Fig. 1.7.

(c) *Conversion coals*, i.e. coals which are used as feedstock in the production of gaseous and/or liquid fuels derived from coal. At present, this form of utilization represents only a small proportion of total coal utilization.

Currently, about 25% of the world's coal production is used for coke making and almost all the remainder is thermal coal. However, about one-third of the thermal coal produced in the world consists of low-rank coals and these, owing to their relatively low calorific values and high moisture contents, are generally usable only in an area fairly local to where they are mined, and thus do not figure in the world export coal trade. International thermal coal trade is, therefore, generally limited to the so-called hard coals, excluding most of those bituminous coals used for coke making.

Figure 1.8 shows the predicted world coal-trade increase until the year 2000, and Fig. 1.9 shows major international coal movements during the mid-1980s. Major importers will remain the European and the Pacific Rim countries.

Although the initial export thrust was coking coal, demand for thermal coals has steadily increased as the rise in oil prices has rendered coal use more competitive. Conversion coals, however, are unlikely to feature in sea-borne trading

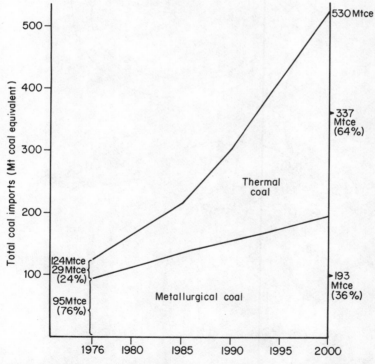

Fig. 1.8 World coal trade (source IEA, Steam Coal, *Prospects to 2000*—November, 1976).

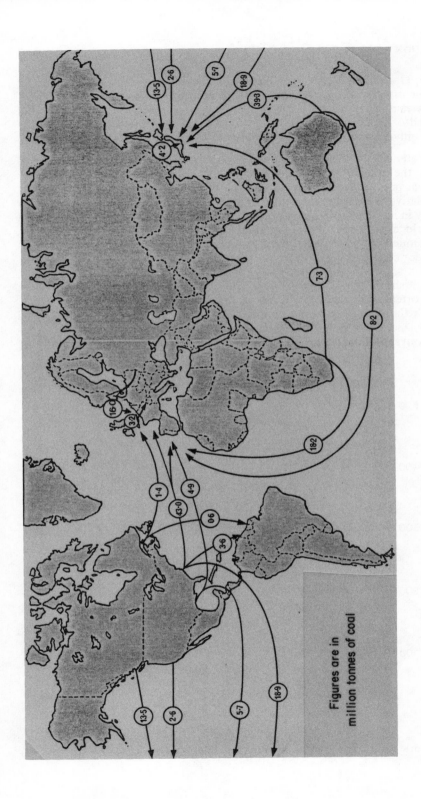

Fig. 1.9 Main coal movements by sea (1982) (courtesy BP Limited).[5]

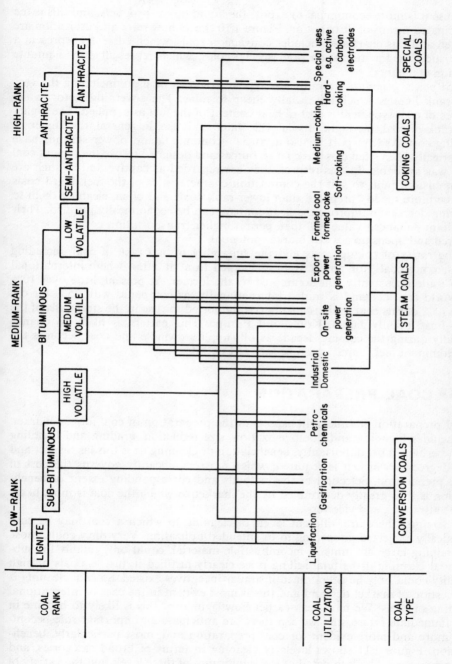

Fig. 1.10 Chart showing relationship between coal rank and utilization.

because it is more economical to export the liquid or gas products, and this is the anticipated means of distribution. Figure 1.10 shows how rank and utilization are related and this chart also includes conversion coals, which currently represent a futuristic area for utilization; and special coals, which represent only a minute portion of the total.

For *power generation*, almost any type of coal can be used, including the very low-rank lignites which are usually cheap to mine. The governing factor in the choice of coal type here is cost of heat content of the fuel as supplied. The higher the rank of coal, the higher its calorific value and also, in general, the higher its mining cost. As transport is also a crucial economic factor, power stations have historically been sited close to coal resources and designed to use the type of coal that was available. In this respect, they are regarded as captive and as such are often custom built to suit the combustion characteristics of the fuel. Hard coals can be more readily prepared than lower-rank coals and often, cleaning them to varying degrees to improve the heat content, can be economically justified. Their resultant enhanced value will then permit higher transportation costs to be considered and opens up a wider market potential.

The rarity of coking coals and the suitability of any type of coal, including non-coking coals, for power stations has led to a situation where international coal trade is currently still dominated by the former. At present, little over 10% of world coal demand is traded internationally, as compared with 75% for oil, and 70% of this coal trade is coking coal imported for use in the steel mills of the Far East, mainly Japan, Korea and Europe. This pattern is likely to change rapidly during the coming decade as oil supplies dwindle and coal becomes the predominant fuel source as can be seen in Fig. 1.8.

1.5 COAL PREPARATION

Coal preparation embraces all aspects of the preparation of coal for the market. It includes blending and homogenization, size reduction, grading and handling and, perhaps most importantly, beneficiation or cleaning. It is this last aspect and the degree to which it is required, which most significantly governs the cost of coal preparation. Selection of the methods and corresponding extent of beneficiation is thus greatly determined by the market to which the coal is most likely to be attractive, and vice versa.

Therefore, there are different levels of cleaning to which a coal may be economically subjected, according to its intended utilization. Very dirty coals, those containing large amounts of incombustible material, could only qualify for substantial cleaning if the final selling price clearly justified it, just as coals of high quality could only be deep-mined if similar incentives existed. So coal cleaning is a question of market demand and this is most evident in the case of international trading of coals. We have seen earlier how significant this is likely to become in the immediate future, and we can therefore anticipate an important role becoming more and more evident for coal preparation and, most particularly, beneficiation. Figure 1.11 shows levels of cleaning in terms of broad categories, and later on we shall look in detail at the significance of these levels and the variety of equipment available for these treatments. But, first, it is necessary to look at the reasons for cleaning and at some current statistics.

INTRODUCTION

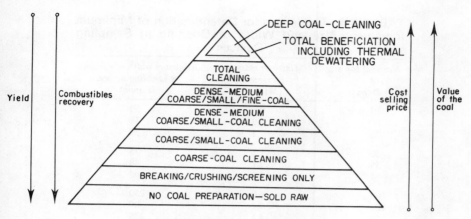

Fig. 1.11 Levels of coal beneficiation.

Only about one-third of the 3.3×10^9 metric tonnes of coal produced every year is at present cleaned. The remainder is simply mined, crushed, graded, possibly deliberately blended to 'iron' of greatly varying quality, and then sold to the consumer. Most cleaned coal is used in coke ovens as what is known as metallurgical coal, for principally ferrous extractive metallurgical applications. Comparatively little coal cleaning is carried out in the large utility steam-coal market, although with the recent sharp rise in international demand and therefore the exportation of steam coal, the proportion of this type of product that is cleaned is increasing rapidly. It is now becoming an increasingly accepted and economically justifiable practice to clean such coal, and several run-of-mine coals which have hitherto been mined and sold raw are now being prepared by cleaning for export markets.

Low-sulphur, underground-mined metallurgical coal, the backbone of the coal-cleaning industry, is mined in Europe, Australia, the southern Appalachians of the United States, South Africa and Canada. Low-sulphur but relatively high ash-content, surface-mined steam-coal for world markets is being produced in Australia, South Africa, Canada, Colombia and the United States in increasingly large tonnages, with single mines producing in excess of 5×10^6 tonnes per year in some cases. If these coals were used domestically, coal cleaning might not be considered economical and probably would therefore not be justifiable. For world markets, however, coal cleaning would improve market potential of the coal and reduce transportation costs. It might also improve the 'environmental' properties of the coal where, in the case of high-sulphur coals containing predominantly pyritic sulphur, coal cleaning can effectively cut down sulphur content to an economically acceptable level. Each of these aspects will be discussed in some detail in later chapters.

Quality and cost of treated coal are highly dependent on the specific coal and the market for which it is aimed. In order to ascertain the amount of cleaning required, run-of-mine coal must be subjected to certain minimum qualitative and quantitative analyses, from which conditions of cleaning and information about ultimate quality can be obtained. Obtaining a representative sample of run-of-mine coal and carrying out such test work is a lot more easily said than done,

TABLE 1.4 Guidelines for Determination of Minimum Increment Size and Width of Opening in Sampling Devices

Nominal top size of coal, D (mm)[a]	Minimum mass per increment, (M) (kg)[b]	Minimum width of opening in sampling device, W (mm)
10 or less	0.6	30
15	0.9	38
20	1.2	50
30	1.8	75
35	2.1	88
65	3.9	163
100	6.0	250
120	7.2	300
150	14.1	375
200	33.3	500

[a] Except in the case of $D = 10$, $W = 2.5\,D$
[b] Based on the appropriate one of the following formulae:
   1. for coals of nominal top size of <120 mm, $M = 0.06\,D$
   2. for coals of nominal top size $\geqslant 120$ mm, $M = 7.2\,(D/120)^3$

and to illustrate this point, Table 1.4 shows minimum amounts of sample which are required to carry out common test and analytical work. Besides the need to satisfy quantity requirements, the representativeness of the sample with relation to future run-of-mine coal is another difficult criterion to achieve. It is usually achieved by a combination of several data sources that are correlatable. Prospective coal reserves are drilled by open-hole and coring methods to a predetermined grid in order to obtain physical and chemical data. A relatively large bulk sample, usually of several tonnes, would be required in order to obtain a coal sample of size-consist similar to that which would eventually be mined. This would then be tested to provide cleaning data, with corresponding quality determined for various fractions, following preparation of the sample to conform to the treatment size-ranges required. This aspect is also covered later on because it makes a vitally important contribution to the profits of the eventual total mining operations. Suffice to say, the analyses of the raw coal and prospective products for such parameters as ash, heat, volatile and sulphur contents, as well as elemental components of carbon, hydrogen, oxygen and nitrogen, will provide a good insight into market suitability. Then, other special parameters relating to specific applications such as swelling index, fluidity and dilatation of coking coals and Hardgrove grindability, ash fusibility and ash analyses for energy coals will further define the overall quality of the coal.

Float–sink tests provide washability data (see Fig. 1.12) vital for assessing the probable performance of separation equipment, which depends upon density differences; and froth-flotation tests simulating the separation characteristics of the density separator provide valuable performance data for fine-coal treatment.

All separation equipment has well-established empirically based performance characteristics which allow its performance to be determined once the washability data have been obtained. This predictability lends itself to computerization and facilitates fairly rapid simulation testing. These predictions can then

INTRODUCTION

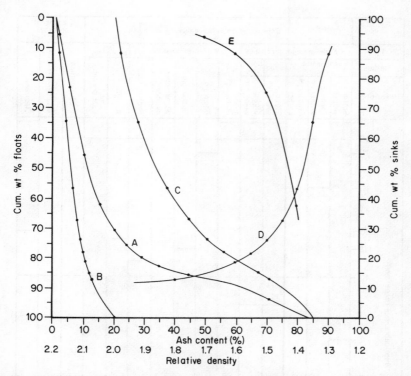

Fig. 1.12 Washability curves: A, primary curve; B, clean-coal curve; C, discard curve; D, relative density–yield curve; E, ±0.1 r.d. distribution curve.

be compared with the performance of other known installations of a similar type, thereby providing for confident selection of the correct cleaning method and accurate estimation of plant performance. Figure 1.13 shows the relationship between the various types of coal-cleaning equipment and the size range of coal for which they are normally applied.

One of the disadvantages of coal cleaning by wet processes lies with the fact that the ash removed by cleaning is to some extent replaced by water. Just as cleaning becomes more difficult with decreasing particle size, so too does moisture removal from the product. In some cases, fine coal sized below 500 μm may contribute a higher level of moisture than incombustible ash,[3] and any additional beneficiation by decreasing either ash or moisture may still not render fines-addition economical. In such cases, the fines are usually rejected as uneconomical, although some countries, keen to conserve resources, are becoming increasingly anxious to eliminate such wastage.

So we shall see that washability data provide a yield value which is akin to almost perfect cleaning conditions, and thus it is often called the theoretical yield. Actual yield of saleable coal, always poorer, will depend upon the degree of efficiency of the total plant when any form of cleaning is used.

First, coal cleaning is a continuous, dynamic process, not batch like float–sink tests, and cannot clean perfectly. In practice, this means that separator per-

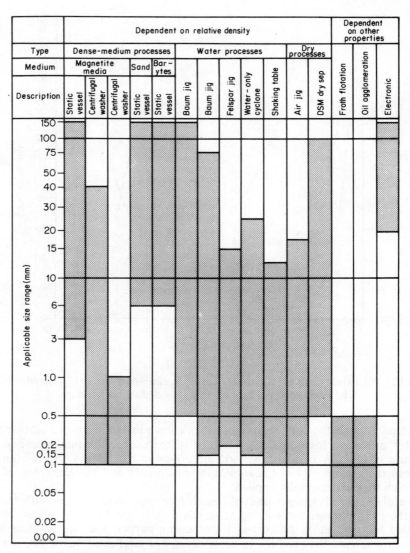

Fig. 1.13 Coal-cleaning methods and corresponding size range.

formance is affected by near-density material which could report to either stream, depending upon the accuracy of the separator. Hence, some misplacement, i.e. coal in discard and vice versa, will always occur.

Secondly, cleaning becomes more difficult with decreasing grain size of the feed. This becomes increasingly acute when fines of minus 500 μm are treated, due to the combined effects of poorer efficiency and increasing moisture content.

Thirdly, ultrafine coal, even when beneficiated, may still be lost during the final dewatering processes.

TABLE 1.5 Coal-cleaning Methods, 1980 (Percentage of Total)

	Soviet Union[a]	United States	China[a]	United Kingdom	Poland	West Germany	Australia	South Africa	France	Japan	Canada[a]	India
Water medium												
jigs	50	49	major	61	51	70	18	23	29	62	3	31
tables	—	11	—	1	—	—	4	—	—	—	1	—
cyclones	4	—	—	1	—	1	10	—	—	—	8	2
other	—	2	—	1	4	—	2	—	—	1	3	—
Dense medium												
vessels	25	31	minor	22	33	21	16	62	54	10	4	44
cyclones	4	—	—	6	5	1	40	12	16	16	12	22
Froth flotation	15	4	minor	7	6	7	10	3	9	11	12	1
Pneumatic	2	3	—	1	1	—	—	—	—	—	—	—

[a] Reliable data are not available
Source: SRI International

As we have already discussed, efficient and effective coal cleaning embraces the use of a number of coal-preparation operations including handling, storage, blending, sizing, beneficiation, comminution and dewatering, sometimes including thermal drying. For each of these operations, there exists a range of equipment which is selected partly out of proven suitability, partly out of personal preference or confidence from previous experience, and partly, of course, for relative cost. For coarse-coal cleaning, Fig. 1.13 shows that Baum jigs and the more efficient dense-medium vessels are used. Dense-medium cyclones, water-only cyclones, tables and felspar Baum jigs are used for medium-size coal cleaning. Froth flotation is the most commonly applied method for fine-coal cleaning.

Table 1.5 outlines the types of coal-cleaning methods used in 1980 by the major cleaned-coal-producing countries. Baum jigs followed by dense-medium vessels and froth flotation are the most widely used methods of coal cleaning. In Australia, South Africa, France and India, dirtier coals with substantial amounts

TABLE 1.6 World Production of Coal and Cleaned Coal for 1977 (tonnes × 10^6 per Year)

	Hard coals	Lignite brown coals	Total coal production	Approximate cleaned-coal production[a]
USSR	587.0	165.1	722.1	300
United States	602.3	13.0	615.3	240
Canada	22.9	5.5	28.4	14
Mexico	2.3	—	2.3	2
South America	9.8	—	9.8	4
Asia				
China	490.0	—	490.0	100
India	99.0	4.0	103.0	15
Japan	18.2	—	18.2	17
North Korea	45.1	—	45.1	NA
South Korea	17.0	—	17.0	NA
Other	7.4	—	7.4	NA
	676.7	4.0	680.7	160
Europe				
West Germany	84.5	173.0	267.5	70
East Germany	0.4	249.0	249.4	—
Poland	186.0	40.7	226.7	80
Czechoslovakia	28.0	43.0	121.0	15
France	21.3	3.2	24.5	19
Spain	11.9	4.5	16.4	NA
United Kingdom	120.8	—	120.8	85
Other	17.4	54.5	71.8	6
	470.3	567.8	1038.1	278
Africa				
South Africa	85.4	—	85.4	37
Other	4.9	—	4.9	4
	90.3	—	90.3	41
Oceania				
Australia	72.4	29.3	101.7	49
New Zealand	2.4	0.2	2.6	2
	74.8	29.5	104.3	51
Total world	2 506	785	3 291	1 090

[a] Mostly hard coal

of near-density material have led to a more widespread adoption of the more accurate dense-medium cleaning methods, whereas, very clean as-mined coals in the United States, Britain and West Germany have resulted in greater popularity of the far less accurate jigs, tables and water-only cyclones as cleaning equipment. Dense-medium methods are most definitely on the upsurgence on a world-wide basis, even in countries which have always traditionally cleaned by the cheapest methods. This is mainly due to the following reasons:

(a) depletion of higher quality coal seams;
(b) increased mechanization in mining, which increases impurities in run-of-mine coal;
(c) high cost of transportation, which makes it uneconomical to transport incombustible material;
(d) market demand for higher-quality coal;
(e) environmental requirements in regard to minimizing pollution;
(f) obsolescence of existing plant and processes and the increasing incentives provided by the foregoing;
(g) higher mining costs which make it imperative to improve coal-washing techniques for optimum yields; and
(h) significant improvement in plant-control equipment and operating capacities.

Of the one-third of the total of world coal production currently cleaned, most is used to produce coke for the iron and steel industry and the remainder is almost entirely consumed in large utility boilers to generate electricity. Table 1.6 summarizes the 1982 world total coal production and estimated clean-coal production by country. Tables 1.7 and 1.8 summarize the anticipated utilization of total coal and cleaned coal for the world and the United States respectively.

TABLE 1.7 Approximate World Coal Utilization in 1982 (tonnes × 10^6 per year)

Sector	All coal	Cleaned coal
Utility steam coal	2 065	240
Industrial metallurgical coal	525	510
Industrial steam coal	500	220
Residential/commercial heating	230	110
	3 320	1 080

TABLE 1.8 US Coal Utilization in 1982 (tonnes × 10^6 per year)

Sector	All coal	Cleaned coal
Utility steam coal[a]	441	70
Industrial metallurgical coal[b]	110	110
Other industrial use (steam coal + direct use)	57	53
Residential/commercial heating	7	7
	615	240

[a] Includes 10 × 10^6 t of export steam coal to Canada
[b] Includes 40 × 10^6 t of exported metallurgical coal

1.6 STEAM-COAL CLEANING

Almost any raw coal can be combusted in a boiler to generate steam. All that is required is a knowledge of the combustion characteristics of the coal or likely range of qualities of feed coals, and furnace conditions can be designed to suit. Brown coals, containing 60–70% water with a heating (calorific) value of only 7–8 MJ/kg, are commonly burned in Germany. Sub-bituminous coals containing up to 30% ash are burned in South Africa, Canada and the United States, and increasing amounts of these low-rank coals will undoubtedly be utilized in the near future. Efficient combustion is achieved with custom-built, captive power stations capable of large generating capacities requiring large amounts of coal of up to 10×10^6 tonnes per year. The major attractiveness of such coal is the low mining cost of extracting shallow, thick deposits. Very few sub-bituminous coals are cleaned at present, and it is unlikely that the amount will perceptibly increase. 'Cleaning' as required can usually be ensured during mining operations and in subsequent treatment, i.e. coal breakers and blending stockpiles. However, it is hoped that an efficient process will be developed in the future which will permit moisture reduction of low-rank coals to well below the normal inherent level.

The most common method of burning coal is in pulverized-coal-fired furnaces, which are used by most utilities and industrial boilers that have more than 100 000 kg/h of steam capacity. The coal is pulverized to 70% minus 200 mesh (0.07 mm) usually, and fed into the boiler by being entrained with a portion of the combustion air. Although most coals can be burned in a pulverized-coal boiler, analysis of the specific coal(s) to be burned determines the exact design and optimum performance. Only 10–13% of the utility coal currently burned is cleaned to any degree beforehand, as shown in Tables 1.7 and 1.8. Even when it is upgraded, only partial cleaning is usually carried out so as to reduce, and render more consistent, the ash level of the fuels as supplied to the furnace.

Ash reduction improves the operability of boilers and electrostatic precipitators. It also becomes important when the coal has high shipping cost or high pyritic sulphur content. A common practice in western Europe is to blend cleaned coal with raw coal so as to control the ash content of the coal supplied to the power station on a continuous basis.

As is the case with metallurgical coal, there are a number of inherent physical differences among thermal coals and these affect the operating and capital costs of conventional coal-fired power plants.

In order to generate electricity using coal, a coal-fired power plant must be designed to achieve steam flow, pressure and temperature sufficient to drive electricity-producing turbines. Plant design is directly affected by coal quality. Coals with difficult quality characteristics, such as the tendency to form mineral deposits on furnace walls and other surfaces in the combustion zone of the boiler (slagging) or the tendency to form high-temperature bonded deposits on the superheater and reheater tubes in the convective section of the boiler (fouling) can be compensated for by building larger units. To do so, however, is very expensive. Thus, the slagging and fouling tendencies of prospective power-plant coals are a major cost consideration for any power plant.

Because it is not practical to design a boiler to accommodate every possible problem, coal-quality characteristics can continue to affect greatly both the operating and the capital costs of power plants. Difficult coals, those that slag and foul or are corrosive and abrasive or that create temperature imbalance in the boiler, can increase fuel costs or act to accelerate the physical depreciation of plant and equipment, thus increasing unit capital and operating costs.

The combustion of coal in a boiler is very rapid, taking two seconds or less. But during this short period, complex chemical reactions occur, primarily between the various mineral components of coal, collectively known as the ash content. How the coal reacts affects the power-plant performance expressed in terms of boiler efficiency and boiler availability. Boiler efficiency is the amount of heat consumed to produce a given quantity of steam, while boiler availability is simply the time the boiler is available to generate steam when the boiler's operation is not reduced due to forced outage or partial outage.

The complex chemical reactions are extremely important, because they affect the formation of chemical compounds that promote slagging, fouling, corrosion and, possibly, temperature imbalance. These are commonly the most troublesome problems caused by coal in a power plant. For example, the most important chemical reactions are those affecting sodium and potassium, the alkali metal salts on superheater tubes. These salts are extremely corrosive and are the principal cause of damage to power-plant superheater tubes.

Such negative impacts can be compensated for in a number of ways. The coal can be washed to reduce the ash quantity or to reduce certain detrimental ash characteristics such as sulphur content; it can be blended with other coals, or have the quality variability reduced; or power-plant flue gases can be cleaned further, etc. The desirability of doing such things, however, depends on an analysis of the economics of how a change in coal quality will affect power-plant performance.

Recent studies conducted by the International Energy Agency[10] have shown, for example, the following economic impact of ash level on power-plant efficiency and availability: a 1% increase in ash (generally after passing the 10% ash level) results in a 1.2–1.5% decrease in boiler availability and a decrease of 0.3% in boiler efficiency. At a capital cost of US$1000/kW, the capital-cost absorption penalty is US$0.95/t and US$0.67/t respectively per 1% ash-content increase, disregarding all but boiler efficiency and availability factors. And this is only one side of the coin. The impact of ash chemistry, as opposed to ash quantity, is recognized as being even more significant.

There is also increasing evidence of the importance of consistency of coal quality and the need to have delivered coal quality that remains within the design parameters of the boiler. A 1983 study of utilities in New South Wales, Australia, reports that a high forced outage rate of 7.2% over the preceding three-year period was due primarily to ash content exceeding design levels and to inconsistency in coal quality. The reduced output resulted in increased fixed capital charges, operating and maintenance costs equivalent to US$6.80/t of coal consumed and a production cost increase due to increased boiler wear of US$4.20/t of coal consumed. The solution to this problem was to install a coal washery to reduce ash content and introduce more consistency in coal quality.

Recent technological developments, such as fluidized-bed combustion and gasification, will enable the use of poorer-quality thermal coals. The qualitative characteristics that are currently of great significance to consumers of thermal coals, such as heat content, ash content, ash composition (slagging and fouling tendencies), volatility (an indication of the ability of the coal to burn efficiently in a boiler), sulphur levels and nitrogen levels (which both contribute to pollution or require expensive stack emission-control equipment), and moisture content, will not be as important. This will result in an increased competitiveness among coals as the range of acceptable coals widens to include coals not used in conventional thermal boilers. Acceptable coals will likely include sub-bituminous coals (lower heat contents due to high inherent moisture contents) and lignites (very high moisture contents). These coals are available in large quantities world-wide, as are poorer, higher-sulphur, higher-nitrogen, heavy-slagging or -fouling bituminous coals.

Some thermal coal prices are also affected by the physical size of the coal provided. Anthracite, in particular, is standardized into 10 or more distinct sizes based upon the coal's inherent physical make-up. Price differences between sizes stem, in part, from the fact that the larger sizes are generally more readily handleable and have higher heat content, but size requirements for particular industrial uses also affect the price levels.

1.7 COKING-COAL CLEANING

Metallurgical coking-coal is an important component in iron and steel production. It is converted into coke for use in blast furnaces as the heat source and reducing agent for converting iron ore into pig iron. Modern by-product coke ovens are batch operation, slot-type ceramic ovens that carbonize or thermally crack the coking coal. A good coal produces large, strong pieces of coke, the two qualities that are required to support the enormous weight of iron ore in the blast furnace.

Only certain high-rank bituminous coals are classified as metallurgical cokingcoals. Generally, only low-sulphur coals are acceptable which further reduces the range available for this special application. In certain cases, medium-sulphur coals are considered for coking because they can be beneficiated by extensive coal cleaning. This, in certain instances, can result in the production of two saleable products: a low-ash coking-coal cleaned to acceptable coking properties, including low sulphur; and a medium-ash steam-coal obtained by cleaning to a higher separating density. Three product separations, i.e. low ash, medium ash and discards, are becoming increasingly common where coking-coals are to be obtained from dirtier run-of-mine coals. Twenty years ago, most coking-coals were of such high quality as mined that only minimal cleaning was required; and in Britain, West Germany and the United States, large amounts of high-quality coking-coal with ash content well below 5% were sent to the coke ovens without cleaning.

Numerous tests have been developed to analyse potential coking-coals, some of which are traditional, such as free-swelling index, Gray–King coke number, coke strength and hardness and petrographic analysis. In addition to these are a number of tests that have become more recently adopted internationally with the onset of blending. The Japanese steel mills, probably more than others, have in recent years exhaustively researched coal blending, and the parameters of fluidity, dilatation, reflectance and coke stability have as a result become important in ascertaining the suitability of blends. Figure 1.14 shows a diagram as produced by Miyazu[9] which is used by the Japanese steel mills and others to determine the possible contribution which a new coal might make to a coking-coal blend. These

INTRODUCTION

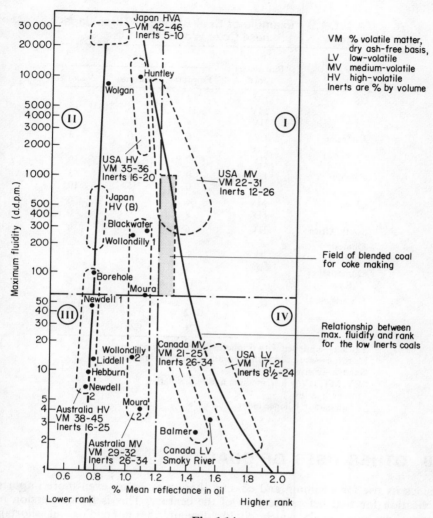

Fig. 1.14

days, not many single-source coals exist which can be regarded as adequate by themselves, and those that do obviously command a very high selling price. The shaded area in the centre of the diagram is seen as representing these rare coking coals. Most coke makers, however, would insist that the proof of the pudding was in the coking, and pilot-scale oven tests are regarded as essential in finally obtaining important coking performance data. Table 1.9 gives an example of coal blends used in three Japanese steel mills employing local and imported coals from many sources.[7]

TABLE 1.9 Example of Coal Blends in Three Japanese Coke Works[7]

	Bituminous coals[a]	Works		
		Yawata	Muroran	Nagoya
US	LV	2	2	4
	MV	11	5	10
	HV	9	6	24
Australia	LV	16	0	8
	MV	24	0	15
	HV	0	0	16
Canada	LV	0	22	13
	MV	0	15	10
China	MV	4	0	0
	HV	9	0	0
South Africa	HV	6	13	0
Domestic (Hokkaido)		0	32	0
Domestic (Kyushu)		17	0	0
US—West		2	5	0
SI[b]		3.92	3.66	4.10
CBI[c]		1.21	1.26	1.21

Yawata—equipped with briquette-blend coking process
Muroran—equipped with preheated coal-charging process
Nagoya—conventional process
[a] LV, MV, HV = low, medium and high volatile
[b] Strength index
[c] Composition balance index

1.8 OTHER USES OF COAL

Besides its use for steaming and coking, there may be an even more important application for coal before the turn of the century. This is for conversion into gaseous or liquid fuels which could alleviate the impending oil shortage.[8] Although countries like Australia, the United States, Canada and South Africa possess huge deposits of coal of various grades and types, no thorough or convincing studies have so far conclusively pointed to any one specific rank which might be the most suitable for conversion. To be economical, such an application would have to consider the ongoing demand for coal for combustion and coking purposes. Several methods of converting the hydrocarbons in coal into liquid or gaseous form are known. All involve some form of hydrogenation or raising the ratio of available hydrogen atoms to carbon atoms in the coal. This is necessary because the greatest proportion of the heat energy of a fuel is derived from its hydrogen content. In combustion, the hydrogen–oxygen combination releases more heat than the carbon–oxygen combination. In fact, the hydrogen produces 4.2 times as much energy per unit mass as the carbon. Thus, coal with a hydrogen–carbon mass ratio of about 8:100 has a heat value of 29 MJ/kg, while

INTRODUCTION

oil with a hydrogen–carbon ratio of 17:100 has a heat value of 43 MJ/kg. Any process which adds hydrogen to coal is clearly useful, if the result is something resembling crude oil.

Attempts to obtain liquid or gaseous fuels from coal have been going on for nearly 200 years. In the early 19th century, some towns in England began to light their streets with 'town gas' produced by the heating of coal. Most of the coal was turned into coke by the process, but up to 10% was driven off as a viscous, black tarry liquid. From this early start, various uses were found for these coal tars (see Fig. 1.15). They were burned in furnaces as fuels, used as feedstock for the making of some synthetics, and eventually, after hydrogenation, found suitable as fuel in internal combustion engines.

The 1920s found intensive research under way, particularly in Germany and Britain, into the large-scale conversion of coal into oil. The Germans were extremely conscious of their lack of natural oil in a world moving rapidly into the era of mechanized warfare, and devoted a great deal of attention to developing the technology. They were successful to the degree that they were able to fight World War II largely on synthetic oil, the production of which reached 3.5×10^6 tonnes in 1944. Of course, the major consideration was survival rather than cost or efficiency, and any attempt today to adopt the war-time German plants would involve prohibitively high capital and production costs. There are at present, however, a number of technically feasible, though not necessarily economical, routes for the conversion of coal to liquid fuels and these are shown in Fig. 1.16.

The first method, pyrolysis, which may ultimately prove to be the most widely applied, involves the heating of coal in the absence of oxygen to produce gas, a solid carbon residue called char, and tar. This process is similar in many ways to the 19th century town gas processing, but new developments involving the rapid heating of finely ground coal to about 500 °C can now produce higher yields of up to 30% of the tarry liquids suitable for use as feedstock in an oil refinery. Bituminous coals are regarded as most suitable for the pyrolysis route, based mainly upon economics, whereby the char is burned directly as fuel in a steam-powered electricity-generating station. As an example, a single 2000-MW power-station would consume all the char from a pyrolysis plant capable of producing about 30 000 barrels of oil per day from coal. With the current world capacity for the refining of crude oil and for producing electricity by steam-generation dependent upon coal as a fuel source, pyrolysis seems a logical step in the transition from the current conventional routes to the more ideal technological solutions of the future.[4]

Another conversion route begins with the production of solvent-refined coal. In this process, the coal is first pulverized then mixed with solvents at a pressure of about 80 bar. The solvents dissolve the combustible hydrocarbons, leaving behind the mineral content of the coal which can amount to as much as 70%. The dissolved hydrocarbons then have to be hydrogenated to produce syncrude for refining.

A third route involves direct hydrogenation of the coal, with the solids being separated out later. Such a process was first demonstrated by Bergius in Germany as far back as 1924. In the Bergius process, a fine suspension of coal in oil is subjected to hydrogen under pressure of up to 250 bar, at temperatures of 400–500 °C. By distillation, a range of liquid fuels including petroleum, kerosene, diesel oil, fuel oil, solvents and gas were obtained, and this was the method devel-

Fig. 1.15 The coal tree (courtesy Coal Association of Canada).[6]

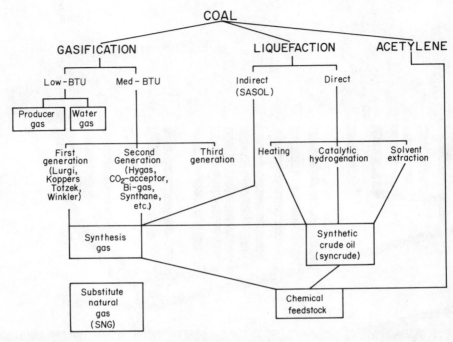

Fig. 1.16

oped through the 1930s which was mainly used by Germany through World War II to maintain vital oil supplies. After 1945, syncrude produced by the Bergius method could not compete economically with the flood of natural crude from the Middle East and its development came to an end. It has been followed by numerous attempts at revival that have taken place in the last decade when it has become increasingly clear that the once cheap Middle East supply would become increasingly costly, more vulnerable and eventually exhausted. New, yet uneconomic, processes have been researched with increasing political justification, but the enormous expense of developing them to commercial scale has eliminated all but the most technologically viable. Of those remaining, the H-Coal, Solvent-refined Coal SRC II, and Exxon Donor Solvent processes are perhaps the most promising and these have reached pilot- or demonstration-scale plant operations in the United States. Simplified flowsheets for each are given in Fig. 1.17. Details can be found in numerous texts.[1, 8]

A fourth route, indirect hydrogenation, has proved a more commercially viable prospect and is receiving attention in a number of countries. This approach involves the complete gasification of coal, and the hydrogenation of some of the gas to yield syncrude. Of the processes available for gasification, only one has featured in a commercial plant, i.e. Lurgi fixed-bed process, but others are available that have found application as commercial low–medium calorific value gas producers and these include the Koppers–Totzek entrained-bed process and the Winkler fluidized-bed process. The hydrogenation process currently in use was first demonstrated by Sabatier in 1905 under pressurized conditions, but was

INTRODUCTION

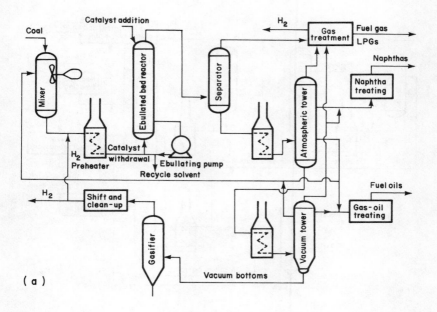

(a)

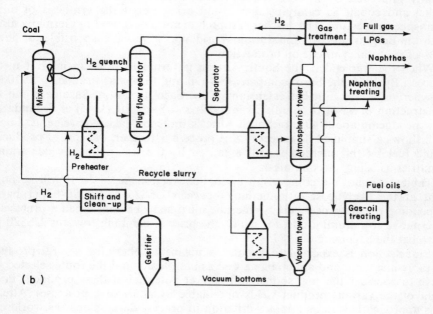

(b)

Fig. 1.17 (a) H-Coal; (b) SRC II; (c) EDS simplified flowsheets.

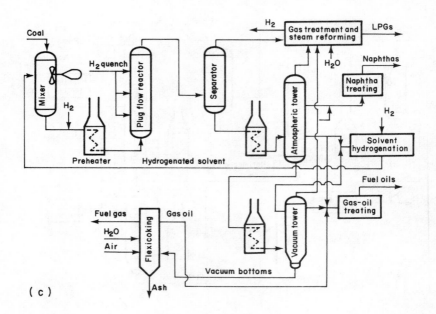

(c)

greatly improved by Fischer and Tropsch in 1925 who found that, by using nickel and cobalt as catalysts, they could bring about the synthesis of oil at atmospheric pressure. The Fischer–Tropsch method was used in Germany along with the Bergius process, and during World War II it was responsible for about 15% of the total synthetic oil production.

When the war ended, the South African government took an interest in the process. By 1955, the Sasol company (SA Oil and Gas) was producing 250 000 t of syncrude a year. Further extensions of the Sasol I plant at Sasolberg and the construction of Sasol II completed in 1982, and Sasol III (1985) at Secunda are now producing about 40% of South Africa's transport-fuel requirements. Figure 1.18 shows a simplified flowsheet of this process. However, the cost of producing liquid fuel by this method is still regarded by the less politically constrained countries as being uneconomical.

Table 1.10 shows a comparison of coal-conversion methods. It will be noted that at the likely cost of something between US$25 and US$40 per barrel, depending on the process used, syncrude must soon be considered a practicable alternative to natural crude, for which the price is climbing towards US$20 per barrel at the beginning of 1987.

Liquefaction is expected to become a means of obtaining several products, notably methanol and ammonia as well as petroleum, and the route selected will have to consider the coal source and market potential of these products, in the light of the various product yields obtainable by alternative processes. Already this approach is seen as a major criterion in process development as world coal resources are constantly reviewed and vast new exploration data become available. This constant reviewing process is necessary in order to establish strategic information regarding the availability of suitable coal reserves as future conversion feedstock. At the end of the day, it will probably be those coal deposits

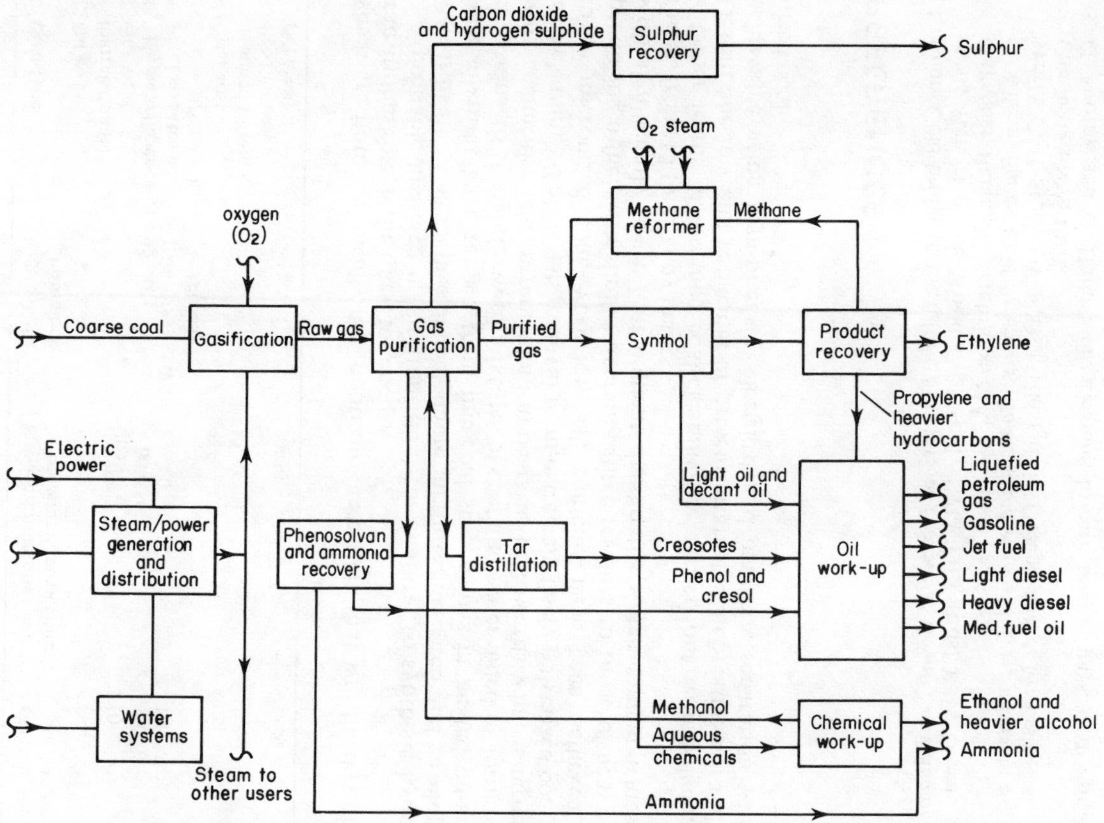

Fig. 1.18 SASOL coal conversion: simplified flowsheet.

TABLE 1.10 Comparison of Coal Conversion Methods[8]

	I Pyrolysis	II Direct hydrogenation	III Indirect hydrogenation gas synthesis
1. Coal type	high-volatile bituminous	low-ash bituminous or sub-bituminous lignites	any (but non-coking for Lurgi)
2. Oil yield (barrels/tonne)	1	<3	1.25
3. Gas yield	small	small	10%
4. Thermal efficiency	65–70%	<60–65%	<50%
5. Plant size (tonnes coal $\times 10^6$/annum)	1.4 + 2000 MW electricity	1.4–4.0	1.4
6. Coal cost, assumed ($/tonne)	15	15	10
7. Product cost ($/barrel)	<18	20–30	>30

which are both technically and politically accessible that will fill this important transitional role in the quest for energy.

A probable first step in the transition from natural to synthetic hydrocarbon fuels is likely to be the use of coal–liquid mixtures. Research and development of such mixtures have ranged from fine-coal suspensions in heated oil-and-water emulsions, to straightforward mixtures of coal and water aided by forms of stabilizing chemicals. The current trend appears to be towards what is termed 'deep coal cleaning'. This is ash removal from extremely finely pulverized coal, referred to as micronized coal. Very low ash contents have been achieved, thereby producing a fuel with similar uncombustible residues to those of heavy oils, but with significantly lower heat content due to the water content which usually ranges from 15 to 25% by mass.

Other forms of ultrafine coal-cleaning are carried out with pulverized coal. Despite the low unit capacities at present attainable, impressive ash reduction is achievable using high-intensity electrical and magnetic separation with micronized coal.

REFERENCES

1. Anon. Coal-bridge to the future. *Future Coal Prospects: Country and Regional Assessments.* Report of the World Coal Study (WOCOL), (Eds) R. P. Greene and J. M. Gallagher, Ballinger, Cambridge, MA, 1980.
2. C. Wilson. *Future Coal Prospects: Country and Regional Assessments*, in Report of the World Coal Study (WOCOL), (Eds) R. P. Greene and J. M. Gallagher, Ballinger, Cambridge, MA, 1980.
3. G. Norton and G. Hambleton. Cleaning to zero seldom pays, *Coal Age* **88**, 50–55, 1983.
4. D. W. Horsfall. A prognosis for coal in an integrated fuel technology, *Optima* **26**, 105–112, 1976. Anglo-American Corporation of South Africa, Johannesburg.
5. Anon. *BP Statistical Review of World Energy*, British Petroleum, London, 1985.
6. Anon. *Coal in Canada*, The Coal Association of Canada, Calgary, Canada, 3rd edn, 1985.

7. Y. Ishikawa. *Technical Aspects of Coal Use in the Japanese Coal Industry*, Nippon Steel Corporation, Tokyo, Japan, 1982.
8. G. J. Pitt and G. R. Millward. *Coal and Modern Coal Processing—An Introduction*, Academic Press, London, 1979.
9. D. E. Pearson. The quality of Western Canadian coking coal, *CIMM Bulletin* **73**, 71–84, 1980.
10. Anon. *Coal Quality and Ash Characteristics*, IEA Coal Advisory Board, Paris, January 1985.

Chapter 2

PROPERTIES OF COAL

2.1 INTRODUCTION

2.1.1 Definition of Coal

Coal is the general descriptive term applied to a group of solid fossil fuels, black or brown in colour, that consist predominantly of altered plant material and usually occur as seams within other consolidated strata.

From a geological standpoint, coal may be classified as a sedimentary rock consisting essentially of organic components and with only a minor proportion of mineral constituents, although in legal and everyday language it is sometimes referred to as a mineral (i.e. a substance within the earth's crust that can be extracted commercially).

2.1.2 Coal Rank

Coals vary in composition and properties in accordance with the extent of alteration, or degree of coalification, of the original plant material from which they were derived. The vast majority of coals are of the banded (or humic) varieties, resulting from the decay of plant material under swampy conditions to form peat, this material then passing through one or more stages of transformation to give one of the varieties of coal commonly known as lignite, sub-bituminous coal, bituminous coal, semi-anthracite and anthracite.

The concept of *coal rank* is used to indicate the stage of alteration attained by a particular coal; the greater the alteration, the higher is the rank of the coal. Thus, lignites and sub-bituminous coals are termed low-rank coals, while semi-anthracites and anthracites are termed high-rank coals.

The high- and medium-rank coals are often referred to collectively as *hard coal*, but this term is rather misleading, as some bituminous coals are quite soft, while some low-rank coals are very hard. The high- and medium-rank coals are also frequently referred to as *black coal* and the low-rank coals as *brown coal*, but this terminology is again not very satisfactory, as low-rank coals which are black in colour are not uncommon.

PROPERTIES OF COAL

The relationship of these various terms is shown below:

Descriptions of banded (humic) coals

High-rank	anthracite	
	semi-anthracite	} hard or black coal
Medium-rank	bituminous	
Low-rank	sub-bituminous	} brown coal
	lignite	

2.1.3 Coal Equivalent

There is a wide variation in energy content or heating power of different coals, since this property is dependent upon both the rank of the coal, and its quality as measured by the amount of incombustible material (moisture and mineral matter) it contains. Thus, high-rank, good-quality bituminous coal has about three times the energy content of a low-rank poor-quality lignite.

In view of this wide variation, it is frequently desirable, particularly in statistics of energy resources, reserves, production and trade, to compare quantities of coal on the basis of their energy contents instead of their actual tonnages, and for this purpose a measure known as *coal equivalent* is commonly used.

The coal equivalent of a particular quantity of coal, or other energy source such as oil or gas, is the equivalent quantity, in energy content, of coal with a specific energy (or, as it is usually termed, calorific value) of 29.3 megajoules per kilogram (MJ/kg), or 12 600 British thermal units per pound (Btu/lb), or 7000 kilocalories per kilogram (kcal/kg).

2.2 COAL PETROLOGY

Coal petrology is concerned with the origin, composition and properties of the distinct organic and inorganic components of different coals. To date, the principal practical applications of coal petrology have been in the specification and selection of coals for carbonization, although the techniques are finding increasing use in connection with coal exploration, preparation, storage, combustion and conversion.

2.2.1 Mode of Origin of Coals

According to the mode of decay of the original plant material, coals can be grouped broadly into the following two categories:

1. *Humic coals*—resulting from plant decay under aerobic conditions. This group includes most of the coal types found in banded coal seams.
2. *Sapropelic coals*—resulting from plant decay under anaerobic conditions. These coals are characterized by a non-banded appearance.

2.2.2 Macroscopic (Megascopic) Components of Coals

Two different systems of nomenclature for the macroscopically distinct varieties of coal, which usually occur as bands within coal seams, have been in extensive use during the last 60 years. The European system, however, has now gained a widespread measure of international acceptance.

European System of Nomenclature
In this system, which was originally based largely on the work of Dr M. C. Stopes on British bituminous coals, the following four components of humic coals are recognized:

(a) *Vitrain*—essentially bright coal, glossy and brilliant in lustre;
(b) *Clarain*—which can be considered as finely interbanded bright and dull coal, bright in overall appearance;
(c) *Durain*—essentially dull coal, often with a suggestion of a slightly greasy overall appearance, and usually harder than bright coal;
(d) *Fusain*—a friable, somewhat fibrous, charcoal-like component that often occurs in pockets and lenticles rather than uniform bands.

These varieties of coal have been invested with the status of separate types of rock, and are therefore termed *lithotypes*.

The sapropelic coals are divided into two groups: the *cannel coals* and the *boghead coals*. The former are dullish black, with a slightly greasy appearance and conchoidal fracture; the latter are more brownish in colour.

American System of Nomenclature
This system is based on the work of Dr R. Thiessen on American coals in general, irrespective of rank or mode of origin. Two main components, one bright and the other dull, were originally recognized and designated *anthraxylon* and *attritus* respectively. Among banded coals, those in which anthraxylon predominates have been termed *bright coals*; those with approximately equal amounts of anthraxylon and attritus, *semisplint coals*; and those in which attritus predominates, *splint coals*.

The ASTM (American Society for Testing and Materials) Standard for megascopic description of coal[1] now describes banded coals in terms of *vitrain* (virtually identical with anthraxylon), *attrital coal* (corresponding broadly with clarain and durain) and *fusain* (as in the European nomenclature).

The American standard recognizes cannel coals and boghead coals as varieties of non-banded attrital coal.

A comparison of the European (in English, French and German) and American terminology is shown in Table 2.1.

2.2.3 Microscopic Components of Coals

The organic particles of coals are classified into different types termed *macerals*, a term intended to indicate analogy with minerals (which are the different types of inorganic particles found in coals and other rocks).

There is a generally accepted international system of nomenclature for macerals and maceral groups and associations. The three groups of macerals in high- and medium-rank coals are as follows:

(a) *Vitrinite*—derived from plant cell substances, varying in appearance from being completely structureless to exhibiting well discernible tissues.
(b) *Exinite*—derived from secretions and waxy coatings of plants, and lower in reflectance than vitrinite.

TABLE 2.1 Nomenclature of Coal Lithotypes

		European classification			Approximate American equivalents		
	English terminology	French terminology	Germany terminology		Earlier terminology		Current terminology
Banded humic coals	Vitrain	charbon brillant	Glanzkohle		anthraxylon	bright coal	vitrain
	Clarain	charbon semi-brillant	Halbglanzkohle		anthraxylon and attritus	bright coal / semisplint coal	attrital coal
	Durain	charbon mat	Mattkohle		attritus and anthraxylon	splint coal	
	Fusain	charbon fibreux	Faserkohle		—	—	fusain
Non-banded (sapropelic) coals	Cannel coal		Attritus		Cannel coal	Cannel coal	
	Boghead		Attritus		Boghead coal	Boghead coal	

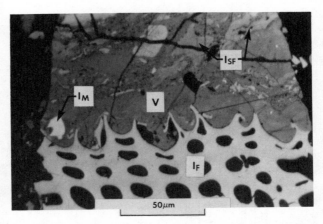

Fig. 2.1 Photomicrograph showing coal macerals in West Virginia Pennsylvania-age coal. V, vitrinite; I_{SF}, semifusinite; I_M, macrinite; I_F, fusinite (courtesy David Pearson and Associates).

(c) *Inertinite*—with or without recognizable plant structures, and higher in reflectance than vitrinite.

These are shown in the photomicrograph in Fig. 2.1.

In low-rank coals, three maceral groups related to the above are recognized, but the groups corresponding with vitrinite and exinite are termed *huminite* and *liptinite* respectively.

2.2.4 Geographical Variation in Petrological Characteristics of Coals

Most of the terminology of coal petrology has been based upon study of coals in Europe and North America, particularly coals of Carboniferous age and of bituminous or higher rank. Some marked petrological differences are apparent between these coals and the so-called Gondwana coals (mainly of Permian age) of India, Southern Africa, Australia, part of South America and Antarctica. The main macroscopic differences are that the Gondwana coals are usually comparatively dull in overall lustre, and occur in seams in which dull bands are commoner and sometimes much thicker than in the Carboniferous coals of the Northern Hemisphere. These differences result from the two groups of coals having been derived from contrasting plant assemblages that grew under different climatic conditions; the plants that gave rise to the Gondwana coals grew in a considerably colder atmosphere.

Microscopically, the same three maceral groups are seen to be present in each of the two groups of coals, and none of the maceral groups shows differences in properties between the two coal groups. However, Gondwana coals generally contain a higher proportion of inertinite and minerals and a lower proportion of vitrinite and exinite, a fact which accounts for the generally dull macroscopic appearance of these coals.

2.2.5 Principal Techniques of Applied Coal Petrology

Maceral and Microlithotype Analyses

Maceral analysis is a technique widely used for providing valuable information on the behaviour of coals during carbonization. In a maceral analysis inertinite is commonly subdivided into the macerals macrinite, micrinite, semifusinite and fusinite, and the analysis summarized in terms of *total reactives* (vitrinite plus exinite plus one-third semifusinite) and *total inerts* (two-thirds semifusinite plus other inertinite macerals plus mineral matter). Microlithotype analysis is sometimes used as an additional aid in the study of coals for carbonization.

Macerals rarely occur at random within any lithotype band; under the microscope such a band can usually be seen to comprise thinner bands distinguished from each other by consisting of different maceral associations, and which are known as *microlithotypes*.

The relationship between the lithotypes of banded high- and medium-rank coals and their principal component microlithotype and maceral groups is shown in Table 2.2.

Determination of Reflectance

As the reflectance of each maceral varies directly with coal rank, reflectance measurement is a valuable technique for determining the latter with precision. However, the macerals of any particular coal differ from each other in reflectance, and determination of coal rank is therefore normally based upon vitrinite reflectance. On a prepared sample of the coal, 100 measurements of the maximum reflectance of the vitrinite in oil are made. From these measurements, the mean maximum reflectance (R_0 max) is calculated, and the frequency distribution of the individual measurements is sometimes reported in the form of a histogram known as a *reflectogram*, which shows the number of measurements within successive ranges of maximum reflectance of 0.1 or 0.05%. These ranges are known as *V-steps* and *half V-steps* respectively, the former being designated by the tenfold multiples of their lower limits, e.g. step V-12 covers the range of maximum reflectance from 1.20 to 1.29%, inclusive. The types of vitrinite within the various V-steps are sometimes referred to as *vitrinoids*.

In coal carbonization technology, reflectance distribution is frequently used to calculate *strength index* (SI) and *composition balance index* (CBI), which are measures of the coking property of a coal or coal blend.[2]

TABLE 2.2 Principal Microlithotype and Maceral Groups in Lithotypes of Banded High- and Medium-rank Coals

Lithotype	Principal microlithotype group(s)	Principal maceral group(s)
Vitrain	vitrite	vitrinite
Clarain	vitrite	vitrinite
	clarite	vitrinite, exinite
	vitrinertite	vitrinite, inertinite
	trimacerite	vitrinite, exinite, inertinite
	durite	inertinite, exinite
Durain	durite	inertinite, exinite
	trimacerite	inertinite, exinite, vitrinite
Fusain	inertite	inertinite

2.3 COAL RANK AND CLASSIFICATION

2.3.1 Coal Rank and Process of Coal Formation

The concept of coal rank in relation to the process of coal formation is also described in brief outline in Chapter 17 of this book. The transformation of plant material to peat is a biochemical process, the earlier stage of this being due largely to the activity of bacteria (aerobic and anaerobic) and fungi. The transformation of peat to coal, or coalification, is a geochemical process, being dependent upon the effects of heat and pressure acting over long periods of time.

2.3.2 Effects of Geological Factors on Coal Rank

Age

A common misconception is that the rank of a coal is a measure of its age. Time is only one factor in coalification, and a relatively young coal may have been subjected to greater depth of burial or more severe earth movements, and consequently attained a higher rank, than many older coals. It follows that coals of the same geological age may show wide differences in rank geographically; for instance, coals of Cretaceous age vary in rank world-wide from lignites to anthracites.

Heat and Pressure in Normal Coalification

Resulting from differences in the depth of burial and intensity of earth movements to which they have been subjected, coals within a particular coalfield often exhibit significant variation in rank. Such variation sometimes occurs in both a vertical direction from one seam to another, and in a lateral direction within each seam. In a seam sequence at a particular locality, a general increase in coal rank from the upper to the lower seams can frequently be demonstrated, and this relationship has been termed *Hilt's law*.

Igneous activity

In some coalfields, the normal course of coalification due to the above factors has been modified locally by the effects of igneous activity.

Where igneous rocks have been intruded through, or extruded in the proximity of, pre-existing coal seams, volatile matter is driven from the coal, and this devolatilization is accompanied by reduction or complete destruction of any swelling and coking properties the coal may have possessed. Such coals are frequently described as *heat-altered coals*, and those which have been severely affected are sometimes known as *cinder coals* or *burnt coals*.

Variation in Rank-related Physical and Chemical Properties of Coals

Changes of major practical importance occur in the physical and chemical properties of all the component macerals of coal as its rank increases. Vitrinite is the maceral usually chosen for study of these changes. Since the amount of mineral matter in a coal is independent of its rank, and since the moisture content, though partly rank-related, is not a component of the organic matter, values of other chemical properties for use as measures of coal rank are usually calculated to the basis of 'pure' (dry, mineral-matter-free) coal, and the values quoted below are expressed on that basis.

The principal changes in physical and chemical properties that occur with

TABLE 2.3 Normal Ranges of Certain Properties in Humic Coals of Different Ranks

Rank of coal	Maximum reflectance of vitrinite (%)	Carbon (d.m.m.f.)[a] (%)	Hydrogen (d.m.m.f.) (%)	Oxygen (d.m.m.f.) (%)	Volatile matter (d.m.m.f.) (%)	Calorific value (d.m.m.f.) (MJ/kg)	Equilibrium moisture (m.m.f.)[b] (%)
Lignite	<0.30–0.37	62–73	5.9–4.3	30–20	65–42	23.7–29.5	>70–31
Sub-bituminous coal	0.37–0.55	70–80	6.0–4.5	22–13	52–37	28.6–31.9	35–16
High volatile bituminous coal	0.50–1.00	76–88	6.0–4.9	16–4.5	49–32	30.9–37.0	21–1
Medium-volatile bituminous coal	1.00–1.55	84–91.5	5.4–4.5	8.5–2	32–20	34.2–37.2	4–1
Low-volatile bituminous coal and semi-anthracite	1.55–2.30	90.5–93.3	4.3–3.7	3.5–1.5	20–9	37.2–36.1	1–2
Anthracite	2.30–>4.00	92.5–>96	3.9–<2.0	2.5–<1	9–<4	36.6–<34.5	1.5–>3

[a] d.m.m.f.—dry, mineral-matter-free basis
[b] m.m.f.—mineral-matter-free basis

increase in coal rank from low-rank to high-rank anthracites are as follows:

(a) a progressive increase in three closely related physical properties—darkness of colour, lustre, and reflectivity; the maximum reflectance of vitrinite (or the corresponding huminite in low-rank coals) increases from less than 0.3 to greater than 4.0%;

(b) an increase in relative density from about 1.2 to 1.7, this change being more rapid in the high-rank coals;

(c) with regard to elemental composition (ultimate analysis), a progressive and almost uniform increase in carbon from about 60 to over 96%, and a corresponding decrease in oxygen from about 30 to less than 1%. Hydrogen remains fairly constant at about 5–6% until the carbon content reaches about 89%, after which there is a relatively rapid decrease in hydrogen to under 2%.

(d) a decrease in volatile matter from about 65% to less than 4%;

(e) an increase in calorific value from about 23.5 MJ/kg (10 100 Btu/lb or 5600 kcal/kg) to about 37.2 MJ/kg (16 000 Btu/lb or 8900 kcal/kg) at the stage where hydrogen begins to decrease, when a related decrease in calorific value commences, continuing to below 34.5 MJ/kg (14 850 Btu/lb or 8250 kcal/kg);

(f) a trend of variation in swelling and coking properties of the coal which is broadly similar to that shown by calorific value. Swelling and coking properties increase from zero values in the low-rank coals to maxima in the higher-rank bituminous coals, decreasing thereafter to zero values in the anthracites;

(g) a trend of variation in the moisture-holding capacity of the coal, under standard humidity/temperature conditions, which is broadly the converse of that shown by calorific value and swelling and coking properties. There is a decrease in this moisture-holding capacity from over 70 to under 1%, followed by an increase to over 3%;

(h) a general decrease in the solubility of the coal in solvents, and in its reactivity with oxidizing or hydrogenating agents.

The normal ranges of most of these properties in the broad rank categories of humic coals are shown in Table 2.3.

2.3.3 Coal Classification

Most of the detailed systems of classifying coals are based upon one or more properties related to coal rank.

The most scientifically refined system is probably that due to the famous British coal scientist, Dr Seyler, in which the two main parameters are the carbon and hydrogen contents of the pure coal (see Fig. 2.2). This system has found little use commercially, because ultimate analysis of a coal is not normally carried out, and because the system was developed largely from data on British coals, which include almost none of low rank, and the names given to some of the coal classes are confusing in a world context.

In the principal coal classification systems that are used nationally and internationally, the main parameters are rank-related properties, such as volatile matter and calorific value, which are determined in routine coal analysis as described in Chapter 17. However, in some of these systems the calorific values have to be expressed on a moist, ash-free basis or a moist, mineral-matter-free basis, a requirement sometimes involving the determination of moisture-holding capacity

PROPERTIES OF COAL

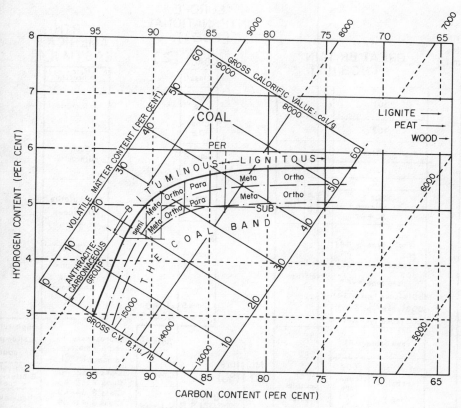

Fig. 2.2 Seyler's chart.[12]

(equilibrium moisture), which is not a routine procedure in coal analysis.

The inter-relationship between the coal classification systems in use in the major coal-producing countries is depicted in Fig. 2.3. The notes on this diagram include a brief description of the basis of each classification system.[10]

In most countries, the classification is still primarily based on the volatile matter content. This term, unfortunately, suggests that coals are a mixture of components, some of which can be distilled off, whereas others cannot. In fact, it means the percentage loss in weight when a sample of crushed coal is rapidly heated to 900 °C under standard conditions in a crucible of a specified type, allowance being made for moisture and mineral matter in the coal. Volatile matter is therefore a measure of the amount of material driven off under standardized decomposition conditions rather than distillation of pre-existing components; it decreases with increasing rank from about 40% to about 5%. In a given coal, exinite has a higher volatile matter content than vitrinite, whereas inertinite has a lower value.

The use of volatile matter as a parameter in coal classification has obvious relevance to the carbonization of coal, giving a rough measure of the yields of tar plus gas which may be expected when a coal is heated in the absence of oxygen.

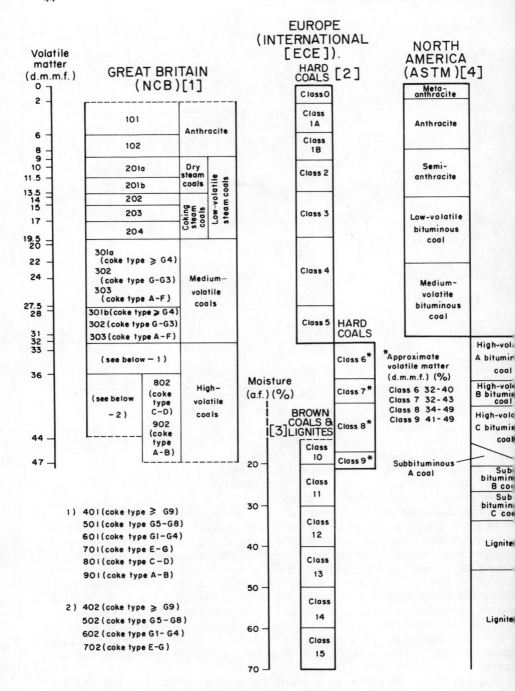

Fig. 2.3 Inter-relationship of coal-classification systems used in various countries

PROPERTIES OF COAL

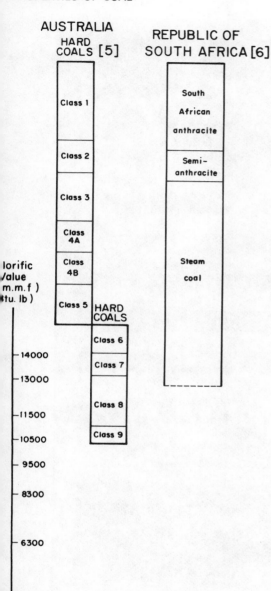

NOTES

NCB classification

Coals with volatile matter (d.m.m.f.) of under 19.6% are classified by this parameter alone. Other coals are classified by the parameters of volatile matter (d.m.m.f.) and Gray-King coke type.

International classification

Hard coals with volatile matter (d.a.f.) not exceeding 33% are divided into classes on the basis of this parameter; class boundaries are shown on this diagram at the approximate equivalent values of volatile matter (d.m.m.f.). Hard coals with volatile matter (d.a.f.) exceeding 33% are divided into classes on the basis of calorific value (m.a.f.); class boundaries are shown on this diagram at the approximate equivalent values of calorific value (m.m.m.f.).

Hard-coal classes are divided into groups on the basis of caking properties, and the groups are divided into sub-groups on the basis of coking properties. The classification of any coal is denoted by a three-figure code number (e.g. 722), the first figure of which denotes the class, the second figure the group and the third figure the sub-group.

Soft coals are divided into classes on the basis of total moisture (a.f.), and the classes are divided into groups on the basis of tar yield (d.a.f.). The classification of any coal is denoted by a four-figure code number (e.g. 1203), the first two figures of which denote the class and the last two figures the group.

ASTM classification

Coals with fixed carbon (d.m.m.f.) of 69% or more, i.e. with volatile matter (d.m.m.f.) of 31% or less, are classified by fixed carbon (d.m.m.f.), i.e. effectively by volatile matter (d.m.m.f.). Other coals are classified primarily by calorific value (m.m.m.f.).

Coals with calorific values (m.m.m.f.) in the range 10 500–11 500 Btu/lb are classified as high-volatile C bituminous coal if agglomerating and as sub-bituminous A coal if non-agglomerating.

Australian classification

Hard coals with volatile matter (d.m.m.f.) not exceeding 33% are divided into classes on the basis of this parameter. Hard coals with volatile matter (d.m.m.f.) exceeding 33% are divided into classes on the basis of calorific value (d.a.f.), the class boundaries being so chosen as to be equivalent to the class boundaries of the International Classification that are based on calorific value (m.a.f.).

Hard-coal classes are divided into groups on the basis of crucible swelling number, the groups are divided into sub-groups on the basis of Gray-King coke type, and the sub-groups are further divided on the basis of ash (d.). The classification of any coal is denoted by a four-figure coal code (e.g. 532(1)), the first figure of which denotes the class, the second figure the group, the third figure the sub-group and the fourth figure (which is placed within brackets) the ash designation.

South African classification

Coals are divided into three broad classes on the basis of volatile matter (d.a.f.); class boundaries are shown on this diagram at the approximate equivalent values of volatile matter (d.m.m.f.). For commercial purposes, coals of each class are graded on the basis of calorific value (a.d.), ash (a.d.) and ash fusibility.

Abbreviations

a.d.	air-dried basis
d.	dry basis
a.f.	ash-free basis
d.a.f.	dry, ash-free basis
d.m.m.f.	dry, mineral-matter-free basis
m.a.f.	moist, ash-free basis
m.m.m.f.	moist, mineral-matter-free basis

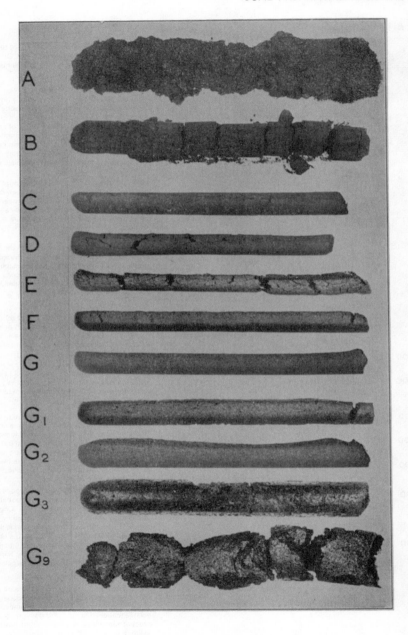

Fig. 2.4 Gray–King coke types (reproduced with permission of the controller of HMSO from *The Efficient Use of Fuel*, 1958, HMSO, London).

Coals with ash of over 10% must be cleaned before analysis for classification to give a maximum yield of coal with ash of 10% or less.

Coal rank code			Volatile matter (d.m.m.f.)[a]	Gray–King coke type[b]	General description
Main class(es)	Class	Sub-class	(%)		
100			Under 9.1	A	
	101[c]		Under 6.1	A	**Anthracites**
	102[c]		6.1–9.0		
200			9.1–19.5	A-G8	**Low-volatile steam coals**
	201		9.1–13.5	A-C	
		201a	9.1–11.5	A-B	Dry steam coals
		201b	11.6–13.5	B-C	
	202		13.6–15.0	B-G	
	203		15.1–17.0	E-G4	Coking steam coals
	204		17.1–19.5	G1-G8	
300			19.6–32.0	A-G9 and over	**Medium-volatile coals**
	301		19.6–32.0	G4 and over	
		301a	19.6–27.5	G4 and over	Prime coking coals
		301b	27.6–32.0		
	302		19.6–32.0	G-G3	Medium-volatile medium-caking or weakly caking coals
	303		19.6–32.0	A-F	Medium-volatile weakly caking to non-caking coals
400 to 900:			Over 32.0	A-G9 and over	**High-volatile coals**
400			Over 32.0	G9 and over	High-volatile, very strongly caking coals
	401		32.1–36.0	G9 and over	
	402		Over 36.0		
500			Over 32.0	G5-G8	High-volatile, strongly caking coals
	501		32.1–36.0	G5-G8	
	502		Over 36.0		
600			Over 32.0	G1-G4	High-volatile, medium-caking coals
	601		32.1–36.0	G1-G4	
	602		Over 36.0		
700			Over 32.0	E-G	High-volatile, weakly caking coals
	701		32.1–36.0	E-G	
	702		Over 36.0		
800			Over 32.0	C-D	High-volatile, very weakly caking coals
	801		32.1-36-0	C-D	
	802		Over 36.0		
900			Over 32.0	A-B	High-volatile, non-caking coals
	901		32.1–36.0	A-B	
	902		Over 36.0		

[a] d.m.m.f.—dry, mineral-matter-free basis
[b] Coals with volatile matter of under 19.6% are classified by using the parameter of volatile matter alone; the Gray–King coke types quoted for these coals indicate the general ranges found in practice, and are not criteria for classification.
[c] In order to divide anthracite into two classes, it is sometimes convenient to use a hydrogen content of 3.35% (d.m.m.f.) instead of a volatile matter of 6.0% as the limiting criterion. In the original Coal Survey rank coding system the anthracites were divided into four classes then designated 101, 102, 103, and 104. Although the present division into two classes satisfies most requirements, it may sometimes be necessary to recognize more than two classes.

NOTES

1. Coals that have been affected by igneous intrusions ('heat-altered' coals) occur mainly in classes 100, 200 and 300, and when recognized should be distinguished by adding the suffix H to the coal rank code, e.g. 102H, 201bH.
2. Coals that have been oxidized by weathering may occur in any class, and when recognized should be distinguished by adding the suffix W to the coal rank code, e.g. 801W.

Fig. 2.5 The coal classification system used by British Coal (formerly the National Coal Board). Source *Energy World*, 1986.

While volatile matter forms the primary basis of coal classification, it is found necessary in practice to use at least one more parameter in order to characterize the behaviour of a coal when heated, because two coals may have the same volatile content but differ from one another in the extent to which they swell up or cake together. In Britain, the parameter is known as the Gray–King coke type and is based on the appearance of a coke specimen prepared under standard conditions (Fig. 2.4). The details of this parameter are discussed later in Chapter 17. Figure 2.5 shows its use in the current British coal classification scheme, particularly to subdivide high-volatile coals into six different degrees of caking

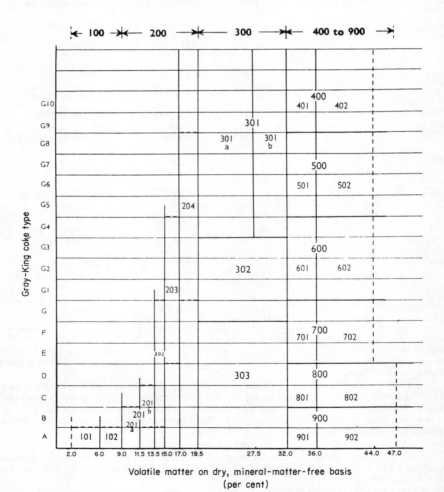

Fig. 2.6 Coal classification system used by British Coal (formerly the National Coal Board). (Courtesy British Coal.)

PROPERTIES OF COAL

capacity. Figure 2.6 shows the same classification graphically. Broken lines define a general limit as found in practice (although this is not necessarily a boundary for classification purposes); solid lines define a classification boundary.[12]

2.4 MOLECULAR STRUCTURE OF COAL

The chemical analysis of coal having been established, it must be admitted that a knowledge of the carbon, hydrogen and oxygen contents does not convey much

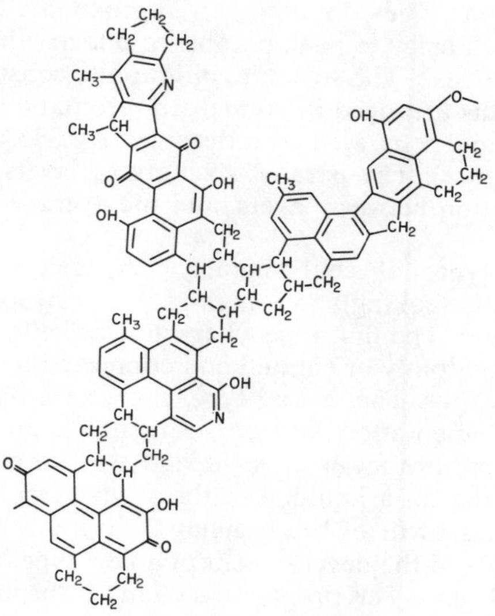

Comparison of properties

Property	Coal	Model
Empirical formula		$C_{100}H_{76}O_{10}N_2$
Molecular weight		1464
Analysis	C81.7, H5.3, N1.9, O11.1%	C82.0, H5.2, N1.9, O10.9
$C-H_{ar}/C-H_{al}$	0.21–0.23	0.20
$H_{CH_3} + H_{CH}/H_{CH_2}$	0.8–0.9	0.74
Hydroxyl content	6.1% O as OH	5.4% O as OH
Carbonyl content	4.0% O as C=O	4.4% O as C=O
Unreactive oxygen	1.0% O	1.1% O as ether
No. of carbon atoms/ aromatic cluster	16 (average)	11–22
Aromaticity	0.60–0.75	0.66

Hydrogen distribution

H_{ar}	H_{OH}	H_{CH}	H_{CH_2}	H_{CH_3}	H_{total}
12	5	13	34	12	76

Fig. 2.7 Example of a molecular model for the maceral vitrinite.

about its chemical make-up. Even in elementary organic chemistry, a substance of molecular weight 46 containing 52% carbon, 13% hydrogen and 35% oxygen might be ethanol or dimethylether, and if the molecular weight is unknown, the number of possible alternatives becomes boundless. Other evidence is necessary before it is possible to define the way in which the various atoms are assembled.

Coal is not a single chemical compound with identical molecules. It can be regarded as a statistical structure made up of small, condensed, aromatic units or layers with a variety of substituent groups around the perimeter and some cross-linking between adjacent units (see Fig. 2.7). Some of the units may not be strictly planar because of the presence of hetero-atoms (oxygen, nitrogen or sulphur) or hydroaromatic portions. They do, however, approximate to planarity and consequently show some tendency to pack parallel to one another, although this parallelism is not very extensive. The structure of graphite consists of extensive parallel layers of carbon atoms arranged in condensed, aromatic arrays, the layers being mutually orientated and separated by a distance of 0.335 nm (the carbon–carbon van der Waals' distance). The parallel stacking of layers in coal is similar, but lacks mutual orientation between layers, and the average spacing between layers is somewhat larger.

The three main types of coal—namely low-rank (non-coking or weakly caking), medium-rank (coking) and high-rank—correspond approximately to three types of structure. The first type, characteristic of low-rank coals, has small layers more or less randomly orientated and connected by cross-links, so that the structure is highly porous. The second type, characteristic of coking coals, shows a greater degree of orientation and consequently a greater tendency towards parallel stacking; there are fewer cross-links, fewer pores and the structure has some similarities to that of a liquid. The third type, seen in high-rank coals and anthracites, shows a growth of the individual layers, a marked increase in the degree of orientation and the development of a new type of pore elongated parallel to the stacks of layers. The progressive changes throughout the coalification can then be more fully understood.

2.5 MINERAL MATTER IN COAL

The amount, mode of occurrence and composition of the mineral matter in coal are factors of great practical importance in determining its market acceptability and economic value. The yields and qualities of products obtainable by cleaning the coal are also dependent upon the characteristics of the mineral matter.

2.5.1 Amount

In contrast to crude oils, in which a mineral matter content of under 0.1% is normal and one of 0.25% is high, few coals as marketed for conventional utilization contain under 5% of mineral matter, and many contain 10% or more. Even a mineral matter content of 60% does not render a fuel incombustible if such developing techniques as fluidized combustion are considered.

2.5.2 Mode of Occurrence

According to its mode of origin, mineral matter within a coal seam may be classified into the following two categories:

1. *inherent mineral matter*—consisting of inorganic constituents that previously formed part of the tissues of the plants from which the coal was derived. Most inherent mineral matter is chemically or colloidally combined with the coal substance.

2. *adventitious mineral matter*—derived from sources outside the swamps in which the coal-forming plants decayed. Adventitious mineral matter occurs in the following four forms:

(a) in chemical or colloidal combination with the coal substance;
(b) as fine particles, often disseminated fairly evenly throughout the coal;
(c) as partings, bands and lenticles of mudstone, 'shale', siltstone, sandstone, etc., within the seam, resulting from deposition of sediments between periods of coal formation;
(d) as material that has been introduced into pores, fissures and planes of weakness in the seam during or subsequent to coalification.

From a practical viewpoint, mineral matter within a coal seam can conveniently be classified as follows:

(a) inorganic material in chemical or colloidal combination with the coal substance, and irremovable by commercial coal preparation processes;
(b) mineral matter in fine, discrete particles disseminated within the coal, much of which cannot be removed except by fine grinding of the coal;
(c) mineral matter occurring as partings, bands, lenticles, veins and nodules within the seam; much of this material (sometimes described as *'free' dirt*) is usually removable by an appropriate commercial coal-preparation process.

Under modern systems of mining, run-of-mine (ROM) coal often contains, in addition to the mineral matter from within the seam, mineral matter derived from strata above and/or below the seam. This additional material, which is known as *extraneous dirt* or *contamination*, mostly consists of mudstone, 'shale', siltstone and sandstone and is usually removable by a commercial coal-preparation process.

2.5.3 Composition

The principal minerals found in coal seams are listed in Table 2.4, together with the chemical composition of each mineral.

In addition to the elements in these minerals, coals contain other elements in very small amounts. These *trace elements* include the metals antimony, beryllium, bismuth, chromium, cobalt, copper, gallium, germanium, lead, molybdenum, nickel, silver, strontium, tin, uranium, vanadium, zinc and zirconium. Some of them appear to be associated with the inherent mineral matter, others with the adventitious mineral matter, and others with both these types of mineral matter. The possibility of commercial recovery of some of the more valuable of these metals, such as germanium and uranium, has been considered. However, attempts to do so have always proved unsuccessful, largely due to the sporadic occurrence in coals of worthwhile concentrations of these elements.

TABLE 2.4 Principal Minerals in Coal Seams

Minerals occurring in mudstone, 'shale', siltstone and sandstone	Quartz, SiO_2
	Minerals of the clay group, principally: Kaolinite, $Al_2 Si_2 O_5 (OH)_2$ Montmorillonite, $Al_2 Si_4 O_{10} (OH)_2 \cdot n\, H_2O$ with some Mg and Na
	Minerals of the mica group, principally: Muscovite, $KAl_2 (Si_3Al) O_{10} (OH)_2$ Illite, $K_xAl_4 (Si_{8-x}Al_x) O_{20} (OH)_4$
	Minerals of the chlorite group, general formula $(Mg,Fe)_3 (OH)_6 \cdot (Fe,Mg,Al)_3 (Si_3Al)O_{10} (OH)_2$
Sulphide minerals	Pyrite ('iron pyrites'), FeS_2
	Marcasite, FeS_2
	Arsenopyrite (mispickel), FeAsS
Carbonate minerals	Siderite (chalybite), $FeCO_3$
	Ankerite, $(Ca, Mg, Fe)CO_3$ with some Mn
	Calcite, $CaCO_3$
	Dolomite, $CaMg (CO_3)_2$
Chloride minerals	Halite, NaCl
	Sylvite, KCl
Other minerals	Fluorapatite, $Ca_{10} F_2 (PO_4)_6$
	Minerals of the felspar group, including: Orthoclase, $KAlSi_3O_8$ Albite, $NaAlSi_3O_8$
	Titanium minerals, including: Sphene, $CaTiSiO_5$ Rutile TiO_2 Ilmenite, $FeTiO_3$
	Tourmaline, $(H, Na, K, Fe, Mg)_9 Al_3 (B,OH)_2 Si_4O_{19}$
	Barytes, $BaSO_4$
	Minerals formed through weathering, including: Limonite, $2Fe_2O_3 \cdot 3H_2O$ Gypsum, $CaSO_4 \cdot 2H_2O$ Melanterite, $FeSO_4 \cdot 7H_2O$

2.5.4 Petrological and Geographical Variation

The lithotypes of a coal differ from each other in both the quantity and the composition of the mineral matter they contain. The amount of mineral matter in most durains and fusains is considerably greater than in the corresponding vitrains, and the mineral matter content of most clarains is intermediate. Qualitatively, the mineral matter in vitrain generally tends to be lower in silica and alumina, and appreciably higher in alkalis, than the mineral matter in durain, while clarain is again intermediate. The mineral matter in fusain varies widely in composition.

With regard to geographical variation in mineral matter, the generally higher mineral matter contents of Gondwana coals, in comparison with Carboniferous coals of the Northern Hemisphere, has been noted. Although the same minerals are found in each of these two groups of coals, the proportions of many minerals

PROPERTIES OF COAL

show marked differences between the coal groups. In particular, pyrite is generally present in much smaller amounts in Gondwana coals, giving rise to lower sulphur contents than those of most Northern Hemisphere coals.

2.5.5 Relationship with Ash

Failure to distinguish between the mineral matter and the ash of a coal is frequent, and the distinction is obscured by the common, but incorrect, use of the term 'ash content'. Coals contain mineral matter but not ash; ash is the solid residue, different from the mineral matter in both amount and composition, that is derived from it on combustion of the coal.

The distinction between mineral matter and ash is of practical importance, since the effects of some of the chemical changes that occur when mineral matter is heated are significant in coal utilization. The principal chemical changes are as follows:[11]

(a) loss of water of hydration from the minerals of the clay, mica and chlorite groups;

(b) conversion of the pyrite into ferric oxide with emission of sulphur dioxide, oxidation of some of the latter to sulphur trioxide, and formation of alkali sulphates by reaction of sulphur trioxide with alkali oxides in the ash;

(c) decomposition of carbonate minerals with emission of carbon dioxide, leaving the corresponding oxides of the metals in the ash;

(d) decomposition of chlorine compounds with emission of hydrogen chloride and leaving, in the case of chloride minerals, the corresponding oxides of the metals in the ash.

2.5.6 Calculation of Mineral Matter Content

As extraction of the mineral matter from coal is a difficult and time-consuming procedure only undertaken exceptionally in coal analysis, mineral matter content is normally calculated from the values for certain determined properties, by means of one of a number of formulae based on the chemical changes that occur when mineral matter is converted to ash. The formulae used in various major coal-producing countries are as follows:

- *Original Parr formula* (North America):
  $MM = 1.08 A + 0.55 S_{tot}$
- *Modified Parr formula* (North America):
  $MM = 1.13 A + 0.47 S_{pyr} + Cl$
- *King–Maries–Crossley (KMC) formula*, as revised by National Coal Board (Britain):
  $MM = 1.13 A + 0.5 S_{pyr} + 0.8 CO_2 - 2.8 S_{ash} + 2.8 S_{sulph} + 0.31 Cl$
- *British Coal Utilization Research Association (BCURA) formula* (Britain):
  $MM = 1.10 A + 0.53 S_{tot} + 0.74 CO_2 - 0.36$
- *Standards Association of Australia formula* (Australia):
  $MM = 1.1 A$
- *National Institute for Coal Research formula* (South Africa):
  $MM = 1.1 A + 0.55 CO_2$

where, as percentages of the air-dried coal:

MM = mineral matter
A = ash
S_{tot} = total sulphur
S_{pyr} = pyritic sulphur
S_{ash} = sulphur retained in ash
S_{sulph} = sulphate sulphur
CO_2 = carbon dioxide
Cl = chlorine

The original Parr formula has probably been the most frequently used on a world-wide basis, but it is based on North American coals and is unsatisfactory for coals high in carbonates and/or chlorides, of which it takes no account. The modified Parr formula is also unsatisfactory for high-carbonate coals. Each of the other formulae shown above was derived specifically for application to coals of the country named.

2.6 COAL BEHAVIOUR ON HEATING

From a practical standpoint, the most important properties of coal are those associated with its behaviour when heated, with or without the presence of air, since carbonization accounts for about 20% of the coal currently used in Europe and North America, and virtually all the rest is burned in one way or another.

When heated, most coals evolve tarry vapours, gas and moisture (Fig. 2.8), and some coals soften and fuse into a coke residue. In the presence of air, the combustible products burn. The tarry vapours normally burn with a smoky flame, the gases burn with a non-smoky flame and the solid residue glows, leaving ash derived from the mineral matter. The calorific value of coal varies with rank in a manner which can be predicted from the changes in chemical composition, so

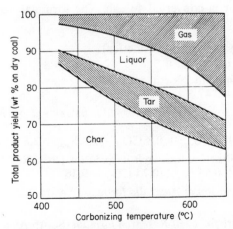

Fig. 2.8 Distribution of products from carbonization at various temperatures.[14]

PROPERTIES OF COAL

that low-rank coals generate less heat because of their higher oxygen content (and because of their higher moisture content), but they are widely burned because of greater availability and low cost. If the combustion conditions lead to the production of smoke, there is not only a pollution problem, but also a reduction in efficiency.

In the absence of air, the sequence of changes which occur when coal is heated forms the basis of the carbonization process from which metallurgical coke, domestic coke, coal tar, town gas (now largely replaced by natural gas) and other products, including ammonia, are derived. The way coals behave when heated is consistent with the coal structural types described earlier. In the first place, the tar is derived from aromatic layer 'molecules' in the original structure which are only weakly held by cross-links and can therefore be evolved as a result of minor decomposition. This process is limited because the breakage of cross-links generates free valencies and these tend to satisfy themselves by recombination. Gaseous products such as methane and hydrogen are derived from the breakage of bonds to peripheral substituent groups and combination of the resulting radicals; this also leaves free valencies and increases the incidence of recombination so that the coke remaining becomes more cross-linked and involatile.

When heated, coking coals soften in the temperature range 400–500 °C, become plastic and agglomerate. It would appear that plasticity depends on the availability of an adequate concentration of liquid decomposition products and a limited degree of cross-linking of the network of layers in the coal. There is less cross-linking in a coking coal than in a low-rank coal, and plasticization is therefore enhanced. In high-rank coals, on the other hand, there is a lower concentration of liquid decomposition products and therefore plasticization decreases again. Evidently there is a delicate balance and it can be shifted to some extent. For example, increasing the rate of heating leads to increasing evidence of plasticity in a low-rank coal.

It is important that the behaviour of coal when heated, and the changes that occur in its properties as a result, should be fully understood when determining the suitability of a particular coal for a specific application.

2.7 COAL UTILIZATION AND THE SPECIFIC PROPERTY PARAMETERS INVOLVED

These aspects are discussed in detail under the subject of coal testing in Chapter 17, but a cursory treatment of them is now given to demonstrate the relationship between certain properties and the utilization of the coal.

2.7.1 Physical Properties

The general physical properties determined from coal samples include moisture, porosity, relative density, hardness, grindability, abrasiveness, size consist, quantities of impurities or non-combustible mineral matter.

2.7.2 Chemical Properties

The general chemical properties determined from crushed coal samples include the components of the proximate analysis (volatile content, fixed carbon content, ash content and moisture content); the components of the ultimate analysis (carbon, hydrogen, nitrogen, sulphur and oxygen), together with other elemental components of which the most common are chlorine, phosphorus and arsenic. Forms of sulphur are also determined.

2.7.3 Properties Relating to Specific Applications

Other specific properties and indices, or factors derived, are examined in order to determine the suitability of a given coal for a specific form of utilization:

Coking coals
Crucible-swelling number, fluidity, dilatation (coke strength), (composition balance index), etc., and the data emanating from movable wall and pilot-scale coke-oven tests.

Thermal coals
Ash analysis, ash fusibility (slagging and fouling ratio), ash resistivity, etc., and the data emanating from burning profile testing and pilot-scale combustion rig tests.

Conversion coals
Fischer assay, petrographic analysis, in addition to all the general chemical properties.

Environmental factors
Environmental factors directly resulting from treatment or utilization of coal can be subdivided into three forms:
(a) solid, e.g. acid-generation potential of mine solid waste, solid contamination of airborne or waterborne particulate, autogenous heating potential;
(b) liquid, e.g. acidity/alkalinity, chemical analysis, leaching potential;
(c) gaseous, resulting from combustion or spontaneous combustion, e.g. SO_x and NO_x emissions, CO and CH_4 emissions, or temperature effects.

The individual properties of each coal, will, to a large extent, dictate the impact of each form.

REFERENCES

1. American Society for Testing and Materials. Standard definitions of terms relating to megascopic description of coal and coal beds and microscopical description and analysis of coal, In ASTM, *1980 Annual Book of ASTM Standards*, Pt 26, ANSI/ASTM D 2796-78, pp. 363–366, ASTM, Philadelphia, 1980.
2. E. Stach, M. Th. Mackovsky, M. Teichmüller, G. H. Taylor, D. Chandra and R. Teichmüller. *Stach's Textbook of Coal Petrology*, 2nd completely revised edn, Gebruder Borntraeger, Berlin–Stuttgart, 1975, pp. 1–4, 54–121, 132–153, 157–172, 176–184, 187–198, 200–263, 281–303, 310–316, 331–335, 341–381, 386–388.
3. National Coal Board. Scientific Control. *Coal Survey: The Coal Classification System Used by the National Coal Board* (Revision of 1964). NCB, London, 1964.
4. Economic Commission for Europe. *International Classification of Hard Coals by Type*, Geneva, ECE, 1956, *E/ECE/247, E/ECE/COAL/110*, Pub. Sales No. 1956, 11E.4.

5. Economic Commission for Europe. *Mining and Upgrading of Brown Coal in Europe—Developments and Prospects*, Geneva, ECE, 1957, pp. 14–15, *E/ECE/247, E/ECE/COAL/124*, Pub. Sales No. 1957, 11E Mim. 20.
6. American Society for Testing and Materials. Standard classification of coals by rank, In ASTM, *1980 Annual Book of ASTM Standards*, Pt 26, *ANSI/ASTM* D 388–77, pp. 223–227, ASTM, Philadelphia, 1980.
7. Standards Association of Australia. *Classification System for Australian Hard Coals*, AS 2096–1977 North Sydney, SAA, 1977.
8. Republic of South Africa. Manner of Grading Coal, *Government Gazette*, February 24, 1961, Pt 111. Pretoria.
9. Classification Standards for South African Coals Developed by FRI. *Coal Gold Base Metals Southern Africa*, June 1978, pp. 67–87.
10. P. L. Rumsby. *BP Coal Handbook on Coal*, Internal Publication, 1982.
11. J. G. King, M. B. Maries and H. E. Crossley. Formulae for the calculation of coal analyses to a basis of coal substance free from mineral matter, *Journal Society Chemical Industry*, 1936, **55**, 277–281T.
12. J. Gibson. Coal—an introduction to its formation and properties, Chapter 1, In *Coal and Modern Coal Processing—An Introduction*, Pitt and Millward, (Eds), Academic Press, London, 1979.
13. J. W. Leonard, (Ed.), *Coal Preparation*, 4th edn, 1979, Society Mining Engineers, AIME, New York.
14. J. Owen. *Residential Conference on Science in the Use of Coal*, Institute of Fuel (now Institute of Energy), 1958, pp. C34–39.
15. M. A. Elliott (Ed.), *Chemistry of Coal Utilization*, 2nd Supp. Vol., Wiley, New York, Chichester, 1981, pp. 20–36, 130–283, 425–521.
16. W. Francis. *Coal—Its Formation and Composition*, 2nd Edn, Edward Arnold, London, 1961, pp. 361–634, 667–753, 771–779.
17. D. W. van Krevelen. *Coal—Typology-Chemistry-Constitution*, Elsevier, Barking, 1961, pp. 10–57, 89–200, 304–477.
18. H. H. Lowry, (Ed.), *Chemistry of Coal Utilization*, Wiley, New York; Chapman and Hall, London, 1945, pp. 38–85, 337–424, 677–773.
19. H. H. Lowry (Ed.), *Chemistry of Coal Utilization*, Supp. Vol., Wiley, New York, Chichester, 1963, pp. 27–118, 232–295.
20. G. Thiessen. Composition and origin of the mineral matter in coal. In H. H. Lowry (Ed.), *Chemistry of Coal Utilization*, Wiley, New York; Chapman and Hall, London, 1945, pp. 485–495.
21. I. A. Williamson. *Coal Mining Geology*, Oxford University Press, London, 1967, pp. 236–256.

Chapter 3

SIZE REDUCTION

3.1 INTRODUCTION

Size-reduction equipment for coal applications ranges from heavy-duty crushers and breakers, capable of accepting lumps of up to a metre cube in size, to coal-pulverization equipment capable of milling coal to a fine powder. The duty required, being known beforehand, will usually dictate the type(s) of size-reduction equipment, and to a large extent it is possible to categorize this equipment according to the upper size limit and degree of reduction required. This in turn will be influenced by the nature of the coal and other rocks which the machine must reduce and the relative proportions of each.

In selecting a size-reduction machine or a flowsheet comprising more than one stage, the following must be borne in mind:[1]

(a) Any size-reduction machine will be limited by:
- size range of feed;
- nature/properties of rocks to be crushed;
- feed rate;
- moisture content.

(b) It may be necessary to employ a number of machines in series and/or parallel due to constraints imposed by the limiting factors. No single machine efficiently reduces large sizes to small sizes by repeated breakage, and therefore size reduction is usually performed in a number of stages.

(c) The comminution mode may be influenced by the limiting factors, e.g. compression, impact, grinding, and the final requirements, e.g. minimized fines.

(d) Size reduction is required for one of the following reasons:
- to facilitate handling: for example, in run-of-mine handling, conveyor-belt size is determined by capacity and maximum lump size. Breakage of large pieces of coal will reduce the belt width required and also potential spillage;
- to provide coal in a form which can be readily utilized. This includes oversize elimination, grading into size ranges and pulverizing;
- to provide liberation of coal and non-coal components to facilitate improved coal recovery. Regarding this, improved liberation creates increased fines and potential loss of coal and hence, there are limitations to consider.

SIZE REDUCTION

(e) Efficient size reduction necessitates that the material requiring size reduction be maximized within the breaking/crushing zone of the equipment, i.e. fines should be screened out or removed rapidly to maximize crushing performance and capacity.

There are four basic phenomena involved in size reductions:[2]

1. Impact—which is a sharp, instantaneous impingement of one moving object against another.
2. Attrition—which takes place as a result of a grinding action caused by two hard surfaces sliding relative to each other.
3. Shear—which consists of a trimming or cleaving effect as opposed to the rubbing action of attrition.
4. Compression—which is created when the material is trapped between two hard, enclosing surfaces applying increasing pressure towards one another.

Figure 3.1 illustrates each of these phenomena.

The *impact* effect is utilized in two different ways. Both objects may be moving such as a bat hitting a ball, or one may be motionless such as a rock being struck by a hammer. These two variations are often referred to as gravity and dynamic impact. In the first, coal is dropped onto a hard surface resulting in some degree of fragmentation. In the latter, coal is struck by some form of hammers, and since the material is unsupported the force of impact accelerates movement causing further impacts to occur with other hammers or the metal surfaces inside the crushing chamber.

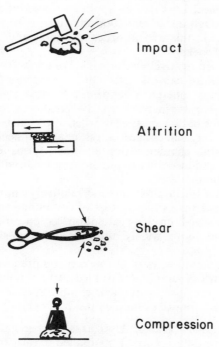

Fig. 3.1 Size-reduction phenomena.

The *attrition* effect reduces material by a form of scrubbing action. This usually occurs as a secondary effect in breakers and crushers whereby the coal is drawn through close clearances between moving and stationary surfaces causing size reducing by the combined effects of attrition and shear. Although attrition consumes more power and can exact heavier wear on crushing surfaces, it is a valuable size-reduction method for the less abrasive materials such as coal and its associated rocks.

The *shearing* effect, like attrition, is a secondary effect encountered (and encouraged) in toothed roll crushers in particular where it exploits the various weaknesses in the structure of the coal by a form of cleaving action.

The *compression* effect is predominantly exploited in crushing hard materials and jaw, gyratory and cone types of crushers are specifically designed to reduce size by this effect assisted by combination of shear and attrition effects.

The contribution of each of these four size-reduction phenomena will become more evident as the various types of size-reduction equipment are discussed.

3.2 MECHANISM OF SIZE REDUCTION

We have seen that size reduction occurs as a result of a combination of four effects and that each effect is exploited to varying degrees by the size-reduction mechanism of the various types of crushing equipment.

Principles of rock mechanics can be applied to describe the behaviour of rock when it is passed through a size reduction machine. It is useful to consider size-reduction mechanisms in this way in order to understand the relative behaviour of coal and associated rocks.

In fact, most coals are relatively weak when compared to other common rocks and minerals, and failure causing size reduction occurs largely as a result of exploitation of the many inherent weaknesses. However, energy requirements of size reduction can sharply increase as the weaker structural characteristics created during coal formation disappear when the coal is reduced to finer and finer size-consists. Under such circumstances, coals can begin to exhibit the more predictable behavioural characteristics of other forms of sedimentary rock. Hence, coal-grinding circuit design requires the specialized design data typical of other milling applications.[5]

The size reduction of brittle substances to relatively fine powders is referred to in mineral processing terms as the process of comminution. In recent years coal comminution has received growing attention as pulverized coal combustion systems have become more widely utilized and coal-slurry technology has emerged.

One of the most common ways of testing the compressive strength of a rock is to produce a symmetrical sample and subject it to a uniaxial compression test (see Fig. 3.2). When pressure is applied, compressive forces are converted to tensile forces and a fracture will eventually result either:

(i) initiated by internal cracks or flaws (as is most commonly the case) or
(ii) caused by plastic deformation from applied tensile stress, or
(iii) from a combination of both.

SIZE REDUCTION

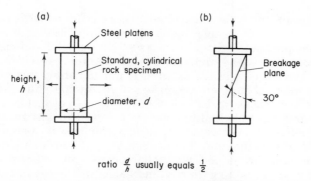

Fig. 3.2 (a) Uniaxial compression testing. (b) Failure or breakage plane.

For a fracture initiated by internal cracks it can be demonstrated[6] that for a brittle substance the compressive strength will be about ten times the tensile strength of the same material. In the event of fracture due to compression, the breakage plane will be situated at an angle of about 30° to the direction of applied pressure (see Fig. 3.2). If plastic deformation is proportional to strain energy, comminution by tensile forces will require only one-hundredth of the energy needed to achieve size reduction by compression. However, experimental results have demonstrated that comminution caused mainly by compressive forces is more economical than that due mainly to tensile stress. This is because, when a rock is severed into two pieces due to tensile forces, the strain energy accumulated in each fragment is suddenly released and largely transformed into useless thermal energy. On the other hand, comminution resulting mainly from compressive forces propagates the initial cracks which ultimately give rise to further fracturing. These cracks do not necessarily separate but lead to further cracks, whilst the fractured parts remain in close contact, as shown in Fig. 3.3. Therefore, the elastic strain within the rock persists and most of the strain energy thus accumulated can be fruitfully utilized in creating more fracture surface.

The observed brittle strength of rocks is variable for every rock type and greatly influenced by the structural characteristics of the specific rock under test. The practical value obtained from testing is usually well below theoretically derived values by a factor of somewhere between one-tenth and one-hundredth.

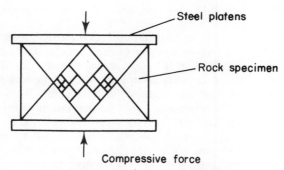

Fig. 3.3 Fragmentation pattern in compressed specimen.

This clearly demonstrates the value of conducting crushing testing of one type or another with actual rock samples requiring crushing. The source of this inherent weakness is the presence of flaws or cracks within the solid, especially at or close to the surface, which act as stress multipliers. According to Griffith[7] the energy required for fracture of a brittle material is not uniformly distributed over the volume of the material but there are regions of energy concentration produced by minute faults and cracks in the material. In the case of metals such flows are often minute and may only lead to failure in extreme circumstances. However, in rocks and minerals the occurrence of such phenomena is common and a very significant aid to comminution. Rocks such as coal and its associated in-seam, roof and floor materials depend greatly upon cracks and other structural anomalies for their ease of crushing. In rock mechanics, the behaviour of rocks *in situ* will also be determined by structural anomalies, but uniaxial testing will not fully define the rock behaviour, and some form of triaxial testing is necessary to cater for the three-dimensional forces involved.

3.3 CRUSHING AND COMMINUTION TESTING

Some discussion of coal grindability and abrasiveness test-work associated with preparing coal for the market is included in Chapter 17. Chapter 5, which deals with size-distribution analysis, includes discussion of the Rosin–Rammler distribution for broken coal.

In addition to uniaxial compressive testing, the data from which are usually only applicable for compression type crushers, there are several non-standard forms of tests used to determine the suitability and capacity of size-reduction equipment.

The most commonly employed tests are those conducted with either small-scale crushers similar to those shown in Fig. 3.4 or special test rigs incorporating smaller sizes of the commercial range of crushers. For such tests to be meaningful the sample mass and representativeness are all-important in the determination of reliable crushing characteristics of the test material:

- Coal mass required for crushing tests is given in Table 3.1 but care must be taken to ensure that the material is packed in containers which will ensure it will retain its original characteristics, e.g. moisture content, size, shape, etc. These containers should normally also be airtight in order to minimize weathering effects.
- The usual procedure is to crush a representative subsample of the test material and then conduct a size analysis of the product. The size analyses of the type shown in the graph in Fig. 3.5 are typical test results obtained for these separate sub-samples of test coal. By comparing these data with others associated with familiar sources of coal, an experienced crusher applications engineer can estimate overall crusher performance and power requirements and select the appropriate machine type and size. The tests can also be used to detect characteristics that require special design considerations or auxiliary equipment. These tests may indicate that two or more types of crusher will adequately handle the test coal. By taking into account other contributing factors such as fines generation potential, available headroom, product shape, economics of operations and

SIZE REDUCTION

Fig. 3.4 Pilot-scale coal-crushing equipment.

maintenance and the adaptability of each alternative to further plant expansions, the crusher applications engineer can determine the best selection. Another, more specialized form of testing is the drop test, employed to determine the size of rotary breaker machine required for size reduction of run-of-mine coal. In this test lumps of coal, shale or other rock likely to be encountered by the breaker are dropped from a predetermined height onto a steel plate and the degree of resultant fragmentation is used to determine the required drop height within the breaker and hence the equivalent machine diameter. The procedure for this test varies from one manufacturer to the next, but each has established empirical data for its own equipment that permit reliable machine sizing. Tests for abrasiveness

TABLE 3.1 Minimum Sample Masses Required for Crushing Tests (Source: Australian Standard)

Nominal particle size fraction[a] (mm)	Recommended mass (kg)				
		Coal preparation plant products			
		Comprehensive test		Control test	
	Raw coal	Clean coal	Discard	Clean coal	Discard
− 125 + 63	400	600	1200	200	400
− 63 + 31.5	100	150	300	50	100
− 31.5 + 16.0	25	35	70	15	35
− 16.0 + 8.0	7	10	20	4	10
− 8.0 + 4.0	3	4.5	9	2	5
− 4.0 + 2.0	1.5	2.5	5	1.5	2.5
− 2.0 + 0.5	1.0	1.5	3	1.0	1.0
− 0.5	0.5	0.5	1	0.5	0.5

[a] The particle sizes shown in the above table may be supplemented or replaced by other particle sizes or gauges in accordance with personal preference.

and hardness assessment have been devised independently by various manufacturers of crushing equipment.

Pennsylvania Crusher Corporation[8] has reported a simple method of determining relative abrasiveness that provides data used to augment previous experience or testing in the semiproduction test rigs described earlier. Four 0.4-kg samples of material to be crushed are individually placed in a drum with a steel paddle and rotated for 15 minutes. The loss in mass from each paddle is obtained by weighing and the average value is used as a relative index of abrasive action. The greater the index the greater the degree of abrasiveness. Based on data accumulated over a period of many years this company has compiled tabulated data of physical characteristics of rocks and ores similar to those shown in Fig. 3.6.

The most common grindability test used in conjunction with coal size-reduction is that first derived by Hardgrove. It is obtained by grinding 50 g of 1 mm × 0.6 mm of dried coal sample in a standardized ball-and-race mill for 60 cycles at a speed of 20 r.p.m. The sample is then removed and sieved at 74-μm (woven wire-mesh) aperture to determine the amount of material, W, passing through the sieve. The Hardgrove grindability index (HGI) is calculated using the formula

$$\text{HGI} = 13.6 + 6.93W$$

According to this formula, when the resistance to grinding increases the value of HGI will decrease. From this formula the value of HGI can be used to predict the grain-size distribution i.e. Rosin–Rammler–Bennet curves based on the formula given in Section 4.2.1 of Chapter 4 and Fig. 4.5. A distribution modulus, m; and size modulus, $d_{63.2}$, must be known to enable the size distribution to be determined. This value of distribution modulus has been found to be related to Hardgrove grindability for Australian coals[9] as

$$\text{HGI} = 35.5(m)^{-1.54}$$

which demonstrated good reproducibility over a range from 48 to 106 HGI.

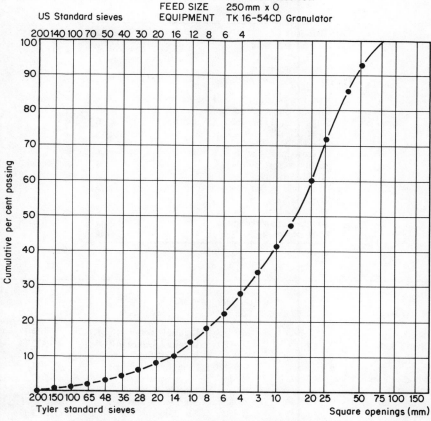

Screen openings	Cumulative % passing	Cumulative % passing	Cumulative % passing
50 mm	92.3		
40 mm	85.5		
30 mm	72.0		
20 mm	60.8		
15 mm	47.8		
10 mm	41.5		
3 mesh	34.2		
4 mesh	28.0		
6 mesh	22.0		
8 mesh	18.7		
10 mesh	14.5		
14 mesh	11.4		
20 mesh	8.32		
28 mesh	6.22		
35 mesh	4.54		
48 mesh	3.30		
65 mesh	2.27		
100 mesh	1.68		
150 mesh	1.0		
200 mesh	.82		

Fig. 3.5 Size-distribution graph for run-of-mine coal (courtesy Pennsylvania Crusher Corporation).

Abrasiveness	Class
Low abrasiveness	1
Mildly abrasive	2
Very abrasive	3
Flowability	
Very free-flowing angle of repose up to 30°	4
Free-flowing angle of repose 30°–45°	5
Sluggish angle of repose 45° and up	6
Special characteristics	
Hygroscopic	A
Highly corrosive	B
Mildly corrosive	C
Degradable when exposed to air	D
Very friable	E
Mildly friable	F
Tough—resists reduction	G
Plastic or sticky	H

Material	Class[a]	Average weight (lb/ft^3)	Material	Class[a]	Average weight (lb/ft^3)
Alumina	3-5-G	60	Lignite—Texas (ROM)	1-5-D-E	45–50
Aluminium Oxide	3-G	70–120	Lignite—Dakota (ROM)	1-6-D-F	45–50
Bagasse	1-6-C	7–8	Lime—Pebble	1-5-E	53–56
Barite	3-5	140–180	Limestone—broken	2-5-F	90–10·
Bark (wood refuse)	2-6-G	10–20	Manganese ore	2-5-G	125–1·
Basalt (broken)	3-G	10–20	Marble—broken	2-5-F	90–10·
Bauxite (crushed)	3-5-F	75–85	Marl—raw and wet	2-6-E-H	130–1·
Bentonite	2-5	40–50	Middlings—Coal	2-5-G-F	—
Brick	3-F	100–125	Phosphate—Rock—Ore	2-5	—
Carbon electrodes (baked)	2-G	—	Potash ore	1-E	—
Carbon electrodes (unbaked)	1-F	—	Potash ore compactor flake	1-E	—
Cement clinker	3-5-F	75–95	Quartz—broken	3-5-E	85–95
Cement rock	2-5-F	100–110	Refuse—household	2-G	45–50
Charcoal	2-5-D-F	18–25	Sand—dry bank	3-5	90–11·
Clay (dry)	3-5	60–75	Sand—foundry	3-5	90–11·
Calcined clay	3-F	80–100	Sandstone—broken	3-F	85–90
Coal—anthracite	1-4-D-E	55–60	Shale—broken	2-5-F	90–10·
Coal—bituminous	1-5-C-D-E	45–55	Shells—Oyster	2-5-E	70–80
Coal—sub-bituminous	1-5-C-D-E	45–55	Dicalcium phosphate	2-5-E-H	43
Coke—petroleum	2-5	35–42	Dolomite	2-5-F	90–10·
Cryolite	1-5-F	110	Slag—Open hearth	3-G	160–1·
Cullet—glass	3-5-E	80–120	Slag—Blast furnace	3-4-F-E	80–90
Diatomaceous earth	2	11–14	Slate	2-E	85–95
Dross—aluminium	3-F-C	—	Soapstone (talc)	1-F	40–50
Fluorspar	2-5-F	90–100	Superphosphate	2-6-F-H	50–55
Fullers earth—raw	2-5	35–40	Traprock—broken	3-5	105–1·
Granite—broken	3-5-G-H	90–100	Triple superphosphate	2-6-F-H	50–55
Gravel	3-5-F	90–100	Trona ore	2-5-F	90–10·
Gypsum rock	2-5-F	90–100	Tungsten carbide	3-4-G	—

[a] Refer to table above for class description

Fig. 3.6 Table of physical characteristics of rocks and ores (source, Pennsylvania Crush Corporation[2]).

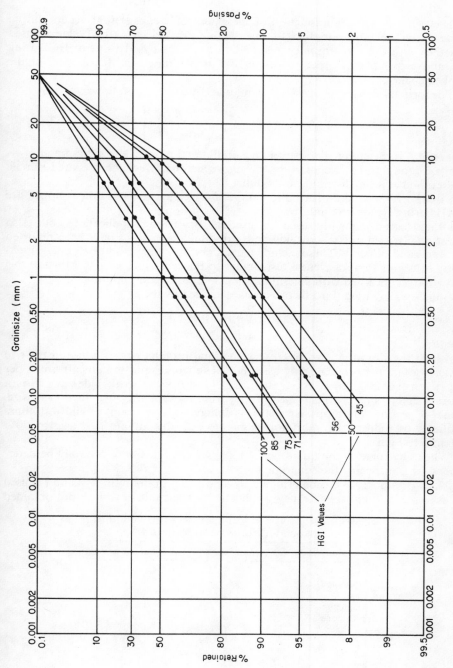

Fig. 3.7 Size-distribution curves for various cleaned coals of different HGI values.

The value of $d_{63.2}$ is a function of the extent of breakage that occurs during mine extraction and handling to the plant. Once selected, this value is used together with the value of m corresponding to the analysed value of HGI using the above formula in order to obtain the appropriate size distribution of the broken coal.

The formula derived by Rosin, Rammler and Bennet is as follows:

$$F(d) = 100\left[1 - \exp\left(-\frac{d}{d_{63.2}}\right)^m\right]$$

where $F(d)$ is the cumulative percentage of material passing size d, in mm; $d_{63.2}$ is the size modulus, i.e. the aperture through which 63.2% of the prepared sample passes; and m is the distribution modulus.

Figure 3.7 shows size-distribution curves for cleaned coals of similar upper size but differing Hardgrove indices.

Other tests, frequently used in mineral processing applications but also occasionally used in coal-crushing and -grinding applications have been reported by Bond and others.[10] These include a crushability test, rod-and-ball mill grindability tests and an abrasiveness test similar to that described earlier. Much of this testwork is concerned with determining work input in comminution.

Bond's widely used third theory is characterized by the equation:

$$W = \frac{10Wi}{\sqrt{P}} - \frac{10Wi}{\sqrt{F}}$$

in which the size in µm which 80% of the final product passes is P and the size which 80% of the original feed passes is F. The work input in kilowatt hours per short ton (i.e. 2000 lb) is designated W, and Wi is the work index which is a parameter defining the resistance of a specific rock or ore to reduction in size, under operating conditions. Hence Wi is defined as the amount of kilowatt hours required to reduce one short ton from theoretically infinite feed size to 80% passing 100 µm.

When any three of the quantities are known, the remaining one may be determined by transposing the equation.

The work index, Wi, is determined by laboratory grindability tests and is used to estimate ball-mill sizes for new applications. The value of work input obtained

TABLE 3.2 Typical Values of *Wi* for Coal and Associated Rocks and Minerals

Material	Relative density	Typical Wi value
Clay	2.23	7.10
Bituminous coal	1.63	11.37
Coke	1.51	20.70
Magnetite	3.88	10.21
Limestone	2.69	11.61
Oil shale	1.76	18.10
Pyrite	3.48	8.90
Quartz	2.64	12.77
Sandstone	2.68	11.53
Shale	2.58	16.40
Blast furnace slag	2.39	12.12
Slate	2.48	13.83

SIZE REDUCTION

represents that transmitted from the motor drive shaft to the mill. Because electrical input to the drive motor is usually measured, it should be reduced by the motor and speed reducer drive efficiency loss, but when such losses are not known, the motor input should be multiplied by 0.95 to correct the test work index. Some typical values of work index for various rocks and ores are given in Table 3.2.

3.4 SIZE-REDUCTION EQUIPMENT

3.4.1 Duty

Coarse or heavy duty	—usually raw coal
Medium duty	—commonly raw coal occasionally product
Light duty	—usually product (cleaned coal)
Fines duty	—milling and pulverizing of product coal or slurry preparation
Ultrafines duty	—micronized, pulverized product coal

Size ranges usually typified by each of the above are as follows:

Duty	Feed size (mm)	Product size (mm)
Coarse	up to 1 m cube	100–150
Medium	up to 150 mm	35–75
Light	up to 100 mm	25–50
Fines	up to 25 mm	75% 0.15
Ultrafines	up to 1 mm	95% 0.074

Coarse- and medium-duty breakers and crushers are used for *primary* size-reduction purposes, i.e. for size reduction of run-of-mine or 'raw' coal.

Medium- and light-duty crushers are used for *secondary* size reduction with the latter often employed in conjunction with a screen to minimize production of fines by pre-screening the coal at an aperture close to or at the crusher setting.

Light-duty crushers are used for *tertiary* size reduction, perhaps in preparing washed coal for subsequent milling or pulverization. Coal milling to produce coal–liquid slurry would probably form a *quaternary* size-reduction step in the sequence of coal-preparation operations if such a step was included in a conventional coal-preparation plant. However, at present, such slurry preparation is still in the developmental stage.

The normal power requirements for the various categories of size-reduction equipment are likely to be as follows:

Usual duty	Type of machine	kW/tonne/hour
Heavy	rotary breaker	0.1–0.3
Heavy	jaw crusher	0.2–0.5
Heavy/medium	roll crushers	0.2–0.6
Medium/light	hammer mill	0.3–8.0
Light	impactor	0.5–20.0
Fines/ultra	ball mill	1.0–30.0
Fines/ultra	roller mill	5.0–40.0

It is clear from the above list that the term 'duty' could be misleading if related in any way to power requirements. It relates only to the lump size and reduction ratio combination of the feed coal. Power requirements increase significantly with decrease in size-consist of the product coal and are most influenced by the hardness of the coal and associated materials requiring size reduction.

3.4.2 Types of Equipment

(a) Heavy duty
- rotary breakers
- feeder breakers
- jaw crushers
- gyratory crushers
- single roll crushers
- open hammer and ring impactors

(b) Medium duty
- cone gyratory crushers
- single-roll crushers
- double- (and triple-) roll crushers
- open hammer and ring impactors

(c) Light duty
- cone gyratory crushers
- hammer, cage impactors
- double- (triple-) roll crushers

(d) Fines duty
- high-speed hammer mill (screen)
- dry, ball-and-rod mills
- roll/race pulverizers

(e) Ultrafines duty
- vibration mills
- high-frequency attrition mills (wet)
- wet-ball mills
- roll/race pulverizers
- jet mills

3.4.3 Rotary Breaker

The forerunner of the rotary breaker was the heavy-duty trommel sizing screen. This type of screen, used extensively a hundred years ago, was observed to break coal down by its tumbling action whilst having little effect on shale and other harder rocks. Hezekiah Bradford of Pennsylvania is attributed with the invention of the first breaker in 1893 and his Bradford breaker design is still in use today.

The breaker consists of an outer fixed shell with an inner rotating cylindrical frame fitted with perforated plates (see Fig. 3.8). This drum rotates at speeds of between 12 and 18 r.p.m. and this speed is expressed as a function of the so-called critical speed which is defined as the lowest speed that will cause rock fragments to be carried around without slippage, i.e. the centrifugal speed. This critical speed decreases with increase in diameter, which means that the operating speed of the drum must also decrease with increase in drum diameter. The perforated plate-lining serves the dual purposes of screening and breakage. The machines are also fitted with lifters to raise the coal for breakage, and the lifter design may

SIZE REDUCTION

Fig. 3.8 Rotary breaker (source, Pennsylvania Crusher Corporation[2]).

also cause transportation of the cascading rocks towards the discharge end. In other designs, special ploughs are provided for this purpose. The rotary breaker is able to maintain a uniform product top-size while producing a minimum amount of fines, due to the fact that the undersized material is removed at the rate it is produced through the screen-plate openings. Most of the naturally arising undersize is removed at a very early stage, almost immediately upon entering the drum, as can be seen in Fig. 3.9. In many installations, however, a fixed or vibrating grizzly screen is installed ahead of the rotary breaker to remove

Fig. 3.9 Photograph of the inside of a rotary breaker handling 800 t/h run-of-mine coal.

Fig. 3.10 Rotary breaker with pre-screening (source, McNally Pittsburg Company[10]).

undersized material first and therefore avoid further comminution (see Fig. 3.10). All of the large-size rocks and any tramp material (wood is commonly encountered for underground mined coal) pass along the breaker, being harder or otherwise more resilient to breakdown than the coal, and are subsequently discharged by some form of peripheral plough. A slow-speed belt conveyor is often used to transport the coarse refuse to a stockpile, but chutes are also employed. Due to the size and mass of this material it is frequently directed to open, accessible piles rather than into bins or hoppers where blockage may occur. The rotary breaker is relatively trouble free, requiring little maintenance, due mainly to the fact that it is rugged in design and also because the drum rings usually ride on rollers or trunnions, there being no centre shaft or shaft bearings. There are, however, one or two shaft-driven designs for smaller, portable machines.

The versatility of the rotary breaker is seen in its ability to handle fluctuations in feed moisture, feed tonnage and size-consist. It has some limitations, too. Perhaps the greatest limitation to its use is in the size reduction of raw coal containing plastic, clayey materials. Such materials often have a tendency to roll into balls and plug up the screen perforations, requiring frequent stoppage for cleaning. The installation of chains and in some cases water sprays at the feed end of the breaker drum, has been tried to reduce this problem, with varying degrees of success.

In general, the rotary breaker can reduce run-of-mine coal of up to one metre cube to a normal minimum of about 50 mm. Some softer coals have been reduced to as low as 25 mm size, and apertures of 32 and 37 mm are common. However, most installations have apertures in the range 75–200 mm.

The size of the breaker is based upon the diameter and length of the drum. Diameters range from about 1.5 to over 4.5 m. Lengths range from 2.5 to over 8.5 m. Length/diameter ratios therefore range from 1.3 to 2.0. The length is a function of the hardness of the coal and the aperture at which separation is desired. Capacities are available up to 2000 t/h per unit.

SIZE REDUCTION

TABLE 3.3 Performance Data for Rotary Breakers[4]

Size[a] Diameter × Length (mm)	(Product top-size (mm)) Approximate capacity (tonne/h)							Approx. power (kW)
	(38)	(50)	(63.5)	(75)	(100)	(125)	(150)	
275 × 485	—	—	—	400	500	600	700	45
275 × 610	—	—	400	500	600	700	800	50
305 × 485	—	—	400	500	600	700	800	55
305 × 610	—	400	500	650	800	1 000	1 200	60
305 × 730	400	550	700	850	1 000	1 250	1 500	75
365 × 670	400	550	700	850	1 000	1 250	1 500	80
365 × 825	700	800	900	1 000	1 200	1 450	1 700	85
425 × 860	800	1 000	1 200	1 400	1 600	1 800	2 000	150

[a] Rounded to nearest 5 mm

Manufacturers' power requirements range from approximately 12 to 200 kW corresponding to a capacity range from 50 to 2000 t/h. Capacity is largely dependent upon coal hardness, grindability, and size differential between nominal upper size of the feed and the desired product upper size. One formula has been derived for calculating approximate power requirements

$$P = d\left(\frac{85 + \pi L}{13.5}\right)$$

where P is power in kilowatts, d is diameter in metres, and L is length in metres.

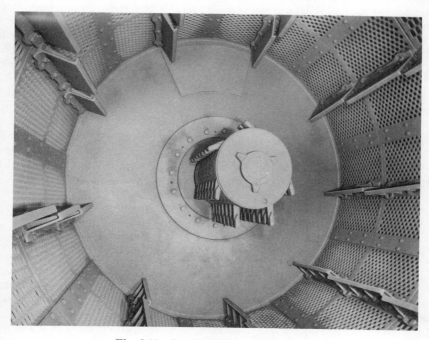

Fig. 3.11 Bradford Hammermill.[2]

Table 3.3 gives capacity and power requirement for some common sizes of rotary breaker.

Several manufacturers have produced a modified form of rotary breaker which contains an impacting device close to the feed end of the breaker drum. The addition of this impacting rotor helps facilitate breakage of harder coal lumps, thereby eliminating any potential loss with the discards. In addition, the capacity of the breaker is increased which means that such units may be of value where space is limited. Increase in capacity over similarly sized conventional breakers may be as much as 20%, but power requirements also increase from 0.1 to 0.3 kW/t/h to 0.2–0.7 kW/t/h. Figure 3.11 shows a typical hammer-mill insert.

3.4.4 Rotary Pick Breakers (Feeder–Breaker)

The basic design concept for this type of machine is that a rotating drum fitted with picks fractures rather than crushes material, thereby delivering size reduction with minimum fines and minimizing wear and tear on the size-reducing elements. Carbide-tipped picks, usually arranged either in a spiral or as parallel discs on a solid alloy–steel shaft, rotate in the same direction as the material flow which is delivered by a flight or scraper conveyor. This is also constructed of solid alloy–steel flights, pin mounted to high-strength chain. The machine drive, which usually drives both the rotor and the conveyor, can be either hydraulic or mechanical. Hydraulic drives, popular in underground installation for which this type of unit was first developed, slow down or even stall when the breaker encounters very hard material, but maintain pressure until the material is broken. To cater for surges in run-of-mine coal delivery the unit is usually accompanied by some form of lined feed-hopper. In order to reduce power requirements, some manufacturers incorporate a fly-wheel to provide the added inertia necessary to overcome surges in power demand caused by operation at high capacity, or for harder rock. However, feeder–breakers are usually selected to operate at about 50% of the design load. Figure 3.12 shows an example of a typical feeder–breaker machine employed in crushing run-of-mine coal in a surface installation. Most of the machines currently used in coal applications are, however, employed in underground coal-handling systems because of the benefit which the unit provides in terms of conveyor-belt width.

A growing application for this type of unit is as a portable raw-coal breaker for in-pit applications. Wheel- or crawler-mounted units incorporating feed hopper, scraper conveyor, breaker and transfer-belt conveyor are available with capacities of up to 3000 t/h. Many such installations result in an all-conveyor handling system to the coal-preparation plant, eliminating the need for trucking.

3.4.5 Roll Crushers

Roll crushers are size-reduction machines which shear and compress the material between two hardened surfaces. The surfaces used can take the form of a rotating roll and stationary anvil, or two rolls of equal diameter rotating at the same speed in towards one another. The surfaces of the rolls may be smooth, corrugated or toothed and for coal, where high reduction ratios are required with minimal fines generation, some form of surface feature is usually selected.

If the rolls are smooth and μ is the coefficient of friction between coal and steel surface, C is the compressive force necessary to crush the coal and θ is the angle

SIZE REDUCTION

Fig. 3.12 Stamler feeder–breaker (courtesy Stamler). Run-of-mine coal (max. lump size 1 m reduced to 205 mm) at 1000 t/h at an installation in the eastern USA.

of nip (see Fig. 3.13), a spherical piece of coal would be just gripped by the roll when

$$C \sin \frac{\theta}{2} = \mu C \cos \frac{\theta}{2}$$

or

$$\mu = \tan \frac{\theta}{2}$$

For most materials of construction this gives θ greater than 30° which means that (if θ exceeds 30°) the roll will not grip the coal. This factor therefore determines

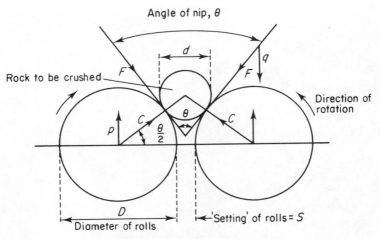

Fig. 3.13 Forces acting in crushing rolls.
C = compression forces
p and q are vertical components of friction forces, F

Hence $\qquad p = F \sin \dfrac{\theta}{2}; \qquad q = \mu C \cos \dfrac{\theta}{2} \quad$ because $F = C\mu$

For the rock to be crushed $\qquad \mu > \tan \dfrac{\theta}{2}$

assuming its weight to be negligible.

$\dfrac{d}{s}$ = reduction ratio which is usually $\approx 4:1$

the size of crushing rolls necessary to crush rock. Table 3.4 gives maximum lump sizes for various smooth-roll diameters calculated from the above relationship. The speed of the rolls depends on the angle of nip being used. The larger the angle of nip (i.e. the coarser the feed) the slower the peripheral speed must be to

TABLE 3.4 Feed Size for Smooth-roll Crushers

Roll diameter (mm)	Maximum lump-sized feed[a] (mm)
230	9
305	12
460	18
610	24.5
760	30.5
915	36.5
1070	43
1220	49

[a] Rounded to nearest 0.5 mm

SIZE REDUCTION

ensure nipping; but the crusher must be designed to take the nominal upper-sized feed material.

By introducing surface features such as corrugations or teeth (or both) the reduction ratio can be significantly enhanced. However, if the rock to be crushed is abrasive the rate of wear may prove excessive. This is why toothed roll crushers are rarely used for ore crushing. Only heavily lined, high-manganese steel, smooth rolls are sufficiently resilient to prove economical for ore crushing. Coal and associated mudstones and shales may qualify for the application of toothed single- or double-roll crushers providing that the correct form of surface is carefully selected beforehand. The speed and diameter of the toothed rolls, the required product size-consist and the feed size-consist, influence unit capacity and are therefore important in crusher selection. The clearance (or setting) between the crushing surfaces can be approximated once the feed and product specifications are known. It must be remembered that toothed double-roll crushers have roll-liner designs with which the teeth on the roll mesh with one another. Roll crushers, therefore, particularly double rolls, have a tendency to produce slabby material in the product when crushing slabby material and this must be considered when selecting the type of roll-liner design.

Roll speed must be selected to be fast enough to ensure rapid nipping of the feed but not so fast as to cause the feed to be thrown back from the crushing zone, which wastes energy and can lead to fines generation.

A single-roll crusher of the type often used for primary crushing is shown in Fig. 3.14. This machine consists of one crushing roll and a curved anvil which has renewable wear plates. The anvil is normally hinged at the top to permit the passage of tramp material without damage to the unit. Most single-roll crushers are also fitted with shear pins or other form of overload tripping device to protect the drive system. For a given reduction ratio (usually between 4:1 and 6:1), single-roll crushers are capable of primary or secondary crushing to produce a top size of between 200 mm and 20 mm. Single-roll machines are available ranging from 450 mm diameter × 450 mm long rolls to 900 mm diameter × 2500 mm long, which represents a single-unit capacity range of 20–1500 t/h, with cor-

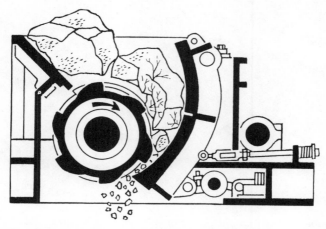

Fig. 3.14 Heavy-duty single-roll coal crusher for primary coal crushing.[3]

TABLE 3.5 Data for Single-roll Crushers[3]

Roll size diameter × length (mm)	Upper-limit feed size (mm)	(Product top-size (mm)) Approx. capacity (tonne/h)								Approx. power required (kW)
		(32)	(38)	(50)	(75)	(100)	(125)	(150)	(200)	
510 × 460	305	45	55	70	85	100	—	—	—	15
510 × 610	355	60	70	80	105	125	150	175	—	15
510 × 760	355	75	85	95	120	130	170	190	—	20
610 × 610	405	60	70	80	105	125	150	175	—	20
610 × 915	405	80	95	110	135	155	190	219	230	22.5
610 × 1220	405	125	150	170	190	240	280	300	330	30
915 × 915	460	125	150	200	240	270	300	330	360	40
915 × 1220	510	200	225	275	325	275	425	475	525	45
915 × 1375	510	235	260	310	370	425	485	545	600	60
915 × 1525	510	250	285	345	410	470	530	590	655	80
915 × 1675	510	285	320	380	440	520	590	660	730	80

responding power requirements of 15 to over 150 kW. Table 3.5 gives data for some of the more common sizes of machine.

Some double- and triple-stage single-roll machines have been developed to perform primary and secondary size reduction in a single unit.

Double-roll crushers like the one shown in Fig. 3.15 are used for primary crushing, being able to reduce run-of-mine coal of up to 1 m^3 size to a product with top size of 350 to about 100 mm, depending upon the characteristics of the raw coal. Double-roll crushers of the type shown in Fig. 3.16 have various alternative crushing-roll interlocking tooth segments, as shown in Fig. 3.17. These machines can be used as secondary raw-coal crushers, middlings crushers or

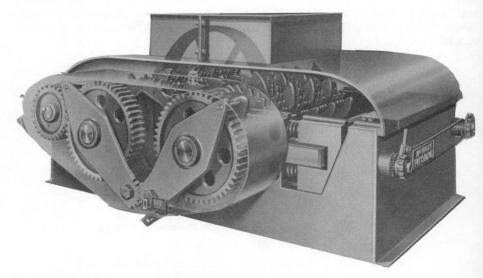

Fig. 3.15 Double-roll coal-crusher for primary crushing (courtesy McNally Pittsburg Company[10]).

Fig. 3.16 An example of a toothed-roll crusher for product coal or middlings crushing (courtesy McNally Pittsburg Company[10]).

product-sizing crushers. They are widely used for producing a stoker product, for which excessive fines generation must be avoided.

From feeds of up to 350 mm in top size, single-stage, double-roll crushers can produce crushed coal with a top size varying between 50 and 20 mm. Capacities of all double-roll crushers range from 10 to 2000 t/h per unit, with corresponding power consumption ranging from 5 to over 100 kW. The setting of these crushers can usually be readily affected by means of a spring compensate screw adjustment which is clearly shown in both Figs 3.14 and 3.15. Table 3.6 gives data for some of the more common sizes of machine.

Double-roll crushers are also manufactured in three- and four-roll, two-stage configuration. The upper stage performs coarse crushing and the lower produces the finer coal. In the triple-roll unit, the upper stage consists of a single-roll crusher while the lower section is a double-roll unit as shown in Fig. 3.16. In the four-roll unit both upper and lower sections are double-roll units.

These two-stage machines can accept material with a maximum size of 600 mm and produce a product with a top size of between 100 and 10 mm in a single pass. Usually the rolls vary in size from 300×600 mm to 450×1800 mm (diameter by length). The largest units can handle up to about 600 t/h, requiring about 175 kW power input.

Typical capacities for double-roll coal crushers, as shown in Table 3.6 are for guidance only and applications are usually determined by means of crushing tests.

As a further method of assessment, particularly for cost estimation purposes, the crusher size may be estimated by the following formula. If the theoretical volume per unit time for a ribbon of coal drawn through the crusher is $\pi dwsn$, where d is the roll diameter, w is the roll width, s is the setting, and n is the revolution per minute, then

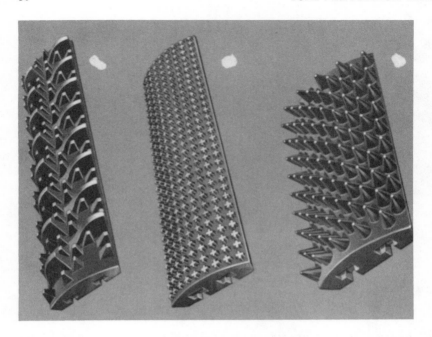

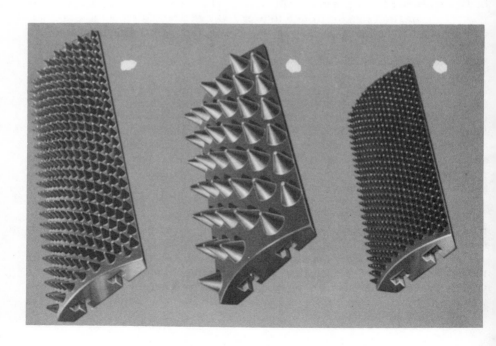

SIZE REDUCTION

TABLE 3.6 Data for Double-roll Crushers[3]

Roll size diameter × length (mm)	Upper-limit feed size (mm)	(Product top-size (mm)) Approx. capacity (tonne/h)					Approx. power required (kW)
		(150)	(200)	(250)	(300)	(350)	
(a) Gearmatic (heavy-duty) breaker							
760 × 1220	750	750	800	800	900	950	75
760 × 1525	1000	900	950	1000	1050	1100	95
760 × 1830	1250	1200	1250	1300	1350	1400	115
(b) Stoker product (medium duty)							
		19	25	32	38	50	
915 × 915	175	90	120	150	180	210	30
915 × 1220	175	120	160	200	240	280	40
915 × 1525	175	150	200	250	300	350	45
610 × 610	150	60	80	100	120	140	20
610 × 915	150	90	120	150	180	210	22.5
610 × 1220	150	120	120	200	240	280	30
460 × 460	100	25	40	50	65	—	10
460 × 610	100	35	55	75	90	—	15

$$\text{crushing rate} = \pi d w s n \rho c \text{ t/h}$$

where ρ is the bulk density of the coal in kg/m^3, and c is an empirical constant.[1]

	Value of c		
Roll diameter, d (mm)	Soft bituminous coal	Medium	Hard
460	—	0.25	—
610	—	0.27	—
660	—	0.30	—
760	1.20	0.37	0.80
910	—	0.50	—
1220	—	0.67	—
1370	—	0.75	—

3.4.6 Jaw Crushers

Although not as widely used in coal primary crushing as single-roll or gyratory crushers, jaw crushers are frequently selected for open-pit mining operations where the coal is hard, or unusually hard out-of-seam rock is likely to be encountered in varying quantities. The distinctive feature of this type of machine is the two crushing plates which open and shut like the jaws of animals.[12] The jaws are set at an acute angle, one jaw being pivoted in order to be able to swing relative to the other. Two types of jaw crusher may be seen in coal-crushing applications:

(a) single-toggle machine,
(b) double-toggle machine,

both of which are often referred to as the Blake crusher as characterized by the moving jaw being pivoted at the top. This ensures variation of the crusher 'setting' whilst maintaining fixed-feed opening dimensions i.e. 'width' and 'gape'.

A single-toggle machine as shown in Fig. 3.18 has the swing-jaw suspended on

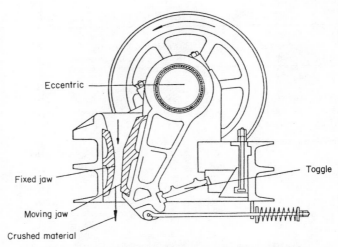

Fig. 3.18 Single-toggle dodge type of jaw crusher.[10]

the eccentric shaft, which allows for a lighter, more compact design than the double-toggle machine. Not only does the swing-jaw move towards the fixed one, by the action of the toggle plate, but it may also move vertically as the eccentric rotates. This elliptical motion aids in moving the rock through the jaws forming the crushing chamber but may also lead to higher wear on the jaw-liner plates and hence greater running costs. However, where relatively friable raw coal and shaley material requires crushing, with resultant lower abrasiveness, the single-toggle machine is the preferable option because of its lower installation and power cost. Another limitation is the fact that the direct attachment of the swing arm to the eccentric creates greater strain on the drive shaft, tending to cause higher maintenance costs than for double-toggle machines.

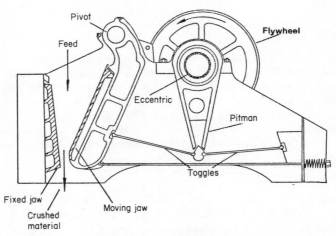

Fig. 3.19 Double-toggle Blake type of jaw crusher.

SIZE REDUCTION

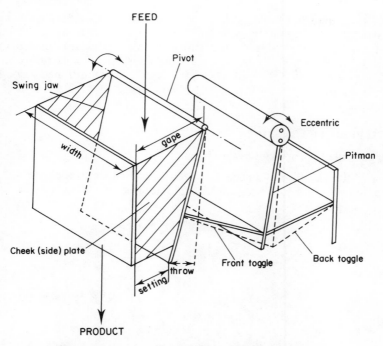

Fig. 3.20 Schematic diagram of double-toggle jaw crusher action.

The double-toggle machine is shown in Fig. 3.19. In this type of crusher the oscillating motion of the swing-jaw is created by vertical movement of the pitman. This in turn causes upward and downward motion of the front toggle connected to the swing-jaw, which moves the swing-jaw back and forward through a predetermined 'throw'. The back toggle plate causes the pitman to move sideways as it is flexed up and down. The schematic diagram in Fig. 3.20 illustrates this mechanism.

TABLE 3.7 Data for Blake-type Jaw Crushers

Feed opening gape × width (mm)	Approximate capacity (tonne/h) for various settings (mm)						Approx. power requirement (kW)
	setting	capacity	setting	capacity	setting	capacity	
180 × 250	25	3	38	4	50	6	5
250 × 500	38	10	50	15	65	20	10
380 × 610	50	17	75	35	90	45	25
710 × 915	75	40	100	65	125	105	65
915 × 1070	100	75	125	110	150	145	95
915 × 1220	100	85	125	115	150	150	100
1070 × 1525	125	165	180	330	230	580	140
1680 × 2200	180	415	280	940	300	1100	225
2150 × 3050	230	1100	280	1600	360	2850	375

All dimensions rounded to nearest 5 mm
Hard coal and shale

Double-toggle machines cost about 50% more than single-toggle machines of similar size and are generally selected for crushing tough, hard or abrasive materials where the rate of wear and tear creates a significant superiority over single-toggle machines. For coal applications, their selection is rarely justified, although they are commonly used when purchased as reconditioned second-hand units, when the price differential may not be so marked.

Table 3.7 gives data for jaw crushers.

3.4.7 Gyratory Crushers

This category of crusher includes two types:

(a) heavy-duty, long spindle; characterized by a small cone angle;
(b) medium-light-duty, short spindle; characterized by a large cone angle.

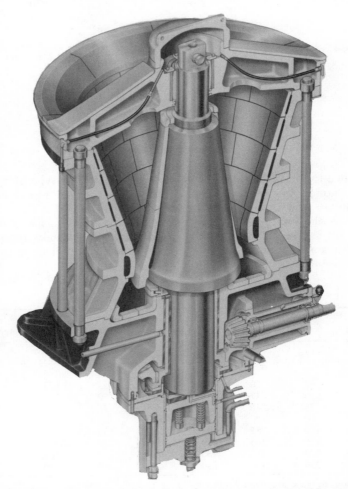

Fig. 3.21 Nordberg heavy-duty gyratory crusher (courtesy Rexnord Incorporated).

SIZE REDUCTION 85

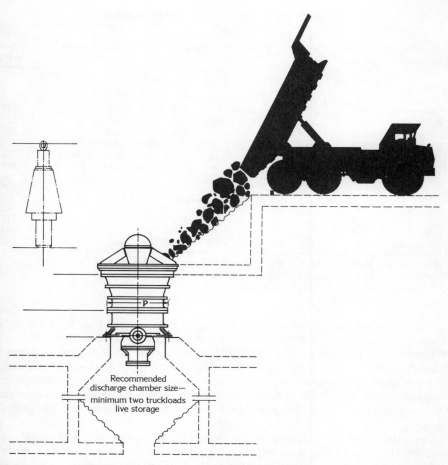

Fig. 3.22 Arrangement of feed system for primary gyratory crushers (courtesy Allis–Chalmers).

The heavy-duty machines, similar to that shown in Fig. 3.21, are primary crushers for run-of-mine coal crushing. They consist of a long spindle carrying a hard-steel, conical crushing element seated in an eccentric sleeve. This spindle is supported from an upper frame called a spider and rotates in a conical motion with respect to the crushing chamber. The spider is in turn supported by the main steel frame encasing the machine. Both components of the crushing chamber are lined with wear-resistant, replaceable liner plates. Those lining the fixed shell are called concaves. As with the jaw crusher, maximum movement of the head occurs near the discharge, which tends to relieve the tendency for chokage caused by swelling of the broken rock.

Heavy-duty gyratory crushers range in sizes up to 1830 mm gape and can crush material sized up to 1350 mm at rates of up to 3000 t/h to produce up to 300 mm top-sized product. In coal applications this type of crusher is best suited to a direct-tip type of feed system. Raw coal is fed directly into the crusher which

TABLE 3.8 Data for Gyratory Crushers

Crusher size[a] (mm)	Gyrations (per min)	Throw (mm)	(Open-sided settings[b] of discharge opening (mm)) Approximate capacity (tonne/h)								Approx. power (kW)
			(50)	(75)	(100)	(125)	(150)	(180)	(200)	(250)	
					Primary crushers						
750 × 1400	175	32	—	510	650	810	—	—	—	—	225
910 × 1400	175	32	—	—	600	760	—	—	—	—	225
1070 × 1650	150	38	—	—	—	1000	1250	1650	—	—	300
1220 × 1880	135	40	—	—	—	1700	2000	2300	2700	—	375
1370 × 1880	135	40	—	—	—	—	1950	2250	2550	—	375
1525 × 2260	125	45	—	—	—	—	2500	2840	3260	—	450
1525 × 2700	110	50	—	—	—	—	—	—	4200	5260	750
					Secondary Crushers						
330 × 910	285	25	200	300	—	—	—	—	—	—	95
405 × 130	225	32	350	430	—	—	—	—	—	—	110
610 × 1525	175	32	—	—	660	—	—	—	—	—	175
750 × 1780	150	38	—	—	1080	1350	—	—	—	—	300

[a] Size designates the size of the receiving opening by diameter of the mantle
[b] Open-side setting is maximum opening at discharge
All dimensions have been rounded to nearest 5 mm

operates very effectively when choke fed in this manner (see Fig. 3.22). However crushers of this type require height which can result in costly structures. This type of crusher has the largest unit capacity: crushing rate capability of all types of primary crushers, mainly due to the fact that, due to its conical action, it crushes on a full cycle as compared to a jaw crusher in which compression is intermittent. To exploit this advantage, gyratory crushers should be choke fed to ensure continuous operation. Table 3.8 gives data for gyratory crushers used for primary crushing.

The other form of gyratory is the cone crusher which was developed as a modified version of the heavy-duty machine to act as a secondary or tertiary crushing unit. Following their development, they quickly emerged as a replacement for smooth, double-roll crushing in ore-treatment applications. There are three common types:

  (i) standard-head conical;
  (ii) short-head conical;
  (iii) hemispherical head or gyrosphere.

Figure 3.23 illustrates the cross-section of each type, demonstrating their close resemblance to the primary crusher type. The essential difference is in the short spindle which is not suspended, but instead supported in a curved universal bearing located below the gyratory head or 'cone'. Unlike the heavy-duty machine the throw is easily adjusted and can be up to as much as five times that of the primary machine as a result of the lighter duty and lower working stresses encountered. Cone crushers range from about 500 mm to over 3000 mm and have capacities of up to 1500 t/h of hard ore. Their use in coal is restricted by their comparative cost with roll-and-hammer types of crusher. Because coal tends to be less abrasive and softer than most ores, cone crushers, which were essen

SIZE REDUCTION

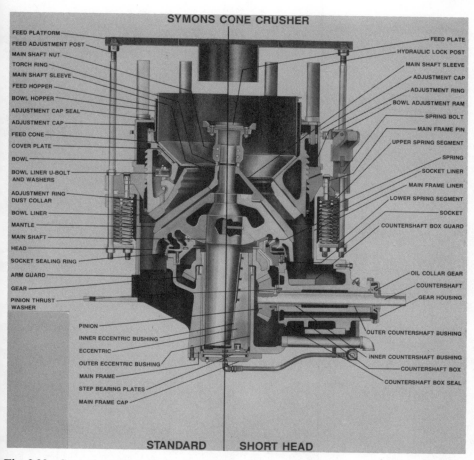

Fig. 3.23 Secondary and tertiary types of Symons cone crusher (courtesy Rexnord Incorporated).

tially developed for crushing hard rocks and ores, can rarely be justified. Nevertheless, they are frequently found in coal applications, particularly as components of semiportable or mobile treatment plants.

3.4.8 Hammer Mills and Impact Crushers

The *hammer mill*, as shown in Fig. 3.24 is a size-reduction machine in which the coal is initially impacted by rotating hammers or rings and then subsequently further reduced by impact against grid plates, the inner lining and other hammer impacts. In such machines, the majority of the size reduction occurring as a result of impact occurs above the rotor by the action of the hammers. Other, subsequent size reduction is by compression and attrition between the hammers and the lining and grid.[3] This attrition component often results in the generation of large amounts of fines, especially for the more friable coals. In such cases a hammer mill with either a large grid aperture, or a machine without a grid, might

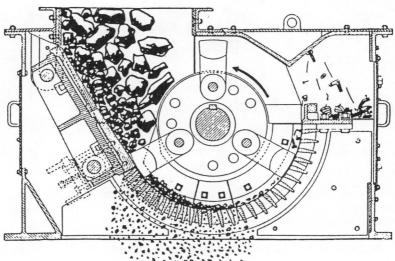

Fig. 3.24 Hammer-mill type of coal crusher.

prove applicable if used in closed circuit with a screen as shown in Fig. 3.25. The ring type of hammer mill as shown in Fig. 3.26 was developed to help minimize fines production. In this type of mill the rings often have alternate toothed and plain perimeters and are free to revolve. This feature tends to concentrate the compressive action and reduce attrition effects. Another method employed to reduce fines production is the use of an air-swept crusher unit. Air-flow regula-

SIZE REDUCTION

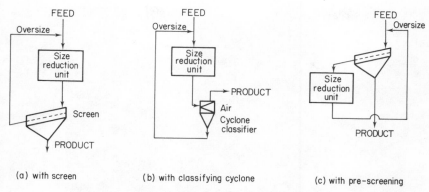

(a) with screen   (b) with classifying cyclone   (c) with pre-screening

Fig. 3.25 Closed-circuit crushing system).[3]

(a) and (b) Capacity of crusher (t/h) $= \dfrac{\text{feed (t/h)}}{E \times p}$

(c) Capacity of screen (t/h) = capacity of crusher + feed (t/h)

Capacity of crusher $= \dfrac{1 - (E \times f)}{E \times p} \times$ feed (t/h)

where E = screen efficiency, and f and p are cumulative fraction of feed and product passing screen opening, respectively.

Fig. 3.26 Ring-type hammer mill.

90 COAL PREPARATION TECHNOLOGY

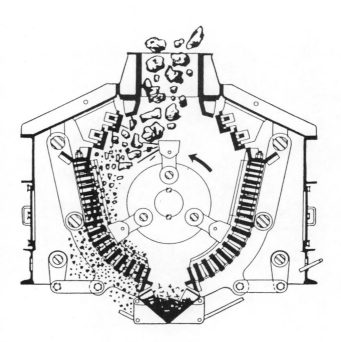

Fig. 3.27 Reversible hammer mill (courtesy Pennsylvania Crusher Corporation).[2]

SIZE REDUCTION

tion with such units, in conjunction with some form of air classifier, usually a cyclone, creates a closed-circuit crushing system with reasonable control over fines production. The hammer mill is a versatile heavy- to medium-duty machine which is employed in size reduction of raw coal, washed coal, middlings and in some cases coarse refuse. Because of the high rotational speed of such machines they are vulnerable to severe damage by tramp iron and steel and must usually be protected by metal detectors and magnets. Tramp iron that does enter the machine is often captured by some form of side pocket or a special trap similar to that shown in the drawing of the reversible hammer mill shown in Fig. 3.27.

Factors that affect the performance of hammer mills are rotor speed, rotor diameter, feed characteristics, machine type, hammer or ring design and clearance and grid apertures.

The most critical of the feed characteristics other than, perhaps, hardness is probably moisture. An increase in moisture content in the feed to the machine tends to reduce capacity and result in greater fines generation because of the longer residence time in the crushing zone.[14] Also, capacity is reduced as upper size in the feed decreases, as with most other crushers.

Rotor speed has been shown to be directly proportional to capacity, and the fineness of the product is largely determined by the intensity of the hammer impact force. This in turn has been observed to be directly proportional to the tip velocity of the hammer.[13] The clearance between the hammer and the grid plates affects the product size-consist. As the clearance is reduced there is an increase in the associated comminution occurring in this zone which most affects the coarser fragments.[15] Grid aperture has been observed to have greatest effect on size reduction and the generation of fines when it exceeds the hammer clearance.[15] To help reduce uneven wear, which leads to higher maintenance costs, most units are fitted with reversible hammers. Hammer wear has been shown to have its greatest influence on mill performance and maintenance requirement with higher rotor speeds.[15] In order to attempt to optimize performance and maintenance, the reversible hammer mill, of the type shown in Fig. 3.27, was developed. This permits rotation in either direction, and by periodic reversal the wear rate is maintained at a more or less even level.

Hammer mills are available for crushing run-of-mine coal as large as 1500 mm, although normally the majority are required to handle top sizes in a range of 450–650 mm. Rotor size is designated by the diameter and width, and sizes range from about 300 × 300 mm to 1850 × 3600 mm (diameter × width). Power requirement ranges from 15 to about 1850 kW and rotor speeds vary from 400 to 1800 r.p.m., the higher speeds being applicable to the lighter-duty (smaller coal-sized feed) machines. Crushing capacities, although greatly dependent upon coal characteristics, will range from as low as 5 t/h for sample and pilot-plant crushers to as much as 200 t/h.

Impact crushers are size-reduction machines which impact the incoming feed causing it to be directed towards a hard surface. The main difference between the hammer mill and the impactor types of crusher is the elimination of the compression/attrition zone created by the grid bars. In many respects, impact crushers are perhaps the best machines for obtaining a cubic-shaped, well-distributed size distribution for a crushed coal because of the predominance of the impacting effect which causes maximum breakage along natural cleavages. An impact mill is believed to apply its shattering power in direct proportion to

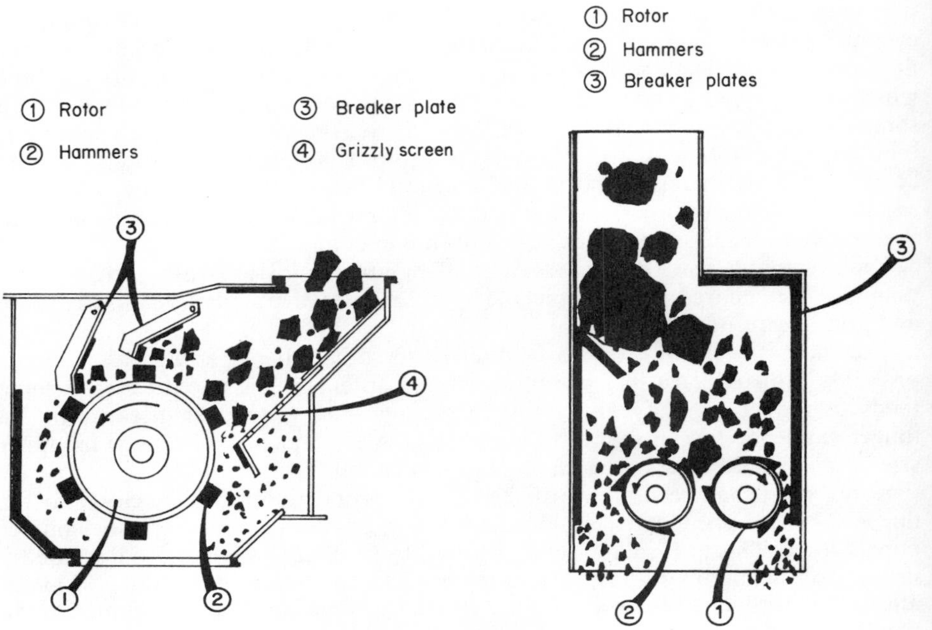

① Rotor
② Hammers
③ Breaker plates

① Rotor
② Hammers
③ Breaker plate
④ Grizzly screen

Fig. 3.28 Rotary-type impact crushers.

the lump or grain size of the feed.[16] Owing to the absence of grid bars, the product obtained from impact crushers is usually distinctly coarser than that obtained from hammer mills, but the risk of producing oversize material is greater and a closed-circuit flowsheet of the type shown earlier in Fig. 3.25 may prove more effective. Both screens and air classifiers are employed as the size controller in such circuits.

The rotor type of impact mill as shown in Fig. 3.28 uses one or two rotors, often equipped with some form of impacting device, to create size reduction. The feed descends into the path of the rotor which impacts and shatters the coal and at the same time directs the fragments into the lined-steel breaker plates. By a succession of rebounds between the two, the fragmented coal gradually passes downwards to the discharge opening. Some machines employ an internal breaker screen through which finer material may pass to avoid further size reduction. Other machines are available which have reversible rotors and therefore offer more favourable wear-control facility. Factors which influence the performance of rotary impact mills include rotor speed, rotor diameter, feed characteristics, coal type and hardness and feed rate. These factors follow the same general relationships as discussed earlier for hammer mills. Feed-size upper limit may vary from 3 to 1500 mm, the lower end being mainly sample preparation and test crushers. Product size will range from 50 μm for laboratory machines to about 150 mm for medium-duty crushers used for secondary raw-coal crushing, or middlings or cleaned-coal applications.

Rotor size normally ranges from 450 × 450 to 1800 × 3000 mm, diameter × width, and rotor speed ranges from 200 to 1500 r.p.m., with higher speeds corresponding to finer-sized feeds.

SIZE REDUCTION

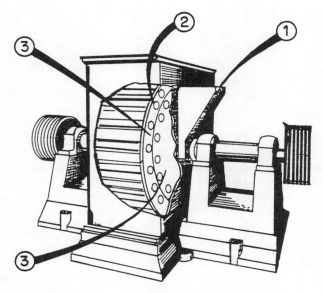

① Feed inlet
② Rods
③ Cages

Fig. 3.29 Two-row cage-mill type of crusher.

Another form of impact crusher is the *cage mill* as shown in Fig. 3.29. These can comprise one, two, four or six concentric cages with each adjacent cage spinning in the opposite direction but at equal speed. Two drives are required for the multistage machines and one motor for the single-cage units. Each cage consists of rows of rods supported at each end by a ring. Rod diameters vary from 100 to 250 mm in the range of crushers from light to medium duty. The feed is introduced into the centre of the inner cage and moves outwards by centrifugal force until it reaches the casing of the machine which is lined with wear-resistant material. The various impacts which occur during the passage to the casing create the size reduction of the coal. The final product discharges through the bottom of the mill and is usually classified in a flowsheet of the type shown in Fig. 3.25. The principal means for controlling the extent of size reduction is speed control of the cages. But product size can also be influenced by rod spacing.

Another version of the cage mill is the stud or peg mill which employs studs or pegs instead of rods or bars. Various sizes of cage mills are available for handling a range of feed size from 10 mm to over 450 mm. Product size can therefore vary from about 35 mm upper limit to below 150 μm. The largest machines can handle capacities of up to as much as 3500 t/h. Power requirements range from as little as 5 kW for laboratory machines to well over 1000 kW, with corresponding cage rotation speeds of 1600–600 r.p.m. Commercial cage mills have cage diameters of 450 mm to about 2500 mm, this diameter being that of the outside of the largest cage.

TABLE 3.9 Data for Hammer-mills

Rotor diameter[a] (mm)	Max. feed lump-size (mm)	Rotor speed (r.p.m.)	(Product size (mm); 85–90% passing) Approximate capacity (tonne/h)					Approx. power required (kW)
			(38)	(32)	(25)	(19)	(12.5)	
380	250	600–720	70–80	50–65	45–50	35–40	25–30	15–20
460	300	600–620	80–90	70–80	60–65	50–55	40–45	20–25
610	450	600–720	140–150	130–140	100–120	80–100	60–70	25–35
760	500	600–720	200–225	175–200	150–170	130–150	80–100	45–55
970	600	600–720	250–300	225–250	200–235	175–200	125–150	55–70
1070	700	600–720	325–350	300–325	285–325	275–300	175–200	70–95
1370	700	600–720	375–400	350–375	330–350	300–325	225–250	95–115
1525	700	600–720	400–450	390–420	360–380	325–340	250–275	120–150

[a] Rounded to nearest 5 mm

Some combination crushers employing an initial hammer-mill stage and a subsequent impact stage have been developed to provide pulverization capacity in a single machine for use in electrical power plant coal-preparation circuits. Such machines are commonly incorporated into a closed circuit with air classification which further acts as a final coal-drying stage. They usually accept a feed with a 50 mm top size and produce a pulverized coal with about 70% minus 75 μm. Capacities range from 700 to 2500 kg/h and require power of up to about 350 kW.

Tables 3.9 and 3.10 give data relating to hammer-mill and impact types of crusher.

3.4.9 Tumbling Mills

Tumbling mills are widely used in ore treatment for fine grinding to create liberation of valuable minerals prior to beneficiation. Most often they are operated with wet slurries of the ore but dry milling is also widely practised.

In coal preparation, coal milling using tumbling mills is more or less confined to the following fields of application:

(a) preparation of dry pulverized coal for power plant cumbustion systems;
(b) preparation of fine, coal slurries for coal transportation by pipeline;
(c) deep cleaning of coal by wet processes such as agglomeration;
(d) deep cleaning of coal by dry processes such as electrical and magnetic separation;
(e) preparation of coal–liquid slurry fuels.

Two types of mill charge (i.e. grinding media) are employed which tend to characterize the form of system. They are:

1. ball mills, which utilize forged steel, cast alloy or ceramic balls or some form of specially shaped medium such as cylpeb;
2. rod mills, which utilize hot-rolled, high-carbon or alloy–steel rods.

Since most coals are relatively soft compared to these forms of media, autogenous grinding, commonly practised with ore treatment, is not a practical pro-

TABLE 3.10 Data for Impact Crushers[3]

Machine size[a] (mm)	Maximum capacity (t/h)	(for product size (mm)) Normal rotor speed (r.p.m.)											Approx. power required (kW)
		(150)	(125)	(100)	(75)	(65)	(50)	(38)	(32)	(25)	(20)	(12)	(3)	
925 × 1220	200	—	—	—	96	90	77	64	57	50	40	32	12	110
1220 × 1270	400	—	100	96	89	75	60	50	44	35	22	18	3	225
1525 × 1830	800	95	88	77	68	56	44	38	32	26	19	15	5	300

[a] Single-rotor impactor

Fig. 3.30 Tumbling mills used for coal comminution.

position for coal treatment. Figure 3.30 shows typical ball-and-rod mills used for coal comminution.

Wet-milling units are further characterized by the discharge arrangement as shown in Fig. 3.31.

Grinding within a tumbling mill, whatever form of charge is used, is influenced by aspects of size and quantity, the type of motion and the voidage characteristics within the mill, as well as by whether the process is being carried out wet or dry. As opposed to crushing, which occurs between relatively rigid surfaces, grinding is a random process employing several mechanisms which may change in relative proportions as size reduction takes place. Principal mechanisms such as impact or compression gradually give way to shear, abrasion or attrition.

The distinctive feature of all tumbling mills is the use of loose crushing media which are usually large in size and mass relative to the coal being crushed. They

SIZE REDUCTION

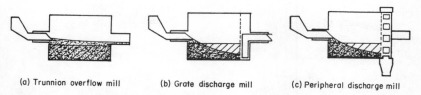

Fig. 3.31 Various types of ball-mill discharge arrangement for wet-milling applications.

are, however, small in relation to the mill dimensions and occupy slightly less than half of the internal volume of the mill. Due to the rotational and frictional effects of the mill shell, the grinding media and material to be crushed, collectively known as the *charge*, is lifted along the rising side of the mill until a position of dynamic equilibrium is reached. Under these conditions the uppermost material is caused either to cascade or to cataract down the free surface of the remaining material, about a dead zone where limited movement occurs. This culminates at what is called the *toe* of the charge, as shown in Fig. 3.32. The speed at which the mill is run is most important since it will govern the degree of comminution achieved and also the amount of wear on the shell liners and grinding media. Various types of lining are available, of which the most common are shown in Fig. 3.33.

The driving force of the tumbling mill is transmitted via the liner to the live charge. At relatively low speeds, or with smooth liners, the charge tends to roll around to the toe and much of the comminution achieved is from abrasion. This effect, known as cascading, is a popular requirement of coal-milling applications. At higher speeds, the charge is projected clear of the revolving mass of grinding media and coal and describes a series of parabolas prior to impacting in the vicinity of the toe. This effect, known as cataracting, is often too severe for coal milling unless relatively coarse material is added to the mill. At a so-called 'critical speed', the theoretical trajectory of the medium is such that it would fall outside the shell; but instead it is pinned to the inner periphery by centrifugal force and therefore revolves with the mill. It is easy to determine mathematically a theoretical value of this speed, N, and then adjust it by using constants to cater for frictional and lining effects.

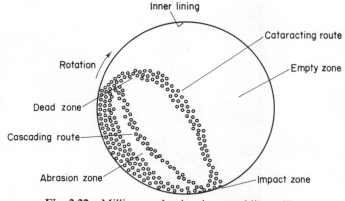

Fig. 3.32 Milling mechanism in a tumbling mill.

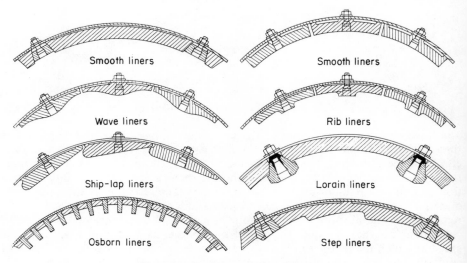

Fig. 3.33 Common mill-shell liners.

If the ball (or rod) which is lifted has a diameter, d, and the mill, a diameter, D, and the shell revolves at a velocity, V m/s, the ball will abandon its circular path for a parabolic one when just balanced by the centrifugal force, i.e.:

$$\frac{2mV^2}{D} = mg \cos \theta$$

where m is the mass in kg of the ball and θ coincides with the point on the mill circumference shown in Fig. 3.34. Acceleration due to gravity is g in m/s² and since:

$$V = \frac{\pi DN}{60}$$

$$\cos \theta = 0.0005 N^2 D$$

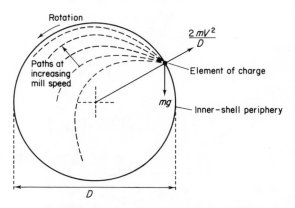

Fig. 3.34 Forces acting in a tumbling mill.

SIZE REDUCTION

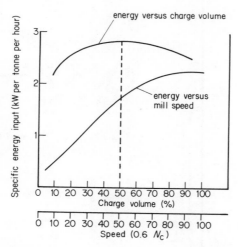

Fig. 3.35 Relationships between energy input and mill-speed and charge volume.

When the diameter of the ball is taken into account, the radius of the outermost path becomes $(D - d)/2$ and

$$\cos \theta = 0.0011 N^2 \left(\frac{D - d}{2} \right)$$

If the critical speed is when $\theta = 0$, then $\cos \theta = 1$ and

$$N_c = \frac{42.3}{\sqrt{(D - d)}} \text{ r.p.m.}$$

This is, of course, a theoretical value, and practical values of operating speeds are usually 50–90% of N_c for wet-coal milling. This often equates to about 20–30 r.p.m. in wet-milling applications where the optimum will be determined as being the point where energy consumption approaches a minimum value. Both speed and charge volume have characteristic relationships with energy, as shown in Fig. 3.35. Both efficiency curves are relatively flat but show a peak value which in the case of the mill charge gives a maximum value at about 50%.

In ore treatment, ball mills are usually operated at higher speeds than rod mills in order that full advantage may be obtained from the cataracting effect as a result of the more 'fluid' charge. For coal-milling applications, rod mills are not often employed, but where they are the speed is similar to that employed for a ball mill of the same size.

Where dry milling is employed, the milled product size is controlled by an air classification system of the type shown in Fig. 3.36. Sweeping air is often preheated in power-plant milling applications using waste heat from the combustion system. This ensures drying of any surface moisture entering with the feed and consequently improves classification. Air-flow regulation controls the product size so that low air-flows result in very fine grinding and longer milling circuit residence times, and vice versa for high air-flows. Flow rate may be determined

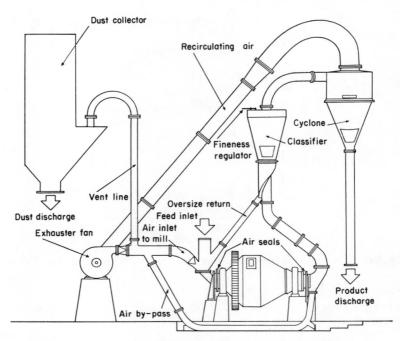

Fig. 3.36 Air-swept, dry-ball mill with closed-circuit classification.

by the moisture content of the feed and a suitable range of adjustment will be designed into the system to cater for drying requirements.

All tumbling mills must incorporate means for adjusting mill rotational speed in order to permit optimum operating conditions to be determined and maintained.

Ball mills are capable of producing finer-sized pulverized coal than rod mills, and are normally fed with coal of up to 50 mm top-size. Product size range may be from 1 mm to as fine as 10 μm in mean size. The characteristic length-to-diameter ratio of a ball mill for coal milling is between 1.0 and 1.5, which is

TABLE 3.11 Ball Mill Performance Data[3]

Diameter × length (mm)	Feed size (mm)	Product size (mm)	Hardgrove GI	kW/tonne	Approx. capacity (tonne/h)	Type of circuit
1525 × 2800	50.0	75%—0.074	50	14.0	4	open/dry
3960 × 6555	12.5	95%—0.50	55	7.5	230	open/dry
3050 × 5640	19.0	85%—0.074	46	35.0	18	open/dry
3660 × 6100	10.0	85%—0.074	48	27.0	34	closed/dry
4450 × 7160	6.4	70%—0.074	41	?	100	closed/dry
2745 × 3175	25.0	90%—0.074	50	38.0	9	closed/dry
2745 × 3175	6.4	75%—0.074	57	23.0	16	closed/dry
2135 × 3660	50.0	90%—0.074	50	1.0	5	closed/dry
2440 × 2860[a]	40.0	85%—0.074	52	21.0	8	closed/dry

[a] Only one which is not bituminous-rank coal

SIZE REDUCTION

somewhat narrower than for ore treatment, where length-to-diameter ratios can range from 0.5 to over 3.5. Charge diameter is usually about 50 mm average but can vary from 20 to 100 mm. Varying proportions of balls of different sizes will be added at the outset but only balls of a fixed size added thereafter. At present ball-mill systems of up to about 300 t/h are employed which utilize power at up to 5000 kW depending upon the fineness of grind required.

Rod mills employed for coal milling tend to have length-to-diameter ratios of 1.3–2.0. Rod lengths are sized in ratio to diameter and the range is 1.25–1.4 to minimize tangling of worn rods. Charge filling in rod mills is usually about 35–45% and rod diameters available range from 35 to 75 mm for coal milling.

Some performance data for dry, air-swept ball mills are given in Table 3.11.[3]

3.4.10 Roller Mills

This type of mill includes roll-and-race, ring-ball mills and bowl mills, all of which are shown in concept in Fig. 3.37. All of these varieties are used exclusively for fine-coal pulverization. Size reduction occurs as a result of the coal being ground or compressed and abraded between two hardened surfaces, whereby one surface rolls over the other, trapping the coal in between. The moving, grinding surface takes the form of rollers in a bowl, rollers in a ring or race or balls in a

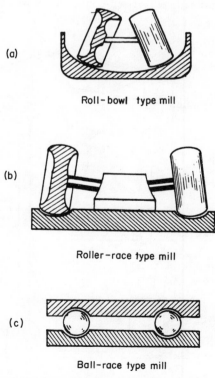

(a) Roll-bowl type mill

(b) Roller-race type mill

(c) Ball-race type mill

Fig. 3.37 Various forms of roller-type mills.

ring or race, and either mechanical springs or hydraulic cylinders are used to provide the appropriate pressure.

In a ball-ring type mill, the diameter of the ball is about 20–30% of that of the ring diameter, whereas for a roll-ring type, the roll diameter is about 25–45% of the ring diameter. As the coal is pulverized it becomes entrained in a stream of hot air which serves to dry and transport the ground coal to an air classifier in a circuit similar to that shown previously in Fig. 3.36. In this type of pulverizer, size-reduction forces are applied to a static bed of relatively fine coal rather than to individual coal grains. In order to achieve significant size reduction a greater amount of energy is required than for other forms of fine-coal crusher. It has been observed that for milling to ultrafine sizes, energy requirements increase exponentially as a result of energy losses created by a cushioning effect produced by the large quantity of very fine material present in extensively milled coal.[17] Other factors which influence the efficiency of roller mills are hardness and abrasiveness, feed rates, air-flow rates and the air temperature. For a given product fineness, the power requirements will decrease as Hardgrove grindability index increases. This is perhaps predictable since the index measures increasing HGI for softer coals[18] but the relationship is almost a proportional one for bituminous coal. Roller mills are capable of reducing feed material with an upper size of 50 mm, although the initial size reduction to about 10 mm is often performed with another type of crusher. The typical product from this type of mill is 70% minus 75 μm, i.e. suitable for pulverized-coal-fired, power-generation boiler firing.

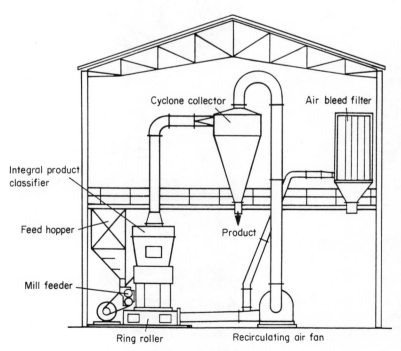

Fig. 3.38 Typical roller-mill installation.

SIZE REDUCTION

Units are available for handling at up to 100 t/h requiring over 750 kW to achieve the desired milled product. Figure 3.38 shows a typical installation.

3.4.11 Other Forms of Crusher (Pulverizer)

A great deal of attention has been focused on fine-coal size reduction in recent years. In particular, ultrafine coal milling has commanded much interest with the upsurgence of interest in deep coal-cleaning and micronized coal-combustion. These new areas of coal utilization have motivated the development of various devices for ultrafine size reduction including the following:[3]

- vibration mills;
- jet mills;
- high-speed hammer mills;
- attrition or stirred mills.

The vibrating mill is a cylindrical mill with its shell almost completely filled with grinding media. Transverse vibration is produced along the axis of the mill to cause high-frequency impact to occur within the charge, rather than rotation around the longitudinal axis. These vibrations cause the charge to oscillate and move longitudinally through the mill like a fluid in a complex spinning helix, thus allowing the grinding medium to reduce it by attrition. The material is fed and discharged through flexible-bellow type hoses in the type of machine shown in Fig. 3.39. The outstanding features of an effectively applied vibration mill are their relatively small size in relation to their capacity and low power consumption. They can mill coal to extreme fineness, i.e. to surface areas of above 500 m^2/g, a degree of fineness well below that of a tumbling mill.[19] Individual units can mill up to 15 t/h with oscillation frequencies of up to 1800 oscillations per minute.

A jet mill is a size-reduction device which has no moving parts. It utilizes a high-velocity stream of air, steam or hot waste gases at pressures of up to 700 kPa, in order to generate interparticle collisions. This type of mill, commonly known as a fluid-energy mill, has been utilized for many years for pulverizing very soft materials. More recently they have been adapted to suit specific materials, including coal, for which they are able to produce average particle sizes of below 5 µm at feed rates in excess of 4000 kg/h. Typical fluid-energy mills used for preparing micronized coal are the Sturtevant micronizer and the Jet-O-Mizer mills. Table 3.12 gives data for the latter mill for coal applications.

High-speed hammer mills or micropulverizers are able to pulverize coal to below an average particle size of 10 µm. Special stirrup or bar types of hammer design combined with close-fitting grinding surfaces permit ultrafine commi-

TABLE 3.12 Data for Jet-O-Mill Pulverizers (source: Fluid Energy Processing and Equipment Company)

Material	Feed (kg/h)	Fluid medium	Pressure (kPa)	Temperature (°C)	Approx. feed size (mm)	+50 micron	Average micron	Maximum micron
Anthracite	1350	steam	1380	400	5–8	0	3.5	35
Bituminous coal	4100	air	690	115	3–4	0.5	4.5	—

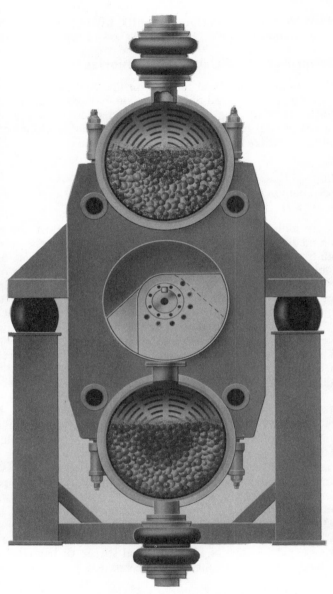

Fig. 3.39 Vibratory mill.

nution, and air elutriation is used both to cool the fine coal and to remove product. Higher speeds are also employed, but power requirements are only slightly higher than for conventional hammer mills. One manufacturer, Micro-Pulverizer, produces a range of machines with power requirements of 2–45 kW, capable of milling bituminous coal to 99% minus 30 μm at rates of 150–2500 kg/h.

SIZE REDUCTION

Attrition mills utilize a paddle-wheel or impeller to stir the grinding media and coal within a totally enclosed chamber. The grinding media may be spherical or disc shaped, and the coal is introduced as a paste or high-concentration slurry. Attrition machines can produce ultrafine coal below 10 μm in size.

3.5 SELECTION OF SIZE-REDUCTION EQUIPMENT

In the foregoing section many data regarding the performance characteristics of more or less the entire current range of coal size-reduction equipment are given. Apart from the broad-based principles which allow for selection into the various duty characteristics, i.e. heavy, medium, light, etc., there is relatively little to assist with the final selection of a specific type of crusher from within each duty category, e.g. hammer versus impact. The process of such a selection must begin with the collection of the information necessary for producing a shortlist of alternatives.

In this regard, most crusher manufacturers provide useful information checklists similar to that shown in Fig. 3.40. These, when completed, will allow the manufacturer to select and quote for a suitable crusher or other size-reduction machine. If certain information is not available, the manufacturer may request sample material to conduct compression, crushing or other forms of test work to ascertain the selection, and also to compare product size-distributions obtainable with various operating conditions, as shown in Fig. 3.41.

It is usually advisable for a potential purchaser of size-reduction equipment to request technical and cost quotations from several potential suppliers that are collectively able to offer the various alternatives under consideration. In several instances, a single manufacturer may be able to supply more than one type of crusher, in which case the eventual recommendation will be of greater value to the purchaser. This should be taken into consideration when preparing bid lists for such equipment.

We have seen in the third section of this chapter what information can be obtained by testing the material requiring crushing, and in the fourth section we have learned the value of empirical data in eventual crusher selection and sizing.

The key factors associated with the final selection are closely related and are as follows:

- specific energy (SE) or the energy per unit of throughput required to reduce the feed coal to the required product size;
- grindability (HGI) as obtained by the standard Hardgrove method of testing;
- reduction ratio (RR) or the ratio of average feed size to average product size obtained.

The first of these factors must be derived either by test or by calculation. The second is determined by testing. The third is a function of the size-reduction unit and the coal requiring crushing.

Obtaining all three is an important step in finalizing the technical requirements for selection. There is a relationship between them which is:

$$SE = f \times HGI \times RR$$

106   COAL PREPARATION TECHNOLOGY

Crushing Machinery
Primary — Secondary — Tertiary
Application Data

ALLIS-CHALMERS
CRUSHING & SCREENING

GENERAL CRUSHER INFORMATION

ITEM 1
Page 1

October, 1963

Company _____

Address _____

Location of Plant _____

Consulting Engineer _____

Address _____

Customer's Inq. or Order No. _____   District Office Inq. or Requisition No. _____   A-C Order Number _____

NECESSARY DATA

	PRIMARY	SECONDARY	TERTIARY
Name of material to be crushed			
Maximum size of feed			
Is material slabby?			
Is material abrasive?			
Is material sticky?			
Does material contain clay? ____%			
Is material damp? ____% moisture			
Crushing resistance			
Type of feeder			
Capacity (tons per hour) Average			
Peak			
Short 2000 lb ☐ Metric 2205 lb ☐ Long 2240 lb ☐			
Size of product(s) required			
Are fines removed from feed?			
Size of fines removed			
Open or closed circuit			
Percent circulating load			
Preceding crusher discharge opening			
Power — electric motor, engine, etc.			
Electrical characteristics			
Maximum hp for full voltage starting			
Is cooling water available? _____ temperature			
Altitude of plant			

(over)
17S8754

Fig. 3.40   Typical information data form (courtesy, Allis-Chalmers).

SIZE REDUCTION

ITEM 1	GENERAL CRUSHER
Page 2	INFORMATION

ALLIS-CHALMERS
CRUSHING & SCREENING

Crushing Machinery

SUPPLEMENTARY DATA

Product(s) Specification (Screen Analysis) — additional information not covered in "Size of product(s) required" on reverse side of this sheet.

Primary	Secondary	Tertiary
_____	_____	_____
_____	_____	_____
_____	_____	_____

Remarks or Additional Data _____
_____
_____
_____

Sketch of plant flow (show existing and proposed equipment with screen separations).

Data Secured:

From _____
      (Name of Individual)

Position _____ Date _____

Data Submitted:

By _____

_____ Sales Office

TABLE 3.13 Hardgrove Grindability Index Ranges for USA Coals (source Ref. 11)

Bed	Grindability index range			Bed	Grindability index range		
	High	Low	Average		High	Low	Average
ALABAMA				**OKLAHOMA**			
America	51	47	49 (3)	Henryetta	—	—	—
Blackburn	—	—	—	McAlester	86	69	78 (2)
Black Creek	47	37	42(10)	Secor	—	—	—
Black Shale	61	47	53 (3)	Stigler	101	87	94 (2)
Brookwood	71	63	67 (2)	Upper Hartshorne	107	52	80 (2)
Corona	—	—	—	**PENNSYLVANIA**			
Helena	50	45	47 (2)	Barnett	85	83	84 (2)
Jagger	55	49	52 (2)	Brookville	107	57	90 (10)
Margaret	—	—	—	Brush Creek	115	97	105 (4)
Mary Lee	70	46	53 (8)	Clarion	107	49	82 (5)
Milldale	69	69	69 (2)	Little Pittsburgh	93	92	93 (2)
Pratt	—	—	—	Lower Freeport	110	53	90 (88)
Thompson	59	54	57 (2)	Lower Kittanning	117	45	94 (152)
ARKANSAS				Middle Kittanning	121	48	89 (27)
Lower Spadra	83	83	83 (2)	Pine Hill No. 2	95	91	93 (3)
Paris	—	—	—	Pittsburgh	94	50	63 (38)
Philpot	—	—	—	Sewickley	97	61	74 (6)
Upper Hartshorne	115	110	113 (2)	Upper Freeport	106	50	77 (129)
COLORADO				Upper Kittanning	111	58	96 (42)
B	53	44	47 (6)	**TENNESSEE**			
Brookside	47	43	45 (12)	Blue Gem (Rich Mountain)	—	—	—
Cameo (Books Cliff)	58	51	54 (15)	Bon Air No. 2	57	49	53 (7)
Collum	56	55	56 (9)	Coal Creek	65	47	53 (3)
C	55	49	52 (9)	Dean	50	45	48 (7)
D	55	47	50 (6)	Glen Mary	63	52	57 (7)
E	56	50	53 (4)	Jellico	55	47	50 (5)
Fox Hill	53	39	45 (6)	Jordan	—	—	—
Hesperus	64	48	53 (7)	Mason	54	52	53 (3)
Jack O'Lantern	50	45	48 (4)	Pee Wee	44	43	44 (2)
	47	45	46 (2)	Red Ash	47	44	46 (7)

Seam				Seam			
Lennox	54	49	52 (4)	Sewanee	98	55	65 (11)
No. 2	58	48	52 (4)	**UTAH**			
No. 3	52	48	50 (3)	Aberdeen	48	46	47 (3)
Palisade	67	44	52 (16)	Blind Canyon	51	45	48 (14)
Pinnacle	52	44	48 (8)	Castlegate A	50	43	57 (10)
Primero	58	57	58 (2)	Castlegate B	49	48	49 (2)
Upper Robinson	44	41	43 (3)	Castlegate D	48	43	45 (4)
Wadge	48	43	45 (7)	Castlegate Sub No. 1	47	44	46 (3)
Walsen	44	44	44 (2)	Ferron J	43	40	41 (3)
Wheeler	—	—	—	Gilson	47	42	45 (4)
Wolf Creek	47	42	45 (2)	Hiawatha	52	41	47 (15)
ILLINOIS				Ivie	51	50	51 (2)
No. 5	67	56	62 (23)	Liberty Sub No. 3	—	—	—
No. 6	66	52	60 (55)	Lower Sunnyside	—	—	—
No. 7	64	58	61 (2)	Rock Canyon	46	43	45 (6)
INDIANA				Sub No. 1	—	—	—
No. 3	66	61	63 (5)	Sub No. 2	—	—	—
No. 4	70	66	68 (2)	Sunnyside	49	47	48 (3)
No. 5	68	52	61 (17)	**VIRGINIA**			
No. 6	67	59	62 (7)	Clintwood	73	57	66 (11)
KANSAS				Dorchester	68	61	65 (3)
Bevier	71	59	66 (6)	Eagle	64	62	63 (2)
Cherokee	70	63	68 (4)	Imboden	57	55	56 (4)
Mineral	65	59	62 (2)	Jewell	104	93	98 (8)
KENTUCKY (East)				Lower Banner	100	43	75 (16)
Alma	44	43	44 (2)	Pocahontas No. 3	109	97	102 (3)
Clintwood	64	58	61 (3)	Raven	99	54	80 (7)
Creech ('E')	53	50	52 (2)	Shannon (Jawbone)	—	—	—
Elkhorn No. 1	44	36	39 (4)	Splash Dam	83	66	75 (10)
Elkhorn No. 2	52	45	49 (9)	Taggart	61	53	58 (3)
Elkhorn No. 3	54	43	48 (4)	Upper Banner	69	58	64 (5)
Flag No. 7 (Hazard No. 7)	53	40	44 (6)	**WEST VIRGINIA**			
Harlan	52	45	47 (6)	Alma	50	44	47 (18)
Hazard No. 5A	47	42	44 (3)	Bakerstown	103	67	83 (15)
Hindman (Hazard No. 9) (High Splint)	63	41	50 (9)	Beckley	106	84	96 (3)
Horse Creek	52	50	51 (3)	Cedar Grove	65	43	52 (28)
Jellico (Straight Creek)	57	44	49 (13)	Chilton	56	43	49 (15)

Seam	Max	Min	Avg (N)
Kellioka (Keckee)	—	—	—
Mason (Mingo)	60	56	58 (2)
Millers Creek (Van Lear)	—	—	—
No. 4 (Hazard No. 4, Fire Clay, Dean—Wallins)	53	41	48 (24)
No. 6	56	54	55 (2)
Pond Creek (No. 2 Gas)	57	43	49 (6)
Thacker	47	44	46 (3)
Whitesburg	—	—	—
Winifrede	47	43	45 (3)
KENTUCKY (West)			
No. 6	56	47	53 (4)
No. 9	63	49	59 (45)
No. 11	63	50	57 (29)
No. 12	63	51	57 (16)
No. 14	66	62	64 (2)
MARYLAND			
Pittsburgh	99	93	95 (5)
Upper Freeport	103	100	102 (3)
MISSOURI			
Bevier	70	58	64 (6)
Tebo	71	52	58 (7)
NEW MEXICO			
Raton	56	50	52 (3)
Sugarite	—	—	—
Yankee	53	53	53 (2)
OHIO			
No. 4	58	50	52 (8)
No. 4A	51	44	48 (12)
No. 5	58	48	52 (4)
No. 6	68	44	52 (43)
No. 7	67	46	54 (11)
No. 8	64	49	57 (42)
No. 8A	56	59	53 (13)
No. 9	55	52	53 (3)
Coalburg	48	41	44 (5)
Eagle	85	50	60 (25)
Fire Creek	103	76	97 (15)
Hernshaw	54	53	54 (2)
Iaeger	—	—	—
Lower Kittanning	78	39	53 (17)
Middle Kittanning	75	48	64 (4)
No. 2 Gas	86	38	59 (25)
Pittsburgh	72	50	60 (181)
Pocahontas No. 3	105	94	100 (5)
Pocahontas No. 4	—	—	—
Pocahontas No. 6	100	97	99 (2)
Powellton	62	51	57 (7)
Red Ash	98	85	91 (4)
Redstone	70	57	61 (39)
Sewell	111	60	84 (38)
Sewickley	66	59	61 (11)
Stockton	—	—	—
Upper Freeport	99	57	77 (61)
Upper Kittanning	77	52	71 (7)
Winifrede	60	43	50 (19)
WYOMING			
Adaville	59	55	57 (3)
Blind Bull	55	53	54 (4)
Frontier No. 1	54	49	51 (3)
Masters	53	49	51 (2)
Monarch	45	37	41 (5)
No. 1	54	48	51 (3)
No. 3	51	49	50 (2)
No. 7	52	45	49 (5)
No. 7½	50	48	49 (2)
No. 15	52	51	52 (2)
Roland-Smith	62	56	59 (3)

Number in parentheses indicates the number of determinations made.

SIZE REDUCTION

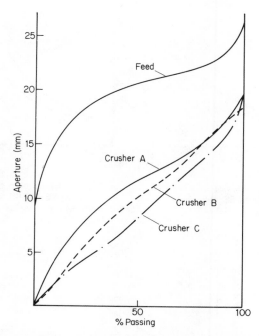

Fig. 3.41 Size-distribution graphs for crusher applications testing.

which is borne out by test data obtained from numerous types of size-reduction equipment.[3] From this expression it can be shown that the total energy requirement increases with increasing throughput for a given type of feed and crusher of specific reduction ratio. This is shown in graph (a) in Fig. 3.42, while the relationship between specific energy and reduction ratio for a crusher is shown in graph (b). This relationship appears to be an exponential one for most types of crusher.[20] Although two similar crushers may have similar energy requirements in a given range of reduction ratio, outside this range one crusher may prove superior to the other. Graph (b) demonstrates that the mode of operation of a size-reduction machine affects the relationship between specific energy and reduction ratio. In practice, observations have been made which confirm that if a crusher is fed correctly, throughput capacities in excess of the design figure can be achieved, particularly when the feed rate is controlled by the energy draw. Feed control is therefore an important consideration in crusher selection in particular. Graph (c) in Fig. 3.42 shows the relationship between the specific energy and grindability which can be seen also to be very dependent upon the type of size-reduction unit.

As for reduction ratio, there are often distinct ranges for which one device is better suited than another. Table 3.13 lists values of Hardgrove grindability as determined for various coals from the USA. Such data are frequently employed by crushing- and grinding-equipment manufacturers in North America to select size-reduction equipment.

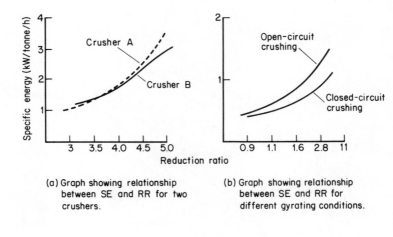

(a) Graph showing relationship between SE and RR for two crushers.

(b) Graph showing relationship between SE and RR for different gyrating conditions.

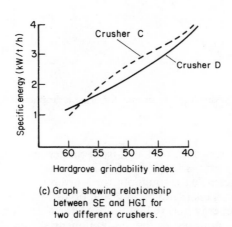

(c) Graph showing relationship between SE and HGI for two different crushers.

Fig. 3.42 Relationships between specific energy, reduction ratio and grindability.[3]

REFERENCES

1. L. G. Austin and J. D. McClung. Size Reduction of Coal, Chapter 7, *Coal Preparation*, (Ed.) J. W. Leonard, Society Mining Engineers, AIME, New York, 1979.
2. L. G. Austin and J. D. McClung, *Handbook of Crushing*, Pennsylvania Crusher Corporation, 1979.
3. R. R. Bevan. *A Review of Industrial Size Reduction Equipment used in the Processing of Coal*, Report FE-2475-9, US Energy Research and Development Administration, July 1977.
4. R. R. Bevan. *Coal Preparation Manual*, McNally Pittsburg Manufacturing Corporation, 1978.
5. A. L. Mular, and G. V. Jergensen. *Design and Installation of Comminution Circuits*, Society Mining Engineers, AIME, New York, 1982.
6. A. H. Cottrell. *The Mechanical Properties of Matter*, Wiley, London, 1964.

SIZE REDUCTION

7. A. A. Griffith. The phenomena of rupture and flow of solids, *Philosophical Transactions Royal Society London* **221**, 163, 1920.
8. B. B. Burbank. *Measuring the Crushing Resistance of Rocks and Ores*, Pennsylvania Crusher Corporation 1975.
9. A. E. LePage and F. Pollard. Methods for providing reliable data for coal preparation plant design, Paper D2. 7th International Coal Preparation Congress, Sydney, 1976.
10. F. C. Bond. Comminution: testing and calculations, Chapter 2, *Unit Operations and Processes*, SME Mineral Processing Handbook. (Ed.) N. L. Weiss, Society Mining Engineers, AIME, New York, 1985. pp. 3A.16–3A.27.
11. F. C. Bond, *Coal Breaking and Crushing*, McNally Pittsburg Company, Handbook 1974.
12. B. A. Wills. *Mineral Processing Technology*, Pergamon Press, Oxford, 1979. Chapter 5, pp. 96–125.
13. B. N. Popelnik, V. G. Shkodkin, M. V. Donde, N. K. Ivanov and P. V. Akulov. Research into hammer crusher performance, *Coke and Chemical, USSR*, **4**, 46–50, 1968.
14. T. G. Callcott. A study of the size-reduction mechanisms of swing hammer mills, *Journal Institute Fuel* **33**, 529–539, 1960.
15. C. E. Bird. *Report on Test Carried out on the Swing Hammer Pulverizer*, International Chemical Metallurgical Mineralogical Society of South Africa, 1951.
16. G. H. Roman, Coal crushers and crushing, *Coal Age*, no. 72, 107–122, 1967.
17. T. G. Callcott. Pulverizing coals, *Proceedings Australian IMM* **183**, 43–76, 1957.
18. A. Fitton and R. Jackson. Some experiences of coal grinding plants tests on a ring-ball and a tube mill, *Journal Institute Fuel* **32**, 520–529, 1959.
19. E. A. Smith. Grinding very hard and very soft materials, *Processing*, 16. November 1974.
20. M. B. Flavel. Power and the way it is applied affects equipment performance and production costs, Proceedings Society Mining Engineers, AIME, Annual Meeting, March 1977, New York.

Chapter 4

SCREENING AND CLASSIFICATION

4.1 INTRODUCTION

Screening is the mechanical separation of mineral fragments by size, which to some extent is influenced by the shape of the fragments.

Classification is the mechanical separation of mineral fragments by size and relative density, through the use of a fluid (air or liquid) medium. The separation is also influenced to some extent by shape.

Screening is used in coal-preparation operations in the following applications:

Scalping or pre-screening (Fig. 4.1)—the removal of oversize or very coarse material from run-of-mine coal. The oversize is then crushed, and the resulting raw coal fed to the plant.

Sizing/grading (Fig. 4.2)—the separation of raw coal into size ranges suitable for each beneficiation process, or the sizing of clean-coal products for the market.

Draining and rinsing (Fig. 4.3)—the removal of magnetite and fines from the products of separation of a dense-medium process.

Dewatering (Fig. 4.4)—the removal of water from wet coal and refuse.

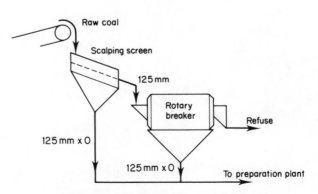

Fig. 4.1 Application of scalping screen.

SCREENING AND CLASSIFICATION

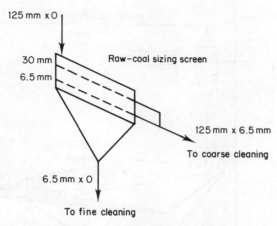

Fig. 4.2   Raw-coal sizing screen.

Screening operations in coal preparation operations embrace the entire range of sizes and occur at all stages in the preparation of coal. In larger-size separations it is carried out dry and at finer sizes wet screening is used. There are so many inter-related variables involved in the screening process that no single workable formula has been developed to allow complete selection simply by the substitution of numerical values in an equation. Familiarity and experience are prerequisites for the correct selection of a particular type and size of screen and screen deck.

The term 'classification' has come to mean the process of fine-coal sizing where screening would be impractical. The term 'size classification' is sometimes used to describe screening. Classification is usually carried out in modern coal-preparation plants by hydrocyclone systems and to a lesser extent by hydraulic settling or upward-current laminar flow systems. Some pneumatic classifiers are still utilized, and dedusting is used in the German coal industry.

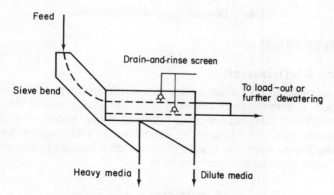

Fig. 4.3   Drain-and-rinse screen for dense-medium recovery.

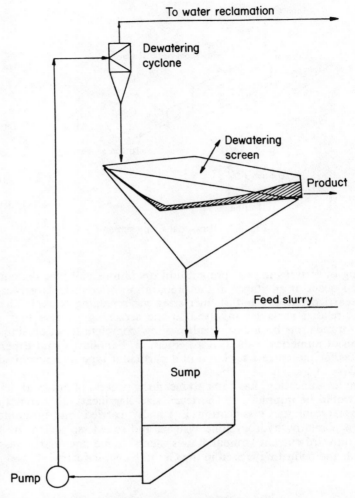

Fig. 4.4 Dewatering screen application.

4.2 SCREENING

4.2.1 Size Distribution

For the correct selection of a screen for industrial application, it is necessary to know the size distribution of the feed material (see Chapter 5). Rosin and Rammler established that broken coal followed a size distribution similar to a Gaussian distribution according to an experimental relationship. Bennet established that the Rosin–Rammler distribution law also applied to run-of-mine coal and put the equation into the common form:

$$R = 100 \exp\left(\frac{-x}{\bar{x}}\right)^n$$

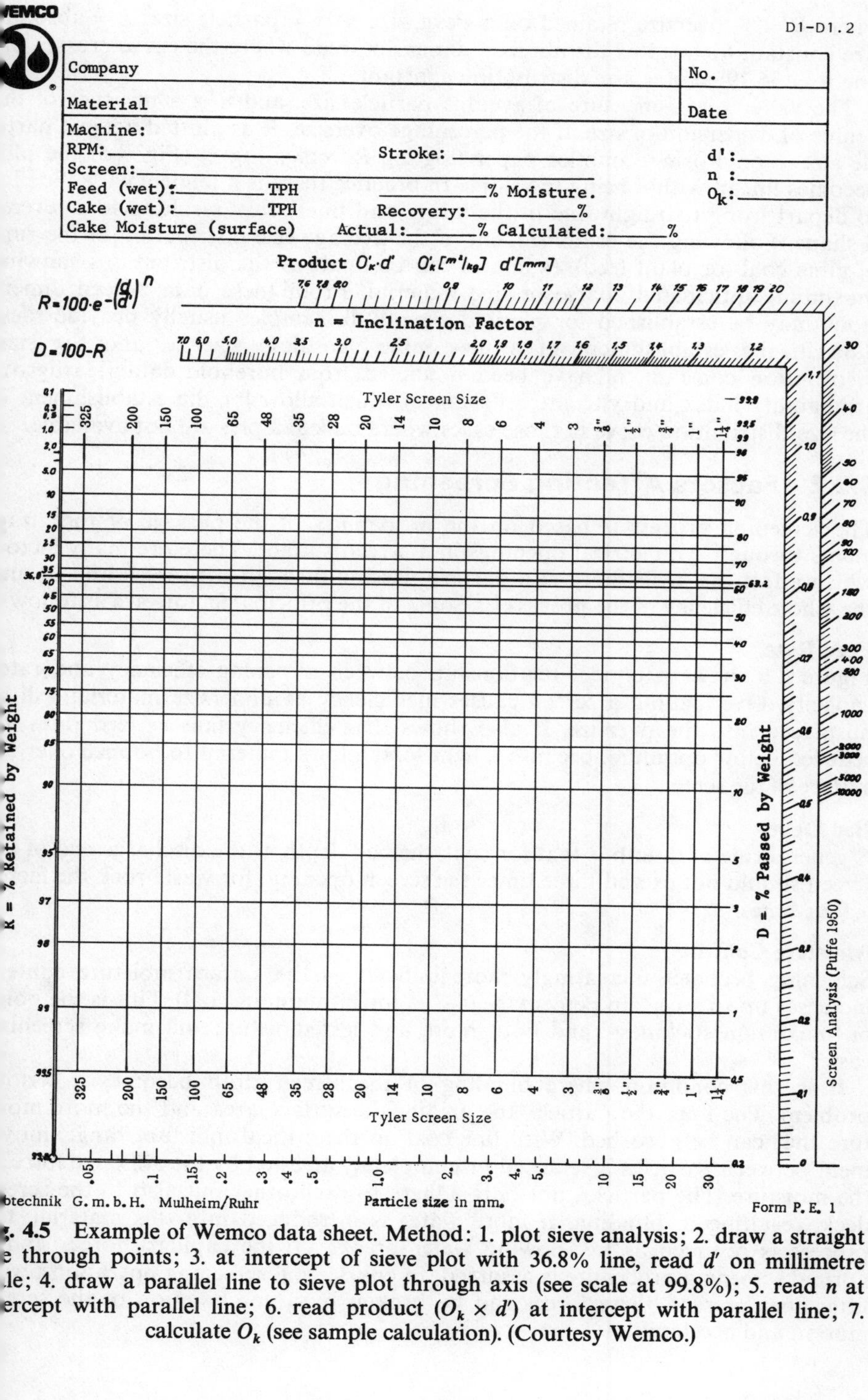

SCREENING AND CLASSIFICATION

WEMCO

D1-D1.2

Company		No.	
Material		Date	
Machine:			
RPM: ___ Stroke: ___		d': ___	
Screen: ___		n: ___	
Feed (wet): ___ TPH ___% Moisture.		O_k: ___	
Cake (wet): ___ TPH Recovery: ___%			
Cake Moisture (surface) Actual: ___% Calculated: ___%			

Product $O'_k \cdot d'$ $O'_k [m^3/kg]$ $d'[mm]$

$R = 100 \cdot e^{-\left(\frac{d}{d'}\right)^n}$

n = Inclination Factor

$D = 100 - R$

Tyler Screen Size

R = % Retained by Weight

D = % Passed by Weight

Screen Analysis (Puffe 1950)

Tyler Screen Size

Particle size in mm.

btechnik G.m.b.H. Mulheim/Ruhr

Form P.E. 1

4.5 Example of Wemco data sheet. Method: 1. plot sieve analysis; 2. draw a straight line through points; 3. at intercept of sieve plot with 36.8% line, read d' on millimetre scale; 4. draw a parallel line to sieve plot through axis (see scale at 99.8%); 5. read n at intercept with parallel line; 6. read product ($O_k \times d'$) at intercept with parallel line; 7. calculate O_k (see sample calculation). (Courtesy Wemco.)

where R = % oversize retained on a sieve, size x; x = particle size; $\bar{x}$ = absolute size constant (defined as the abscissa dimension read where the curve crosses the line $R = 36.79\%$); n = size distribution constant.

The value x is a measure of average particle size, and n is a measure of the degree of dispersion of size. If the percentage oversize, R, is plotted against particle size x on Rosin–Rammler paper (log log R versus log x) (Fig. 4.5) the plot becomes linear, with n being the slope. In practice there is a tendency for the plot to depart from a straight line in the larger and finer sizes, resulting in a reverse S-shape. Following the establishment of the average-size distribution of the run-of-mine coal (or plant feed), estimates can be made of the distributions showing maximum anticipated coarse or fine material. From these data, screen dimensions may be established for plant design. Bulk samples usually provide these data in the establishment of a new mine. Absolute size, x, and the size distribution constant, n, have been predicted from borehole data (Hardgrove grindability index and vitrinite reflectance) which allow for the establishment of the size-distribution curve in those cases where bulk samples are not available.

4.2.2 Factors Affecting Screening

The screening process is based on the probability of the passage of rock fragments through symmetrical openings in a screen surface. There are many factors which affect the probability of a particle passing through the screen and in turn the whole efficiency of the process.[1,2] Some of the principal factors are as follows:

Feed Rate

Figure 4.6 shows a typical relationship between screening efficiency and rated capacity. Overloading a screen causes inefficiency as undersize material will be misplaced into the oversize. It also shows that efficiency falls as feed rates are reduced below optimum, because a light load allows material to bounce over the surface of the screen.

Bed Depth

A general rule of thumb is that for coal, the bed depth at the discharge end of the screen should not exceed three times the screen opening; for waste rock the factor is four times.[3,4]

Moisture Content

Screening becomes increasingly more difficult as the surface moisture content increases up to a certain percentage (4–7% for bituminous coal). This is the point of 'maximum stickiness' and both more and less moisture will make screening easier.

It is this condition where blinding of the screen cloth becomes a serious problem. The finer the particle, the greater the surface area and the more moisture that can be absorbed. With fine coal, in the critical moisture range, movement between particles is retarded or completely arrested by the adhesive force of the moisture. The particles not only adhere to each other but also to the screen deck, resulting in blinding. If more water is introduced into this material, the process is reversed as free-flowing water breaks up the capillary forces and normal screening procedure is assumed. In coarse-coal screening operations, with little fines present, these forces can be broken by a combination of the screen motion and acceleration.

SCREENING AND CLASSIFICATION

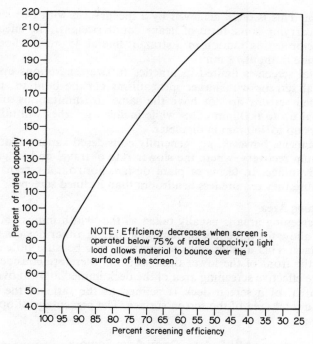

Fig. 4.6 Typical relationship between screen efficiency and feed rate.

Screen Angle and Mechanism

The capacity and efficiency of an inclined screen is affected by the angle at which it operates. This angle is usually related to the material being screened and the aperture size of the screening surface.

As a general rule, for slopes less than 20°, the mechanism rotates with the flow, and for slopes greater than 20°, it rotates contraflow. Capacities of sloping screens are reduced with a decrease in slope as shown:

Slope	%age of rated capacity
Optimum	100
less 2.5°	90–92.5
less 5°	80–85
less 7.5°	70–75
less 10°	60–65

Sloping screens have a circular motion, the amplitude being between 6 and 9 mm. The amplitude should be sufficient to lift the bed from the screen surface at each revolution and while the material flows down under the effect of gravity. Speed, amplitude and slope are determined from empirical data by the screen designer.

The mechanism used for sloping screens with a circular 'throw' is unsuitable for horizontal screens, where gravity is a hindrance rather than a help. The mechanism of a horizontal screen must fulfil the double duty of lifting the bed from the surface at each stroke and imparting movement to convey the material along the screen. This is often achieved by a mechanism with two shafts geared together, each carrying some form of heavy counterweight. The effect is to give the screen a reciprocating motion in a straight line at 45° to the horizontal. An average amplitude is about 12 mm.

The horizontal screen is limited in practice to size separations up to 60 mm, although special screens with larger mechanisms can be built to screen up to 100 mm. Inclined screens do not have the same size limitations and can take run-of-mine coal up to 600 mm cube, while inclined grizzlies will take feed containing boulders up to 750 mm in diameter.

Horizontal screens, however, are generally considered superior for drain-and-rinse and medium recovery, where the slower rate of travel and longer residence time are an advantage. In terms of plant design, horizontal screens have some attraction because they require less headroom than inclined screens.

Effective Screening Area

The effective screening area is usually taken as the 'width minus 150 mm, times the length', to allow for the effect of clamping bars or other devices for securing the screen surface. This applies only to the top deck because little screening is performed on the front of the lower deck which is therefore considered to have only 90% of the effective screening area of the deck immediately above it.

The 'open area' of a screen deck is defined as the ratio of the area of the apertures to the total area of the screen surface. The percentage of open area can

TABLE 4.1   Open Area Factors

Aperture	Formula
1. Rectangular opening	$F_{oa} = \dfrac{a_1 a_2}{(a_1 + d_1)(a_2 + d_2)} \times 100$ where F_{oa} is the open area, expressed in %; d is the diameter of the wire (or the horizontal width of a bar (for plates)); a is clear opening or aperture.
2. Square opening (equivalent to a rectangular opening with $a_1 = a_2$ and $d_1 = d_2$)	$F_{oa} = \left(\dfrac{a}{a+d}\right)^2 \times 100$
3. Square opening expressed in terms of the mesh size m	$F_{oa} = a^2 m^2 \times 100$ since $m = \dfrac{1}{a+d}$
4. Wedge or parallel opening	$F_{oa} = \dfrac{a}{a+d} \times 100$
5. Sloping screen: for a sloping screen the effective aperture is the horizontal projection of the actual screen aperture.	$A_{horiz} = a_s \cos \alpha$ where a_s is the slope aperture and α is the slope angle.

SCREENING AND CLASSIFICATION

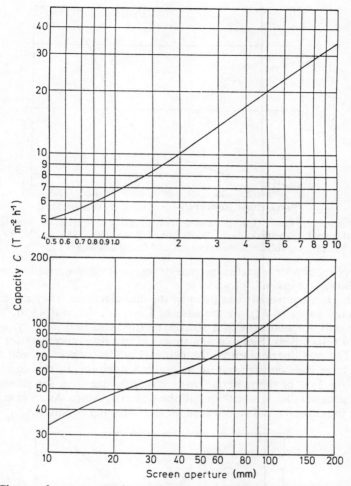

Fig. 4.7 Charts of screen capacity against aperture. (Reproduced from L. Svarovsky (Ed.), *Solid–Liquid Separation*, 1981, p. 225, by permission of the publishers, Butterworth and Co. (Publishers) Ltd ©)

vary from 75% for woven wire down to 12% for wedge wire. It is important that calculated areas be adjusted in accordance with the open area of screening surface to be used. If a heavier cloth is substituted for that supplied or recommended by the manufacturer, in order to prolong screen life, then capacity and efficiency may be sacrificed to some extent. Table 4.1 gives the open area relationship for various types of screen surface.

As an example, the percentage open area for a woven-wire 10 mm square aperture screen constructed from 2.25 mm diameter carbon-steel wire, using the formula $F_{oa} = 100[a/(a + d)]^2$, is

$$F_{oa} = \left(\frac{10}{10 + 2.25}\right)^2 \times 100 = 67\%$$

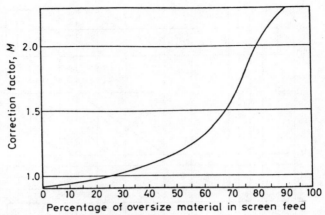

Fig. 4.8 Correction factor for percentage of oversize material in the screen feed. (Reproduced from L. Svarovsky (Ed.), *Solid–Liquid Separation*, 1981, p. 226, by permission of the publishers, Butterworth and Co. (Publishers) Ltd ©)

The screen capacity for a particular type of screen is often derived from charts of the type shown in Figs. 4.7, 4.8 and 4.9.

These charts are normally compiled by the manufacturer. They are constructed for a selected value of F_{oa}, in the case of Figs. 4.7, 4.8 and 4.9 60%, and are applicable only for materials of a specific relative and bulk density, in this case 2.6 and 1.6 respectively. Suitable adjustment is therefore necessary for any other material. The size distribution of the material is also important and the design chart will have been constructed on a specified standard. In this case the charts are based on 25% of the material being larger in size than the screen aperture and 40% being smaller in size than half the screen aperture. As an example, let us determine the capacity and screen area for the following conditions:

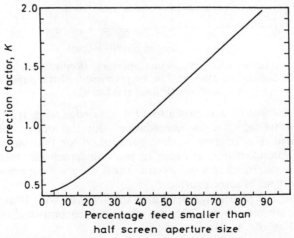

Fig. 4.9 Correction factor for the percentage of material smaller than half the screen aperture size.
(Reproduced from L. Svarovsky (Ed.), *Solid–Liquid Separation*, 1981, p. 226, by permission of the publishers, Butterworth and Co. (Publishers) Ltd ©)

SCREENING AND CLASSIFICATION

$F_{oa} = 64.5\%$; $a = 10$ mm
Feed size range has 50% oversize and 38% less than half aperture
Material: coal of relative density 1.3
Required feed rate: 130 t/h
Number of decks: 1

From Fig. 4.7, which is based on $F_{oa} = 60\%$, the capacity, C_i, for a screen of aperture 10 mm is 33. To obtain the corrected capacity, C_{ii}, for $F_{oa} = 64.5\%$ we use $C_{ii}/C_i = 64.5/60.0$

$$\therefore C_{ii} = \frac{64.5}{60} \times 33 = 35.5 \text{ t/m/h}$$

A correction is now required for the fact that the percentage of oversize particles is not 23% as standard. For 50% oversize, Fig. 4.8 gives a correction factor of 1.18, hence the capacity, C_{iii}, corrected for 50% oversize is:

$$C_{iii} = 1.18 \times 35.5 = 41.9 \text{ t/m/h}$$

The correction for 38% of the coal being less than half the screen aperture size instead of the standard 40% is given in Fig. 4.9 as 0.96. Thus C_{iv}, the capacity corrected for the percentage less than half the screen aperture size

$$C_{iv} = 0.96 \times 41.9 = 40.2 \text{ t/m/h}$$

The capacity for a material of relative density, ρ, different to the standard 2.6, is given by

$$\frac{C_\rho}{C_{2.6}} = \frac{\rho}{2.6}$$

Thus the capacity corrected for a density of 1.3 is

$$C_v = \frac{1.3}{2.6} \times 40.2 = 20.1 \text{ t/m/h}$$

The area of screen required is given by the ratio of the feed rate and the capacity

$$\text{area} = \frac{\text{feed rate}}{\text{capacity}} = \frac{130}{20.1} = 6.47 \text{ m}^2$$

A further modification to this value obtained for the area must be made to compensate for the number of screens in a multideck arrangement. Figure 4.10 illustrates the characteristic flow mechanism of material on a screen surface. If this mechanism applies each underlying screen surface will receive passing material a little farther along the deck (in the direction of the material flow) than the one above. It is thus normal to apply a correction to allow for this and assume that only 90% of the effective area is utilized on the lower of two screens. This phenomenon is compounded as the number of screens increases. Thus, for the lowest of three screens the area is only 90% of 90% = 81%. Thus for a multiscreen system the screen area required will be the area calculated for a single screen $+ 0.9^{(n-1)}$ where n is the number of screens counting from the top.

A final contingency allowance is usually applied to compensate for surge capacity and other unpredictable factors such as mutual shape or stickiness. For

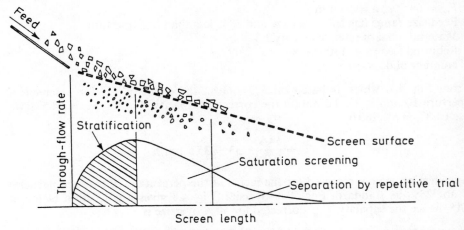

Fig. 4.10 Characteristic flow of material on a screen surface. (Reproduced from L. Svarovsky (Ed.), *Solid–Liquid Separation*, 1981, p. 227, by permission of the publishers, Butterworth and Co. (Publishers) Ltd ©)

the purpose of an example a contingency factor of 20% is applied; thus:

$$\text{area} = 6.47 \times 1.20 = 7.76 \text{ m}^2$$

At this stage it is possible to select a particular sized screen from what is normally a range of standard manufactured sizes. A screen 1.75 m wide by 5 m long is therefore available, which has an area of 8.75 m² and it is now necessary to see if the dimensions are acceptable. It is wise to remember at this stage that the effective screening area is not the product of width and length. Some area is taken up by fixing strips and clamping plates and this must be taken into account.

A rule of thumb here is that the effective area is $(W - 0.15)L$ where W and L are the width and length respectively. Hence the effective area for the example above is

$$(1.75 - 0.15)5 = 8 \text{ m}^2$$

which would still be acceptable for the calculated area requirement of 7.76 m².

It is now necessary to consider whether the length is adequate.

Experience has indicated that the bed height should not exceed three times the screen aperture for material of relative density of 1.3; and not more than four times the screen aperture for material of relative density 2.6. Often this relationship is shown graphically.

Bed depth is determined using:

$$D = \frac{50T}{3BSW_1}$$

where D is the bed depth, mm; T is the tonnage of oversize, t/h; B is the bulk density, t/m³; S is the material travel rate, m/min; W_1 is the effective width, m.

For the screen type under consideration in the example

$$S = 15 \text{ m/min for horizontal application}$$

SCREENING AND CLASSIFICATION

which obviously would increase with the sloping application. Therefore, if coal has a bulk density of 1.13 t/m³

$$D = \frac{50 \times (130 \times 50/100)}{3 \times 1.13 \times 15 \times (1.75 - 0.15)} = 40 \text{ mm}$$

As the screen aperture is 10 mm, the bed depth is $40/10 = 4$ times the screen aperture and therefore the machine under consideration is not acceptable.

It is therefore necessary either to repeat the calculations for an alternative screen which has a larger area or else make use of the fact that the rate of travel of the material across the screen can be increased by increasing the slope of the screen. A further alternative would be to use a double-deck screen with say, a 20 mm upper deck and the 10 mm as the lower deck. The calculations involved would be the same but the burden on the lower deck would now be substantially reduced.

Screen Surface (Deck) Materials

Screen deck types may be woven wire, wedge wire, parallel rods, punched plate, perforated hard rubber or polyurethane.

Woven-wire screen cloths are usually constructed from carbon- and stainless-steels and are by far the most commonly used screening surfaces, accounting for 80% of normal applications. About 15% of screen decks are constructed of closely spaced bars or wires. These are usually wedge shaped, although a 'rod-deck' screen uses rods held loosely on each in 'combs' within which they are free to rotate as a result of the motion imparted to the screen deck. The application of this type of deck is the screening out of moist, small-sized material usually about 5×10 mm in top size. Its drawback is that it is not a positive sizing screen in that oversize appears in the undersize in the form of elongated particles slipping through the rods.

A wedge-wire screen deck consists of wedge-shaped stainless-steel bars which taper from approximately 2 mm wide to approximately 1 mm, over a length of approximately 4 mm. Apertures range from approximately 1 mm to below 0.25 mm. Some typical profiles are shown in Fig. 4.11.

Approximately 5% of screen decks are plates of metal or rubber perforated with square or round holes. One form of perforated surface, which has been increasingly used because of its wear-resistance properties, is polyurethane. Embedded reinforcements of commercial mild steel are sometimes used, or in some cases stainless-steel or spring-steel in order to facilitate tensioning and provide the necessary intrinsic strength. Another commonly used non-metallic surface is wear-resistant hard rubber. Such surfaces are applicable over a wide range of apertures from 300 mm to below 20 mm. However, all non-metallic surfaces appear to have the common disadvantage of decreasing percentage open area with decreasing aperture.

Feed Distribution

The coal feed should be distributed over the total width of the screen. The screen capacity is dictated by width rather than length, but poor distribution affects capacity and therefore, efficiency. The initial positioning of the screen relative to the feed point, i.e. height, direction of flow, speed of the coal travel, etc., must all be taken into consideration. Attempts at a later date to improve distribution by inserting baffle plates often cause blockages and maintenance problems.

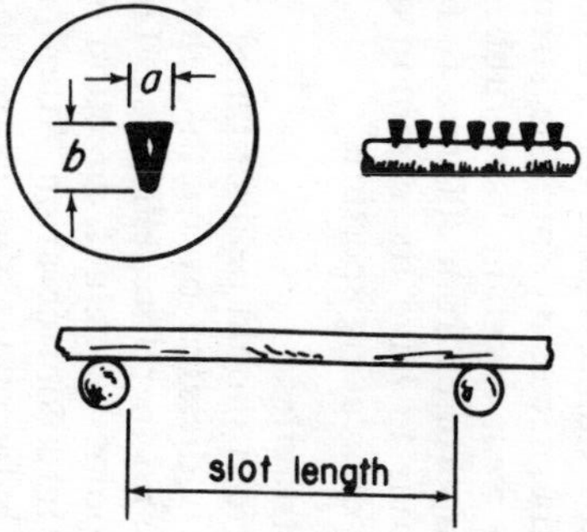

	Aperture size (slot width)							
	0.315 mm	0.4 mm	0.5 mm	0.63 mm	0.8 mm	1.0 mm	1.25 mm
Looped profile wire sieve								
OPEN AREA % for wire size $a = 1.8$ mm; $b = 3.1$ mm slot length = 70 mm	12.4	16.6	19.8	23.7	28.1	32.6	37.4
OPEN AREA % for wire size $a = 2.2$ mm; $b = 3.8$ mm slot length = 70 mm	10.4	13.8	16.6	20.0	24.0	28.1	32.6
OPEN AREA % for wire size $a = 2.5$ mm; $b = 4.5$ mm slot length = 70 mm	9.2	12.4	15.0	18.1	21.8	25.7	30.0
OPEN AREA % for wire size $a = 3.0$ mm; $b = 5.5$ mm slot length = 70 mm	—	—	—	15.3	18.6	22.0	36.9
Welded profile wire sieve								
OPEN AREA % for wire size $a = 1.5$ mm; $b = 4.0$ mm slot length = 40 mm	17.4	21.0	25.0	29.6	34.8	40.0	45.4
OPEN AREA % for wire size $a = 2.2$ mm; $b = 4.5$ mm slot length = 50 mm	12.5	15.4	18.5	22.3	26.7	31.2	36.2
OPEN AREA % for wire size $a = 2.8$ mm; $b = 5.0$ mm slot length = 50 mm	10.1	12.5	15.2	18.4	22.2	26.3	30.9
OPEN AREA % for wire size $a = 3.4$ mm; $b = 8.5$ mm slot length = 50 mm	8.5	10.5	12.8	15.6	19.0	22.7	26.9

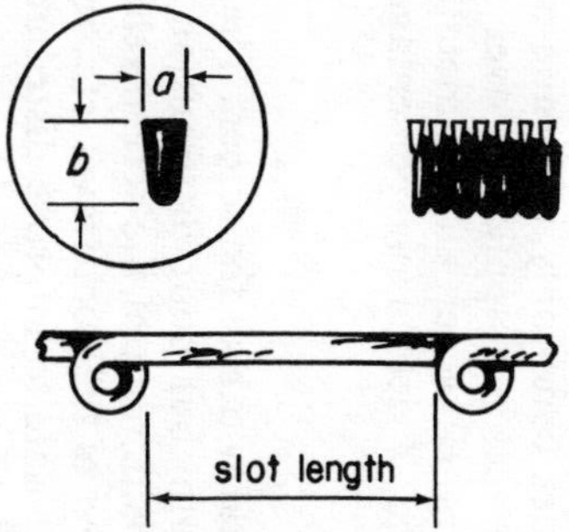

Fig. 4.11 Some typical wedge-wire profile data.

4.3 SCREENING APPLICATIONS IN THE PREPARATION PLANT

4.3.1 Scalping and Prescreening

This operation usually takes place outside the preparation plant, situated in a separate 'breaker station'. Typically, the screen is situated ahead of the primary size-reduction equipment which is a rotary breaker or a roll crusher. Many coal-preparation plants employ a rotary breaker for top-size reduction prior to treatment. The rotary breaker often serves both a screening and a crushing function, in treating coal directly from the pit. This machine is more fully described in Chapter 3 (Section 3.3). If a screen is used prior to the breaker, it is either a fixed-screen grizzly or a vibrating screen.

For the latter application, the raw coal is conveyed to a screen box located above the screen and it absorbs the impact of the coal and spreads it over the screen for more efficient operation and reduction in decking replacement costs. The scalping screen is a rugged, single-deck vibrating screen inclined at an angle of 17–25° (20° typical). The aperture may be up to 150 mm and the deck can be woven, perforated plate or grizzly bars. A scalping screen is typically a single-deck inclined screen with four bearings, a single shaft and operated with a circular motion. The deck is usually fitted with manganese grizzly (skid) bars or perforated plate with skid bars welded on top. Extra-high side plates are sometimes fitted to contain the raw-coal surge.

4.3.2 Sizing and Grading

Coal beneficiation processes are size related, and therefore sizing is usually carried out on the raw coal. The term 'grading' is referred to as the sizing of the clean coal.

Screening of coal above 12 mm is usually carried out dry. Commonly, the sizing screen is a double-deck vibrating type mounted at 17–20°. The decks are woven wire with a load-breaking top deck.

If sizing is carried out at 12 mm or below, then wet screening would normally be used. The water for the sizing operation may be added as a spray, via a flood box. For fine sizing operations at 0.5 mm and below, a coal–water slurry is delivered to the screen. An arbitrary size for dry screening is 10 mm, but often dry screening is practised down to 5 mm aperture. The practical size limit depends on the 'stickiness' of the coal which is influenced by the moisture content (both surface and residual), shape factor and the amount of fines present. Dry screening can be used below 10 mm if a rod-deck screen is used. The screen deck consists of a series of rods suspended across the width of the screen and supported in side combs. The screen is inclined, and the movement of the coal over the screen causes the rods to turn and effectively avoids blinding even with fine, sticky coal. The gap between the rods is the screening aperture, and adjustments can be made by changing the rod diameter or by 'sleeving' the original rods.

When raw coal has been dry screened, it is usually found necessary to rinse off adhering fines from the large coal prior to dense-medium cleaning. A 'prewet' or feed preparation screen is used for this function using low-pressure spray water at 0.8–1.4 $m^3/t/h$. The screen is normally a heavy-duty, vibrating screen, of double-

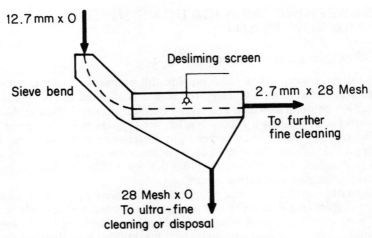

Fig. 4.12 Application of desliming screen.

deck, horizontal type with a lined feed box and water spray bars directed to both decks. The double deck is used to improve feed stratification and, therefore, assist the fines to be washed through the bottom deck. Top decking is of wire cloth, polyurethane or rubber with an approximate 25 mm opening. The bottom deck is usually of T-304 stainless-steel profile decking or polyurethane slotted with a one-millimetre clear opening.

Current coal-preparation practice is for the dense-medium process to be used down to a size of 0.5 mm. The standard sizing practice on dense-medium plants

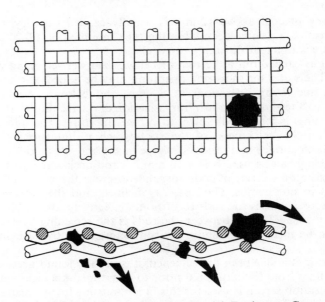

Fig. 4.13 Derrick sandwich screen (courtesy Derrick Equipment Company).

SCREENING AND CLASSIFICATION

is to use a combination of sieve bends and wet desliming vibrating screens with an aperture of 0.5 mm (Fig. 4.12).

Desliming is one form of sizing prior to beneficiation. The name 'desliming' comes from the function it performs since it removes very fine material. This operation is normally carried out on horizontally mounted, single-deck screens. To promote good separation, the depth of material is kept to a minimum. The feed passes through a curtain of water supplied from a flood box and through water sprays.

An interesting screening problem in some Canadian plants is the separation of minus 150 μm material from the 0.5 mm × 0 clean-coal product. Initially this separation was made on the plants using classifying cyclones. The separation was not accurate enough and it was found that 150 μm material was being misplaced and reporting to froth flotation where its recovery was low.

One mine solved the problem by replacing the classifying cyclones with Derrick screens. The Derrick high-speed vibrating screen is fitted with a 'sandwich' deck of two fine-mesh screens supported by a backing wire screen. The two screen cloths are held together in such a manner that wires of the bottom cloth create an interference in the free openings of the top cloth. This prevents small fragments from lodging in the openings of the upper surface while at the same time providing a size separation via the combination of cross-wires from both screen surfaces (Fig. 4.13).

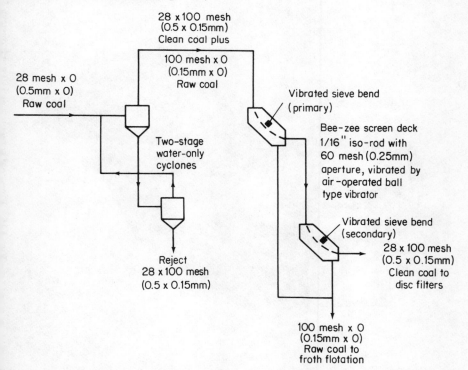

Fig. 4.14 Westar Mining's Elkview fine-coal screening operation at 100 mesh (0.15 mm) (Elkview Coal Preparation Plant, Westar Mining Limited, Sparwood, B.C., Canada).

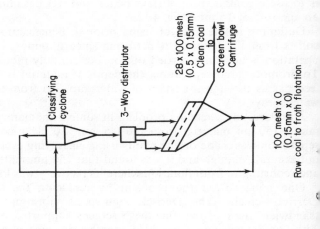

SCREENING AND CLASSIFICATION

Another mine replaced classifying cyclones with rapped sieve bends of 0.25 mm aperture. Rubber-tipped pneumatic rappers strike the back of the profile screen, two rappers to each sieve bend operating intermittently. These are not always totally successful as minus 150 μm material can be displaced to the oversize. This necessitates increase in the screen aperture to 0.4 mm (effectively cutting at 250 μm). The rappers can also cause excessive damage to the sieve bend screen decking. As an alternative, vibrators can be used to replace the rapping mechanism (Fig. 4.14).

A further approach to screening at aperture of 150 μm size is by combining classifying cyclones with Derrick screens (Fig. 4.15). The cyclones are adjusted so that the overflow contains virtually no plus 150 μm. The underflow, which contains all the 0.5 mm by 150 μm fraction plus quantities of minus 150 μm, is fed to the Derrick screen at a pulp density of 15–20% (volume) and the final separation made at 150 μm. The Derrick screen underflow is combined with the classifying cyclone overflow and reports to froth flotation.

4.3.3 Draining and Rinsing

Drain-and-rinse screens are positioned in the preparation circuit as the means for recovery of medium from the clean-coal product and the rejects from any prior equipment in which a dense medium, commonly magnetite–water suspension, is the vehicle of separation. The object of this operation is three-fold: (a) to remove

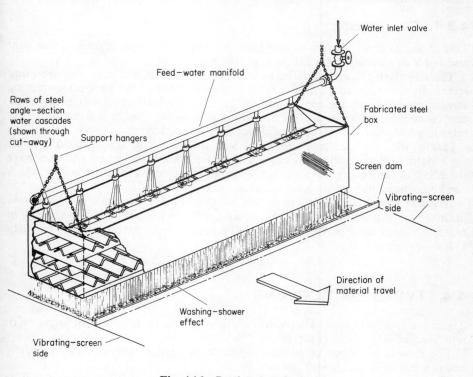

Fig. 4.16 Bretby spray box.

magnetite from the products as it would be a contaminant; (b) to recover the magnetite for re-use; (c) to recover water for re-use.

This process is normally accompanied on a horizontally mounted single-deck vibrating screen. The first section of the screen simply allows the magnetite to drain through. On the second section, the material passes beneath a low-pressure shower or spray of water which rinses off the magnetite clinging to the coal and refuse. The underflows from the first and second section report to separate underpans located beneath the screen. To aid the drainage process and to reduce the screen size, the material normally first passes over a sieve bend as shown in Fig. 4.3.

The vibrating motion of the screen promotes the rinsing action of the spray water. Rinsing is carried out by means of flood boxes and/or low-pressure sprays. As the rinsing water is usually reclaimed water, it can have some solids in it. This often causes problems at the spray nozzle. One simple solution to this is the Bretby spray box which comprises layers of angle iron, each layer mounted at 90° to the last (Fig. 4.16). When water passes through this configuration an even curtain of water is produced, with no chance of blocking occurring.

A key design feature is to size the screen properly so that the bed depth is kept to a minimum. Sometimes, operators will mount retarding strips of baffles on the screen to slow down the flow of solids in order to achieve more rinsing. This should only be done with caution since it increases the bed depth and *may* in fact reduce the rinsing effect.

4.3.4 Dewatering

This aspect is dealt with in more detail in Chapter 10, with regard to fine coal and reject dewatering applications in particular.

The dewatering screen may be an inclined or horizontally mounted vibrating screen. In many instances, a sieve bend is placed ahead of the screen. Dewatering screens operating on fine coal (minus 0.5 mm) will be unlikely to reduce the moisture content much below 25%. If horizontal screens are used, it is common to use retarding baffles to increase the residence time on the screen.

Ideally, for minus 0.5 mm slurry, the feed should not be less than 30% solids (by weight) and a densification stage is often used in this operation and they have the effect of thickening the feed and also removing slimes in the overflow which can be rerouted to another form of dewatering.

In many cases, vibrating screens are used in conjunction with sieve bends in drainage applications in order to cut down on the number of vibrating screens. In other instances sieve bends operating in tandem provide extra drainage capacity at low operating cost.

4.4 TYPES OF SCREENS

There are two main types of screen: rigid-bar screens or grizzlies (for large coal); and fixed sieves (for fine coal).

The former are often of simple, rugged construction, either set horizontally above a bin or inclined, with the coal flow.

Fixed sieves are used for dewatering and/or size separation of slurried coal or

SCREENING AND CLASSIFICATION

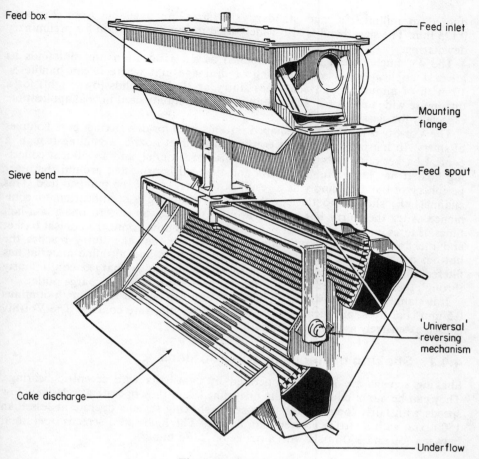

Fig. 4.17 Sieve bend.

discard. The sieve bend, the Vor-Siv, the Wilmot riffle and the cross-flow screen fall into this category.

The screening device most commonly used in coal-preparation plants is the sieve bend (Fig. 4.17). Coal slurry is fed tangentially via a feed box onto a fixed, curved wedge-wire screen deck. The solids are separated at a predetermined particle size, delivering a partially dewatered overproduct and a dilute underflow. A series of baffles spreads the feed evenly across the width of the deck. The slurry drops from the last baffle, passes through the feed spout and is then fed tangentially to the screen. It is important that the feed should contact the surface immediately at the front end of the sieve bend. It is possible to lose as much as 25% of screening area with a poorly fitted feed box.

The full stream of coal slurry flowing over the sieve bend decreases in depth, in increments of approximately one-quarter the slot width each time it passes a slot. Sieve bends make a size separation at half to two-thirds the slot width. This feature virtually eliminates blinding of the screen surface. Slot sizes are normally quoted in millimetres, and standard slot sizes are available from 2 mm to 0.125 mm. The sieve bend is also widely used in coal preparation for preliminary dewatering.

The 45° angle screen is widely applied as a classification unit but finds its greatest application as a dewatering or drainage unit where it can handle 3–10 m^3/h of minus 0.5 mm coal fines slurry at 30–50% solids (by weight) for a one-metre wide unit. A 60° angle bend is also commonly used in coal application, especially for classification.

The Vor-Siv is a fixed, bowl-shaped screen,[1] used for dewatering large volumes of slurry. In its simplest form, it consists of an inlet nozzle, a guiding trough, a conical basket of radially slotted wire, a discharge outlet, and an effluent collecting box. It has no mechanical parts and the feed is projected around the inner periphery of the unit simply as a result of the pressure of the pumped feed. This channels the slurry into the circular raceway causing a stratification to commence. After the slurry has rotated around the raceway one to one-and-a-half times, it loses enough of its energy so that it drops down onto the conical basket and the flow becomes a spiralling vortex. By the time the slurry reaches the bottom quarter of the basket, almost all of the fluid and ultrafine material has filtered through the screen. The dewatered solids are discharged continuously through the central outlet while the effluent passes through a discharge outlet.

It is claimed that the two-metre diameter unit can handle 12–15 m^3/h of minus 0.5 mm coal fines, producing cakes with 35–50% moisture content. The Vor-Siv is used extensively as a size classifier.

4.4.1 Shaking or Reciprocating Screens

Shaking screens are normally operated by camshafts with eccentric bearings. They can be mounted horizontally or inclined, and they operate at relatively low speeds with fairly long strokes. Low-speed shaking screens operate at less than 150 r.p.m. with a stroke greater than 85 mm. The high-speed screens operate at between 150 and 300 r.p.m. with a stroke of 25–75 mm.

4.4.2 Vibrating Screens[5]

Mechanically vibrated screens are mainly of two types, namely, the low-speed vibrator (300–700 r.p.m. with strokes of 25–100 mm), and high-speed vibrators (700–2000 r.p.m. with strokes of less than 12 mm).

High-frequency vibrating screens usually employ an out-of-balance weight arrangement to provide the vibration. Mechanically this can take several forms, of which the most common are:

(i) *Four-bearing type*, employing an eccentric, counterbalanced drive shaft. The stroke of these machines is predetermined by the amount of drive shaft eccentricity and cannot be changed except by substitution of the proper shaft for the desired stroke. In this application the counterweights on each end of the shaft are used as balancers to provide smooth operation. Because of these balancing requirements no additions to the vibrated weight can be made without careful calculation of the transportation effect.

(ii) *Two-bearing type*, normally actuated by a concentric drive shaft which can be adjusted in operation, within the limits of bearing capacity, by varying the

SCREENING AND CLASSIFICATION

amount of counterweight used, and by alteration of the speed. These vibrators produce a circular action and depend on gravity for their feed rate.

(iii) *Horizontal screens* actuated by off-centre shafts or weights geared together. These are also of the four-bearing type, and the motion produced is virtually in a straight-line throw with a slight ellipse. The shafts or weights are geared together in a certain fixed relationship to produce a certain direction of action which can be varied between the vertical and the horizontal. The most commonly used throw is at 45° to the horizontal. Such screens are often employed where headroom is at a premium, or where sharp sizing, dewatering or feeding of a product is desired.

Positive stroke, four-bearing screens start and stop smoothly, while two bearing units of any type of 'full-floating' screens have a characteristic stop-start bounce. However, from an application standpoint it may be said that, with only few exceptions, there is no job that a four-bearing type screen can do that a two-bearing unit, properly designed and applied, cannot do equally well. For the most part, the two types of unit have identical stroke characteristics, size for size, and the lump of material on the deck behaves more or less the same in each case.

Slowly vibrating screens are usually suspended on hangers or cables and are driven through electric cams with little facility for adjustment. Such units are most commonly applied for fine solids or slurry dewatering.

Various methods of mounting are employed, such as massive springs or air cushions for floor mounting, or alternatively, suspension hangers, cables or springs.

The general tendency with recent coal-preparation plant design has been to install reliable large-capacity equipment.[6] Until recently, the largest-width vibrating screen available was 3 m wide. New developments in drive mechanisms now allow for screens up to 5.5 m wide and 6.5 m long. Combined with large throughput, screens have been developed (Fig. 4.18) with varying slope. The first section of the screen has a 34° slope allowing the coal to travel at 1.5–2 m/s, the middle section has a slope of 23° allowing a speed of 0.75–1.25 m/s, and the final section has a slope of 12°, resulting in coal travel at 0.25–0.5 m/s. This concept has been applied to coal-sizing screens resulting in efficient screening at very high capacity, i.e. 1000 t/h on one screen. For desliming and drain-and-rinse applications, a large horizontal screen is used. Many of these high-capacity screens are operating successfully in coal-preparation plants in West Germany, South Africa and Canada.[7]

Typical applications of vibrating screens in coal-preparation plants are shown in Table 4.2.

4.4.3 Resonance Screens

Resonance screens have been designed to save on energy consumption. The screen deck is mounted on flexible hanger strips and attached to a balance frame, three or four times heavier than the screen itself.

Movement is imparted to the screen by an eccentric drive and connecting rod. A rubber pad connects the rod to the screen. Rubber trippers restrict the movement of the screen and serve to store up energy which is returned to the screen frame. Any movement given to the screen is transmitted to the balance frame. The throw of the balance frame is less than that of the screen because its mass is

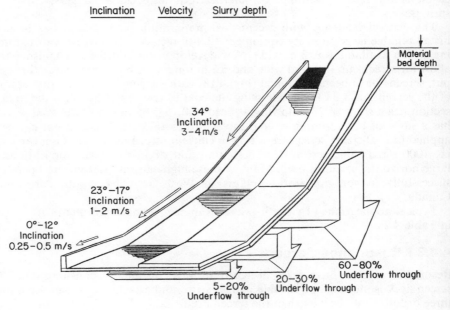

Fig. 4.18 PennTechnic multislope or 'banana' screen. (Courtesy Pennsylvania Crusher Corporation.)

TABLE 4.2 Vibrating Screen Applications in Coal-preparation Plants

Type	Number of decks	Installation angle	Aperture[a]	Screen deck type	Accessories
...-of-mine ...alper	single	17–25°	150 mm	manganese skid bars, AR perforated plate with skid bars	feed box with liners, extra-high side plates, drive guard enclosures
... coal ...zing ...reen	double	17–25°	25 mm	AR steel perforated plate, polyurethane, rubber	dust enclosures, drive guard enclosures
			8 mm	polyurethane, wire 304 stainless steel profile deck, rubber	feed box with liners
...-wet ...reen	double	horizontal	25 mm	wire, polyurethane, rubber	water spray bar, side-plate drip angles, drive guard enclosures, feed box liners
			1 mm	stainless-steel profile deck, polyurethane	
...se-medium ...rain-and-...nse screen (...oarse coal)	double	horizontal	25 mm	wire, polyurethane, rubber	side-plate drip angles, spray bars, shower box cross-flow screen or sieve bend, drip-lip angles, drive guard enclosures
			1 mm	304 stainless-steel profile deck, polyurethane	
...atering ...reen (...oarse coal)	single	horizontal	1 mm	304 stainless-steel profile deck, polyurethane	sieve-bend or cross-flow screen, dam, discharge drip-lip angles, drive guard enclosures
...iming ...reen	single	horizontal	0.5 mm	304 stainless-steel profile deck, polyurethane	sieve-bend or cross-flow screen, spray bars, shower box, drive guard enclosures
...sifying ...reen (...ne coal)	single	28°	100 mesh	Stainless-steel woven-wire 'sandwich' screens	three-way slurry distributor and feed system
...se-medium ...rain-and-...nse screen (...ne coal)	single	horizontal	0.5 mm	304 stainless-steel profile deck polyurethane	sieve-bend or cross-flow screen, spray bars, shower-box, drip-lip angles, drive guard enclosures
...atering ...reen (...ne coal)	single	horizontal or 27–29°	0.5 mm	304 stainless-steel profile deck or woven wire, rubber polyurethane	sieve-bend or cross-flow screen, dam, drip-lip angles, drive guard enclosures

[a] ...pical application

...uch greater. Therefore, any motion given to the balance frame sets up vibra-...ons which, instead of being wasted, are imparted back to the screening frame, ...educing loss of energy to a minimum. In certain types of resonance screen, the ...eck movement is restricted by a series of stops which are attached to the deck ...nd operate between buffers attached to the frame. In addition to storing energy, ...he sharp return motion of the deck imparts a lively action to the particle bed ...nd therefore promotes good screening.

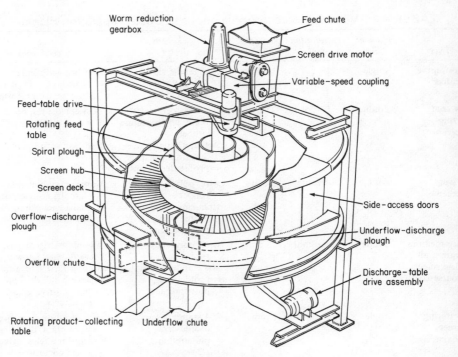

Fig. 4.19 Rotating probability screen.

4.4.4 Electromechanical Screens

The electromechanical type of screen operates with a high-frequency motion of very small throw. The motion is usually caused by a moving magnet which strikes a stop. There are other types in which an electromagnet attracts an armature (in this case impact is avoided because of obvious damage to the armature).

While the principle is currently widely applied to vibrating feeders, it is not often used for screens and only applications to fine-particle screening and dewatering are likely to be encountered.

4.4.5 Rotating Screening

A new process for the screening of raw coal of high moisture content at fine sizes has been developed by British Coal.[8] The 'rotating probability screen' (Fig. 4.19) has the advantage that the effective screening aperture may be changed while the screen is running. The screen 'deck' is made up of small-diameter stainless-steel rods radiating from a central hub. The hub is rotated and the coal falls onto the rotation spokes. The undersize passes through the spokes, and the oversize is deflected over them. The speed of rotation dictates the screening aperture. The ability to tolerate high levels of moisture in the coal has taken precedence over the accuracy of size separation. A screening operation by which the proportion of underflow product can be controlled while the machine is in operation represents an important advance in the preparation of blended coals in treatment plants

SCREENING AND CLASSIFICATION

where the fines are not cleaned. Numerous screens with 2.4 m diameter decks are at present installed in plants and can handle up to 100 t/h when accepting feeds sized 20 mm × 0 and separating at 5 mm.

4.5 SCREENING EFFICIENCY

Within any given mix of particles, perfect screening would require the sorting of all larger particles from all smaller particles at some given size of separation. In practice this perfect separation is seldom attained, but it may be approached if the screening system used is effective. Numerous methods of calculating screen efficiency are used. In practice, screening efficiency is measured by sampling the feed, overproduct and underproduct. Screen efficiency should always be viewed within the context of the amount of fine material in the feed. Perhaps the most widely applicable formula for calculation of screen efficiency is that suggested by Leonard.[9] This approach is based on the concept of misplaced sizes and is suitable for the unqualified calculation of screening efficiency. It treats efficiency on the basis of the percentage of the total misplaced fine material in the screen overproduct and the misplaced coarse material in the screen underproduct. The total percentage of misplaced material is then subtracted from 100%. The formula is:

$$E = (100 - b) - \left(\frac{a-b}{c-b}\right)(100 - b - c) \tag{4.1}$$

where E is %age screen efficiency, a is %age undersize in the screen feed, b is %age undersize in the screen overproduct and c is %age undersize in the screen underproduct.

Three other formulae are:

$$E = \frac{100(c-a)}{bc} \tag{4.2}$$

where a is the %age of coarse material in the feed, b is the %age of fine material in the feed and c is the %age of coarse material in the overproduct.

$$E = \frac{100(a-b)}{a(100-b)} \tag{4.3}$$

where a is the %age of undersize material in the feed and b is the %age of undersize material in the overproduct.

$$E = \frac{100(a-b)}{a(c-b)} \tag{4.4}$$

where a is the %age of undersize material in the feed, b is the %age of undersize material in the overproduct and c is the %age of undersize material in the underproduct.

These three formulae define efficiency as the ratio of the amount of material that actually passes through a screen to the amount which should pass through the screen on a basis of the square-hole screen analysis of the feed. The first two

TABLE 4.3 Comparison of Formulae for Calculation of Efficiency

Test no.	Moisture	Feed	Passing screen mesh		Formula			
			Overproduct	Underproduct	(4.2)	(4.3)	(4.4)	Leonard's (4.1)
1	3.9	75.7	8.2	93.4	97.1	97.1	105	93.1
2	6.7	60.6	17.5	95.1	86.2	86.2	91.3	89.5
3	8.3	70.5	12.4	98.0	94.0	94.0	96.3	94.7
4	10.8	66.0	36.8	94.0	70.0	70.0	77.5	78.9

are fundamentally the same and differ only in the definition of the terms. They were meant for evaluating square-hole screens and therefore do not compensate for the oversize material in the screen underproduct that occurs with slots or parallel-bar screen surfaces.

The third formula was likewise meant for evaluating square-hole screening efficiencies but does, however, account for a quantity of oversize material in the screen underproduct. Nonetheless, the formula is not really applicable to rectangular openings because coarser material will be passed, and thus the comparison using square-hole testing can result in efficiencies greater than 100%. The first formula, that derived by Leonard, treats efficiency on the basis of the percentage of the total misplaced fine material in the screen overproduct and the misplaced coarse material in the screen underproduct. The total percentage of misplaced material is then subtracted from 100%. Table 4.3 compares all four formulae, giving results obtained from four separate tests for a rectangular-hole screen using a square-hole test screen.

4.6 CLASSIFICATION

4.6.1 Principles of Classification

Classification is used in coal preparation for the fine sizing, where it is impractical or too expensive to use screening. The density difference between the particles becomes significant when classification is employed to separate particles into two size groupings, because of the following considerations.[10] Size, relative density, and consequently relative mass determine the behaviour of the particles in water, and two important laws will apply. In general terms, Stokes' law is valid for particles below about 50 μm and Newton's law is valid for particles above 5 mm.

i.e.
$$v = k_1 d^2 (D_s - D_f) \quad \text{Stokes' law}$$
$$V = k_2 d (D_s - D_f)^{1/2} \quad \text{Newton's law}$$

where v is terminal velocity, d is particle diameter, D_s is particle density, D_f is fluid density and k_1, k_2 are constants.

All classifiers consist of a classifying column in which water is rising at a uniform rate. Particles contained in the water, when introduced to the column, either sink or rise according to whether their terminal velocities are greater or less than the rising velocity of the water. The classifying column therefore separates the particles into two groups:

- an overflow product consisting of the finer fragments having terminal velocities below the upward flow rate;
- an underflow or spigot product with terminal velocity greater than the rising velocity.

In coal preparation, sizing classifiers are usually most commonly utilized for dilute suspensions of relatively finely sized fragments (i.e. less than one millimetre in size). As such the classifier is likely to operate under predominantly free-settling conditions, and for such conditions it can be shown that:

$$\frac{d_{s_2}}{d_{s_1}} = \left(\frac{D_{s_1} - D_f}{D_{s_2} - D_f}\right)^n$$

This is called the free-settling ratio, where n is 0.5 for small particles obeying Stokes' law and 1.0 for larger particles obeying Newton's law; D_{s_1} is the relative density of one variety of rock, e.g. coal, with a mean diameter of d_{s_1}; and D_{s_2} is the relative density of another variety of rock, e.g. shale with a mean diameter of d_{s_2}.

This simple rule signifies that the density difference between the two rock types has a more pronounced effect on classification at coarser size ranges. Hence its acceptable performance for fine sizes.

Similarly, if the solids concentration of the feed to the classifier was to increase so as to create hindered settling conditions instead of free settling, the resultant effect would be to reduce the effect of size while increasing the effect of density on classification; again a counterproductive effect. It can be shown that a corresponding hindered settling ratio is defined as:

$$\frac{d_{s_2}}{d_{s_1}} = \left(\frac{D_{s_1} - D_p}{D_{s_2} - D_p}\right)$$

4.6.2 Types of Classifier

Sizing classifiers, operating with fairly dilute slurries under free-settling conditions, can be subdivided into three groups:

(a) settling classifiers;
(b) mechanical classifiers;
(c) centrifugal classifiers.

Settling Classifiers

Settling classifiers are the simplest and oldest form of classifier. They employ a quiescent reservoir of fixed volume in order to create the correct settling conditions so as to effect the separation. This occurs in a fairly narrow size-range embracing the desired cut-point. Conditions are regulated by varying the underflow rate either continuously or intermittently. Settling cones and spitzkasten classifiers are the most widely employed in coal preparation. Applications include the desliming of fines and recovery of magnetite but true sizing applications are nowadays fairly uncommon.

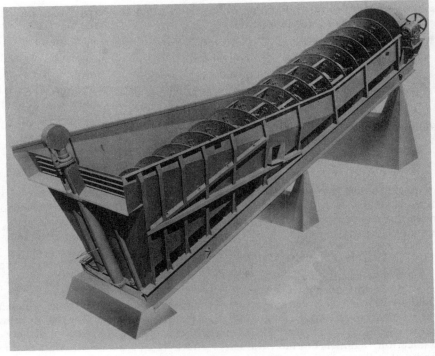

Fig. 4.20 Mechanical classifier.

Mechanical Classifiers

Several forms of this type of classifier exist, including horizontal-current classifiers such as the rake and spiral type and vertical-current or hydraulic classifiers such as multispigot hydrosizers.

A spiral type of mechanical classifier is shown in Fig. 4.20. The slurry feed is introduced into the inclined trough and forms a settling pool, in which particles of high terminal velocity settle rapidly to the bottom. Above this sediment layer is a quicksand zone where hindered settling occurs.[11]

The depth and shape of this zone is a function of the classifier action and feed solids concentration. Above the hindered settling zone is another formed predominantly by free-settling material, comprising a stream of flowing dilute suspension moving in a horizontal direction to an overflow weir. The sediment is removed continuously by a mechanical scroll or by rakes. The fines are carried by the water flow over the edge of the weir. Hydraulic classifiers are more commonly used for mineral processing applications, but occasionally use is made in them for coal treatment as size classifiers to prepare feed slurries for gravity-concentration units such as shaking tables, spiral concentrators and, less frequently, water-only cyclone and froth-flotation circuits.

The Stokes' hydrosizer,[12] shown in Fig. 4.21 is an example of this type of classifier. Each teeter chamber is provided at its bottom with a supply of water

SCREENING AND CLASSIFICATION

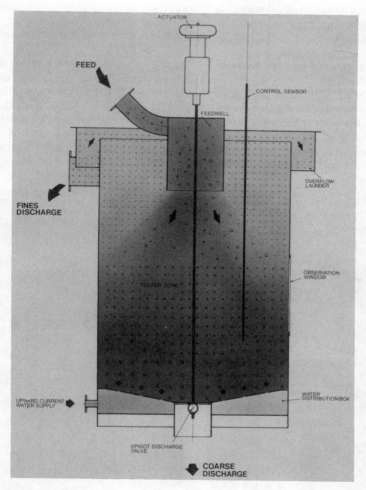

Fig. 4.21 Stokes' single-cut hydrosizer (courtesy R. O. Stokes and Company Limited).

under constant head which is used to maintain teetering condition in the solids which percolate downwards against the interstitial rising flow of water. Each chamber is fitted with a discharge spigot connected to a pressure-sensitive valve so that the desired cut-point can be regulated.

Centrifugal Classifiers

Two forms of centrifugal classifier are employed in coal treatment, of which by far the most important is the hydrocyclone. The other is the decanter solid-bowl centrifuge which is discussed later in the book in Chapter 10 which deals with solid–liquid separation. Since the real function of the solid-bowl centrifuge is in fact to effect dewatering of solids, its ability to classify out ultrafine solids is more often undesirable than desirable in coal treatment. The hydrocyclone, on the other hand, is most definitely an important size classifier in many coal-preparation applications.

The flow pattern in the common design of cyclone is that of a spiral within a spiral. The entering slurry commences a downward flow in the outer regions of the cylindrical section of the cyclone (see Fig. 4.22) thereby forming the outer spiral. The provision of an upper central outlet and the inability of all the slurry to depart through the conical apex outlet results in the gradual upward migration of some of the slurry. This migration increases as the conical apex is approached, creating a reversal of fluid in its vertical-velocity direction resulting in an upward-flowing inner spiral towards the overflow outlet.[13]

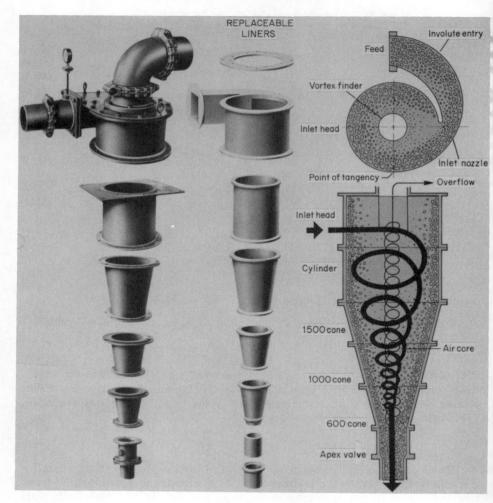

Fig. 4.22 Hydrocyclone.

SCREENING AND CLASSIFICATION

Some imperfections in operating conditions occur in practice. Most result from wear and tear within the body of the cyclone. Others are the result of poor operation or poor finishing of the cyclone casting, etc.

Common imperfections are:

- short-circuit flow caused by the formation of a flowpath created by some form of obstruction or wear of the vortex finder, or a combination of both. This is one of the main reasons why a vortex finder is used;
- eddy flows, which are vertical flows in the region outside of the outer wall of the vortex finder caused by inability of the normal opening to cope with the natural upflow to the vortex.

Research into the internal conditions which occur when a hydrocyclone is operating have shown that a number of peculiarities exist. These have a significant bearing upon the effectiveness of the unit to perform at a given size separation.[13,14]

When the feed is introduced under pressure through the tangential inlet, a swirling motion is imparted to the slurry when it enters the involuted feed entrance. This generates a vortex in the cyclone, with a low-pressure zone forming along the vertical axis. Consequently, an air core forms along this axis, normally connected to the atmosphere through the apex opening, but in part created by dissolved air emanating from the slurry within the zone of low pressure. Formation of the air core is an indication of vortex stability. The centrifugal force developed accelerates the settling rate of the particles, thereby separating them according to their size and relative density. There is evidence to demonstrate that Stokes' law does apply with reasonable accuracy to cyclones of conventional design. Faster-settling grains move to the inner wall of the cyclone where the velocity is lowest and migrate to the apex opening or underflow discharge. Slower-settling particles move towards the zone of low pressure along the axis and are carried upwards through the vortex finder to the overflow discharge. The existence of an outer region of downward flow and an inner region of upward flow creates the existence of a position at which there is no vertical-velocity component at all. This applies throughout the functional part of the cyclone, and an envelope or locus of zero vertical velocity can be traced. Eddy flows tend to centre around this locus.

The most commonly used method for representing cyclone efficiency is by means of the grade efficiency or Tromp curve as shown in Fig. 4.23. This relates the mass fraction as a percentage reporting to the underflow, of each particle size in the feed, to the particle size of the underflow. The cut-point, usually referred to as the 'd_{50} size' is the point at which 50% of particles in the feed, of that size, report to the underflow, i.e. particles of this size have an equal chance of migrating to either product. The grade efficiency curve provides an indication of the sharpness of cut which the cyclone under consideration is able to provide. The slope of the curve and the d_{50} therefore provide the measure of efficiency or imperfection, I:

$$I = \frac{d_{75} - d_{25}}{2d_{50}}$$

The lower the value of I, the more efficient the cyclone is. It should therefore be borne in mind that the test which provides the determination of I is purely a

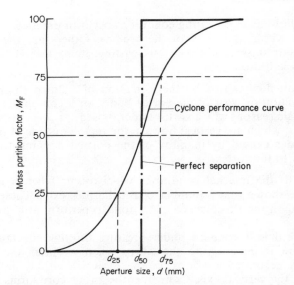

Fig. 4.23 Tromp type, grade efficiency curve for hydrocyclone classifying performance.

measure of the efficiency of a particular hydrocyclone, and as such is therefore independent of the characteristics of the feed slurry. The slurry is merely providing the means of determining individual cyclone performance. In coal preparation applications the hydrocyclone is employed for both classifying and thickening. Classifying cyclones are usually of conventional design, i.e. long slender cones normally ranging in diameter from about 25 to 750 mm. In general the feed to classifying cyclones will be 2–15% solids. Multiple units consisting of several (10–30) small-diameter units can be employed and these will be fed by either a straight or a circular manifold arrangement, as shown in Fig. 4.24. Cyclones of small radii develop very high centrifugal force and are therefore able to cut at fine sizes. A 75-mm diameter cyclone has a nominal rating of 1.5–4 m^3/h at 70–275 kPa feed pressures, but the orifice size in the underflow determines flow volume.[15]

At the other end of the range the large-diameter cyclones are usually used as primary thickeners or for coarse classification. A 610-mm diameter cyclone can vary from 220 to over 550 m^3/h at 70–250 kPa.

Smaller units are made from cast aluminium or stainless-steel metallic bodies or ceramic, or polyurethane for the smallest types.[17]

Replaceable, moulded, natural or synthetic rubber or polyurethane liners are usually fitted into the metallic body type. The overflow orifice and top plate will also be moulded in natural or synthetic rubber or made from stainless-steel whereas the underflow orifice may also be made from ceramic material. The larger units are available in all-cast Ni-Hard, cast-steel or aluminium shells fitted with ceramic lining or bonded rubber, elastomer and urethane lining. Underflow orifices are often made from similar wear-resistant materials.[16]

Figure 4.25 shows an application of a hydrocylone operating in conjunction

Fig. 4.24 Straight- and circular-manifold feed arrangements for hydrocylone assemblies (courtesy Richard Mosley Limited).

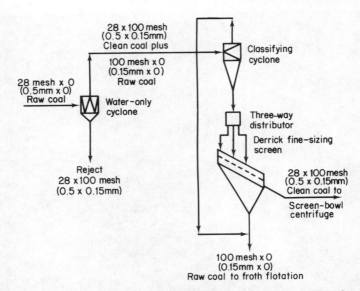

Fig. 4.25 Application of a hydrocyclone operating in conjunction with a fine-coal sizing screen (100 mesh or 0.15 mm) (from Cardinal River Coal Preparation Plant).

with a fine-coal sizing screen, and Fig. 4.26 shows a more general circuit.

Finally, Fig. 4.27 shows a multicomponent cyclone design which is assembled by means of 'Victaulic' type ring couplings to facilitate rapid dismantling. Such types are of value in multicyclone assemblies, as shown where rotational maintenance is essential for good operating conditions.

4.6.3 Pneumatic Separation

Separation at fine sizes can also be achieved by pneumatic centrifugal classifiers (dedusters) (Fig. 4.28). The deduster works according to a closed-air principle whereby the material is effectively separated with the aid of gravitational and centrifugal effects.[18] The separator operates on a coal feed of about 5 mm and makes separations in the 0.25–0.5 mm range. The raw coal is fed through a hollow shaft to a rotating distributor, giving a centrifugal effect. The coal is dispersed by this distributor radially outward into a rising air stream, where the fines pass upward. A set of impellers ejects the fines to the outer shell, where they drop out into a large chamber. A rotating scraper at the bottom then moves the fines to an opening which discharges them to a conveyor. The large air chamber provides for differential settling of the dust, with coarser particles moving through vanes into the inner chamber, where they combine with the coarse coal. The coarse-coal particles drop down into the centre chamber and discharge by gravity to the coarse-coal conveyor. In order to avoid water condensation, a portion of 10–15% of the circulating air volume is exchanged with fresh air which is introduced with the feed.

The deduster is often employed to remove dry fines which will then be added back to the washed coarser fraction. In removing the fines by classification, the

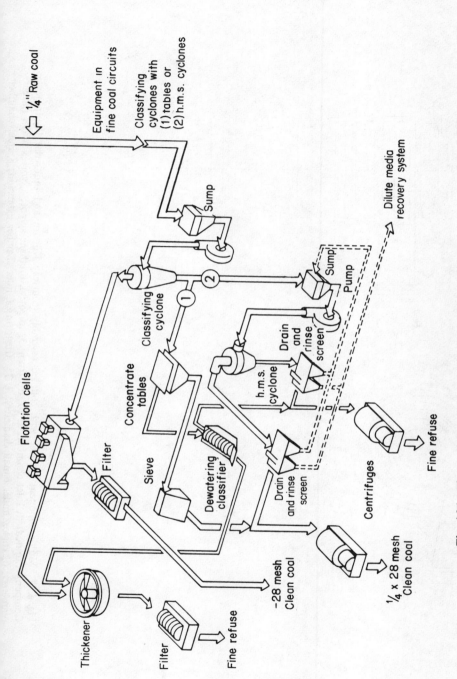

Fig. 4.26 Fine-coal cleaning circuit using classifying cyclones.

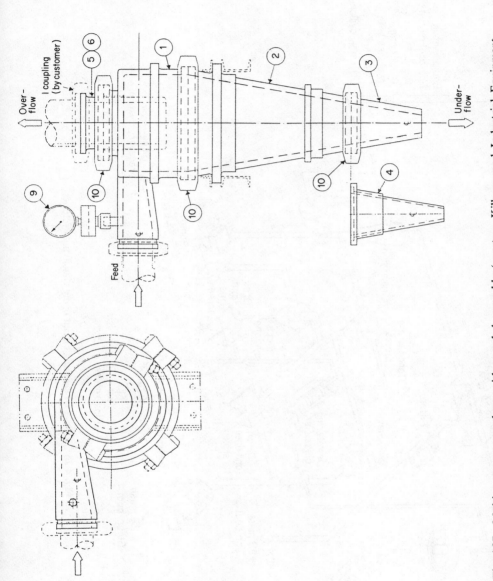

Fig. 4.27 Multicomponent cyclone with coupled assembly (courtesy Kilborn and Industrial Equipment Company Limited). 1, Inlet chamber; 2, 20° frustrum; 3, spigot; 4, urethane spigot; 5, vortex finder (larger diameter); 6, vortex finder (small diameter); 7, 75° frustrum; 8, inlet chamber extension; 8, inlet chamber

SCREENING AND CLASSIFICATION

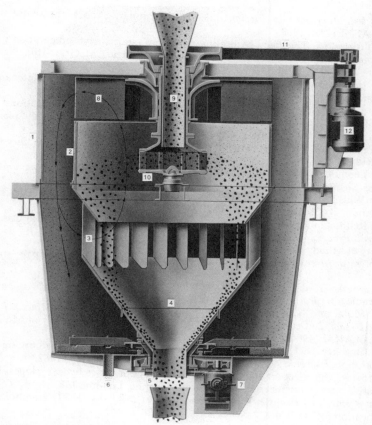

Fig. 4.28 Pneumatic centrifugal classifier (deduster). 1, Outside cylindrical mantle; 2, inside ring; 3, hopper cover; 4, grit outlet; 5, grit outlet; 6, dust discharge; 7, drive of dust discharging device; 8, impeller; 9, hollow shaft with product inlet; 10, distributor; 11, belt drive; 12, driving motor (courtesy Humboldt-Wedag.)

density effect comes into play and fine particles of coal are removed and fine particles of high-density, high-ash material remain with the larger particles which report to the beneficiation process. This results in the dedusting process performing a fine-size separation and some particle cleaning.

REFERENCES

1. D. G. Osborne. Screening, in *Solid–Liquid Separation*, (Ed.) L. Svarovsky, Butterworth, Guildford, England, 1981, Chapter 8, pp. 213–241.
2. A. C. Partridge. Principles of screening, *Mineral Engineering Handbook, Mine and Quarry*, Dec. 1977, pp. 33–38.
3. Anon. Vibrating screen. Theory and selection, Bulletin 26M5506, Allis-Chalmers Company, Wisconsin, USA.

4. Anon. *Vibrating Screens—Design Bulletins*, Vibrating Screen Manufacturers Association, ANSI 121.1, 1968.
5. G. E. Groseclose. The role of vibrating screens in modern coal preparation plants, in *Coal Age Handbook of Preparation*, (Ed.) P. C. Merritt, McGraw-Hill, New York, pp. 109–113.
6. R. A. Irvine. Large vibrating screen design—manufacturing and maintenance considerations, *Mining Engineering*, **36**, pp. 1341–1346, 1984.
7. W. R. Leeder, J. W. Hogg, E. M. Jacobs and D. G. Osborne. Applications of high-capacity multislope screens for coal desliming applications in heavy media plants, 10th International Coal Preparation Congress, Edmonton, 1986.
8. M. P. Armstrong. Demonstration of the NCB/MAGCO rotating probability screen at Hickleton, NCB–MRDE, June 1980.
9. J. W. Leonard. Determination of industrial screening efficiency, *Transactions Society Mining Engineers of AIME* **256**, 185–187, 1974.
10. B. A. Wills. Classification, in *Mineral Processing Technology*, 2nd edn, Chapter 9, Pergamon, Oxford, pp. 215–251, 1982.
11. R. E. Reithmann and B. M. Bunnel. Application and selection of spiral classifiers, in *Mineral Processing Plant Design*, 2nd edn, (Eds) A. L. Mular and R. B. Bhappu, Society of Mining Engineers of AIME, New York, 1980, pp. 362–385.
12. Anon. *Hydrosizer*, R. O. Stokes and Company Limited Publication RPS-500/3/3/60, 1965.
13. L. Svarovsky. *Hydrocyclones*, Holt Saunders, Eastbourne, England, 1984.
14. D. Bradley. *Hydrocyclones*, Pergamon, Oxford, 1965.
15. A. L. Mular and N. A. Jull. The selection of cyclone classifiers, in *Mineral Processing Plant Design*, 2nd edn, (Eds) A. L. Mular and R. B. Bhappu, Society of Mining Engineers of AIME, 1980, pp. 376–403.
16. G. F. Brookes, N. J. Miles and M. G. Ayat. The possible use of small cyclones in coal preparation, *University of Nottingham Mining Magazine* **XXXIV**, 1982.
17. R. Mozley. The selection and operation of high performance hydrocyclones, Filtration Society/Minerals Engineering Society Joint Meeting, London, 1983.
18. Anon. *Centrifugal Deduster with Mechanical Dust Discharge*, KHD Humboldt Wedag, Publication 3.430, 1974.

Chapter 5

SIZE ANALYSIS AND FLOAT–SINK TESTING

5.1 INTRODUCTION

The two most important quantitative analyses in coal preparation are the size-distribution analysis and the float–sink analysis. Their combined impact on both plant design and the day-to-day running of plants is of paramount importance and yet many data obtained cannot be regarded as being truly representative of the raw coal to be treated. Perhaps the greatest need of all is that of relating laboratory data to the practical situation, and in this regard both types of analysis present some difficulties.

First, it is very important to obtain truly representative samples of adequate mass to permit the required analytical work to be carried out. The analytical data can only be as good as the sample is representative of its source. Secondly, the sample point location must be selected to provide samples which truly represent the coal in the required state, e.g. raw-coal feed as it actually enters the plant rather than when it leaves the raw-coal stockpile. Thirdly, the samples must be used for carrying out a required minimum of work in analysing the sample to contribute analyses which cover most of the forseeable needs for the coal, e.g. low- and high-density separations to produce two saleable products: clean coal and middlings. Finally, sample data may be needed for developing the means for scaling-up from laboratory testing to obtain data indicative of plant conditions.

Size analyses are normally carried out to provide the following information:

(a) capacity around a plant flowsheet and the variations likely to occur as a result of variations in size consistency;
(b) the proportions of various sized or graded products;
(c) the amount of fines present in the coal.

Some form of qualitative analysis is normally carried out with each of the size fractions. Usually, at least ash content is analysed.

Float–sink analyses are normally carried out to provide:

(a) float–sink (theoretical) yield of a specified product or products;
(b) an indication of the degree of difficulty which might be encountered in cleaning the coal, e.g. the amount of near-density material;
(c) an indication of the effectiveness of a cleaning process, or of the total plant,

e.g. the amount of misplaced material in the product;
(d) an indication of the distribution of coal quality characteristics with regard to various density fractions, e.g. ash and sulphur content, calorific value.

Both kinds of analyses are used in a variety of circumstances, such as:

(a) preliminary and detailed evaluation of new mine development or modification and/or expansion of existing mines—drill-core sample, bulk sample, etc;
(b) day-to-day plant control in existing operations for different seams, different sources, combined plant feed, different parts of the plant circuit;
(c) Plant design purposes such as flowsheet development.

5.2 OBTAINING THE SAMPLES REQUIRED

The analytical data are only as good as the sample is representative of the coal. Despite good intentions at the outset, much analytical work is eventually found to be futile, when it is later established that the sample material obtained was not truly representative of the particular source of coal from which it was sampled.

Key factors in sampling are, therefore, as follows:

(a) Select a location which:
   (i) lends itself to convenient sampling using specified methods;
   (ii) provides coal which typifies the requirement.
(b) Use a sampling method which:
   (i) avoids bias;
   (ii) cuts the entire range of coal fragments.
(c) Sample under steady-state conditions.
(d) Take enough sample to carry out all of the work, and use accepted methods for reducing sample mass to obtain subsamples.
(e) Store the sample material properly and handle carefully to avoid changes in properties.

Some sampling standards exist (discussed in Chapter 22) which specify the amount required for sizing and float–sink analyses. At the present time, the most comprehensive national standards for testing coal are arguably those produced by Australia. Other countries which have large coal production have developed working standards which are specified in varying degrees of detail, e.g. Britain, West Germany and the USA. Formulae have been developed which can be used to determine the amount of sample required both for size distribution and float–sink analyses. These are generally empirically based to ensure adequate sample material is obtained. However, as yet no firm specification exists for the determination of the mass of sample or the number of increments required when taking samples for float–sink analysis.

The International Standards Organization (ISO)[1] recommends that for screen (size) analysis, a minimum of 40 increments should be taken for each sample, but this is dictated by the nature and source of the sample, so that in practice such recommendations are really only guidelines.

The sample for float–sink analysis is usually eventually subdivided into a number of subsamples without further crushing or degradation and therefore calculation of the total sample required should be based on the minimum recommended mass required for each test.

A general rule which has been proven to give satisfactory results is that a minimum of 1000 fragments are required for each float and sink test. Such work is usually carried out with samples of fairly closely graded coal and therefore a minimum mass can be calculated for each size fraction.

Based on the assumption that the coal fragments are of fairly uniform spherical shape and that the coal has a relative density of 1.5, the following formula has been derived for sample mass:[2]

$$P = 5.24D$$

where P is the mass in kg and D is the mean diameter in mm (e.g. the midpoint of the screens defining the size of the coal).

Using this formula, the following minimum rounded-off quantities are obtained:

Size (mm)	Mean size (mm)	Mass (kg)
150 × 100	125	700
100 × 75	86.5	450
75 × 50	52.5	330
50 × 35	42.5	220
35 × 20	27.5	140
20 × 12	16.0	85
12 × 5	8.5	45
5 × 3	4.0	21
3 × 0.5	1.8	10

It should perhaps be stressed that these are minimum quantities for run-of-mine coal and must be increased for raw coals containing large amounts of stone and shale or for secondary material from waste-pile sources, etc.

When sampling discards, e.g. during performance trials, the above amounts should probably be increased by 50%.

The Australian standard,[3] widely used even outside Australia, suggests that the number of discrete particles in any size fraction should not be less than 2000. It also suggests that because some coals give low yields in the middle fractions of

		Coal-preparation plant products (kg)			
		Comprehensive test		Control test	
Size (mm)	Raw coal	Clean coal	Discard	Clean coal	Discard
125 × 63	400	600	1200	200	400
63 × 31.5	100	150	300	50	100
31.5 × 16	25	35	70	15	35
16 × 8	7	10	20	4	10
8 × 4	3	4.5	9	2	5
4 × 2	1.5	2.5	5	1.5	2.5
< 2	1.0	1.5	3	1.0	1.0

relative density, there may be insufficient sample for analytical requirements (i.e. less than 20 g) and therefore there should be at least 10 fragments in each fraction.

This standard suggests that when the bulk sample is being taken it is better to over sample to avoid obtaining insufficient material. For tests including the coarser size-fractions, samples of over 5 t may be needed, and rather than use a formula the standard suggests pro-rata adjustment using the table to obtain other size ranges. A similar approach is suggested in the ASTM Standard used in North America.[4]

5.3 SIZE ANALYSIS AND PRESENTATION OF DATA

The most common method for carrying out size-distribution analyses is to utilize selected sets of standard screens and/or sieves which can be fitted as a nest into a shaking machine (see Fig. 5.1). This type of approach is applicable down to a sieve aperture of about 35 μm, depending upon the standard selected. Below this size, i.e. in the subsieve range, where mechanical methods become unreliable, elutriation, sedimentation or other methods of size analysis are often necessary.

Subsieve sizing methods most commonly employed therefore include:

(a) Elutriation/sedimentation —Andreason pipette
—Sedimentation balance
(b) Centrifugal —Warman cyclosizer
(c) Microsieves —Electroformed sieves used in conjunction with ultrasonic bath
(d) Microscopic/Image analysers—Quantimet
(e) Electronic pulse —Coulter counter
(f) Surface-area determination

When the size-analysis sample is predominantly fine material and a subsieve analysis is required, the entire sizing analysis is usually conducted wet. In such cases, if a sedimentation or elutriation method is selected for the subsieve analysis, it becomes necessary to determine a correlation for the transition from physical sizing to classification. This is usually achieved by overlapping the two ranges, e.g. sieving 100 μm to 50 μm: elutriation 100 μm and below.

5.3.1 Sieve Analysis

The method and reporting of results of a sieve analysis are set out in published standards (e.g. ISO 2591, British Standard Specification (BSS) 1796, American Society for Testing and Materials (ASTM) E-11-70) and one such standard should be rigorously adopted for good practice. These standards call for the use of one of several standardized series of testing sieves, and Fig. 5.2 shows a comparison table of standards for various commonly employed international sieve series. Recently, an international standard series has been recommended (ISO) in an effort to reduce the number of different standards in current usage and assist

SIZE ANALYSIS AND FLOAT–SINK TESTING

Fig. 5.1 Test sieves for size analysis (source, Tyler Handbook[5]).

USA (1)		TYLER (2)	CANADIAN (3)		BRITISH (4)		FRENCH (5)		GERMAN (6)
Standard[a]	*Alternate*	*Mesh designation*	*Standard*	*Alternate*	*Nominal Aperture*	*Nominal mesh No.*	*Opening (mm)*	*No.*	*Opening*
125 mm	5 in.		125 mm	5 in.					
106 mm	4.24 in.		106 mm	4.24 in.					
100 mm	4 in.		100 mm	4 in.					
90 mm	$3\frac{1}{2}$ in.		90 mm	$3\frac{1}{2}$ in.					
75 mm	3 in.		75 mm	3 in.					
63 mm	$2\frac{1}{2}$ in.		63 mm	$2\frac{1}{2}$ in.					
53 mm	2.12 in.		53 mm	2.12 in.					
50 mm	2 in.		50 mm	2 in.					
45 mm	$1\frac{3}{4}$ in.		45 mm	$1\frac{3}{4}$ in.					
37.5 mm	$1\frac{1}{2}$ in.		37.5 mm	$1\frac{1}{2}$ in.					
31.5 mm	$1\frac{1}{4}$ in.		31.5 mm	$1\frac{1}{4}$ in.					
26.5 mm	1.06 in.	1.05 in.	26.5 mm	1.06 in.					
25.0 mm	1 in.		25.0 mm	1 in.					25.0 mm
22.4 mm	$\frac{7}{8}$ in.	0.883 in.	22.4 mm	$\frac{7}{8}$ in.					
19.0 mm	$\frac{3}{4}$ in.	0.742 in.	19.0 mm	$\frac{3}{4}$ in.					20.0 mm
									18.0 mm
16.0 mm	$\frac{5}{8}$ in.	0.624 in.	16.0 mm	$\frac{5}{8}$ in.					16.0 mm
13.2 mm	0.530 in.	0.525 in.	13.2 mm	0.530 in.					
12.5 mm	$\frac{1}{2}$ in.		12.5 mm	$\frac{1}{2}$ in.					12.5 mm
11.2 mm	$\frac{7}{16}$ in.	0.441 in.	11.2 mm	$\frac{7}{16}$ in.					
									10.0 mm
9.5 mm	$\frac{3}{8}$ in.	0.371 in.	9.5 mm	$\frac{3}{8}$ in.					
8.0 mm	$\frac{5}{16}$ in.	$2\frac{1}{2}$	8.0 mm	$\frac{5}{16}$ in.					8.0 mm
6.7 mm	0.265 in.	3	6.7 mm	0.265 in.					
6.3 mm	$\frac{1}{4}$ in.		6.3 mm	$\frac{1}{4}$ in.					6.3 mm
5.6 mm		$3\frac{1}{2}$	5.6 mm						
	No. $3\frac{1}{2}$			No. $3\frac{1}{2}$			5.000	38	5.0 mm
4.75 mm	4	4	4.75 mm	4					
4.00 mm	5	5	4.00 mm	5			4.000	37	4.0 mm
3.35 mm	6	6	3.35 mm	6	3.35 mm	5			
							3.150	36	3.15 mm
2.80 mm	7	7	2.80 mm	7	2.80 mm	6			
2.36 mm	8	8	2.36 mm	8	2.40 mm	7	2.500	35	2.5 mm
2.0 mm	10	9	2.00 mm	10	2.00 mm	8	2.000	34	2.0 mm
1.70 mm	12	10	1.70 mm	12	1.68 mm	10	1.600	33	1.6 mm
1.40 mm	14	12	1.40 mm	14	1.40 mm	12			
							1.250	32	1.25 mm
1.18 mm	16	14	1.18 mm	16	1.20 mm	14			

USA (1)	Tyler (2)	Canadian (3)	British (4)	British mesh	French (5)	French No.	German mm (6)	German No.	German μm
1.00 mm	18	16	1.00 mm	18	1.00 mm	16	1.000	31	1.0 mm
850 μm	20	20	850 μm	20	850 μm	18	0.800	30	800 μm
710 μm	25	24	710 μm	25	710 μm	22	0.630	29	630 μm
600 μm	30	28	600 μm	30	600 μm	25	0.500	28	500 μm
500 μm	35	32	500 μm	35	500 μm	30			
425 μm	40	35	425 μm	40	420 μm	36	0.400	27	400 μm
355 μm	45	42	355 μm	45	355 μm	44	0.315	26	315 μm
300 μm	50	48	300 μm	50	300 μm	52	0.250	25	250 μm
250 μm	60	60	250 μm	60	250 μm	60			
212 μm	70	65	212 μm	70	210 μm	72	0.200	24	200 μm
180 μm	80	80	180 μm	80	180 μm	85	0.160	23	160 μm
150 μm	100	100	150 μm	100	150 μm	100	0.125	22	125 μm
125 μm	120	115	125 μm	120	125 μm	120			
106 μm	140	150	106 μm	140	105 μm	150	0.100	21	100 μm
									90 μm
90 μm	170	170	90 μm	170	90 μm	170	0.080	20	80 μm
75 μm	200	200	75 μm	200	75 μm	200			
63 μm	230	250	63 μm	230	63 μm	240	0.063	19	71 μm
									63 μm
									56 μm
53 μm	270	270	53 μm	270	53 μm	300			
45 μm	325	325	45 μm	325	45 μm	350	0.050	18	50 μm
									45 μm
									40 μm
38 μm	400	400	38 μm	400			0.040	17	

(1) USA sieve series—ASTM Specification E-11-70
(2) Tyler standard screen scale sieve series.
(3) Canadian standard sieve series 8-GP-1d.
(4) British Standards Institution, London BS-410-62.
(5) French Standard Specifications, AFNOR X-11-501.
(6) German Standard Specification DIN 4188.
^a These sieves correspond to those recommended by ISO (International Standards Organization) as an International Standard and this designation should be used when reporting sieve analysis intended for international publication.

Fig. 5.2 Comparison Table of USA, Tyler, Canadian, British, French and German standard sieve series.

with cross-referencing in technical publications. It is hoped that eventually the ISO sieve series will be accepted as a world standard.

Sizing test results can be graphically represented in numerous ways with emphasis placed upon the amount of material retained on the screen or passing, or both. Size distributions are usually calculated on direct and cumulative bases, or both, and presented in tabular form.

Histograms of the percentage passing or retained per fraction are occasionally used to compare the size distributions of various samples, but the most common approach is to plot cumulative distributions on logarithmic graph paper which assists comparison.

The distributions are therefore given as either frequencies or cumulative frequencies, which are related because mathematically the frequency curve can be determined by differentiation of the cumulative curve or, vice versa, the cumulative curve can be obtained by integration of the frequency, hence:

$$f(x) = \frac{dF(x)}{dx} \quad \text{and} \quad F(x) = \int f(x)dx$$

so that the area beneath the frequency curve equals unity, and $F(x)$ ranges from 0 to 1 (or 0–100%).

Also
$$F(x) \text{ oversize} = 1 - F(x) \text{ undersize} \tag{5.1}$$

With computers becoming more and more widely used, programs for curve-fitting the analytical function to the experimental size-distribution data are readily available to categorize or quantify the characteristics of the distribution for a given sample source. In addition to the many curve-fitting techniques which are available, use can be made of special graph papers produced for several of the more common analytical functions. This latter option is still the most common practice.

The log-normal distribution is probably the most widely used type of function and graph papers are available which require the cumulative values, $F(x)$, to be plotted against size (ln x). For distributions which more or less follow the log-normal law, a fairly straight line is obtained. Most coal samples, however, exhibit a characteristically wide size range. This factor, combined with natural breakage, mechanical comminution and the heterogeneity of the constituent rock and mineral components, results in a complex distribution which rarely obeys any kind of mathematical law.

The Rosin–Rammler distribution function is arguably the closest law for the representation of coal size-distributions within a size range of upper limit 50 mm. It was first reported in 1933[8] and is usually expressed in terms of cumulative percentage oversize:

$$F(x) = \exp\left(-\frac{x}{x_r}\right)^n \tag{5.2}$$

where $F(x)$ is the residue, or percentage of oversize grains of minimum size, x, i.e. the mass percentage retained on a sieve of x aperture; x is the grain size, i.e. sieve aperture; x_r is a constant giving a measure of the range of grain sizes present; while n is another constant characteristic of the source material, and is a measure of the degree of dispersion of the grain size. It is often called the size-distribution constant and is dimensionless.

Equation 5.2 can be reduced to:

$$\log [\ln 1/F(x)] = n \log x - n \log x_r \qquad (5.3)$$

which gives a straight line if $\log [\ln 1/F(x)]$ is plotted against $\ln x$. This forms the basis for Rosin–Rammler graph paper.

The value of the constant, x_r, can be obtained from the graph plotted on this paper because it is the size corresponding to

$$\frac{100}{e} = 36.79\%$$

and the size-distribution constant, n, is represented by the slope of the line: it is, therefore, the degree of dispersion of the grains. Thus, if all the grains of coal were spread out in a straight line in order of size, the value of n would be a measure of the size of each grain divided by the amount each one varies in relation to its immediate neighbours.

Figure 5.3 shows size-distribution plots for various raw-coal samples plotted on Rosin–Rammler graph paper. From this graph it is evident that, when size-distribution graphs covering a wide range of size-consist are plotted, the linear law is usually applicable only for certain parts of the size-distribution curve. This demonstrates that great care must be taken in applying the law, in particular to run-of-mine coal, because of the different size-degradation mechanisms involved during mining, handling and in preparation for washing.

Bennet[9] found that although the Rosin–Rammler law was intended to define relatively fine, crushed coal, it could often be legitimately applied to run-of-mine coal, and it is the Bennet form of the Rosin–Rammler size distribution which is now most commonly used, i.e.

$$R = 100 \exp -\left(\frac{x}{x_r}\right)^n \qquad (5.4)$$

where R is the residue or percentage of oversize grains of minimum grain size, x. More practically, this is the percentage mass retained on a sieve which has an aperture of x. As mentioned earlier, the other two components are both constants: x being called the absolute-size constant, and n, the size-distribution constant.

If logarithms of this equation are taken twice, and both constants are combined, the equation becomes:

$$\log \log R = -n \log x + c \qquad (5.5)$$

This equation is linear, with $\log \log R$ and $\log x$ being directly proportional to each other, and the slope is equivalent to n. The limits for n are zero and infinity.

As n approaches infinity, the curve will approach a vertical line, i.e. all grains will be similar in size. When n equals unity, the law simplifies to ideal breakage in which absolute random distribution of comminution forces is assumed. When n approaches zero, this size-consist will envelop a wide range and it is therefore usual for actual coal size-distributions to fit into this category.

Hypothetical Rosin–Rammler plots are shown in Fig. 5.4. The double logarithmic approach is adopted and the line representing x_r is shown, i.e. $R = 36.79\%$, together with a nomogram for rapid determination of the value of n. This type of graph is used as follows.

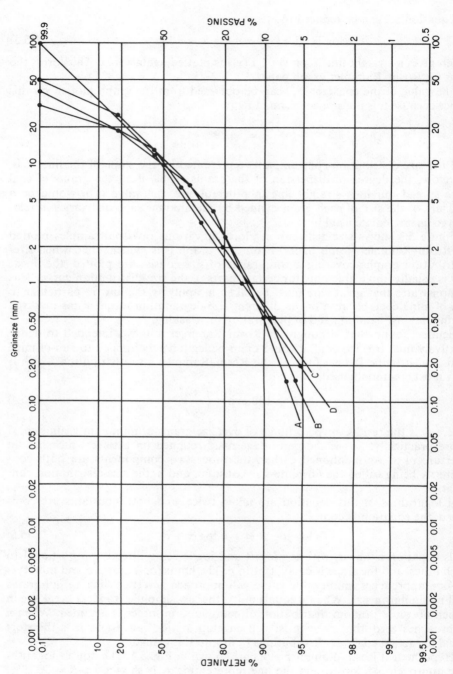

Fig. 5.3 Size-distribution plots for various raw-coal samples.

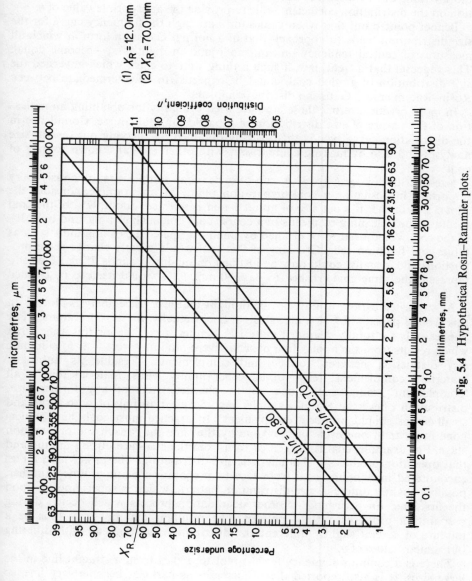

Fig. 5.4 Hypothetical Rosin–Rammler plots.

(1) $X_R = 12.0$ mm
(2) $X_R = 70.0$ mm

The size distribution for a particular coal sample is drawn and then a line is constructed parallel with the plotted line and the cross shown immediately below the 1.4 value on the abscissa. The extension of this line will produce an intersection on the distribution coefficient scale to provide the appropriate value of n.

Bennet pointed out that when values for n are high the frequency curve for the size distribution tends to approach that of a normal Gaussian form in which all measures of central tendency—mean, mode and median as x_r—become equal. This suggests that a 'real' law, if such a thing were to exist, which governed the size distribution of graded coal, could be expected to be intermediate between Rosin–Rammler and Gaussian distribution forms.

In more practical terms, these laws are used primarily for obtaining an indication of the nature of size distribution of a particular coal source. Confidence in the distribution-curve type is then usually obtained by carrying out several size analyses and also by making comparisons with other, similar, known sources of coal.

Size-distribution graphs of a run-of-mine coal sample are obviously very important in designing a coal-preparation plant. As an example, consider the Plot 2 in Fig. 5.3. If a cut size of about 5 mm is selected, about 97% of the coal would require cleaning by means of a coarse-coal unit such as a jig, and only 3% of the coal would need to be cleaned by small- or fine-coal cleaning units such as tables and flotation cells. Crushing of this coal would greatly change these proportions and, usually, could only be justified on quality grounds. If plot 2 was to represent the same coal crushed to pass 80 mm, 92% would need to be cleaned by the jig.

As suggested earlier, it must not be thought that every size analysis plotted on this type of paper will give a nice straight line. Actually, the Rosin–Rammler law applies only to 'totally degraded material'. Bennet et al.[10] define this as '... that which results from the application of such severe random force ... as to eliminate any distinction between residue and complement'. 'Residue' is unbroken material; 'complement' refers to all other lumps. Thus, material that is all 'complement' and retains none of the characteristics of any former size-distribution obeys the Rosin–Rammler law. Severe crushing of both large and small lumps should produce 'complement'. In practice most coals have undergone more than one cycle of breakage by the time they reach the preparation stage. For example, coal is broken or torn from the coal-face during mining and that operation produces a characteristic size-distribution. The rather mild forces encountered in handling coal from face to preparation plant produce another, possibly slightly different, size-distribution pattern, which is superimposed upon the first one. Any crushing or other degradation process which occurs during preparation produces a third pattern. The resultant material may thus be a mixture of several slightly different size-distribution patterns, all with differing but similar values of n.

There is a common tendency for the plot to depart from a straight line in the large sizes. It is also reported that n increases as particles become very fine, so that at the fine end there may also be a departure from linearity. This departure is in the opposite direction from that at the coarse end. Thus, over a very wide size range, the Rosin–Rammler plot of an actual cumulative coal-particle size-distribution on Bennet diagram paper tends to be S-shaped. This result is not surprising, as the law is an exponential one and assumes a finite number of par-

SIZE ANALYSIS AND FLOAT–SINK TESTING

ticles larger or smaller than any arbitrarily assigned or naturally imposed limit. In fact, an upper limit is set physically by coal bed thickness, and a lower one by micellular size, and no particles larger or smaller do, in fact, exist. Furthermore, if a linear frequency distribution curve is plotted, it usually is bimodal. The larger mode belongs to the distribution of the complement produced by the principal comminution event, and the smaller to the size distribution of the residue and resembles the distribution produced by the original mine breakage or other previous event.

Sieve analyses of carefully taken samples obtained from strategic points in a preparation plant may be used to determine the extent of degradation produced by each sizing, washing, and transporting operation, as well as by any deliberate comminution step. Combined with a size-distribution analysis of the plant feed, these data can be used to control operations for optimum size-consist and probably to predict changes in size distribution which could occur as a direct result of contemplated changes in the preparation system.

The Rosin constants have been shown by various researches to be related to other properties of the coal. Bennet[11] suggests that Rosin numbers may be related to rank and composition of coals and shows that they can be correlated with shatter tests for friability. Vogel and Quass[12] related friability of South African coals of the absolute-size constant, x_r. Shotts[13] has also related size-distribution factor, n, to friability index and to free-swelling index for Alabama coals. Friability index and free-swelling index (FSI) are both related to coal rank so that a relation with rank for values of n is implied for the limited range in rank (medium- to high-volatile bituminous range only) of the Alabama coals examined.

5.3.2 Subsieve Analysis

The table which follows[6] provides an idea of the laboratory methods of particle size measurement which are currently available, together with the approximate range for which they are usually applied.

As can be seen, there is an abundance of methods available and much has been written on the subject. However, because of the limited scope of this chapter, only a short review of the methods relevant to coal applications is given. Further information may be readily obtained from numerous sources.[7,14] Many interrelated factors influence the selection of the sizing method, including:

(a) physical characteristics such as shape or relative density of the grains;
(b) quantity to be measured and amount of sample available;
(c) number of points provided and number of analyses required;
(d) operator involvement and experience required;
(e) cost of equipment and accessories, including sample preparation;
(f) sophistication of measurement devices and result evaluation.

In reviewing the methods available for subsieve analysis it is clear that they can be divided into two basic groups. The first group involves some form of inspection and results in actual measurement of dimensions. These methods are basically microscopic in nature. The second group, and the one of most interest in coal analysis, utilizes a relationship between the particle behaviour and size. This often implies some form of assumption of an equivalent size representing

TABLE 5.1 Classification of Laboratory Methods of Particle-size Measurement

Method	Approximate size range (μm)	Equivalent diameter	Type of size distribution
Sieving (wet or dry)		aperture	
woven wire	37–4000		by mass
electroformed	5–120		
Microscopy		length/area	
optical	0.8–150		by number
electron	0.001–5		
Gravity sedimentation		Stokes	
increment and cumulative	2–100		by mass by surface[a]
Centrifugal sedimentation		Stokes	
two layer (incremental and cumulative) homogeneous (incremental only)	0.01–10		by mass
Flow classification		Stokes	
gravity elutriation (dry)	5–100		by mass
centrifugal elutriation (dry)	2–50		by mass
impactors (dry)	0.3–50		by mass or number
cyclonic (wet or dry)	5–50		by mass
electronic pulse (wet)	0.8–200		by number
particle counters (wet or dry)	0.3–10 2–9000		by number
Surface-area determination		surface/ volume	mean only surface/volume
Permeametry hindered settling gas diffusion gas absorption adsorption from solution flow microcalorimetry			

[a] By photosedimentation

each particle. Consequently, most of the methods employing this concept involve a suitable equivalent-diameter approach. For example, sedimentation is used to measure size because of Stokes' law, light scattering is used to measure the size of Mie theory, etc.

The field-scanning methods involving optical and electron microscopy are sophisticated methods which are normally only encountered in research and development work. In optical microscopy the image of a particle seen in a microscope is two dimensional. From this image, an estimate of particle size must be made, and various accepted forms of diameter have been defined, e.g. Martins Feret, maximum chord, projected area, etc. In order to reduce the tedium of manual counting, automated systems have been developed, turning optical

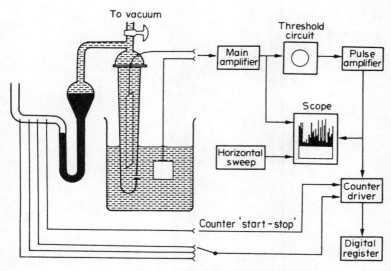

Fig. 5.5 Coulter counter.

microscopy into a sophisticated but usually costly option. Such systems include the Quantimet image analysing computer which is commonly used in coal research and petrography. Electron microscopy methods include transmission electron microscopy, where an image is produced on a fluorescent screen or photographic plate by means of an electron beam (particles 0.001–5 μm) and the scanning electron microscope, when a fine beam of electrons is caused to scan across a sample in a series of parallel tracks (for even finer material). Both of these methods are employed in coal research work.

Stream-scanning techniques, including electrical sensing, of which the Coulter Counter shown in Fig. 5.5 is an example, employ scattering attenuation to register a particle size and count.

The Coulter technique is arguably the most common one used in coal analysis and is a method for determining the number and size of particles suspended in an electrolyte. Pulses are registered when particles pass through a small sapphire orifice on either side of which is located an immersed electrode. The changes in resistance which occur as particles pass through the orifice generate voltage pulses with amplitude proportional to the volume of each particle. Hence, by a computerized system, a size distribution can be automatically produced once the total sample has passed through the tube. The method is suitable for subsieve range down to about one micrometre.

Sedimentation methods are based upon the Stokes equation:

$$U_{st} = \frac{(p_s - p_f)gd_{st}^2}{18\eta} \quad (5.6)$$

where U_{st} = terminal free-settling velocity; p_s and p_f are the relative densities of the solid and liquid respectively; d_{st} is the Stokes diameter; η is the fluid viscosity; and g is gravitational acceleration.

Fig. 5.6 Andreason pipette.

The size distribution is obtained by repeated examination of powder sample immersed in a fluid, usually water.

In one group of methods, the incremental type, changes with time in the concentration or density of the suspension at known (measured) depths are determined, and size distribution is calculated from these data.

In the other group, the cumulative type, the rate at which the individual particles settle out of the suspension is determined.

One of the most traditional methods was the Andreason's pipette shown in Fig. 5.6. Concentration changes occurring within the settling suspension are followed by drawing off definite volumes using the pipette. The initial suspension is usually between 0.5 and 1.0 by volume. After thorough mixing and the addition of a dispersing agent, samples are withdrawn from one minute onwards and the resultant data are used to obtain the various points in the size distribution.

Photosedimentation is a more sophisticated sedimentation technique which lends itself to automatic data processing by means of a coupled computer, as with the Ladal Photosedimentation system shown in Fig. 5.7. A narrow horizontal beam of parallel light is projected through a suspension at a predetermined height, h. As settlement proceeds the emergent light flux increases and hence the concentration of particles in the light beam at any time, t, will be the concentration of particles smaller than d_{st}, the Stokesian diameter, i.e.

$$d_{st} = \left[\frac{18\eta h}{(p_s - p_f)gt} \right]^{1/2} \qquad (5.7)$$

The particle size distribution is calculated from the relationship between th

SIZE ANALYSIS AND FLOAT–SINK TESTING

Fig. 5.7 Ladal wide-angle scanning photosedimentometer (WASP) for sedimentation size analysis and surface area determination.

attenuation of the light beam to the projected surface area of the particles. X-Ray sedimentation involves the use of X-rays instead of the light beam, and instruments such as the Sedigraph are commonly used for size-consists of 1–50 μm.

The sedimentation or specific gravity balance is another instrument in the same group and involves the use of a bob which is immersed in the suspension and thereby responds to changes in density created in the settling suspension. Changes in buoyancy may be varied by means of solenoids which are connected to a chart recorder from which a graph is obtained for the determination of the size analysis.

Varying centrifugal techniques, although not commonly used with coal powders, are available and popular for ultrafine suspension because the centrifuging process speeds up the settling rate, and to some extent eliminates bias caused by convection, diffusion and Brownian movement. Centrifugal sedimentation may be carried out by either two-layer or homogeneous techniques with results obtained either cumulatively or incrementally. The Ladal X-ray centrifuge

Fig. 5.8 Ladal X-ray centrifuge for particle sizing within the range of 0.01–5 μm.

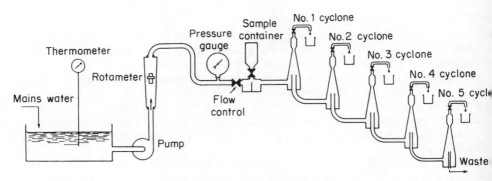

Fig. 5.9 Warman cyclosizer.

shown in Fig. 5.8 is highly automated to produce size distributions with good reproducibility. This device uses a mechanism which allows radial scanning of a sedimentation disc saving analytical time and facilitating rapid data processing. The application range claimed for this instrument is 0.01–5 μm but it cannot be used for substances which are not opaque to X-rays. Two other centrifugal devices are used in coal size analysis, in the range 5–50 μm. One is the Cyclosizer, shown in Fig. 5.9, which is a cyclonic device comprising five small-diameter cyclones arranged in series for analysis of aqueous suspensions. The other is the Infrasizer, a similar device involving larger cyclone units for the analysis of dry-powder-in-air suspension. One advantage of such methods is that small quantities of sample for each size-fraction (each cyclone) can be obtained for analysis.

The size-distribution charts given in Fig. 5.10 show comparative data obtained using a sample from a single source analysed by the various techniques described

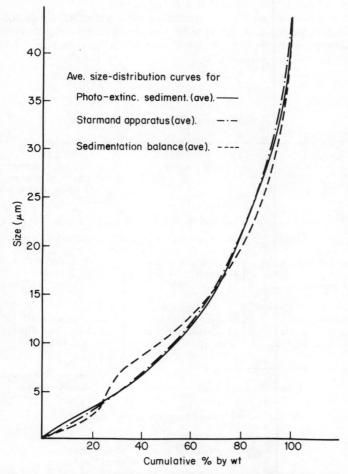

Fig. 5.10 Graph showing size-distribution curves obtained using various subsieve size-analytical methods.

in the previous paragraphs. This graph serves to demonstrate the difficulty which can often be encountered in correlating sieve and subsieve size-distribution results.

5.4 FLOAT–SINK ANALYSIS

When a coal sample is separated into relative-density fractions with defined limits using liquids of different relative densities, this is known as a float-and-sink analysis. The proportions of the fraction are usually shown tabulated as percentages of the total sample, with associated ash content and occasionally other properties also being included.

Coal coarser than 0.5 mm (28 mesh) is tested by consecutively immersing the sample of coal in a series of liquids of different relative densities and by recovering the floats at each of the different relative densities.

The range and number of density intervals used are decided upon according to the information required and the type of sample available, e.g. drill core, bulk etc. A preliminary run should establish whether a certain part of the range

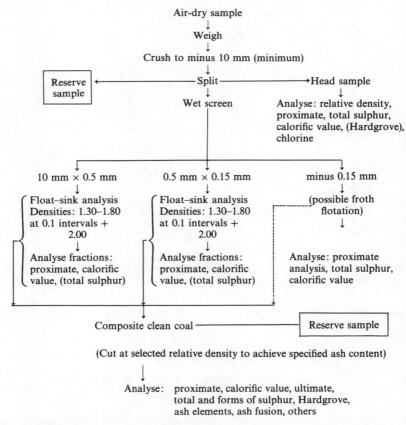

Fig. 5.11 Flow diagram I: drill core analysis for a thermal coal deposit.

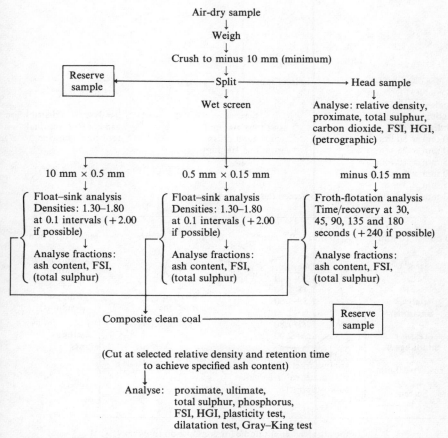

Fig. 5.12 Flow diagram II: drill core analysis for a metallurgical coal deposit.

requires a greater degree of examination than the remainder. Figures 5.11, 5.12, 5.13 and 5.14 give examples of flowsheets commonly adopted for float–sink analysis.

For washability, the intervals are usually 0.05 over the density range 1.25–1.7, and 0.10 for denser fractions. For determinations of washing efficiency, the analysis is normally done at 0.02 intervals over a range of 0.10 relative density on both sides of the anticipated cut-point (perhaps indicated by the preliminary run). For day-to-day plant control, washing samples are often subjected to float-and-sink analysis at only one relative density, and this would normally be a prespecified density of separation or the average density of separation recorded over the period during which the sample was recovered.

The results obtained must be evaluated in accordance with the purpose or reason for the analysis. However, common to all is the transformation of the mass of the individual float-and-sink fractions into percentages of the total sample. The ash content for each relative-density cut-point is then determined and also expressed as a percentage of the total sample.

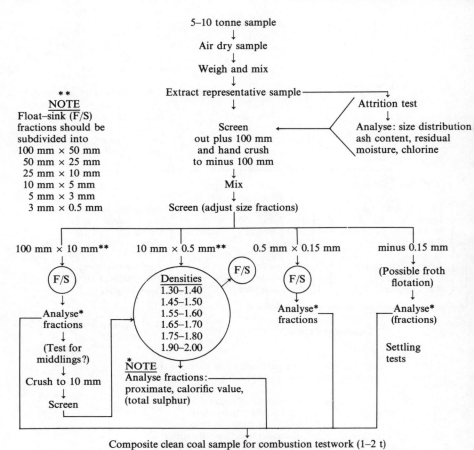

Fig. 5.13 Flow diagram III: bulk sample analysis for a thermal coal deposit.

5.5 EQUIPMENT AND PROCEDURES USED IN FLOAT–SINK TESTING

The main criteria involved in obtaining satisfactory testing apparatus for float sink testing are:

(a) it must be unaffected by the test media;
(b) convenient and safe to use;
(c) simple in operation.

In effect, the test process is simulating the performance of an almost perfect separator, and the apparatus therefore must be capable of providing the most effective operating conditions for dividing a coal sample into two or more relative-density fractions.

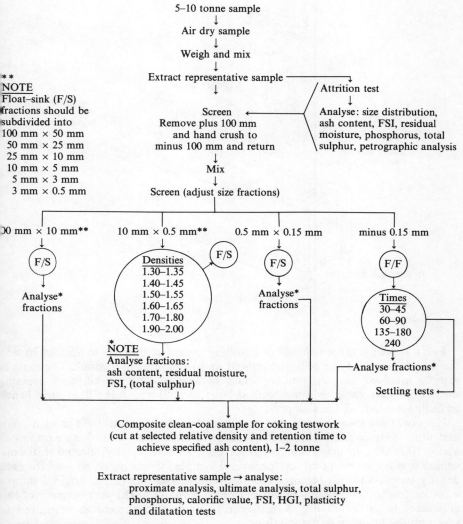

Fig. 5.14 Flow diagram IV: bulk sample analysis for a metallurgical coal deposit.

For coal coarser than 0.5 mm, the test apparatus usually comprises a metal basket with a stainless-steel frame fitted with a fine-mesh base and side walls. This basket fits inside a tank in which the test liquid is stored. Floats are skimmed off the basket as the test proceeds, and when no more floats remain, the basket is removed and, after drainage (and occasionally washing with a suitable solvent), the basket is transferred to another tank containing a liquid having a higher relative density. Figure 5.15 shows a typical float–sink bench employing a powerful down-draught fume-extraction fan which ensures that the operator is safe from the toxic fumes which are commonly encountered with most organic liquids used in float–sink test work.

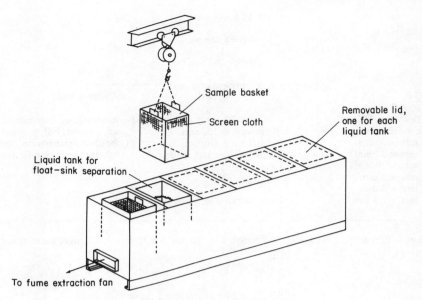

Fig. 5.15 Float–sink bath.

For a test apparatus capable of handling coarse coal of up to 75 mm in size the tank volume may be required to hold up to 100 litres of medium. The mass of sample involved and consequent mass of sample with basket and draining medium usually requires a mechanical hoist located above the float–sink bench to facilitate handling the sink products.

For coal finer than 0.5 mm (but usually coarser than about 0.075 mm), a float and-sink analysis employing glass beakers is often used with fresh samples of about 100–200 g in mass being used for each relative density selected. For convenience, it is common practice for several sample increments to be used for each density tested, and this requires accurate sample splitting of the original sample in order to eliminate sampling errors from the analysis. The arrangement of this apparatus is shown in Fig. 5.16. Centrifuging is also sometimes employed to speed up the recovery of the fractions.

The individual densities of the medium used for the float–sink test are adjusted before and after each test by means of a suitable series of standard hydrometers within the range anticipated for the test. The most commonly used organic liquid employed in float-and-sink testing is perchloroethylene which has a relative density of 1.60. It may be diluted with either petroleum spirit (r.d. = 0.7), white spirit (r.d. = 0.77), naphtha (r.d. = 0.70) or toluene (r.d. = 0.86) for lower densities; or bromoform (r.d. = 2.9) or tetrabromoethane (r.d. = 2.96) can be added for higher densities.

Other liquids such as carbon tetrachloride, acetylene tetrabromide, pentachloroethane and centigrav are also used extensively for laboratory float-and-sink testing, but some exhibit potential health hazards, and special ventilation precautions are required to safeguard the user.

SIZE ANALYSIS AND FLOAT-SINK TESTING

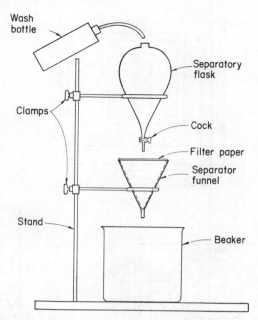

Fig. 5.16 Sketch of a fine-coal float-sink apparatus.

A common inorganic compound used for larger-scale test work, especially, is zinc chloride. The effective range of zinc chloride is from 1.30 to about 1.75, above which the solution viscosity becomes a problem.

When the test is commenced, an increment of the coal being tested, sufficient to form a thin layer, is introduced into the container and gently agitated by stirring. Care must be taken not to overload the container as this is liable to interfere with the separation of float-and-sink material. After sufficient time for settling has elapsed, the float material is removed by a suitable scoop and placed on a draining counter or tray. The time for separation to occur may vary from as little as one minute for coarse size to about 30 minutes for plus 0.5 mm samples. For separation below 0.5 mm, the time required is much longer because of the slow settling rate, and with such analyses extreme care is required to avoid overloading. If no separation is apparent after four hours, it may be necessary to separate the funnel from the flask without disturbing the contents of either, and by means of a filter or centrifuge, obtain the settled and suspended solids representing each fraction. This procedure often introduces error in the test work, and it is normally advisable to leave the suspension for up to 12 hours to obtain an accurate result. Centrifugal separation has been used for ultrafine float-sink testing down to 10 μm particle size, but the practical value of these washability data is questionable.[19] Gravity coal-washing equipment has very poor separation efficiency when operating with material finer than 100 μm in size, and the benefits of such an analysis are therefore questionable.

There are a few published standards for float-sink testing.[3,4] Most specify procedures for carrying out float-sink tests on coarse and fine coals and outline the apparatus and materials to be used. The most comprehensive standard is prob-

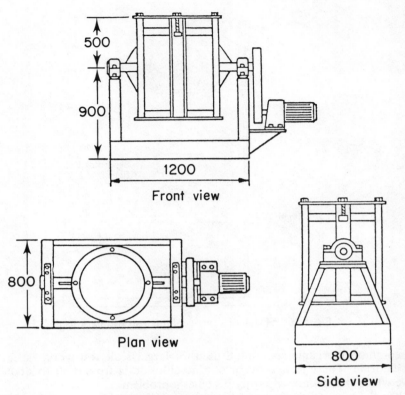

All dimensions in millimetres

Fig. 5.17 Typical drum tumbler.

ably that produced by the Australian Standards Institute which has addressed numerous specialized aspects appertaining to experience with Australian coals. Of particular interest is the appendix dealing with the breakdown of shales in water and also with attrition of coal during washing. The change in size-consist and its resultant effect on the washabilities of certain Australian coals prompted the introduction of a special standard for the quantification of this effect. A test involving subsieve size-analysis of clay by the Andreason pipette method, and a further test involving a drum-tumbler apparatus, are specified in the standard.

In the drum-tumbler method, a 50-kg coal sample is added, together with 150 litres of water, and 18 steel cubes sized 50 mm are added into the tumbler (shown in Fig. 5.17). Tumbling is commenced at a speed of 20 r.p.m. for a time determined from the graph (shown in Fig. 5.18) and based upon the Hardgrove grindability index. When tumbling is complete, the contents are emptied onto a 0.5 mm wedge-wire screen, and the plus 0.5 mm fraction is air dried for subsequent size-distribution float–sink analysis. The minus 0.5 mm fraction is recovered by filtration and air dried for further testing as required.

The Australian standard has been demonstrated as being applicable in suitably modified forms to similar types of coal in other countries. In Canada, a modified

SIZE ANALYSIS AND FLOAT-SINK TESTING

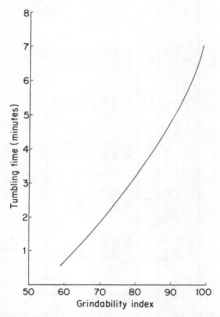

Fig. 5.18 Relationship between Hardgrove grindability index and tumbling time for sample for run-of-mine size distribution.

version of this method has been applied to the determination of the size distribution of highly friable Rocky Mountain low- to medium-volatile coking coal.[15]

5.6 CONSTRUCTION AND USE OF WASHABILITY CURVES

Run-of-mine coal comprises fragments of material which have densities from the lowest to the highest in a continuous range, but the proportions of each vary. Those with the lowest density have the lowest ash but the highest heat value, and those with the highest density vice versa.

Washability curves show the relationship between ash content and the amount of float and sink produced at any particular relative density.

Because the test is conducted under very controlled conditions, the results obtained relate to an almost perfect separation.

From the float-and-sink masses, together with the individual ash content of each relative density fraction, the washability data are calculated.

All the data for the construction of the washability curves can be derived from the weight percentages and ash contents of the relative-density fractions. It is convenient to adopt a standard system of tabulation for these data, and an example is given in Fig. 5.19 which lists washability calculations obtained from a float–sink analysis on a South Durham coal from England. These results were first reported in Ref. 16. The method of computing this table is as follows:

Relative density (r.d.) fraction	Direct		% Wt of ash of total	Cum wt. of ash %	Cumulative floats		Sink wt. of Ash %	Cumulative sinks		±0.1 r.d. distribution	
	Wt %	Ash %			Wt %	Ash %		Wt %	Ash %	r.d.	Wt %
1	2	3	4	5	6	7	8	9	10	11	12
−1.30	39.11	1.4	0.55	0.55	39.11	1.4	27.07	60.89	44.5	1.30	59.61
1.30–1.40	**20.50**	**5.1**	**1.05**	**1.60**	**59.61**	**2.7**	**26.02**	**40.39**	**64.4**	**1.40**	**25.16**
1.40–1.50	4.66	13.8	0.64	2.24	64.27	3.5	25.38	35.73	71.0	1.50	7.05
1.50–1.60	2.39	23.1	0.55	2.79	66.66	4.2	24.83	33.34	74.5	1.60	3.73
1.60–1.80	2.73	37.5	1.02	3.81	69.39	5.5	23.81	30.61	77.8	1.70	2.73
1.80–2.00	1.82	50.7	0.92	4.73	71.21	6.6	22.89	28.79	79.5	1.80	2.03
+2.00	28.79	79.5	22.89	27.62	100.00	27.6	—	—	—	1.90	1.82
Total	100.00		27.62								

Ash content of original sample (direct determination) 26.4%

Fig. 5.19 Washability data for −2 mm + 0.8 mm raw coal, from float-and-sink analysis.

Column 1 Relative-density fraction; limits results directly from test; e.g. 1.30–1.40.

Column 2 Weight percentage of the fraction. This is a direct calculation of percentage, the weight of each fraction being referred to the summation of the weights of *all* the fractions and *not* to the weight of the original sample. The latter provides a check against excessive losses or inaccurate weighings; e.g. weight percentage of the 1.30–1.40 r.d. fraction = 20.50%.

Column 3 Ash percentage of the fraction; from ash determinations on each fraction; e.g. ash %age of the 1.30–1.40 r.d. fraction = 5.1%.

Column 4 Percentage of ash of the total. This column gives the weight of ash in the fraction as a percentage of the total sample; e.g. for the 1.30–1.40 r.d. fraction:

$$\frac{\text{Column 2} \times \text{Column 3}}{100} = \frac{20.5 \times 5.1}{100} = 1.05\%$$

The summation of these values, Column 4, 27.62%, gives the ash content of the sample. It can be compared with the direct ash determination on the original sample and provides a check on the ash contents of the various fractions.

Column 5 Cumulative weight of ash percentage; cumulative of Column 4. Values of this column give the weight of ash, as a percentage of the total sample, which will be obtained in the floats at each separating density; e.g. ash in floats at 1.40 r.d. = 1.60%.

Column 6 Cumulative weight percentage of floats; cumulative of Column 2. This column gives the weight percentage of the floats at each separating density; e.g. weight percentage floats at 1.40 r.d. = 59.61%.

Column 7 Ash percentage of cumulative floats; the ash content of the floats at each separating density; e.g. at 1.40 r.d.:

$$\frac{\text{Column 5}}{\text{Column 6}} \times 100 = \frac{1.60}{59.61} \times 100 = 2.7\%$$

Column 8 Percentage weight of ash in the sinks. The weight of ash as a percent of the total sample contained in the sinks at each separating density; e.g. at 1.40 r.d:

$$\text{Column 4} - \text{Column 5} = 27.62 - 1.60 = 26.02\%$$

Column 9 Cumulative weight percentage of the sinks; the weight percentage of the sinks at each separating density; e.g. at 1.40 r.d:

$$100 - \text{Column 6} = 100 - 59.61 = 40.39\%$$

Column 10 Ash percentage of cumulative sinks; the ash content of the sinks at each relative density; e.g. at 1.40 r.d.:

$$\frac{\text{Column 8}}{\text{Column 9}} \times 100 = \frac{26.02}{40.39} \times 100 = 64.4\%$$

Column 11 Relative density; normally the relative densities at which the examination was conducted. Further points may be interpolated from the r.d.–yield Curve (Curve (d), Fig. 5.21).

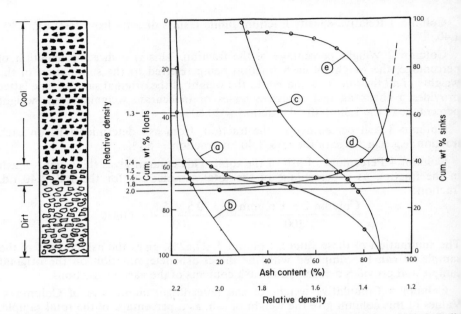

Fig. 5.20 Washability curves (from *A Study of Washability Characteristics*, Vol. 8, Bulletin No. 1. Series: Coal Cleaning No. 16).

Column 12 Weight percentage of the sample occurring between the limits ± 0.10 r.d.; e.g. at 1.40 r.d. (± 0.10) Column 6 (r.d. 1.50) − Column 6 (r.d. 1.30) = 64.27 − 39.11 = 25.16%.

The washability characteristics of the raw coal are, in most cases, sufficiently indicated by five curves shown in Fig. 5.20, which are as follows:

(a) Primary curve, elementary curve or instantaneous ash curve: left ordinate scale, midpoints of the r.d. layers (i.e. the mean of the values corresponding to the upper and lower r.d. limits of each fraction as given in Column 6); abscissa scale ash content—Column 3.

(b) Cumulative sinks curve or discard curve: right ordinate scale—Column 9; abscissa scale, ash content—Column 7.

(c) Cumulative sinks curve or discard curve: right ordinate scale—Column 9; abscissa scale, ash content–Column 10.

(d) Relative density–yield curve: left ordinate scale—Column 6; abscissa scale, relative density—column 1 (upper limit of relative density range).

(e) Plus or minus 0.1 relative density distribution curve: left ordinate scale—column 12; abscissa scale, relative density—Column 11.

5.6.1 Interpretation and Use of the Washability Curves

Referring to Fig. 5.20 and, assuming that an ideal separation is possible, i.e. that all particles of relative density less than a certain value are recovered and all particles of greater relative density are rejected, the cumulative floats curve (b

SIZE ANALYSIS AND FLOAT-SINK TESTING

gives directly the theoretical percentage yield of clean coal of any desired ash content while the cumulative sinks curve (c) gives directly the ash content of the corresponding percentage of discard. For example

Desired ash content of clean coal—6%
From cumulative floats curve (b)
  theoretical yield of clean coal—70%
From cumulative sinks curve (c)
  amount of discard—30%
Ash content of discard—80%

Similarly, it will be appreciated that for any theoretical yield, the ash percentage of the floats can be determined.

For any theoretical yield, reference to the relative density–yield curve (d) will give the relative density at which the perfect separation should be made.

From r.d.–yield curve (d)
  Separating relative density—1.88%

The primary curve (a) reveals the proportions of the raw coal lying within various limits of ash content, and so shows the proportions of free impurities and middlings present. The relative ease or difficulty of cleaning a raw coal to give the theoretical yield and ash content depends on the proportion of middlings present, which in turn determines the characteristic shape of this curve. In addition, curve (a) will give the coincident values of ash content of the dirtiest clean-coal particle and the cleanest discard particle.

From primary curve (a)
  Ash content of dirtiest particle in clean coal—50.5%

Owing to its method of construction, the instantaneous ash curve is often inaccurate, especially in the region where the curve is rapidly changing direction. In this range, extra points may be needed, and may be calculated from the cumulative floats curve, drawn on a large scale for easier reading.[17]

The plus or minus 0.1 relative density distribution curve (e) shows directly the percentage of raw coal lying within the limits $(S + 0.1)$ r.d. and $(S - 0.1)$ r.d. for any particular relative density, S, and hence for any chosen separating density, the amount of near-density material, which is a further measure of the difficulty of obtaining the theoretical cleaning results, particularly with the Baum jig. If, for the desired ash content of the cleaned product, the amount of near-density material at the indicated separating density is appreciable, then, in order to effect an efficient separation, it would be necessary to accept a higher clean-coal ash and so permit a higher separating density where the amount of near-density material was reasonably low. Alternatively, consideration may be given to a dense-medium separating process, capable of an almost ideal separation and a near approach to the theoretical cleaning results in spite of a moderate amount of near-density material. In this case, a plus or minus 0.02 relative density distribution curve would give a comparative measure of the difficulty of cleaning.

From ± 0.1 r.d. distribution curve (e)
  amount of near-density material—1.8%

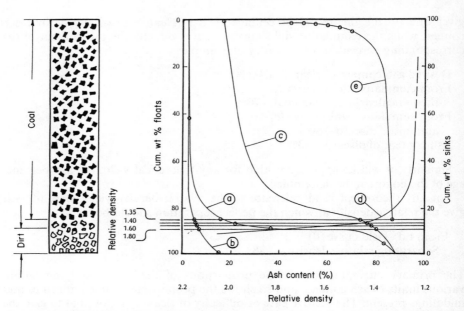

Fig. 5.21 Washability curves for coal with easy washability characteristics (courtesy the Department of Mining Engineering, University of Newcastle upon Tyne).

The near-density material (1.88 ± 0.1 r.d., i.e. between 1.98 and 1.78 r.d.) is given directly by curve (e) from the abscissa 1.88 r.d.

The amount of near-density material is reasonably low and a near approach to these cleaning results should be possible on a Baum felspar jig.

The curves, showing only a moderate amount of middlings, are typical of a coal with normal washability characteristics. The ash contents of the lighter fractions are, however, lower than average and a straightforward two-product separation could yield a clean coal of 6–7% ash, as shown in the example.

The washability curves for a coal with easy washability characteristics are shown in Fig. 5.21. The primary curve for this coal is almost a right angle, indicating a very small middlings content, and reference to the ±0.1 r.d. distribution curve shows that a very efficient separation could be made at any relative density above 1.5.

The washability curves in Fig. 5.22 for a coal with difficult washability characteristics show a very marked difference from those in Fig. 5.21. The primary curve exhibits only a gradual change in slope, revealing a large proportion of middlings. This is also evident in the high values of the ±0.1 r.d. distribution curve over the range of relative densities investigated.

Investigating the possibility of extracting a clean coal of 8% from the coal shown in Fig. 5.22:

Desired ash content—8%
Theoretical yield of clean coal—60%
Theoretical discard—40%
Ash content of dirtiest particle in clean coal and

SIZE ANALYSIS AND FLOAT-SINK TESTING

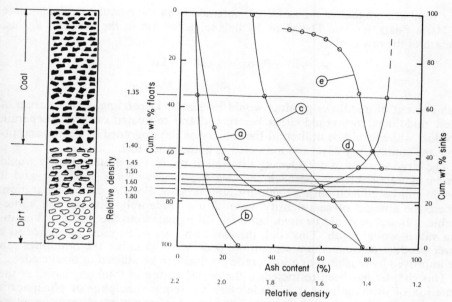

Fig. 5.22 Washability curves for a coal with difficult washability characteristics (courtesy the Department of Mining Engineering, University of Newcastle upon Tyne).

    cleanest particle in discard—16.8%
Ash content of discard—50%
Separating relative density—1.41%
Near-density material—51%

The low ash content of the discard means a high loss of combustible material, and consideration must therefore be given to extracting the middlings fraction.

    Minimum permissible ash content of discard—65%

From cumulative sinks curve (c):
    Theoretical discard—22%
    Yield of intermediate middlings product (40–22%)—18%
    Second separating relative density—1.79%
    Amount of near-density material—6.8%

The ash content of the particles in the middlings will range between 16.8% and 49%, but the theoretical ash content of this product cannot be obtained directly from these curves and must be computed using an ash balance.

Let $x\%$ be the ash content of the middlings. Then the weights of ash, as a percentage of the total raw coal, in the clean coal, middlings and discard products are, respectively:

$$60 \times \frac{8}{100}\%, \quad 18 \times \frac{x}{100}\%, \quad 22 \times \frac{65}{100}\%,$$

i.e. 4.8% $0.18x\%$ 14.30%

The total ash content of the raw coal, obtained from the cumulative floats curve at 100% floats, is 25%. Therefore, by balancing the ash in the products against the ash of the raw coal:

$$4.80 + 0.18x + 14.30 = 25.0$$

$$\text{and } x = 32.8$$

Ash content of middlings product would be 32.8%. Depending on the nature of these middlings, they could either be crushed and re-treated or form a separate product which could be utilized at the colliery or else upgraded to a more acceptable ash content, say 25%, and marketed as a low-grade fuel.

The large amount of near-density material at the first separative density would render it impossible to approach the cleaning results with a Baum jig, and it would still be difficult with a dense-medium process. For the second separation, where the amount of near-density material is considerably less, but still appreciable, a Baum jig could be used, but a fair divergence from the theoretical cleaning values would result. This coal, therefore, appears to be most amenable to a three-product dense-medium separation, provided that either the middlings can be improved by crushing, or a low-grade product can be utilized or marketed.

This chapter has been restricted to the consideration of total impurities as the criterion of washability, but in some cases the required sulphur or phosphorus content of the cleaned product, or its calorific value, may warrant separate investigation. If the relative-density fractions are examined for the particular impurity or property concerned, then washability curves can be constructed, in the manner described, on that basis.

Although economic factors and market requirements have a considerable bearing on the treatment of the raw product, a complete examination of the washability characteristics enables the best compromise to be achieved.

5.7 THE M-CURVE

The mean-value curve, or M-curve, is a different type of washability curve obtained by the graphical (vectorial) summation of the ash content of a raw coal. The M-curve originates from Germany and was devised by Dr F. W. Mayer of Bochum. It has several advantages over the three classical washability curves which it is intended to replace, and also offers wide application in the solving of many general problems of coal-cleaning technology—particularly the mixing of raw coals and the blending of cleaned products.

This description, which is based mainly on a paper by Dr Mayer,[18] is intended only as an introduction to the M-curve method for the evaluation of a raw coal for washing.

It will be found, when using the M-curve method, that the description of the mode of construction is more complicated than the actual preparation and use of the curve. In the diagrams, used to illustrate the construction and use of the curve, parallel displacements are indicated by a normal line between the parallels. The arrow on this line indicates the starting line and which parallel is due to displacement. In other cases, arrows on a straight line indicate the direction in which the line is to be followed. Given values on the ordinate and abscissa axes

SIZE ANALYSIS AND FLOAT-SINK TESTING

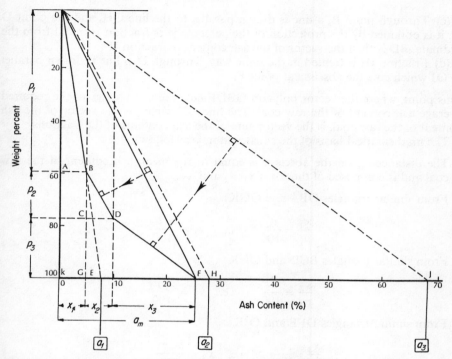

Fig. 5.23 Graphical method for obtaining a mean-value from partial values (courtesy the Department of Mining Engineering, University of Newcastle upon Tyne).

are entered on the side inset in a square and the determined values are set in a circle.

Before describing the construction of the curve, it will assist in the understanding of the vector composition of the M-curve if some consideration is first given to the vectorial summation of partial ash contents in order to obtain a mean or average ash content.

It is possible to obtain a mean-value from partial values by graphical construction as illustrated in Fig. 5.23. In this example, a raw coal is considered, for simplicity, as comprising three different fractions, I, II and III, with ash contents a_1, a_2 and a_3, and occurring in the percentages p_1, p_2 and p_3 respectively. The average ash content of the raw coal is to be determined vectorially.

In the diagram (Fig. 5.23) which has an ordinate scale of weight percentage from 0 to 100, and an abscissa scale of ash content from 0 to 100, the procedure is as follows:

(a) The ash contents of the fractions are marked off on the abscissa scale: points G, H and J. These points are connected to the zero point O of the diagram (top left): lines OG, OH and OJ.

(b) The vector OB is taken off along the line OG by noting the weight percentage of fraction I ($p_1\%$) on the ordinate and projecting on to the line OG (point B).

(c) Through point B, a line is drawn parallel to the line OH, and the point D on it is obtained by the projection of the percentage of fraction II ($p_2\%$) from the ordinate. BD is then the vector of the ash content of fraction II.

(d) Fraction III is treated in the same way. Through D, a line is drawn parallel to OJ which cuts the abscissa at point F.

This point, where the vector polygon OBDF cuts the abscissa, gives the required average ash content of the raw coal. The line OF, which is the vector of the ash content of the raw coal, is the vector sum of the ash contents of the fractions.

The mathematical basis of the construction is as follows:

The distance a_m on the abscissa is equal to the mean ash content of the raw coal and is composed of the parts x_1, x_2 and x_3.

From similar triangles OBA and OGK:

$$\frac{x_1}{p_1} = \frac{a_1}{100} \quad \text{i.e.} \quad x_1 = \frac{a_1 p_1}{100}$$

From similar triangles BCD and OHK:

$$\frac{x_2}{p_2} = \frac{a_2}{100} \quad \text{i.e.} \quad x_2 = \frac{a_2 p_2}{100}$$

From similar triangles DFE and OJK:

$$\frac{x_3}{p_3} = \frac{a_3}{100} \quad \text{i.e.} \quad x_3 = \frac{a_3 p_3}{100}$$

$$a_m = x_1 + x_2 + x_3$$

$$= \frac{1}{100}(a_1 p_1 + a_2 p_2 + a_3 p_3)$$

$$= \frac{1}{100}(ap)$$

Coordinates of B are (x_1, p_1) where

$$x_1 = \frac{a_1 p_1}{100}$$

Coordinates of D are $(x_1 + x_2, p_1 + p_2)$ where

$$x_1 + x_2 = \frac{a_1 p_1}{100} + \frac{a_2 p_2}{100}$$

5.7.1 The Computation of Washability Data for the M-Curve

From the ash contents of the various relative-density fractions obtained by float-and-sink analysis, the mean-value polygon for the raw coal can be drawn. It may

SIZE ANALYSIS AND FLOAT-SINK TESTING

Relative density (r.d.) fraction	Direct		% Wt of ash of total pa/100	Cum. Wt of ash % Σpa/100	Cum. Wt % floats Σp
	Wt % p	Ash % a			
1	2	3	4	5	6
−1.30	39.11	1.4	0.55	0.55	39.11
1.30–1.40	20.50	5.1	1.05	1.60	59.61
1.40–1.50	4.66	13.8	0.64	2.24	64.27
1.50–1.60	2.39	23.1	0.55	2.79	66.66
1.60–1.80	2.73	37.5	1.02	3.81	69.39
1.80–2.00	1.82	50.7	0.92	4.73	71.21
+2.00	28.79	79.5	22.89	27.62	100.00
Total	100.00	26.4[a]	27.62		

[a] Direct ash determination on sample.

Fig. 5.24 Washability data for the M-curve, from float-and-sink analysis. Raw coal fines (−2 mm + 0.8 mm) (courtesy the Department of Mining Engineering, University of Newcastle upon Tyne).

be constructed graphically, as already shown, or more simply by calculating the coordinates of the various points at the angles of the polygon.

The computation of the required data and the construction of the M-curve will be illustrated by using the same analytical data, for a sample of raw-coal fines, as was used in the previous examples to illustrate the computation of the normal washability data and the construction of the normal washability curves. The computed data for plotting the M-curve appears in the table in Fig. 5.24 which is identical with Columns 1–6 of the table in Fig. 5.19, so that the method of obtaining these figures will be easily followed. It will also be appreciated that while the calculations for the normal type washability curves require seven columns (Fig. 5.19, Columns 4–10 inclusive), the calculations for the M-curve are reduced to three columns (table in Fig. 5.25, Columns 4–6 inclusive).

It is convenient to adopt a standard graph sheet for constructing the M-curve, and a suitable type is shown in Fig. 5.25 for raw coals having a mean ash content below 35%.

The left-hand ordinate scale of cumulative weight percent floats has a length of 25 cm so that 2.5 mm corresponds to 1%, and 0.2% can be estimated. The abscissa scale of ash content is 35 cm in length, one centimetre corresponding to one percent ash, and the 0.1% can be read off with ease up to 35% ash. In the M-curve diagram, the ash content appears as a direction, and each line radiating from the zero (upper left) into the area of the diagram corresponds to a particular ash content which is given by the intersection with the abscissa.

It is possible to calculate where these polar or reading-off lines for ash contents greater than 35%, intersect the right-hand ordinate so that it can be divided off for ash contents from 35% to 100%.

Referring to Fig. 5.26, the calculation for the division of the ordinate for reading off values is as follows:
From similar triangles DEB and OEA

$$\frac{y}{(a-L)} = \frac{H}{a}$$

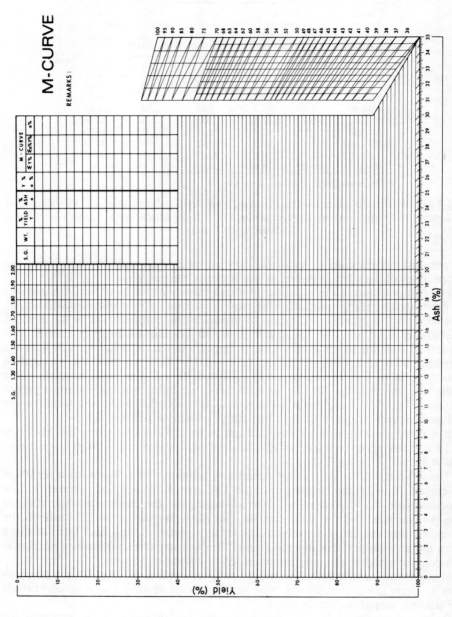

Fig. 5.25 M-Curve diagram (courtesy the Department of Mining Engineering, University of Newcastle upon Tyne).

SIZE ANALYSIS AND FLOAT-SINK TESTING

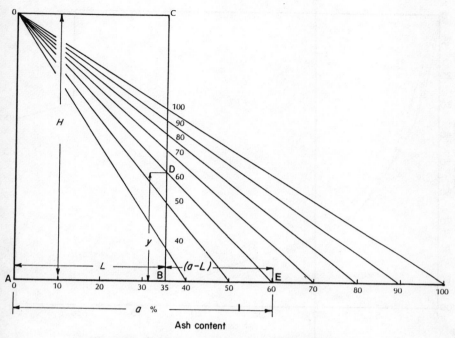

Fig. 5.26 Basis for graphical construction of M-curve.

$$y = H\left(1 - \frac{L}{a}\right)$$

Examples for 60% ash content: $H = 25$ cm; $L = 35$ cm; $a = 60$ cm.

$$y = 25\left(1 - \frac{35}{60}\right)$$

$$= 10.4 \text{ cm}$$

The accuracy with which the ash content can be read off decreases gradually from 0.1% at 35% ash to 1% at 100% ash, but still remains sufficient for all practical purposes.

In the event of the mean ash content of the raw coal exceeding 35%, the abscissa scale will have to be extended accordingly so that the M-curve will intersect it within the diagram.

5.7.2 Construction of the M-Curve

Each side of the mean-value polygon, shown in Fig. 5.27, represents the mean ash content of a particular relative-density fraction in a raw coal. The raw coal, however, contains particles of ash content ranging in a continuous series, and in a relative-density analysis, the particles are separated into various ranges of ash content. If each of the three fractions were further subdivided at closer densities

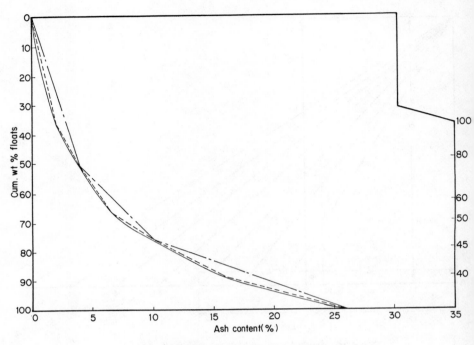

Fig. 5.27 M-Curve showing mean-value polygon.

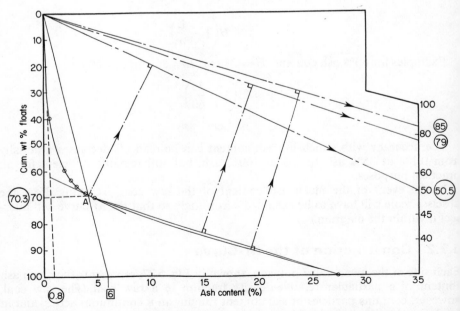

Fig. 5.28 Mean-value curve corresponding to the table given in Fig. 5.24. (Courtesy the Department of Mining Engineering, University of Newcastle upon Tyne.)

SIZE ANALYSIS AND FLOAT-SINK TESTING

into narrower limits of ash content, then mean-value polygons could be drawn for each original fraction. The closing line of each polygon would correspond to a side of the original polygon. It will be seen that as the ash-content limits are made closer, the polygon approaches a smooth curve passing through the angular points. This continuous curve (the mean-value curve), having gradients of all values between the end gradients, represents the continuous range of ash contents, between the lower and upper limit, in the raw coal.

If the relative-density analysis of the raw coal has been conducted at the recommended intervals, then the M-curve may be drawn with sufficient accuracy, through the points of the mean-value polygon by graphical interpolation. The M-curve in Fig. 5.28 has been drawn through the points given by the coordinates in Columns 5 and 6 of the table in Fig. 5.24.

This curve is being used to illustrate the first three parts of Section 5.7.3.

5.7.3 Interpretation and Use of the M-Curve

It is essential to bear in mind that the ash content appears in the M-curve diagram as a direction, steep lines indicating low ash contents, and flat lines indicating high ash contents.

Determination of the Theoretical Yield of Clean Coal of Given Ash Content

Any straight line through the zero point which cuts the M-curve gives by its intersection with the abscissa the ash content of the clean-coal product for a yield, given by the ordinate at the point of intersection with the curve. For a desired ash content of the clean coal, the procedure is to pick off the value of the ash content on the abscissa and connect it to the zero point; where this line cuts the curve (point A), the ordinate gives, from the left-hand scale, the yield of clean coal.

> e.g. desired ash content of clean coal—6%;
> theoretical yield of clean coal—70%

Determination of the Ash Content of the Discard

By joining point A, obtained by the previous construction, with the point at which the curve cuts the abscissa (point B) and paralleling through the zero point, the ash content of the discard is read off on the right-hand ordinate,

> e.g. theoretical percentage discard—30%;
> ash content of discard—80%

Determination of the Boundary Ash Content

Each tangent to the M-curve gives, by its parallel through the zero point for reading off, the ash content at the point of contact. Thus, the tangent at the boundary of the clean coal and the discard will give the boundary ash content which is the value of the highest ash content in the clean coal and the lowest ash content in the discard.

By constructing the tangent at the point A on the curve and paralleling through the zero point, the boundary ash content can be read off on the abscissa,

> e.g. boundary ash content—50.5%

Conversely, where a separation is to be effected at a certain boundary ash content, the boundary point and hence the yield are first found and then the ash contents of the products determined.

The tangents to the M-curve at the zero point and the point of intersection with the abscissa show respectively the lowest and highest ash contents contained in the raw coal,

e.g. lowest ash content in the raw coal—0.8%;
highest ash content in the raw coal—85%.

It is sufficient for all practical cases to estimate tangents by sight without reducing the accuracy obtainable with the M-curve.

Determination of the Ash Content of a Middlings Product
This construction is being illustrated using the M-curve (Fig. 5.29) for a raw coal containing a high production of middlings. The normal type washability curves for the same coal appear previously in Fig. 5.22.

Construction A is used to determine the theoretical yield of clean coal with an ash content, say 8%, indicating a separation at point X on the curve.

desired ash content of clean coal—8%
theoretical yield of clean coal—60%

Construction B is then applied to give the ash content of the discard.

theoretical percentage discard—40%
ash content of discard—50%

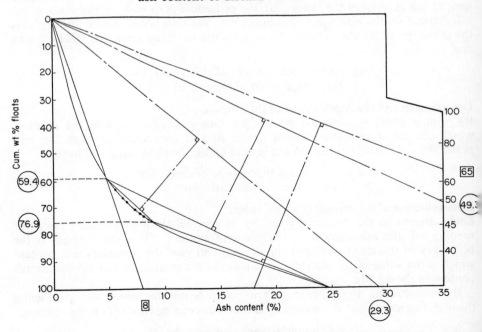

Fig. 5.29 M-Curve for a raw coal containing high middlings content. (Courtesy the Department of Mining Engineering, University of Newcastle upon Tyne.)

SIZE ANALYSIS AND FLOAT-SINK TESTING

The low ash content of the discard necessitates the extraction of a middlings product to reduce the loss of combustible material and raise the ash content of the discard to a minimum value of 65%.

To determine the percentage discard with an ash content of 65%, a line is drawn through point Z (the intersection of the curve with abscissa) parallel to the reading-off line for 65% ash. This line cuts the curve at point Y and the percentage discard is obtained by projecting on to the left-hand ordinate scale. The second separation will be made at point Y.

$$\text{theoretical percentage discard} = (100 - 77) = 23\%$$
$$\therefore \text{ yield of middlings} = (77 - 60)$$
$$= 17\%$$

With the M-curve, the ash content of the middlings product can be obtained directly by simply drawing chord XY, constructing a parallel line through the zero and reading off on the ash-content scale:

$$\text{ash content of middlings—}29.5\%$$

The boundary ash contents at points X and Y may then be determined using construction C. This has not been done in order to avoid complicating the illustration.

Alternatively, if it is desired to make the second separation at a certain boundary ash content, then the point Y may be fixed by constructing the tangent to the curve parallel to the reading-off line for the required ash content. The ash content of the middlings and discard products can then be obtained by paralleling the chords XY and YZ respectively through the zero and the reading-off line.

Again, if it is necessary to produce a middlings product of a certain ash content, a line can be drawn through X, parallel to the reading-off line for the required ash content, which will give, by its second point of intersection with the curve, the position of the second separation.

In order to obtain from the raw coal a given quantity of middlings of certain ash content, it is only a matter of constructing a chord, parallel to the reading-off line for the required ash content, the projection of which on the left-hand ordinate scale gives the required quantity.

The construction lines have been shown in Figs 5.28 and 5.29 merely to illustrate the method. In practice, points on the curve can be fixed and readings obtained without actually drawing these lines.

5.7.4 Comparison of Washability Characteristics

By replacing the three classical washability curves with a single curve, it becomes possible to represent several raw coals on one diagram, thereby making it easier to compare their washability characteristics. This is particularly useful in comparing the curves for the various size-ranges of raw feed or the same size-range from different seams.

In Fig. 5.30, curves A, B and C represent raw coals with easy, normal and difficult washability characteristics respectively. The following should be kept in mind when assessing the washability characteristics of a raw coal from the M-curve:

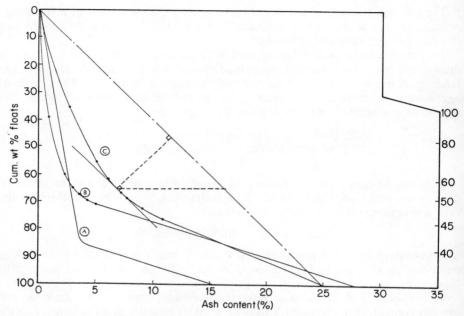

Fig. 5.30 Washability curves for easy (A), normal (B) and difficult (C) coals. (Courtesy the Department of Mining Engineering, University of Newcastle upon Tyne.)

(a) The steeper the line connecting the zero point to the lowest point on the curve, the lower the ash content of the raw coal.

(b) The farther the M-curve is from the connecting line, the greater is the range of variation of ash content in the coal. If a tangent is drawn parallel to the connecting line (e.g. curve C, Fig. 5.30) and through the point of contact, P, a line is drawn, parallel to the abscissa, intersecting the connecting line at Q, then the length of PQ is a measure of the distribution conditions of the ash content.

(c) A sudden change in the slope of the curve, e.g. curve A, indicates easy washability characteristics, while the more gradual the change in slope, e.g. curve C, the more difficult are the washability characteristics.

The M-curve can be used in conjunction with the relative density–yield curve and the ±0.1 r.d. distribution curve, in the same manner as the classical washability curves, so that the cut-points can be related to separating relative densities, and a measure of the ease or difficulty of separations at particular densities can be obtained. An abscissa scale of relative density ranging from 1.2 to 2.2 has, therefore, been included on the recommended M-curve sheet, Fig. 5.25, an interval of 0.1 r.d. being represented by 2 cm. The M-curve from Fig. 5.28 is reproduced in Fig. 5.31, together with these two additional curves.

Compared with the three classical washability curves, the M-curve possesses the following four advantages:

(a) *decreased calculation of data*—only three columns have to be calculated and no division is involved.

(b) *decreased graphical work*—only one curve to be drawn instead of three.

SIZE ANALYSIS AND FLOAT-SINK TESTING

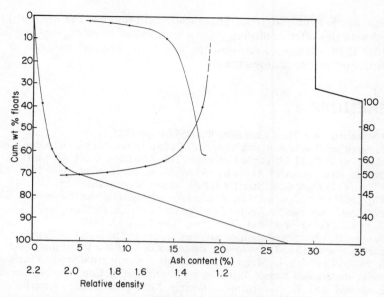

Fig. 5.31 M-Curve with additional 'classical' washability curves.

(c) *increased accuracy in determining ash contents up to 60%*—in the 250-mm abscissa scale for the classical curves, 1% ash is represented by 0.25 cm, while in the M-curve diagram, 1% ash corresponds to 1 cm up to 35% and gradually falls off to 1% equal to 0.25 cm at about 60% ash, and above this value the accuracy still remains sufficient for practical purposes.

(d) *provides more information directly*—particularly with respect to middlings extraction.

5.8 IMPORTANT DOS AND DON'TS

Good sampling is of paramount importance.
Duplicate samples provide a means of correlation.
The cost of a size analysis or float–sink test usually exceeds the cost of obtaining a sample in an operating mine, but is only a small cost component in sampling in new areas. The danger with the latter is that sample size may be determined by financial constraint, and it therefore becomes essential to remain realistic about what the analytical programme can achieve. Often, a very comprehensive analytical programme is conducted on an inadequately sized sample and money is wasted which could be more gainfully used elsewhere.

5.9 SPECIAL ACKNOWLEDGEMENT

The foregoing sections dealing with the construction and use of washability curves, including the M-curve, were first published over 20 years ago in the *Kings College Mining Bulletin*, and are reproduced by permission of the University of

Newcastle upon Tyne, England. The author believes that the manner in which they deal with this often confusing topic is unequalled in subsequent publications, and rather than attempt to reinvent the wheel, the original bulletins are reproduced with only minor changes made.

REFERENCES

1. ISO Standard 1988. *Hard Coal Sampling*. ISO 1988–1975.
2. S. F. Streicher. Float and sink analysis of coal, Paper No. 3, *Advanced Coal Preparation Course*, Rand Afrikaans University, Johannesburg, South Africa, 1979.
3. Standards Association of Australia (SAA), Standard Number 1661–1977. *Float and Sink Testing of Hard Coal*, Standards Association of Australia.
4. ASTM D4371-84. *Standard Test Method for Determining the Washability Characteristic of Coal*, American Society for Testing and Materials, Philadelphia, USA.
5. C. E. Tyler, *Testing Sieves and Their Uses*, Handbook 53, 1980.
6. L. Svarovsky. Characterization of particles suspended in liquids, Chapter 2, in *Solid–Liquid Separation*, (Ed.) L. Svarovsky, Butterworth, Guildford, England, 1981.
7. S. G. Malphan and A. L. Mular. Measurement of size distribution and surface area of granular materials, Chapter 4, in *Design Installation of Comminution Circuits*, (Eds) A. L. Mular and J. P. Jorganson, Society Mining Engineers, American Institute Mining Engineers (SME-AIME), New York, 1982.
8. P. Rosin and E. Rammler. The laws governing the fineness of powdered coal, *Journal Institute Fuel* 7, 29–35, 1933.
9. J. G. Bennet. Broken coal I, *Journal Institute Fuel* 10, 22–39, 1936.
10. J. G. Bennet, R. L. Brown, and H. G. Crane. Broken coal II, III, IV, *Journal Institute Fuel* 14, 111–143, 1941.
11. J. G. Bennet, R. L. Brown and H. G. Crane. Broken Coal, *Proceedings, South Wales Institute Engineering* 47, 1941.
12. J. C. Vogel and F. W. Quass. The friability of South African coals, *Journal Chemical, Metallurgical and Mining Society of South Africa* 37, 469–478, 1937.
13. J. B. Sedgman. Prediction of raw coal size analysis from drill core data, Course notes, Coal Preparation Course, University of British Columbia, Department of Mining and Mineral Processing, 1980.
14. T. Allen. *Particle Size Measurement*, 2nd edn, Chapman and Hall, London, 1975.
15. S. G. Butcher, D. G. Osborne and A. D. Walters. The design, commissioning and operation of plants treating highly friable canadian coals, VIII International Coal Preparation Congress, Donetsk, USSR, 1979, Paper G3.
16. Coal cleaning, Nos 16, 17 and 18. Construction and use of washability curves, *Kings College Mining Society Bulletin*, Newcastle upon Tyne 8, (1).
17. R. M. Horsley and J. O. Young. The representation of coal cleaning results, NCB Scientific Department, CRE Report 1214, January 1954.
18. F. W. Mayer. A new washing curve (die Mittlewertkurve, eine neue Verwachsungskurve), *Gluckauf*, 86, 498–509, 1950.
19. D. J. Brown, B. A. Dockter and M. S. Mitchell. Washability data base of very fine western coal, Proceedings 114th Annual General Meeting, Society AIME, New York, 1985, published by SME-AIME, New York, p. 144.

Chapter 6

DENSE-MEDIUM SEPARATION

6.1 INTRODUCTION

The origin of the concept of dense-medium separation is not clear, but it is likely to have emerged as a result of an observation frequently encountered in operating early forms of jig washer. The phenomenon observed was that a concentration of fine grains of heavy mineral produced a semistable suspension which behaved like a heavy fluid, thereby causing material of low relative density to float. This occurred irrespective of size and was subsequently deliberately encouraged by the operators as being a desirable component of jig operation. Hence, it is still widely accepted that water used for jig-washer operation should contain a minimum amount of semisuspended fines.

When the same phenomenon was encountered with the early trough-and-cone washers, it was gradually incorporated into operating practice until eventually the first commercial dense-medium washing process emerged. This was the Chance process or 'the sand flotation process of coal washing', as it was called. Chance[1] stated that, '... any relatively finely comminuted, insoluble solid material (such as sand), if mixed with a certain quantity of liquid (such as water), can be maintained suspended in the liquid by continuous agitation and that the mixture, so long as agitation is maintained, will form a mass exhibiting physical properties similar in every respect to those of a fluid of relatively high specific gravity'. This rather long-winded description serves to explain the origin of this concept of separation, which is one of the most widely applied in modern coal-preparation technology.[2]

Following on from Chance's separator were others[3] employing separating media other than sand, for example:

- the Conklin process, which used fine magnetite and water; the magnetite being finely ground (minus 75 µm) was maintained in suspension by a rising water current. Ironically, this process was never developed to a commercial stage;
- the de Vooys process, which made use of very finely ground barite and loess (a form of clay) in water, which could be maintained in a semistable state for a density range of 1.25–1.65 without agitation. This process was developed for cleaning German anthracite;
- the Wuensch process, which was similar to the Chance process in that the pulp of solids and water was used in a cone equipped with a stirring device.

Instead of using sand, the natural fine clay and slate from the raw coal were employed as the medium.

In addition to the suspended-particle concept, some attempts had previously been made to commercialize the use of saturated salts and organic liquids as dense media:

- the Lessing process was the earliest attempted commercial dense-medium coal-cleaning process. It was based on the separation phenomenon first reported by Sir Henry Bessemer, over 50 years beforehand in 1858. The separating medium consisted of a saturated solution of calcium chloride having a relative density of 1.4. A refined version of this washer, the Belnap Calcium Chloride Washer, was first installed in 1935 for small-coal cleaning. This was followed by the Du-Pont process, introduced in 1936, which was the first to employ chlorinated hydrocarbons.

Because all of these new processes were considerably more expensive to install and operate, they were developed for and used almost exclusively in anthracite coal-cleaning. Dense-medium 'baths' or 'vessels' continued to become more widely used in both coal and mineral dressing applications, particularly following the economic depression of the early 1930s. The most active area of development appears to have been Holland where the de Vooys process was improved upon to become the world-renowned Barvoys process, and Tromp developed a process using finely ground magnetite or pyrite to form a semistable suspension. The most significant work of all emerged from Staatsmijnen in Linburg (Dutch State Mines (DSM)). It was Tromp, working for DSM at Dominiale anthracite mine in Kerkrade, who designed and encouraged the first dense-medium separator to be installed in a commercial plant (1937). This installation triggered the interest of the world in using dense-medium methods.

Simultaneously, the Barvoys separator emerged with equal popularity, and various modifications of both types of washer emerged in other European countries, then in the United States and elsewhere. From these two designs, and with the Chance cone also being widely utilized, dense-medium vessels rapidly became adopted in place of troughs and upward-current washers, for treating coarse coal too difficult for jig washers, which remained in prominent use for 'easy' coals.

By the mid-1950s, dozens of different designs were being used in the many parts of the world where coal cleaning was practised. The major limitation of dense-medium vessels was found to be an alarming drop in cleaning efficiency with decrease in size of coal fed to the washer. Below 5 mm, the majority of dense-medium vessels fared little better than other types of separators, and it was the DSM which once again came up with the solution to this problem.

The DSM found that two factors made dense-medium separation of smaller sized coal more difficult.[4] These are:

(a) The medium used must have a much lower viscosity to enable accurate separation at reasonable throughput capacity.
(b) The amount of medium removed with the washed products is much greater, and consequently a much larger amount has to be regenerated.

The solution to the problem came with the development of the dense-medium

DENSE-MEDIUM SEPARATION

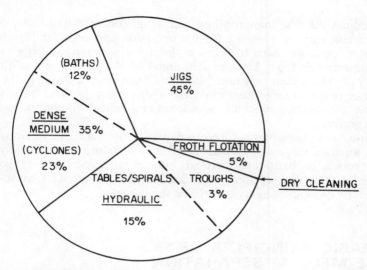

Fig. 6.1 Dense-medium coal washers compared with other types of washer in current usage (1980, based on tonnage treated in the USA and Canada only).

cyclone washer, first pioneered by M. G. Driessen and his colleagues at DSM and reported in 1945.

In the ten years that followed, the DSM cyclone technology became the single most important development in coal washing in the 20th century, and will probably remain so.[5] Hundreds of plants now utilize the DSM cyclone system and many hundreds more use cyclone types of washers, many of which are direct descendants of the original DSM design.

In 1957, the next important DSM development occurred. A flowsheet for a fine-coal dense-medium cyclone circuit was developed by DSM–Stamicarbon and Evence Coppée,[6] and the circuit subsequently installed in the Tertre plant in Belgium. While the technology was clearly proven by this installation, subsequent progress has been disappointingly slow, and despite the apparent demand for the process, relatively few installations are in current operation.[7]

It is difficult to determine the exact number of dense-medium washers now in use throughout the world, but estimates have been made of their use relative to other coal-cleaning methods, as shown in Fig. 6.1.

To summarize the reasons for the growth in popularity of dense-medium separation:

• Dense medium is a highly accurate method for the separation of coal and discard. It can effect separations at much greater efficiencies than any other form of gravity concentration. Significant increases in yield can be achieved and losses of good-quality coal in the discard avoided. The efficiency of dense medium is reflected in the range of probable error, which is between 0.025 and 0.075, depending on coal size and density of separation. Dense medium is particularly effective where there is a significant quantity of near-density material. However, where a relatively high separation density is required, or if the washability indicates that a fairly clear-cut separation is possible between coal and rock, then dense medium may be inappropriate and a jig, concentrating table, or water-cyclone circuit may be adequate for the operation and considerably cheaper in capital and operating costs. In terms of density of separation, dense medium is able to operate effectively between 1.30 and 1.80. At the low end of this range, instability of the medium may take place, and often small changes in the medium density can result in large losses in yield and/or poor product quality. At the higher end, the medium becomes too viscous and difficult to control and recover from the products.

- Because net gravitational force on a particle is related to its mass, smaller particles separate more slowly than large, and therefore less efficiently. Dense-medium vessels or baths are therefore only effective down to approximately 5 mm. Below this size, cyclonic type dense-medium separators must be used, which greatly enhance the gravity force on the particles.

6.2 BASIC PRINCIPLES OF DENSE-MEDIUM SEPARATION

In the next chapter, we shall look at the behaviour of coal and other rock fragments in water and air, i.e. without the benefit of a dense medium, and the reader is directed to Section 7.2 in Chapter 7, which deals with the basic principles of free and hindered settling and their contribution to gravity concentration. A basic understanding of this behaviour is therefore assumed in the following discussion of dense-medium separation.

Consider a spherical piece of coal (assuming that such a thing existed) of mass m and volume, v. If the mass of this piece of coal is less than that of the same volume of liquid, then it will not sink in the liquid (see Fig. 6.2(a)).

The force on the piece of coal is mg, where g is the acceleration due to gravity which can also be represented by gdv, where d is the density of the coal. This is because $m = dv$. The liquid that volume v displaced, having mass M, would be equal to the volume of the piece of coal, v, if the coal sank, but the liquid i exerting an upward force—called buoyancy—which is proportional to M. Hence the effective mass of the coal sphere being pulled downwards towards the earth whilst within the liquid, is the difference in mass of the sphere and the buoyancy of the liquid thrusting upwards. The sphere will therefore not sink if $gdv = gDv$ and hence the velocity will be zero. As the difference between gdv and gDv grows the sphere will settle at a faster and faster rate. Therefore, velocity, s, is proportional to this difference, i.e. $s \propto gdv - gDv$. In 'real' systems, coal fragments are not spherical, nor even smooth, and the piece of coal will meet with some form of resistance other than buoyancy effect.

In attempting to move downwards, the coal will encounter a resistance, R which must also be subtracted from the downward force, to give velocity $s \propto gdv - gDv - R$. Under the action of the net force, the coal sphere will continue to accelerate until the increase in R (occurring as the sphere increases in velocity) reduces the net downward force to zero. At this stage, the sphere reaches the terminal velocity, at which it travels until it reaches the bottom of the vessel containing the fluid (assuming the fluid to be either homogenous or a stable suspension).

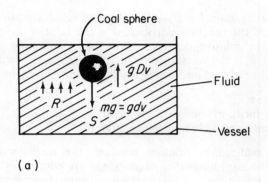

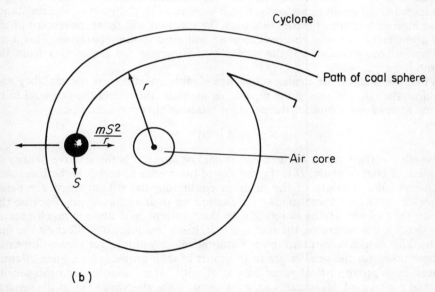

Fig. 6.2 Forces affecting settling behaviour. (a) Forces governing the settling velocity, s of a coal sphere in a static bath:

$$s \propto gdv - gDv - R$$

N.B.! When s is negative, coal sphere floats. (b) Forces governing the settling velocity, s, of a coal sphere in a cyclone:

$$s \propto \frac{s^2}{r}vd - \frac{S^2vd}{r} - R$$

Organic liquids and aqueous solutions are rarely used in commercial dense-medium separation and therefore the properties of fine solid-particle suspensions are of great significance in the commercial use of the process.

The relative density of the medium solids is most important in determining the range of separating density which can be used. The particle size of the solids will determine the stability of the suspension to be used as a medium.

Hence, if settling velocity $\propto gv(d - D) - R$, where g is acceleration due to gravity, v is the volume of the medium particles, d is the relative density of the medium particles, D is the relative density of the fluid containing the media solids, i.e. water $= 1$, and R is the resistance factor, then this relationship shows that settling velocity can be reduced and hence stability improved by having:

(a) small particles, i.e. low v;
(b) low relative-density medium solids, i.e. low d; or
(c) high liquid resistance to particle motion, i.e. high R.

These conditions are realized in practice because: the medium solids are usually ground to very fine size; magnetite suspensions are commonly stabilized by maintaining a proportion of clay slimes; and these slimes, besides reducing v, also increase the medium viscosity, hence increasing R. If, however, the resistance is too high as a result of slimes build-up, the medium will resist movement of the coal and shale fragments and inefficiencies will arise in the separation. This must be rectified by 'cleaning up' the medium to eliminate the excess clay from the circuit.

While the size of the media solids has significant effect on the stability and therefore the effectiveness of the separating medium, the size of the raw coal to be separated *need not*. Consider the previous relationship once again, i.e.:

$$S \propto gv(d - D) - R$$

for which v is the volume of the piece of coal or shale; d is the relative density of the piece of coal or shale; D is the density of the medium; and R is the resistance (viscosity). The viscosity of the medium containing the solids which are being separated, influences these solids by its action on their surface areas. Because the surface area of fine grains in relation to their volume and mass is much greater than that for coarse grains, the resistance, R, has a much greater effect on the fine grains. This causes a resultant lower terminal velocity and larger separation time, i.e. time taken for the smaller grains to report to their correct place. Such circumstances may prove critical when raw coal with large amounts of near-density material are treated. Hence, this *need not* occur if the size range fed to the separator meets the specified criteria for the type of separator proposed, or vice versa. Where only gravitational forces are involved in providing the downward or high-density separating force, the type of dense-medium separators employed treat only relatively coarse solids, i.e. 150–5 mm size range. In fact, the nominal range of treatment is usually narrower, i.e. 100–10 mm.

As the grain size becomes smaller and v gets less, the force expressed by the term $gv(d - D)$ also decreases until it becomes equal to R and settling approaches zero. The magnitude of g cannot be altered or removed, but it can be reduced to a negligible amount by replacing it with a much larger force—centrifugal force. The grain, when subjected to the effects of this new force, will move in the direction which the force is applied. This concept is utilized by dense-medium cyclonic separators, of which the DSM cyclone was the first to make use of this phenomenon in coal separation (Fig. 6.2(b)). Because the force potential of this separator is great, it is possible to treat relatively large quantities of raw coal in a unit of small physical size when compared to dense-medium baths. The corresponding relationship applicable to this form of separation is:

DENSE-MEDIUM SEPARATION

$$\text{settling velocity, } S \propto \frac{S^2}{r} \times v(d - D) - R$$

where v, d, D and R are as before, S is the velocity of the coal or shale grain, and r is the radius of the path of the grain. This relationship explains why, as the coal grain spirals towards the centre of the cyclone, both the settling velocity and the forces causing it, become greater, i.e. the value of r is becoming smaller and V is getting larger. Although gravitational force continues to make a contribution to the separating effect, its value is so small compared to that of the centrifugal force, that it can almost be ignored when considering the normal practical size range for which cyclones are currently used to clean coal. Separation of coal from shale occurs rapidly as a result of the magnified forces of separation involved.

So far, we have only discussed 'relationships' between the settling velocity and the parameters of the solids to be separated and the medium. It is outside the intended scope of this book to describe the detailed research work[8] which has subsequently led to further definition of operating conditions in practical dense-medium bath and cyclone separators. There has been much detailed research

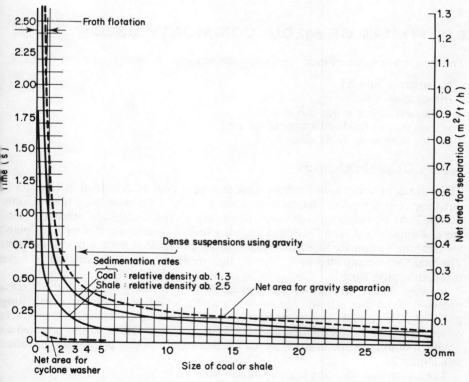

Fig. 6.3 Graph showing time in seconds for coal and shale particles to sink 25 mm in water and net area for separation in square metres per tonne per hour feed. Taken from the paper, Driessen 'The use of centrifugal force', *Journal Institute Fuel*, December, 1945. (Ref. 5.)

work, and it is possible to develop empirical formulae for the estimation of such things as:

- separator residence times for specific feed size ranges;
- separator capacities for specific size ranges.

Figure 6.3 shows a graph developed in 1945 by Driessen[5] to illustrate the effects of grain size on settling velocity using the general formula for static-fluid vessels:

$$s = f\left[\frac{gp(d-D)}{v}\right]$$

where p is particle diameter; v is the kinematic viscosity; f is a constant for the separator. This equation is modified to the following for cyclone separation:

$$s = f\left[\frac{S^2 p(d-D)}{r^n v}\right]^n$$

where n has been shown to be about 0.5.

6.3 FORMS OF MEDIA COMMONLY USED

The separating medium used for dense-medium separation can take five forms:

(a) organic liquids;
(b) aqueous solutions;
(c) aqueous, fine solids suspensions;
(d) aerated or fluidized suspensions; and
(e) low-temperature liquids.

6.3.1 Organic Liquids

These are commonly used for float–sink testing of coal as described in Chapter 5, but their commercial use has proved uneconomical. Of numerous attempts, only the DuPont or Nagelvoort process, developed in the mid-1930s, achieved commercial application. The DuPont Company built a 50-tonne-per-hour pilot plant for cleaning anthracite in Pennsylvania, using chlorinated hydrocarbon medium. The plant was completed in 1936, but high medium losses are believed to have resulted in early closure.

A more recent attempt to commercialize a true heavy liquid is the Otisca process[9] which uses a bath-type separator and a liquid predominantly consisting of CCl_3F. This liquid evaporates readily and is recovered by heating the products of separation in indirectly fired conductive evaporators. The medium is stated as having a boiling temperature of 24 °C. Further details of this process are included in Chapter 23.

The advantages of using organic liquids are:

- low viscosities;
- highly stable and virtually immiscible with water;
- easily regulated and precise separating densities.

The disadvantages are high evaporative loss, health hazard and cost.

6.3.2 Aqueous Solutions

Three salts have been used for producing dense media: the chlorides of zinc, sodium and calcium. The first of these has been confined in its use to float–sink tests; the second has limited potential in terms of relative density range; the third has been used in commercial applications and much interest was devoted to its potential when dense-medium separation was in its infancy. The Lessing and Bertrand processes, both of which emerged in the early part of this century, used calcium chloride in saturated solution to achieve a relative density of approximately 1.4. The Wemco–Belnap Calcium Chloride Washer, based on the Lessing process, was developed to commercial application in the United States in 1935, but its use in coal cleaning was short lived.

The main disadvantages of using such solutions are:

- low relative density;
- highly corrosive to steel and other metals;
- difficult to recover and regenerate;
- quickly reaches high viscosity at saturated-solution strength.

In present-day applications, there are really no advantages favouring the use of aqueous solutions.

6.3.3 Aqueous, Fine-solid Suspensions

With very few exceptions, the vast majority of all dense-medium separation processes used world-wide employ suspension-type media, with finely ground magnetite being the industry standard.

Mixtures of solids and a liquid can take distinct forms. The term 'colloid' is generally applied to those particles which are smaller than 1 μm and a dispersion of these particles is called a 'sol'. Particles of less than 0.2 μm are called supercolloids. Dispersions of larger particles are called 'suspensions' and such dispersion may be classified into either lyophilic or lyophobic. In the former, the solid shows a marked affinity for water or some other dispersing medium, and sols (or suspensions) are formed spontaneously upon mixing. Lyophobic suspensions or sols exhibit low affinity for the host medium, and must be formed by chemical means or by mechanical mixing. This group of suspensions are particularly sensitive to the addition of electrolytes or polyelectrolytes. A common example of a lyophobic colloid is clay, which is a familiar ingredient in most dense-medium circuits in coal-preparation plants. This aspect is dealt with in more detail in Chapter 12.

The rheological properties of the dense medium are largely defined and determined by the following:

(a) composition of the suspension;
(b) relative density of the solids;
(c) particle size distribution of the solids;
(d) solid–liquid concentration;
(e) stability of the medium;
(f) viscosity of the medium;
(g) nature of the solids (including contaminants);
(h) amenability to recovery and regeneration.

Since this category of dense medium assumes the suspensoid of the medium to be water, its density is already defined, but several of the above factors may be further influenced by:

(i) temperature; and
(j) ionic concentration.

There are several materials which when ground or prepared in a fine state would prove a suitable solids ingredient for a dense-medium suspension. Those in current usage are magnetite, sand and crushed shale. As we have already seen, others have been used in coal preparation, including the minerals *barytes* (r.d. 4.5) which is barium sulphate, *pyrites* (r.d. 5.0) which is iron sulphide, and *galena* (r.d.7.5) which is lead sulphide. Of these, barytes was the only one which was extensively used. It was used for many years in Europe, and in particular with the Barvoys process (developed by de Vooys), where it was ground to an average particle-size of 0.025 mm and stabilized with naturally forming clay sols. Current use of barytes is limited.

Silica sand was first used in the Chance process and numerous Chance cones are still operational in various parts of the world. Sand, which is predominantly the mineral quartz, has a relative density of about 2.6, and alluvial or beach sands, with rounded grains tend to provide the most effective form of medium. Normally, the particle size should be 0.5–0.15 mm, with the majority of the grains lying between 0.3 and 0.2 mm to give an average of about 0.25 mm (mean grain size). Sand suspensions have a maximum range of between 1.3 and 1.8 relative density, and the resultant medium has very low viscosity due to the constant upward current of water. It is easily washed from the products and then reclaimed by a simple settling cone. Even with the close size specification, sand provides a comparatively cheap medium, its only real limitations being instability, the need for upward-current water and the abrasive properties of the sand.

Shale Several early dense-medium processes utilized finely ground colliery shale as either the total solids component or as a stabilizing agent with a higher density material, e.g. Barvoys washer. Where shale or clays are used as stabilizers, special types such as loess and occasionally bentonitic clays have been employed, but generally, the cheapest source available is utilized to minimize consumption cost. The most common approach has proved to be the use of froth-flotation tailings, occasionally augmented by crushed shale (crushed to below 0.25 mm). Most shale and clay minerals associated with coal seams are characterized by rapid oxidation rates. This causes them to decompose rapidly in water rendering them ideal suspension components. Many commonly encountered shales decompose so thoroughly that they form sols rather than suspension. Their only real drawback is that the medium so formed may prove too viscous. The maximum relative density of a shale medium is therefore about 1.6, but this may be extended to some degree by use of an upward current and by centrifugal-force (cyclone) separation. Because shale medium is relatively cheap, losses from the circuit are high. Media recovery is effected by classification and thickening usually by means of hydrocyclones.

Magnetite The naturally occurring mineral magnetite has become the standard medium for coal-cleaning operations, whereas ferrosilicon or mixtures of magnetite and ferrosilicon are used for the higher relative-density separations in ore-dressing operations.

Both physical and rheological properties are important in determining the effectiveness of the magnetite medium. Mineralogical and chemical properties may affect such things as magnetic properties or grinding characteristics, and as such are secondary in their influence upon the suitability of a particular source of magnetite for dense-medium application. As the relative density of magnetite is so high (about 5), the medium solids concentration can be quite low, so that the medium is not so prone to viscosity effects arising from slimes contamination. It is also durable and does not significantly break down with use. Because of its high density, it can be used over a wide range of operating densities, which usually extends from about 1.35, below which stability is rapidly lost, to about 1.80, above which viscosity becomes a problem. Further, by virtue of its magnetic properties, it is undoubtedly amenable to recovery and regeneration which usually means that very high efficiencies are attainable in both media recovery and density control.

Generally speaking, the magnetite in bath-type washers is slightly coarser than that used in cyclonic washers, particularly in the higher density range.

Many plants exist which incorporate magnetite milling circuits, some of which operate on a continuous basis, while others are batch-milling operations. Most countries have independent producers and/or suppliers which offer bulk, pre-prepared magnetite to conform with a general specification from the user. In South Africa, Australia and the United States, suppliers offer various grades ranging from relatively coarse for dense-medium baths, to ultrafine for low-density or fine-coal cleaning applications.

5.3.4 Aerated or Fluidized Suspensions

The first work on dry, fluidized suspensions was carried out in the mid-1920s,[10] but the first commercial plant was installed near Pittsburgh in 1930. The medium used was fine, closely sized silica sand, and raw coal up to 100 mm was effectively cleaned in a relative-density range from 1.40 to 1.60. The lower size limit for the coal feed was found to be about 5 mm. More recently, mixtures of limestone and hematite, and limestone and magnetite, have been used with a fluidized-bed cascade separator which resulted in good separating efficiencies being reported.[11] Both coarse (20 × 6 mm) and small (6 × 0.8 mm) coal feeds have been treated with a medium sized between 125 and 600 μm.

5.3.5 Low-temperature Liquids

Water, although cheap and abundant, has disadvantages as the washing medium for coal. At ambient temperatures, it has relatively high viscosity, a high surface tension in contact with air and a relatively low vapour pressure. To use a medium of lower viscosity would enable mechanical separation of much smaller coal, and if the medium also possessed a low surface tension in contact with air or its own vapour, together with high vapour pressure, its removal from 'wetted' surfaces would be easy. These desirable properties are possessed by certain liquefied gases, in particular by liquid carbon dioxide. Carbon dioxide has a critical temperature of 31 °C and can be readily liquified by pressure alone at normal temperatures. Several attempts have been made to utilize this low-temperature liquid for cleaning fine coal.[12]

6.4 PROPERTIES AND METHODS FOR TESTING MAGNETITE

6.4.1 Properties

Magnetite is a naturally occurring mineral of cubic crystal structure. It belongs to the spinel family, having the chemical formula $Fe^{2+}Fe_2^{3+}O_4$. The chemical structure is based on a close-packed, cubic oxygen lattice. The oxygen ions (132 pm or 1.32 Å) are larger than the Fe^{2+} (80 pm or 0.80 Å) and the Fe^{3+} (67 pm or 0.67 Å) and consequently the Fe^{2+} and Fe^{3+} fit into the holes (or sites) within the oxygen lattice. Each 'molecule' of Fe_3O_4 has eight tetrahedral and four octahedral sites, the positions of which can be seen in Fig. 6.4. One of the Fe^{3+} ions occupies an octahedral site and the other Fe^{3+} a tetrahedral site, while the Fe^{2+} ion occupies another octahedral site. This leaves two of the octahedral sites and seven of the tetrahedral sites unoccupied. Many cations (Ti^{4+}, Mg^{2+}, and Mn^{2+} etc.) are of similar size to the iron cations, and it is common to find natural substitution of the iron, yielding magnetite with inferior magnetic properties.

The general specifications for magnetite, as generally agreed upon by consumers and producers throughout the world, are as follows:

particle-size distribution	maximum 5% by weight larger than 45 μm and 30% by weight smaller than 10 μm;
relative density	4.9–5.2 g/cm^3;
magnetic content	not less than 95% by weight.

It should be noted that these specifications are based on those originally set by the British coal-mining industry. As such, they may not be strictly applicable to all other conditions, and they may also have to be related to plant practice. These specifications do not include any reference to rheological properties, such as vis-

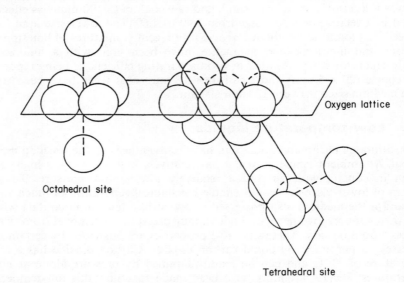

Fig. 6.4 Crystal structure of magnetite.

cosity and settling rates at specific slurry densities, which are important in plant performance and separation, and standard values relating to plant performance are usually established by the operator.

6.4.2 Physical Properties and Test Methods

Physical properties used to specify the quality and suitability of a particular source of magnetite must be assessable by standard techniques, and several recognized procedures and equipment are used in the determination of the physical properties of magnetite. These are summarized in Table 6.1. Some of the apparatus referred to in this table is described in Chapters 4 and 17 of this book.

Size analysis Figure 6.5 shows the particle-size analysis capabilities of the various analytical methods referred to in Table 6.1. The cyclosizer is at present the most widely preferred technique.

Relative density is measured by either a special form of pycnometer instrument or a standard density bottle. Of the two, the former is often regarded as being the more reliable, but both techniques are widely used.

The most common pycnometer instrument is the Beckman type. This apparatus works on the principle that a sample displaces a volume of air equal to its own volume. Two cylinders of identical volume are used, one as a reference and the other for the measurement, each having a movable piston. While both pistons are fully withdrawn, a weighed amount of sample is introduced into the sample cup and placed in the measuring cylinder. Each instrument has a specific calibrating number and the measuring piston is advanced until this number is read

TABLE 6.1 Methods used in the Determination of the Physical Properties of Magnetite[13]

Property	Method or equipment	Limitations
Particle size	dry screening	valid only above 53 μm
	wet screening	many variables introduced
	cyclosizer	time consuming; only limited points obtained
	Coulter method	sensitive to auto-coagulating properties
	sedigraph or other sedimentation analyser	high density of particles causes rapid settling
	Bahco or Haultain: pneumatic classification	time consuming; potential high loss of very fine material
	Microtrac or other microscopic methods	irregular shapes may introduce bias
	optical sizing	sophisticated equipment needed
Magnetic content	Davis tube	many variables
	ISO apparatus	simple method with good reproducibility
	magnetic chute	variable magnetic field
	hand magnet	accuracy depends on operator; only approximate
Density	Beckman pycnometer	slightly operator dependent
	density bottle	often difficult with fine powders

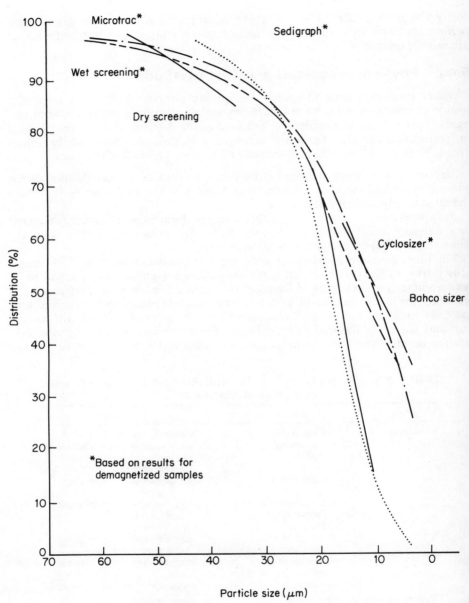

Fig. 6.5 Particle-size distribution of magnetite.

on the meter. The sample cup is then locked in place to give an air-tight seal, and the two cylinders are isolated from each other. The two pistons are then advanced in such a way that the pointer remains on the scale until the reference piston is in the foremost position and comes to a stop. The measuring piston is advanced until the pointer is in the centre of the scale and the valve isolating the

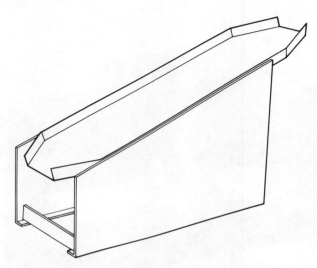

Fig. 6.6 Boxmag-Rapid Magnachute.

cylinders from each other is opened. The value on the meter is then read, being the volume of the sample in cubic centimetres. The relative density of the sample can then be calculated from the formula:

$$\text{density} = \text{mass}/\text{volume}$$

which will give the density in g/cm^3.

The density-bottle method most commonly used is that recommended and described in the recently introduced standard ISO 8833—*The Testing of Magnetite for Use in Coal Preparation—1986*. This standard method involves using a 50-ml capacity density-bottle and the volume of distilled water displaced by a 20-gram sample is measured.

Magnetics content There are numerous forms of apparatus used for determining the magnetics content of magnetite powders, of which the most common are magnetic chutes, the Davis tube, and numerous types of glass-tube devices including the one specified in ISO 8833. Several magnetic chutes have been designed for determining efficiency of wet magnetic separators used in dense-medium plants. One such device is shown in Fig. 6.6. It comprises a sloping non-magnetic frame which contains a permanent magnet unit, above which a hinged non-magnetic tray is fitted. Slurry containing the magnetic particles is poured onto the tray and the magnetic content is held by the magnet. The hinged tray is lifted so that the magnetics are removed from the magnetic zone and are washed off for analysis.

The Davis Tube Tester as shown in Fig. 6.7 is a laboratory unit used specifically for determining the ferromagnetic and non-magnetic fractions of small samples of crushed magnetic iron ore, magnetite, pyrrhotite, etc. It was developed for determining mesh of liberation, selectivity index, grade and recovery of ferromagnetic compounds, but is now widely used for assessing magnetite quality for coal-preparation dense-medium circuits, especially in the United States.

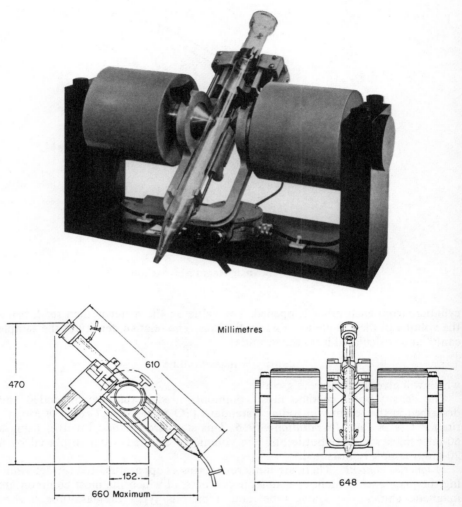

Fig. 6.7 Eriez–Davis tube apparatus (courtesy Eriez Magnetics).

With the magnet energized at a pre-set value of between 400 and 800 gauss (G) or 40 and 80 millitesla (mT), a dry magnetite sample is introduced to the water-filled tube. The tube is then oscillated while a gentle flow of water through the tube washes out non-magnetics. Magnetics are caught and held in the tube between the magnet poles. After sufficient washing, the tube is removed from the separator and the magnetics are flushed out, collected and dried. The same procedure is usually repeated for a different setting, e.g. 400, 550 or 700 G (40, 55 or 70 mT).

The ISO 8833 apparatus, shown in Fig. 6.8, consists of a vertical glass tube of 8 mm nominal diameter and 500 mm in length. Two ceramic permanent magnets of more than 60 mT are located at the centre point of the tube, and both the tube

DENSE-MEDIUM SEPARATION

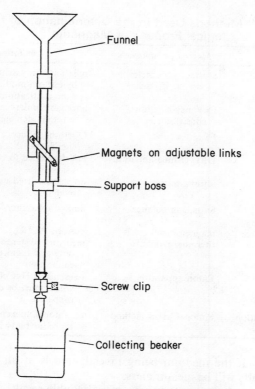

Fig. 6.8 Isomagnetics analysing apparatus.

and the magnets are supported on a laboratory retort-stand. The apparatus is filled with the suspension (2 g) containing the magnetite. The contents are then drained into a 100-ml beaker. The magnets are moved up and down for cleaning (upgrading) the arrested magnetite during drainage. When the tube is drained, the magnets are removed and the captured magnetite is washed into another beaker.

6.4.3 Rheological Properties and Test Methods

These include viscosity, stability, ease of demagnetization, and magnetic susceptibility. Equipment most commonly used in testing these properties in coal industry usage is listed in Table 6.2.

Viscosity Various rotational and capillary viscometers have been used by previous researchers in studies of dense-medium viscosity. The majority have received critical comment, and several unsuccessful attempts have been made to cross-correlate the results obtained. Of the various capillary viscometers, the DeVaney–Shelton consistometer appears to be the only one to give reasonable reproducibility. It is basically an efflux viscometer with an internal stirrer to maintain homogeneity. Being an efflux viscometer, the viscosity is determined at a single point with an unknown shear rate, by comparison with a known

TABLE 6.2 Methods Used in the Determination of the Rheological Property of Magnetite[13]

Property	Method or equipment	Limitations
Viscosity	Stormer viscometer	reliability only achieved by custom modification; demagnetization essential
	DeVaney–Shelton consistometer	unreliable unless magnetite is totally demagnetized
Stability	F5 index tube	Operator influence very significant
	DRL U-tube apparatus	not developed to commercial usage
	settling tests	no standardized method available
Magnetic susceptibility/ coercive force	Satmagan balance	limited reliability
	magnetometer	used by ACIRL
	thermogravimetric	measures hysteresis curve only and requires calibration
	Eaton apparatus	reference is $HgCo(SCN)_4$ which must be carefully prepared
Demagnetization	Simon–Carves method	operator influence, but reproducibility is reasonable

Newtonian liquid. If the medium being investigated is at all non-Newtonian in character, the results will be meaningless.

Rotational viscometers appear to give more reliable results, but some concerns have been expressed regarding the application of conventional rotational types.

A Stormer viscometer, in contrast to a conventional rotational viscometer, applies a constant shear stress to the fluid in the annular gap and measures the speed of rotation (proportional to shear rate). The modified Stormer consists of a multibladed rotor, rotating in the opposite direction to and also beneath the line cylinder. This propels the fluid downward past the inner cylinder and returns it by a series of baffles. Under these conditions, the dense-medium suspension is maintained in a homogeneous state by turbulence. The viscometer is calibrated by comparing known Newtonian fluids to different shear stresses. The viscosity of a Newtonian fluid in a turbulent state does not obey the basic laws of viscosity, and the same applies to a non-Newtonian fluid in this flow condition.

Stability Two methods are commonly used. The F5 Index test is widely used in both Britain and South Africa, but rarely used anywhere else. Jar-settling tests involving large glass cylinders to reduce wall effects are used elsewhere.

The laboratory F5 Index method was developed by the National Institute for Coal Research (NICR) in South Africa.[13] The apparatus is shown in Fig. 6.9 and is in general use in South Africa and Britain. The apparatus consists of a copper tube (the tube must be made of copper or another completely non-magnetic material) with an overall length of 1.225 m. The main portion, which has an outer diameter of 31.75 mm and an inner diameter of 28.6 mm, is 1.14 m in length. The common portion has an inner diameter of 12.7 mm and a length of 70 mm. The reducing piece between the main and bottom portions is 14 mm in

DENSE-MEDIUM SEPARATION

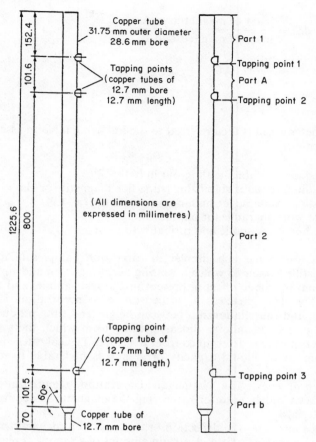

Fig. 6.9 F5 Index apparatus.

length. There are three points in the side of the main portion, all of which are made of copper tubing with an inner diameter of 12.7 mm.

A sample of the medium is placed in the copper tube, adjusted to a constant height by the use of tapping point 1, and the medium in Part 1 of the tube is run off quickly. After a settling time of five minutes, the medium in Part A of the tube is run off into the beaker and weighed. The medium in Part 2 of the tube is discarded, and that in Part B of the tube is retained and weighed. The mass ratio of the medium in Part A to the medium in Part B of the tube is the calculated to give the F5 Index.

The F5 Index is calculated as follows:

If the mass of the magnetite in Part A of the tube (after subtraction of the mass of the beaker and water)

$$= 40 \text{ g}$$

and the mass of the magnetite in Part B

$$= 120 \text{ g}$$

$$\text{the F5 Index} = \frac{\text{mass of magnetite in A}}{\text{mass of magnetite in B}} \times 100$$

$$= \frac{40}{120} \times 100$$

$$= 33.3$$

The value determined is then referred to the following table to assess the condition of the medium:

<10   medium unstable, settling-out in bath
10–20 medium requires stabilizing (superfine magnetite or clay)
20–30 stability suitable for high-density systems (1.6–1.9)
30–60 best working range for bath medium
>60   medium very stable and probably too viscous

Jar-settling tests are much simpler to carry out. Samples of magnetite are mixed with distilled water to which a wetting agent has been added. It is essential that no residual magnetic effect is present in the test sample, and therefore the sample must be demagnetized. A suspension of concentration 1500 kg/m^3 is usually tested, and the relationship between height and time produces a settling curve with a period of uniform (linear) settling from which the settling rate in millimetres per minute is determined from the slope. Typical rates are 10–20 mm/min. A settling jar of 500-ml capacity or greater is preferable to minimize wall effects.

Magnetic properties The Satmagan (saturation magnetization analyser) Balance has been widely used for measuring the magnetic susceptibility of ferromagnetic materials.

This balance measures the deflection in the mass of a magnetic material upon being placed in a magnetic field. A certain amount of a sample is placed in a glass tube on a balance pan. There is a tare control knob with which the mass of the sample can be balanced exactly so that the deflection indicator rests at zero. The magnets are then swung into place, one on either side of the sample, and the measuring knob is rotated until the deflection indicator once more rests at zero. The susceptibility can then be read off the dial on the measuring knob.

Another apparatus used for determining magnetic properties is the Vibrating Sample Magnetometer. This apparatus was used by Parry and Caddy at the University of New South Wales in Australia to characterize magnetite from different sources.[14,15] Measurement of magnetic susceptibility, saturation moment, remanence and coercive force is available from each test sample.

The demagnetization capacity of magnetite is an important, although not a frequently analysed, property. A method was devised by Simon Carves.[13]

A slurry of relative density 1.6 g/m^3 is prepared and placed in a 100-m^3 graduated measuring cylinder equipped with a stopper. The flask is thoroughly agitated and then placed on a level surface, a stopwatch being started at the same moment. The settling time required for all the solids to settle through a given number of gradations is then measured and the settling rate calculated. The cylinder is passed through a magnetizing field of 64 kA/m (800 oersted), and the settling rate is measured again. The cylinder is then passed through a demagnetizing field, and the test repeated. The recommended limits for a sample that will behave well in a plant are as follows:

initial settling rate	2.0–6.5 mm/min
magnetized settling rate	200–250 mm/min
demagnetized settling rate	2.0–7.5 mm/min

6.4.4 Chemical Properties

Chemical analysis, particularly for iron content, is often used for distinguishing magnetites for different sources. Two forms of analysis are common:

(a) determination of total iron content, which is determined using the volumetric method described in ISO 2597 (Method 1b);

(b) determination of ferrous iron content, which is determined by dissolving the sample in hydrochloric acid in an atmosphere of nitrogen and titrating ferrous iron from the resultant solution using potassium dichromate.

More-detailed chemical analysis may be obtained by instrumental methods such as Inductively Coupled Plasma (ICP), to determine the following:

SiO_2, Al_2O_3, TiO_2, P, S, Ba, Cd, Zn, Ca, Mg, Ni, Co, Cu, Au and Ag

Normally, an analysis of this type would not be necessary for known sources, and this type of examination would be carried out for potential new sources of media, including fly-ash-derived material.

6.4.5 Current Practice in Magnetite Testing

With the recent introduction of the ISO 8833 standard, it is hoped that dense-medium plant operators will adopt the recommended methods on a world-wide basis. Although there are some differences in the methodology of magnetite testing, there is almost universal agreement in current practice regarding acceptable standard for delivered magnetite for dense media. This size distribution specified is almost always closely related to the original Dutch State Mines' specification for dense-medium cyclone operation, i.e. 95% minus 40 μm. Figure 6.10 shows size distributions for magnetite used in fine-coal cleaning circuits. Other specified parameters are based upon practices established by the early European users and typified by the current British Coal magnetite specification, i.e.:

(a) relative density	≥ 4.85;
(b) magnetic content	$\geq 97\%$;
(c) size fraction below 60 μm	$\geq 85\%$;
(d) size fraction below 10 μm	$\leq 10\%$;
(e) moisture content	$\leq 10\%$

The ISO 8833 standard was specifically prepared for the purpose of identifying the essential requirements of a magnetite to be used in a dense-medium suspension, and setting down suitable tests using simple apparatus where possible.

Although moisture content, size distribution, magnetic content and relative density are generally considered adequate for routine checking of regular

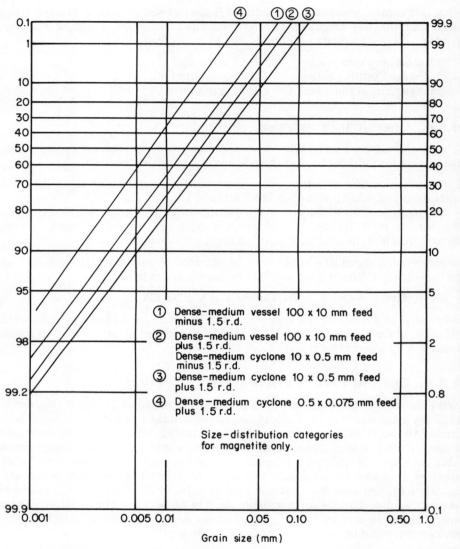

Fig. 6.10 Generalized size-distribution chart for ground magnetite for dense-medium separation.

supplies, extra tests have been included in the document, as they may be necessary for evaluating new sources of magnetite.

The standard specifies procedures for both sampling and testing of magnetite, and the specified tests are intended primarily for the testing of milled magnetite, the top size of which is assumed to be below 250 μm. It is possible to apply the tests to unmilled magnetite with a topsize limit of about 500 μm.

The specified tests should ensure that the properties which make the magnetite suitable for dense-medium usage are acceptable. These properties are:

(a) moisture content;
(b) particle-size distribution;
(c) magnetic content;
(d) relative density;
(e) fundamental magnetic properties, i.e. coercive force, remanence, susceptibility and saturation magnetic moment;
(f) total iron content;
(g) ferrous iron content.

The ISO 8833 standard recommends that all of the tests included for the determination of the above properties should be conducted in order to assess magnetite from a new source, but for routine checking of regular supplies, only the tests for (a) to (d) should be necessary.

The specific test methods recommended for each determination are as follows:

- *Sampling*—should be carried out in accordance with the general recommendations of ISO 3081, *Iron Ores—Increment Sampling—Manual Method*. The mass of sample required to perform the tests is obtainable from ISO 3083, *Iron Ores—Preparation of Samples*.
- *Moisture content*—is obtained by drying either one kilogram of sample from a large consignment, or 100 g (minimum) of a laboratory test sample, using a laboratory air oven capable of being maintained within a temperature range of 105–110 °C.
- *Particle size*—analysis is divided into two ranges:
  (a) sieve;
  (b) subsieve.

The distinction is stated as being 38 μm woven-wire aperture. Sieve analysis is performed in accordance with ISO 565, 'Woven Wire Cloth Test Sieves', and ISO 3310, *Test Sieves—Technical Requirements and Testing*. The number of sieves required would be determined by the grain size range which will usually be between 250 and 38 μm. The mass of each test portion would not be less than 25 g and not more than 100 g, and wet sieving is the preferred method.

Subsieve analysis techniques recommended are those employing centrifugal classification, e.g. Warman cyclosizer. Sedimentation and elutriation methods are affected by the agglomeration tendency exhibited by magnetite particles in suspension. The mass of test portion would not be less than 10 g and usually not more than 50 g (depending on the method).

Magnetic content—the preferred test method employs the special apparatus described earlier or, alternatively, the Davis magnetic tube tester may be used. Sample portions of between 2 and 5 g are suggested.

Relative density—should be determined using the relative-density bottle method in which the volume of liquid displaced by a known mass of sample (15 g) is measured. The preferred liquid is distilled water as this results in a more direct method less liable to measurement error. Alternatively, methylated spirits may be used.

Fundamental magnetic properties—the standard suggests that the determination of suitable parameters requires specialized apparatus and suggests that such work be carried out by a university or research institute. There was apparently a lack of concensus as to which magnetic parameters are relevant in the context of the standard. It was therefore left open for users to decide for themselves, but the standard does suggest that coercivity is a guide to the ease of recovery, and saturation magnetic moment is a measure of the purity of the individual grains.

Total iron content—should be determined using the volumetric method specified in ISO 2597 (Method 1b), *Iron Ores—Determination of Total Iron Content—Volumetric Method*. The test portion is about 0.25 g.

Ferrous iron content—should be determined by dissolving the test sample (0.25 g) in hydrochloric acid under a nitrogen atmosphere and titrating the ferrous iron taken into solution with standard potassium dichromate.

Much technical input in formulating the standard has come from Australia and in particular from the Australian Coal Industry Research Laboratories (ACIRL) which represented the Australian coal industry. For many years, this institution has concerned itself with magnetic testing.[15–17]

Australia's coal industry is almost totally supplied from one source of supply of magnetite, i.e. Mount Biggenden in south-east Queensland. This supplier provides five grades of material based on the following size analysis, with prices increasing with fineness:

Grade	Percentage passing 53 µm
Coarse	50–60
Medium	65–75
Fine	85–90
Superfine	92–95
Super-superfine or ultrafine	94–97

Owing to medium-stability problems, the superfine and ultrafine grades comprise by far the majority of sales. The supplier has only one deposit, and the different grades are produced by altering the operation of the supplier's product ball mill. Because of this, the majority of magnetite testing in Australia is devoted solely to magnetite size analysis in order to determine whether the supplied magnetite is of the specified fineness.

The other general magnetite specifications are:[18]

- Relative density   4.9 minimum
- Magnetic content (Davis tube 700 G)   96%
- Typical chemical analysis:
  SiO_2 1.4%
  Al_2O_3 0.8%
  Total Fe   69.5%
  Fe_3O_4 95.0%
  S   0.4%
  P   0.02%
  Cu   0.005%
  H_2O (110 °C)   10% maximum
- Magnetic properties:
  initial susceptibility, K_{30} (emu/g)   > 0.050
  susceptibility at 800 oersteds, K_{800} (emu/g)   > 0.053
  saturation moment, M_{sat} (emu/g)   > 80
  coercive force, H_c (oersteds)   < 50

DENSE-MEDIUM SEPARATION

TABLE 6.3 Typical Magnetite Quality for North American Sources

	Principal ore value	Magnetite quality					Comments
		Fe (%)	SiO$_2$	r.d.	% − 325 Mesh	% Magnetics	
Canada							
Craigmont	Copper	65–68	2.5–3.0	5.0	90	>95	mine closed
Griffith	Iron	69	3	4.9	90	>95	mine closed
Sherman	Iron	67.5	2.5–5	4.9	—	>95	producing mine export to Australia
Westfrob	Iron	69	5	5.0	90–95	>95	mine closed
USA							
Foote	?	62	1.5	—	92	91–94	—
Tawahus	Ilmenite	67	1.5	5.1	90	>95	tailings source
Reiss Viking	Iron	63	1.5	5.0	95	91	tailings source

Table 6.3 gives some typical magnetite analysis for North American sources.
Table 6.4 gives recent statistics for magnetite annual consumption and working specifications for world-wide usage.

6.5 MAGNETITE PREPARATION AND RECOVERY

6.5.1 Relative-density Calculations

The relative density of a suspension is governed by the proportions of water and magnetite. Hence, if a sample of the medium is composed of 2 ml of water and 1 cm^3 of powdered magnetite of relative density 5, the relative density of the medium is calculated as follows:

mass of 2 ml of water of relative density 1 is $2 \times 1 = 2$ g;

mass of 1 cm^3 of powdered magnetite at relative density 5 g per cm^3 is 5 kg;

∴ mass of the suspension is 7 g and volume is 3 cm^3

TABLE 6.4 Coal-cleaning Dense-medium Magnetite Statistics (1983)

Country	Estimated annual consumption (tonnes)	Working specification			
		Moisture content (%)	Particle-size distribution (μm)	Magnetics content (%)	Relative density (kg/m^3)
Canada	75 000	<10	<5% + 45	>95	>4800
United States	250 000	<10	<5% + 45	>95	>4800
South Africa	115 000	<10	<5% + 45 <30% − 10	>95	4900–5200
United Kingdom		<10	≥85% − 60 ≤27% − 10	>97	>4850
Australia	80 000	<10	<5% + 45 <30% − 10	>95	>4800

Relative density* of the suspension $= \dfrac{\text{mass}}{\text{volume}} = \dfrac{7}{3}$

$= 2.33 \text{ g/cm}^3$

* Density and relative density are taken as being numerically equal in SI units, and 1 cm³ is the true volume of the magnetite (assumed solid).

In order to determine the amount of magnetite of relative density 5 g/cm³ that must be added to 1 m³ of water to obtain a density of 1.5, the following calculation applies:

Suppose x dm³ is the volume of magnetite required—

If density is mass divided by volume

mass of magnetite	$= 5x$ kg
mass of water	$= 1000$ kg
mass of medium	$= 1000 + 5x$
volume of medium	$= 1000 + x$

$$1.5 = \dfrac{1000 + 5x}{1000 + x}$$

x	$= 142.9$ dm³
and mass of magnetite	$= 5 \times 142.9$
	$= 714.3$ kg

If the magnetite was replaced with crushed shale of 2.4 relative density, the corresponding amount of shale required to achieve a relative density of the medium of 1.5 would be:

$$1.5 = \dfrac{1000 + 2.4x}{1000 + x}$$

$$= 555.5 \times 5$$

$$= 2777.7 \text{ kg}$$

Converting these to solids concentrations (by weight) gives

magnetite, % solids $= 41.7\%$
shales, % solids $= 73.5\%$

which demonstrates the potential problems of high viscosity likely to be encountered with the shale slurry.

Similarly, the stabilizing effect obtainable with clay addition can be examined.

Suppose the same relative density is obtained by response to a density controller which cannot distinguish between clay and magnetite (i.e. nuclear gauge). Over a period of time, the circuit builds up to 5% clay (i.e. 95% magnetite).

$$\text{let } x = \text{volume of magnetite}$$

$$y = \text{volume of shale}$$

$$1.5 = \dfrac{1000 + 5x + 2.4y}{1000 + x + y}$$

DENSE-MEDIUM SEPARATION

and

$$\frac{5x}{(1-0.05)} = \frac{2.4y}{(1-0.95)} = \text{total mass of solids}$$

Hence

$$1.5 = \frac{1000 + \left[\dfrac{5x}{(1-0.05)}\right]}{1000 + x + \left[\dfrac{5x(1-0.95)}{2.4(1-0.05)}\right]}$$

$$x = 138.94 \text{ dm}^3$$

$$\text{Mass of magnetite} = 5 \times 138.94 = 694.7 \text{ kg}$$

$$\text{Total mass} = \frac{694.7}{0.95} = 731.3 \text{ kg}$$

$$\text{Mass of shale} = 731.3 \times 0.05 = 36.6 \text{ kg}$$

which converts to a solids concentration of 42.24% by weight.

6.5.2 Magnetic Separation

The key to medium recovery and control in magnetic dense-medium systems is the magnetic separator. Three types of units are used, these being:

(a) permanent magnet wet drum, which is the most common;
(b) permanent magnet disc or magnedisc, which is employed as a scavenger to the drum type; and
(c) horizontal disc, used in conjunction with a settling cone where preconcentration or thickening of dilute medium is employed.

These units and the associated medium control circuits are responsible for ensuring that the dense-medium washer performs at peak efficiency. The medium is the life's blood of the process, medium-pumping plays the part of the heart and the magnetic separator performs a role analogous to liver and kidneys combined.

Wet-drum Separators

Two types of wet-drum separator are employed for coal-cleaning circuits:

(a) cocurrent type—which is the most common (see Fig. 6.11);
(b) counter-current type—which is normally only used as a secondary machine (see Fig. 6.12).

Occasionally, double-drum or two-stage machines are used to save space (see Fig. 6.12).

Drum separators differ little from one manufacturer to the next and all comprise the following:

• *Magnet unit*—made from high-permeability steel and ceramic elements of high-energy anisotropic barium or strontium ferrite. These magnets are characterized by high coercivity, lightness and high resistance to demagnetization. The

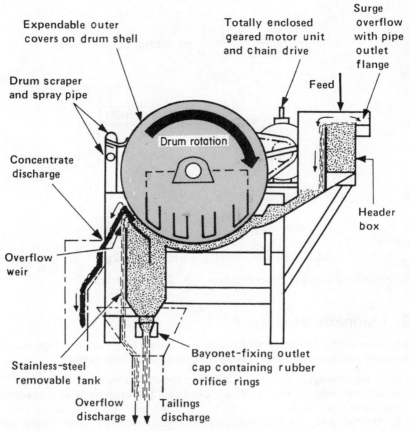

Fig. 6.11 Typical single-drum cocurrent wet magnetic separators (courtesy, Boxmag-Rapid).

magnets are arranged with alternate polarity and are arranged at an optimum distance from the slurry to provide the correct field-strength gradient. Figure 6.13 shows a typical field-strength chart of the type commonly used to determine the correct operating arrangements for a separator. The magnet unit or yoke is mounted on the stationary drum shaft and its position can be adjusted ($\pm 30\text{--}40°$) by a lever. The gap between the tank and the drum, which determines pulp flow rate, can also be adjusted vertically, by shims inserted under the bearing housing and horizontally by slotted holes or guides.

• *Drum*—comprises a stainless-steel main shell which is carried on non-magnetic, corrosion-resistant end flanges fitted with self-aligning sealed ball bearings. An expendable outer cover may be of rubber, polyurethane or stainless steel, although the latter is the most commonly used.

• *Tank*—constructed of stainless steel with easily removable tailings outlet caps for replacement or interchangeable rubber orifice rings or valves. The rubber rings are preferable because they are easily changed and flow adjustment

DENSE-MEDIUM SEPARATION

Fig. 6.12 Counterflow single-drum and cocurrent double-drum wet magnetic separators (courtesy, Boxmag-Rapid).

is simple. As for the drum, the feed launder and all wearing surfaces can be rubber covered to improve life.

- *Feed and discharge box*—the most effective feed boxes are those incorporating a steady-head condition, i.e. two-compartment spill-over type as shown in Fig. 6.11.
- *Drive*—most machines are either belt or chain driven by a totally enclosed geared motor.
- *Ancillaries*—drum scrapers, spray-water pipes and spill boxes are all frequently used, the former two providing assistance in discharging of concentrate and the latter for double-drum machines as shown in Fig. 6.12.
- *Capacities*—wet-drum separator capacities can vary according to the magnetic field strength of the magnet unit, the direction of rotation and the particle

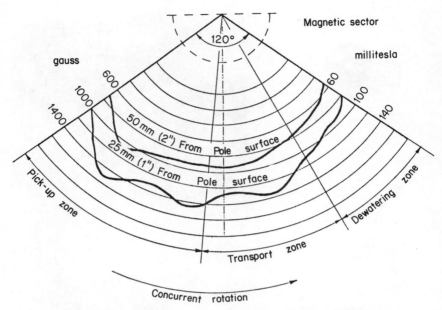

Fig. 6.13 Magnetic-field strength chart (Sala International).

size of the magnetite. Normal machines used for primary- and secondary-stage duty range from 55–80 mT (550–800 G), but some higher strengths of up to 140 mT (1400 G) and their potential benefit may be seen in Fig. 6.13. Normal capacities are given in Table 6.5.

Dutch State Mines recommended[24] the following operating parameters for magnetic separators:

(a) Cocurrent operating drum separators should be selected with permanent ceramic magnets.

(b) The feed concentration to the separator should be a maximum of 250 g/litre.

TABLE 6.5 Magnetic-drum Separator Throughputs and Settings (source Boxmag-Rapid)[19]

Operating criteria *Nominal speed of rotation = 11 r.p.m.* *Maximum solids content of feed = 20% by volume* *Maximum magnetic content of slurry = 250 g/litre*				
Drum diameter (mm)	610	765	915	1070
Operating gap				
normal	14	25	32	38
maximum	25	32	41	50
Maximum throughput m³/h/m^a	57	70	87.5	105
Maximum output concentrate t/h/m^a	12.4	16.5	19.8	23.1

^{a} per metre of drum width.

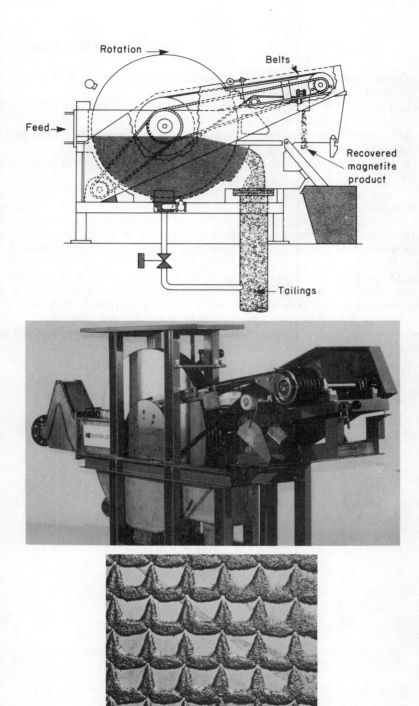

Fig. 6.14 MagnaDisc magnetic separator (courtesy Asea). (a) Diagram. (b) Five-disc unit. (c) Detail of magnetized disc surface.

(c) The amount of magnetic and non-magnetic solids in the feed should not exceed 16% by volume.
(d) The magnets of the separator must be able to give a minimum flux density of 75 mT (750 G) in air at a distance from the drum of 50 mm.
(e) The relative density of the recovered magnetite concentrate should be 2.1 if the separator is operating properly.
(f) A separator with an evenly distributed feed will normally not exceed a loss of 0.3 g/litre.

Disc Separators

As an alternative to drum-type separators, the vertical-disc separator known as the MagnaDisc may be used for dilute slurries (as low as 5%), and contains only a small amount of magnetite (less than 1%). Details of a typical machine are given in Fig. 6.14.

The MagnaDisc developed in the 1970s in Sweden by Asea, consists of 1220-mm diameter discs set at 20-mm spacing. Each disc contains 5000 permanent magnet inserts, covered by stainless steel 0.5 mm thick. Magnetite recoveries of up to 30 kg/h from feed rates of 20 m³/h per disc are possible.

The other type of disc separator, the horizontal disc, is normally used in conjunction with a settling cone, as shown in Fig. 6.15.

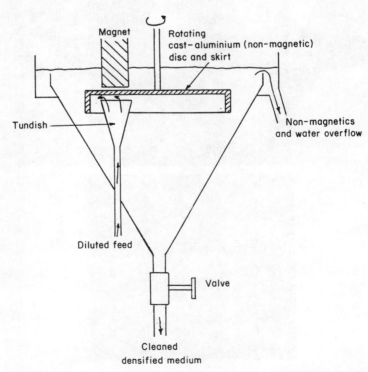

Fig. 6.15 Simcar desliming cone (courtesy Simon-Carves Limited).

DENSE-MEDIUM SEPARATION

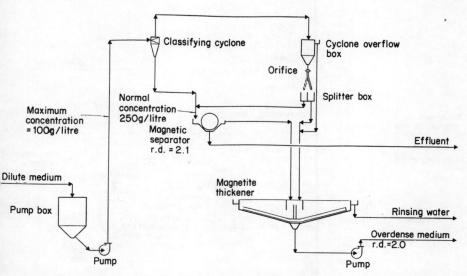

Fig. 6.16 Dense-medium recovery system employing a magnetite thickener.

6.5.3 Magnetite Recovery Systems

Much of the current technology of medium recovery used in modern plants is based on that originally developed by the Dutch State Mines (DSM). The original design concepts for media recovery circuits as recommended by DSM[23,24] were:

(a) recovery system employing magnetic separators and magnetite thickening, as shown in Fig. 6.16;

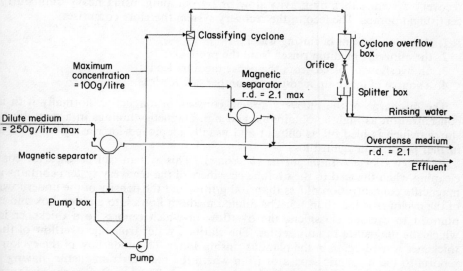

Fig. 6.17 Dense-medium recovery system employing magnetic separators.

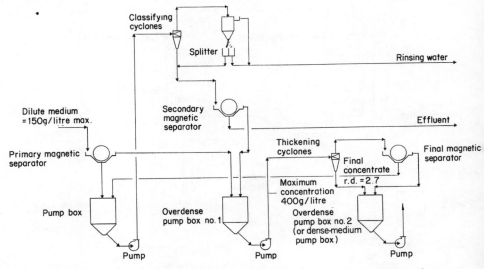

Fig. 6.18 Dense-medium recovery system for very low-feed concentrations.

(b) recovery system without magnetite thickening, as shown in Fig. 6.17;
(c) recovery system with cyclone thickening, as shown in Fig. 6.18.

More detailed versions of (a) and (b) are shown in Figs 6.19 and 6.20. In these flowsheets, typical recovery circuits for dense-medium cyclone and dense-medium bath are given.

In all dense-medium plants, the medium adhering to the products is rinsed away with water. The diluted medium thus formed is then directed to the recovery system either by gravity flow or by pumping, using heavy-duty slurry centrifugal pumps. The feed to the recovery system therefore comprises:

1. the total amount of rinsing water;
2. the adhering medium rinsed from the products;
3. an overflow or bleed, from the dense-medium feed;
4. a source of make-up medium to compensate for loss.

The discharge of the recovery system consists of magnetite, normally with a concentration as high as obtainable and non-magnetics tailings in a very dilute slurry, which is bled off as effluent and usually employed for rinsing water as a further precaution against loss.

Recovery system with magnetite thickeners This system can be used for coal cleaning when the feed to the cyclone classifiers of the recovery system contains a magnetite concentration of less than 100 g/litre, and the density of the underflows of the cyclones is less than 1.3. The diluted medium flows into a pump box and is pumped to cyclone classifiers, the overflow of which reports to a thickener in which the magnetite rapidly settles. The clarified water from the overflow of the thickener is used again in the plant as rinsing water. The underflow of the cyclone reports to the magnetic separator from where the recovered magnetite, having a relative density of over 2.1, reports to the thickener. The overflow of the cyclones

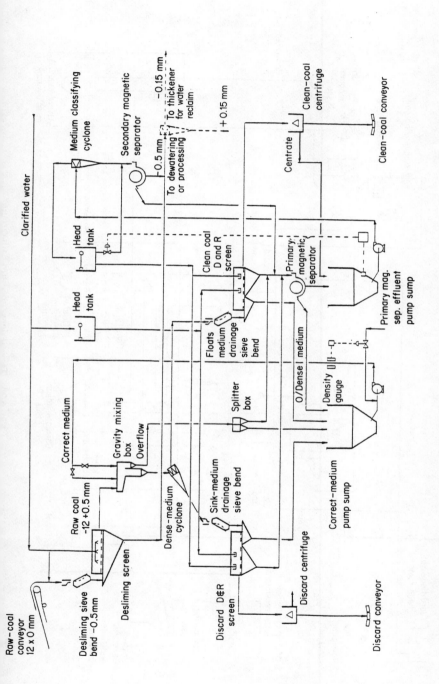

Fig. 6.19 Typical dense-medium cyclone circuit.

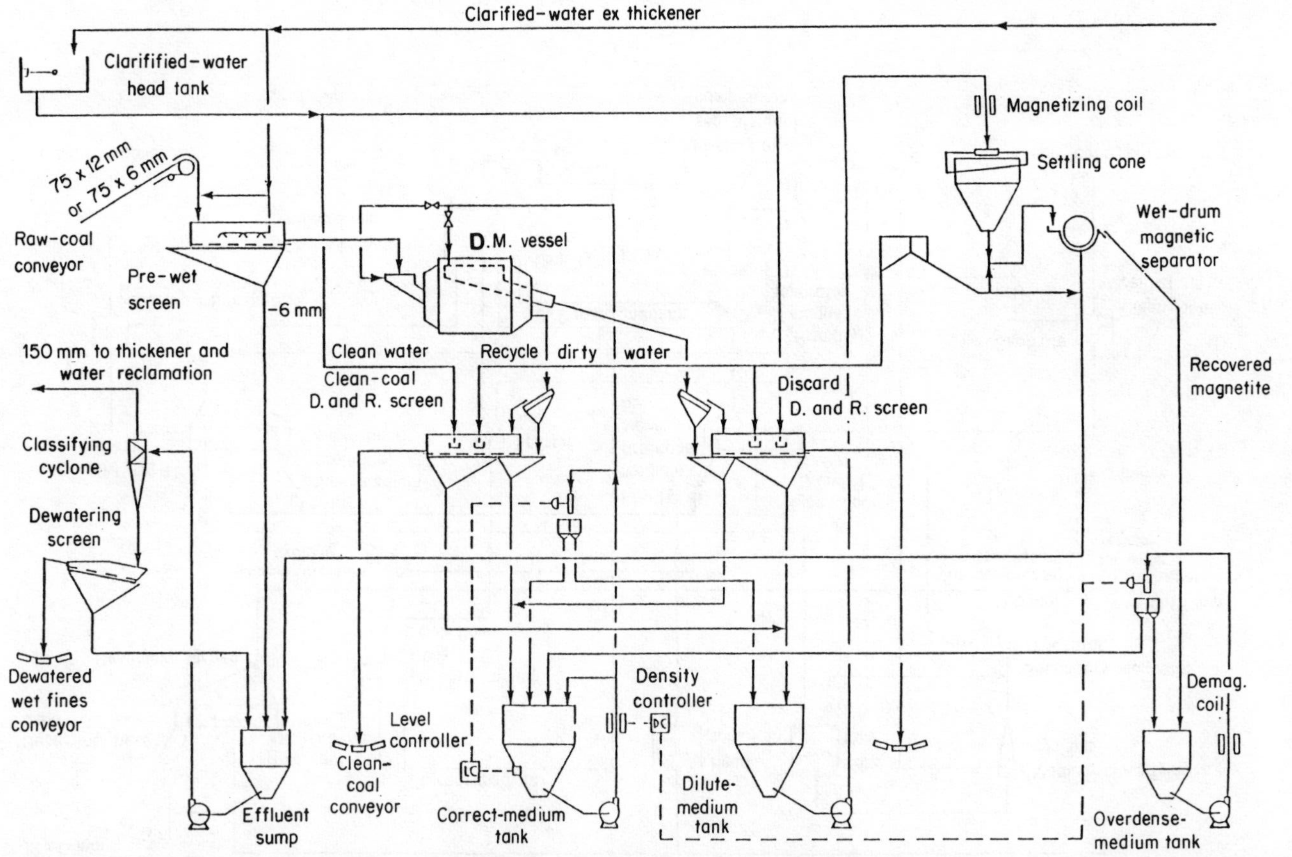

Fig. 6.20 Typical dense-medium bath circuit

containing finer magnetite particles than the underflow, is mixed with the latter in the thickener to obtain magnetite having the original grain-size distribution. The effluent of the magnetic separator is sent to the normal tailings thickener.

Recovery system without magnetite thickening This method of magnetite recovery is applicable where the dilute-medium feed to the recovery system contains a concentration of magnetite up to a maximum of 250 g/litre. The dilute medium reports to magnetic separators where the magnetite is recovered with a relative density of 2.1. The discharge of the magnetic separators containing slurry and a small amount of magnetite is pumped to cyclone classifiers, and the cyclone overflow returns to the rinsing system as clarified water. The cyclone underflow, being the effluent water of the system, reports to a secondary magnetic separator, the width of which is determined by the volume of this underflow.

The slurry discharge from the secondary magnetic separator is removed from the plant. If there is sufficient rinsing water available from other sources, the classifying cyclone step may be eliminated. In these cases, the secondary magnetic separator can be used as desliming water or be sent directly to the slurry section. The amount of effluent is equal to the amount of fresh water fed to the system.

Recovery system with cyclone densifier When a magnetic separator is fed with a product with a relative density too low to effect the required separation and a thickener cannot be used (cost, space, etc.), the relative density can be raised by using a cyclone densifier as the concentrator. The feed to the cyclone concentrator is the magnetite concentrate from the magnetic separator which must be diluted to about 8% by volume before pumping it to the cyclone densifiers. This provides a medium concentration of up to 400 g/litre.

6.5.4 Relative-density Regulation

The control objective of the system is to keep the relative density of the medium constant. In Fig. 6.21 the dense-medium system is indicated by a rectangle and

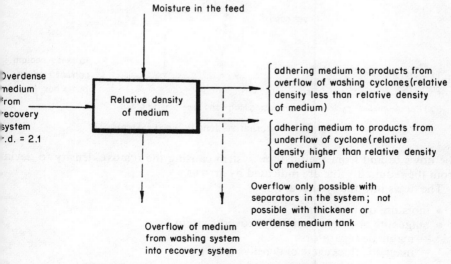

Fig. 6.21 Flow in and out of a dense-medium system.

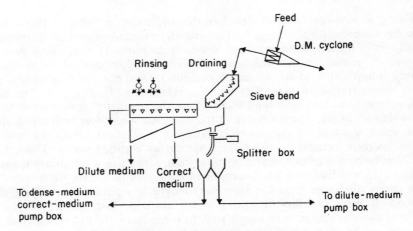

(a) Dense-medium cyclone recovery without magnetic thickener.

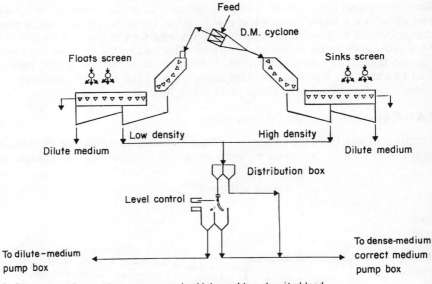

(b) Dense-medium cyclone recovery using high- and low-density bleed.

Fig. 6.22 Alternative-density bleed systems.

the flows to and from the medium system causing the relative density to deviate from the required value are indicated by arrows.

The flows into the system are:

- moisture on the feed;
- magnetite obtained from the recovery system via
  — magnetic separators;
  — magnetic thickeners of densifying cyclone;
  — overdense pump box.

DENSE-MEDIUM SEPARATION

The flow from the system is:

- medium adhering to the products.

The relative-density regulating systems which are employed are:

(a) From the mass balance calculations it almost invariably appears that an overflow of medium from the washing system to the recovery system is necessary. This overflow can be taken from: the medium in the mixing tank (A); overflow of the washing cyclone (B). The method using the cyclone overflow is shown in Fig. 6.22(a). This method can be applied when no thickener is used. A signal from the controller is transmitted to a regulatory device which controls the amount of overflowing medium. When the level in the dense-medium pump box drops below a certain value, as shown by a level indicator, an amount of make-up magnetite has to be added to the system.

(b) If a thickener or magnetite storage tank is used in the recovery system, the amount of concentrated magnetite going into the system is controlled by a relative-density regulating device mounted in the magnetite discharge line from the thickener or storage tank to the dense-medium pump-box tank. The level in the pump box is controlled by regulating the amount of overflow from the washing system into the recovery system, with medium from the mixing tank. When a higher relative density than desired is being obtained in the washing system (relative density of medium in the pump box), then the density control of the system may be achieved by means of regulating a flow of process water into that system. Regulation of medium from the washing system to the recovery system can be achieved by using both the dense-medium overflow and the underflow as shown in Fig. 6.22(b).

Measurement, indication, recording and control in the dense-medium circuits involves the control of the density, the medium, and the level and control of the pump boxes. To obtain an optimum separation in the plant, the density of the medium has to be kept constant within certain limits by automatic means, and to prevent overflow or emptying of the pumps, the levels of liquid in these pump boxes must be controlled.

The density with which the washing medium enters the washing vessel is the relative density that must be measured. However, the medium also contains raw coal and therefore measurement of the density is not possible at this point. The relative density of the medium without coal should therefore be measured at a point in the medium flowstream, prior to the addition of the coal and as far as possible from the point where the last flow enters the medium stream. This will ensure adequate mixing of the various flows.

In the washing system where product and medium are pumped together to the washing cyclone, this point is before the division of the feed launder to the pumpbox. This point is marked A in Fig. 6.23(a).

In the system where the dense-medium cyclone is gravity fed, the recommended points for measurement are shown as A, B or C in Fig. 6.23(b).

For many years, densities were measured by dip tubes recording differential pressure and/or Dp cells. These methods have now been almost totally replaced by nuclear density gauges. Nuclear density gauges actually measure the proportion of solids contained in a slurry, and the signal is then converted to a density reading based on prior calibration to a 'known' standard. If the medium is

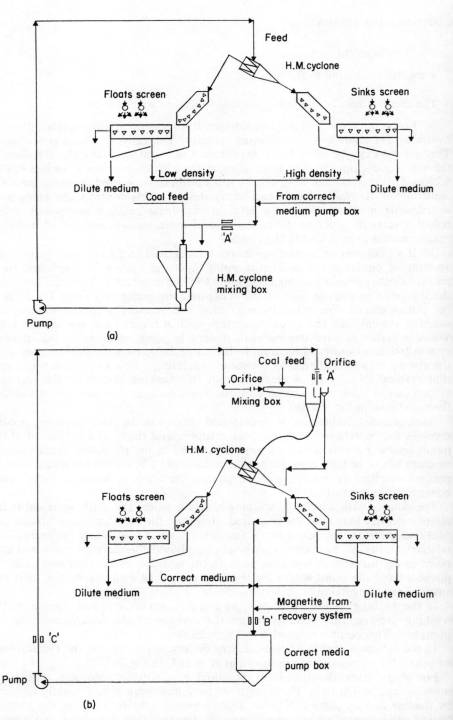

Fig. 6.23 Dense-medium cyclone recovery system: alternative locations for a density gauge: (a) density measurement point for pump-fed circuit; (b) density measurement point for gravity-fed circuit.

heavily contaminated with non-magnetics, then a false reading may arise. Some attempts have been made to introduce a magnetic-coil type of density gauge into the same line as the nuclear gauge. The magnetic device measures only the proportion of magnetic solids contained in the slurry so that by comparing the two signals obtained it is possible to determine the amount of non-magnetics. These combined systems are not often employed in coal preparation at present.

Chapter 13, which deals with instrumentation and control, includes description of some of the more widely used density-measuring instruments and associated equipment.

Because of the unavoidable build-up in residual magnetism of the circulating medium, the fine magnetic particles tend to flocculate unless a demagnetizing coil is used. A demagnetizing coil is therefore often used to prevent magnetics discharged from a wet-drum separator from settling too rapidly. In many plants treating plus 0.5 mm coal, sufficient demagnetization occurs naturally by pump action and pipe friction often to eliminate the need for a coil, but for fine-coal cleaning, where medium stability is extremely important, a demagnetizing coil is probably advisable. An adequately long coil length (usually 350–400 mm) is required for 3 to 4 complete flux reversals to occur at maximum magnetic intensity (750–1200 G; 75–120 mT). Despite the higher centrifugal force required for effective fine-coal cleaning by dense-medium cyclone, there is still a tendency for magnetic flocculation to cause the density differential to increase when a demagnetizing coil is not employed.

In recent South African and Canadian designed plants, a refined and highly accurate density-control system has been installed which greatly reduces the risk of loss and provides for automatic adjustment to 0.005 r.d. or better.[21] This system is shown in Fig. 6.24 and employs nuclear types of density gauges. The nuclear density gauges are located on the discharge of the circulating-medium pumps and provide a reading in the central control room for the dense-medium plant whenever the pump is running. However, they will only control automatically when the recycle by-pass valve is closed and the medium is flowing to the mixing box. A 'manual over-ride' can be used to control density at any time, but when the pump is on 'by-pass' no medium will go to the magnetic separators and the density can only be reduced using water injection. Each dense-medium stream has a fully independent density-control system.

The control is based upon the 'rising density' principle. Medium is allowed to recycle continuously through the 'bleed' splitter box back to the circulating-medium pumpbox (correct-medium sump). To increase the density a manual, air-operated actuator moves the splitter box-feed pipe over to the bleed compartment and a proportion of the medium flows to the primary magnetic separator via a product screen underpan. The separator recovers the magnetite from the medium and returns it to the correct-medium sump at a density of 2.1. Since the correct-medium density will be less than 1.75 the net effect of the 'bleed' is to increase the correct-medium density, and maintain a permanent 'rising density'.

The density must be reduced to the required (set-point) density, say 1.70. If the sump density is higher than 1.70 then water is added at the pump suction by the special injection arrangement, until the density reaches 1.70. It is clearly unwarranted to bleed off more medium than is necessary to give a slight rising density; this minimum bleed is arranged by automatically balancing the amount of bleed with the water injection quantity. It is therefore important to operate the

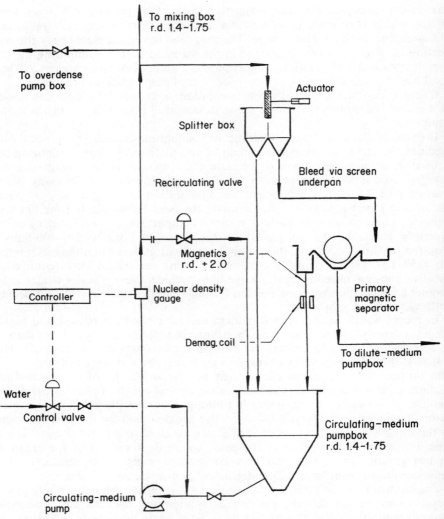

Fig. 6.24 Rising-density control system (courtesy, Kilborn Limited).

density control on 'automatic' since the control system is then optimized to give maximum control and minimum loss of magnetite.

For the purpose of cleaning the correct medium (i.e. removing some of the contamination by the minus 0.5 mm coal) the splitter box must always bleed some medium, say for example, 15% of the total volume is always bled off to the magnetic separator regardless of the operating density. A cleaner medium will generally give better control of the cut-point.

In addition to the density the dense medium has two other properties to be controlled. One is the 'viscosity' which plays an important role in relation to the accuracy of separation, and the other is the 'stability' which is important from an

DENSE-MEDIUM SEPARATION

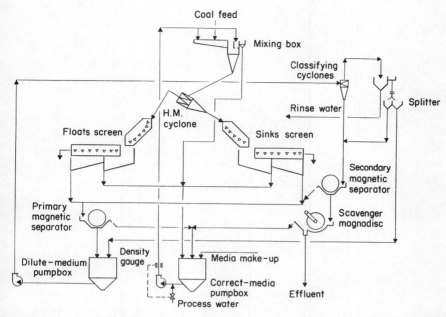

Fig. 6.25 Gravity-fed dense-medium cyclone system employing a tertiary or scavenger magnetic separator.

operational standpoint to improve the settling–plugging tendency of the medium in the pipeline.

The viscosity and stability are increased due to contamination by the slime (fine coal, clay, shale). Various degrees of contamination are found in the medium at the same density. The optimum viscosity and stability at the required density level are determined by sample testing. The volume of bleed is also decided by these test results. It should be noted that a certain amount of slime may be required at start-up to add necessary viscosity and stability to the clean medium.

Each density gauge is usually recorded on a strip chart recorder to obtain a permanent record of operations. A manual over-ride for control is also usually incorporated.

This type of control circuit has the advantage of being very responsive to changes in control density, and also less likely to encounter the condition where an operational problem creates a massive depletion of magnetite from the circuit, causing the density to fall to a very low level. Several hours would normally be needed to restore the required operating condition following this type of problem. The circuit as described applies to gravity-fed dense-medium cyclones is shown in Fig. 6.25. For pump-fed circuits two correct-medium sumps are required, but the control system is essentially the same.

The 'rising density' form of control system represents one of the most effective of the various types currently in use. The main advantages of this system are:

1. only a small quantity of 'top-up' medium is required to provide adjustment in response to demand by the control circuit;

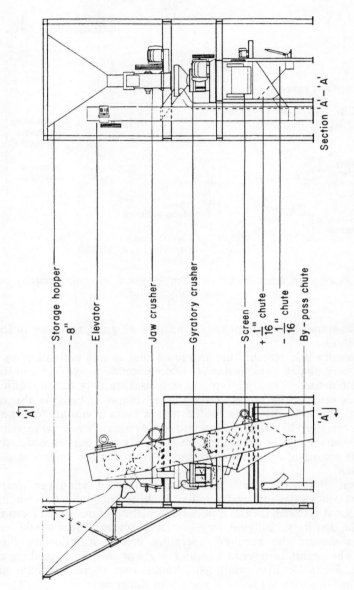

Fig. 6.26 Raw-magnetite crushing plant (source, Simonacco Limited[22]).

DENSE-MEDIUM SEPARATION

2. fast response is obtainable to corrective signal;
3. rapid indication is received when sudden loss occurs as a result of emergency or leakage;
4. in a multistream plant, most streams may be able to continue cleaning despite operating problems encountered in one stream;
5. exact make-up requirements can be measured, enabling media make-up system to be readily automated.

6.5.5 Magnetite Medium Preparation

As discussed earlier (Section 6.4.5) it is possible in most countries where dense-medium separation is employed to obtain 'ready-made' magnetite powder of dense-medium grade in dry, bulk form from suppliers who produce it specifically for this purpose. However, there are varying practices in these same countries ranging from this direct form of procurement to the installation of preparation of

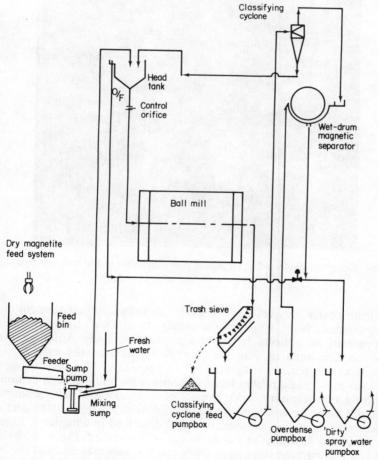

Fig. 6.27 Continuous-milling circuit for magnetite size-distribution control.

Fig. 6.28 Electromagnet used to transport raw magnetite to the ball-mill feed hopper.

medium from coarse magnetite ore. Figure 6.26 shows an arrangement of a raw-magnetite crushing plant, which would usually be followed by a wet ball milling system operated on either a batch or a continuous basis. Although the latter could be run simultaneous with the plant it would normally only be operated when the stock of prepared magnetite fell to a low level. The final sizing step by means of wet milling in a rubber batch or continuous mill has been demonstrated as permitting the necessary strict control of final sizing of the medium. Other forms of mill have, however, been successfully used for both batch and continuous operations, including the vibration mill described in Chapter 3. Figure 6.27 shows a simplified flowsheet of a continuous-milling circuit. Figure 6.28 shows an electromagnet being used to remove finely crushed raw magnetite to the ball mill feed hopper.

DENSE-MEDIUM SEPARATION

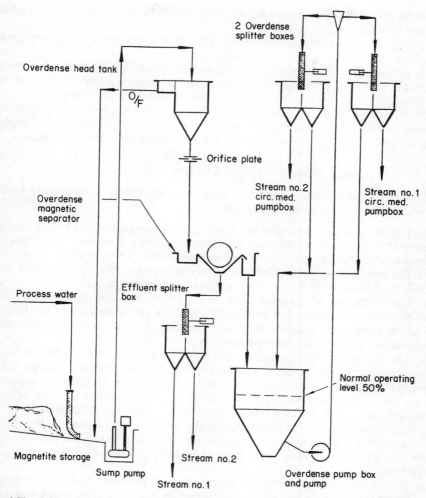

Fig. 6.29 Magnetite mixing and distribution system for a two-stream dense-medium cyclone plant (courtesy, Kilborn Limited).

Figure 6.29 shows a magnetite-mixing system of the type commonly employed for bulk delivery of magnetite.

The dry, bulk magnetite is delivered by truck and stored in an isolated section of the plant. When required the magnetite is moved with a front-end loader. The dry magnetite is washed with water into a floor sump and pumped via a sump pump to a head tank located in the plant. The overflow from the head tank returns to the floor sump. The underflow from the head tank, controlled with an orifice plate, gravitates to the overdense magnetic separator.

The high-density lip product, or overdense, from the magnetic separator flows through a demagnetizing coil into the overdense pumpbox. The overdense pump discharges into two two-way splitter boxes and the two discharge hoses are controlled to divert the overdense medium from recirculation to the two circulating-medium pumpboxes.

The overflow from the splitter boxes then gravitates back to the overdense pumpbox.

The tailing from the magnetic separator discharges into a splitter box which directs the tailing to either the sump pump in the dense-medium cyclone circuit, or stream 1 or stream 2 where it will be treated with the dilute medium.

The overdense pumpbox should only be filled up to 50% with overdense medium, so that circulating medium can be bled off into the overdense pumpbox when a quick reduction of the density of the circulating medium is required.

6.5.6 Magnetic Consumption Levels

The consumption (and subsequent loss) of magnetite is governed by several factors, of which the most important are:

(a) The fineness of the magnetite—finely ground magnetite has process advantages in that it improves efficiency and therefore yield. However, it has the disadvantage that the finer particles are more susceptible to loss from the circuit.

(b) The fineness of the coal being treated—it is more difficult to remove magnetite from fine coal and refuse material than from coarse material.

(c) The efficiency of magnetic separators—there are numerous factors that effect efficiency—magnetic flux density rating of the permanent magnets, the position of the permanent magnets, the feed volume to the separators, the percentage of magnetics in the feed to the separators, etc.

(d) The design of the magnetite recovery circuit—there are numerous design alternatives for magnetite circuits incorporating such variations as single- versus two-stage separators; countercurrent versus concurrent rotation of separator drums, use of dilute-medium cyclones versus magnetite thickeners; and need for secondary and even tertiary magnetite-recovery systems. The design of the magnetite-recovery circuit should match the dense-medium process being used and the density-control system employed. Poorly designed circuits lead to losses which it is almost impossible to correct operationally, without major redesign of the system.

(e) Poor operations and maintenance—removal of solids build-up in magnetite separators, maintaining the correct level in the separator, worn separator surfaces, leaks in medium lines, blocked or damaged rinse-screen surfaces causing the returned medium to be contaminated with fine (and sometimes coarse) coal and refuse material.

The magnetite losses commonly encountered in modern plants will normally range from 0.50 to 3.0 kg/t, quoted in tonnes of feed to the dense-medium section of the plant. It should be noted that many operators quote magnetite consumption based on either a clean-coal tonnes or total raw-coal tonnes basis. This can sometimes be misleading because the quoted figure can vary considerably due to the influence of yield change and fluctuation in raw-coal size analysis. Quoting consumption on the basis of tonnes of feed to the dense-medium section (or sections) avoids this problem. An example of this follows.

Consider a plant operating at 500 t/h with 25% minus 0.6 mm reporting to froth flotation and 75% reporting to the dense-medium section. The yield of the

DENSE-MEDIUM SEPARATION

dense-medium section is 80% and that of the froth-flotation section is 70%. If magnetite consumption is 1.5 kg/t of feed to the dense-medium section, then consumption based on raw coal and clean coal is as follows:

Raw coal

$$\frac{1.5 \times 75}{100} = 1.125 \text{ kg/t raw-coal feed}$$

Total clean coal

Consumption for clean coal produced =

$$\frac{1.5}{0.8} = 1.875 \text{ kg/t clean coal}$$

$$\text{Overall yield} = (75 \times 0.8) + (25 \times 0.7)$$
$$= 77.5\%$$

$$\text{Consumption per tonne } total \text{ clean coal} = \frac{1.125}{77.5} \times 100$$

$$= 1.45 \text{ kg/t}$$

Consider now, if the percentage of minus 0.6 mm changed to 30% and the yields of dense-medium and froth-flotation clean coal changed to 60% and 73% respectively. The consumption rates now become:

Raw coal

$$\frac{1.5 \times 60}{100} = 0.9 \text{ kg/t raw-coal feed}$$

Total clean coal

$$\text{Overall yield} = (70 \times 0.73) + (30 \times 0.6)$$
$$= 69.1\%$$

$$\text{Consumption for total clean coal produced} = \frac{0.9}{69.1} \times 100$$

$$= 1.30 \text{ kg/t}$$

This assumes a constant consumption rate of 1.5 kg/t of feed to the dense-medium section, which is a reasonable one, because magnetite is lost mainly on clean coal and refuse products. Although it is true that some magnetite losses occur even when the plant is idling, the practice of relating the overall losses to the dense-medium feed is more realistic than relating losses to clean coal, which is influenced by yield, or to total raw coal, which is influenced by the size analysis.

The consumption of magnetite is also related to the mean grain-size of the coal being cleaned in that more magnetite will adhere to fine coal (or refuse) on the drain-and-rinse screen than to coarse coal.

As we have already seen, the Dutch State Mines (DSM) contributed much to the early development of the original magnetite-based, dense-medium processes.

Dutch State Mines related adhering magnetite to mean grain-size as in equation 6.1.[23]

The amount of medium adhering to the products on a sieve bend plus drain screen (i.e. before the rinsing section of the screen) is given by:

$$Q = \frac{950}{d_{ave} \times \rho_{pr}} \qquad (6.1)$$

where Q = medium in litre/t product, d_{ave} = average grain-size of the product in mm, ρ_{pr} = average relative density of the grain of the product (i.e. coal or shale).

As an example, one source of coal has a mean grain-size of 1.8 mm for a 10 m × 0.6 mm size-range while another source has a mean grain-size of 2.7 mm for the same size-fraction.

For the finer coal (first source) the adhering medium at the end of the drainage section of the clean-coal drain-and-rinse screen for equation 6.1 would be:

$$Q = \frac{950}{1.8 \times 1.5} = 352 \text{ litres/t product}$$

$$= 0.352 \text{ m}^3/\text{t product}$$

which is equivalent to $0.352 \text{ m}^3 \times 5 \text{ t/m}^3$ or 1760 kg/t of product.

The coarser-coal source would be:

$$Q = \frac{950}{2.7 \times 1.5} = 235 \text{ litre/t}$$

$$= 0.235 \text{ m}^3/\text{t}$$

which is equivalent to 0.235 × 5 or 1175 kg/t of product.

If recovery only depended upon drainage, the losses would obviously be severe. However, we have seen that the second section of the drain-and-rinse screen is intended to recover as much of the adhering medium as possible.

If the rinsing section of the screen is working effectively (i.e. shower boxes and/or sprays are efficient) and the screen itself is correctly sized, the rule of thumb is that the adhering magnetite is reduced by a factor of 1000 prior to the discharge of the product from the end of the screen.

Hence, for the above examples the adhering medium *still* remaining on the screen at the point of discharge would be 1760 g/t for the fine-coal source (i.e. 1.76 kg/t and 1175 g/t for the coarser coal (i.e. 1.18 kg/t). This amount of loss may, however, be reduced further by subsequent passage through a centrifugal dewatering machine, and the centrate so obtained would then normally be routed to the dilute-medium circuit as shown earlier in Fig. 6.19.

A generally accepted upper-limit figure for European, eastern US, Australian and South African coals would be 1.5 kg/t for small coal (10 mm × 0.5 mm) and 0.50 kg/t for coarse coal (>10 mm).

For the finer, Rocky Mountain coals of Canada which resemble the first source in the example, these levels are generally a little higher i.e. 2.0 and 1.0 kg/t of feed to the dense-medium circuit, respectively.

Normal loss (consumption) levels achievable with DSM dense-medium process circuits are stated as being 1 kg/t for the small coal and 0.5 kg/t for the coarse

DENSE-MEDIUM SEPARATION

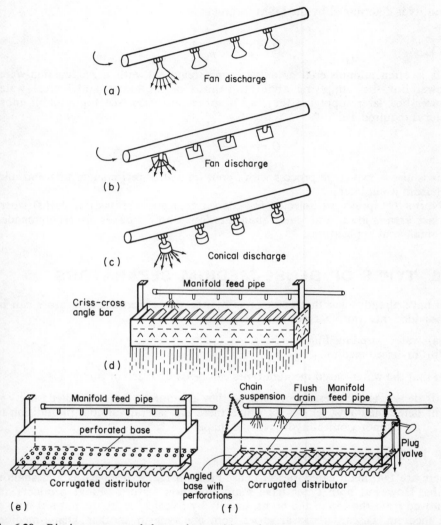

Fig. 6.30 Rinsing sprays and shower boxes: (a) spade type sprays; (b) plate type spray; (c) nozzle type spray; (d) Bretby shower box; (e) Dutch State Mines 'rainmaker' shower box; (f) Kilborn shower box.

coal, and it is these figures which are normally used to determine magnetite make-up requirements for most new plant designs.[23]

A key element in the medium-recovery system is the rinsing of adhering medium from the drained products passing along the drain-and-rinse screens. The quantity of rinsing water required for this purpose depends on the surface area of the material needing rinsing, and varies from 20 m^3/h/m width of screen to about 65 m^3/h/m width of screen.

Where shower boxes are used (see Fig. 6.30(a)) less water is required and the quantity is determined by the DSM formula.[23]

$$Q = \frac{3}{d_{ave} \times \rho_{pr}} \text{ m}^3/\text{t}$$

This is often administered as a two-stage treatment with a recirculated-water shower box first supplying about two-thirds of the above and a fresh-water shower box later supplying the rest. If sprays are used (see Fig. 6.30(b)), more water is required, i.e.

$$Q = \frac{9}{d_{ave} \times \rho_{pr}} \text{ m}^3/\text{t}$$

This is also a two-stage process with about four-fifths recirculated first and one-fifth fresh water later.

Normally, sprays are selected for coarser-coal applications (i.e. baths) where surface area is lower and water quantities low. Shower boxes are recommended for small-coal applications.

6.6 TYPES OF DENSE-MEDIUM SEPARATORS

We have already seen that commercially used dense-medium separators can be subdivided into various types and categories:

(a) water-based medium; and
(b) air-based medium

and that the water-based medium types can be categorized into:

(i) dense-medium vessels, or baths as they are more commonly called; and
(ii) dense-medium cyclones, of which there are several varieties in addition to the cylindroconical hydrocyclone shape.

6.6.1 Dense-medium Baths

There are two main categories of dense-medium baths; deep baths and shallow baths. However, both types have many common features. Float products are removed from the top of the baths, usually by paddles, or by the natural flow of the medium. Discard removal varies from one type to another. Dense-medium baths usually treat coal in the 300 mm × 6 mm size-range.

There have been many proprietary designs of dense-medium baths, particularly since the use of magnetite as a medium became internationally adopted. However, because this chapter is not intended to provide an historical review of their development, the types which are described are those which are currently widely used in several different countries. The published literature,[25,26] will provide further details of other specific types.

The two types can be readily differentiated as follows.

Deep baths have the advantages of a fairly quiescent pool area and long residence times, and are consequently less prone to gravity changes by accidental addition of surplus water. Several deep baths incorporate a second stage by which middlings may be extracted. Baths in this category are the Barvoys, the

DENSE-MEDIUM SEPARATION

Tromp Deep Bath and the Chance Cone. As we have seen, the Barvoys was originally designed to operate on barytes or shale but has since been converted to magnetite use. The discard is removed by a bottom elevator and middlings are removed by a hydraulic middlings tube. The Tromp Deep Bath now also uses magnetite, and the discard and the middlings are both removed by elevator. The Chance Cone, which was the original dense-medium separator, uses relatively coarse sand as a medium but it is not a true dense-medium separator as it uses upward currents (elutriation principle) to stabilize the medium. Although there are many Chance Cone separators still in operation throughout the world, no new installations have been built for many years. The last unit was probably installed in the early 1960s.

Shallow baths do not require as much medium in circulation and take up considerably less area. There are three subcategories within this group. The first category is wedge- or dish-shaped baths from which the discard is removed by conveyor belt within the bath (as for the Ridley–Scholes); or a flight conveyor (e.g. Tromp and Daniels) or a common scraper removes floats and on its return, discard (e.g. Dutch State Mines shallow bath). The second category is the circular-pool type bath where the discard is removed by a vertical or inclined wheel (e.g. Drewboy and Norwalt). The third category is the horizontally mounted drum or disc in which the discard is picked up by lifters on the shell and expelled from the vessel by a chute (e.g. Wemco and Teska).

Fig. 6.31   Two-stage Teska vessel installation (courtesy, K. H. D. Humboldt-Wedag).

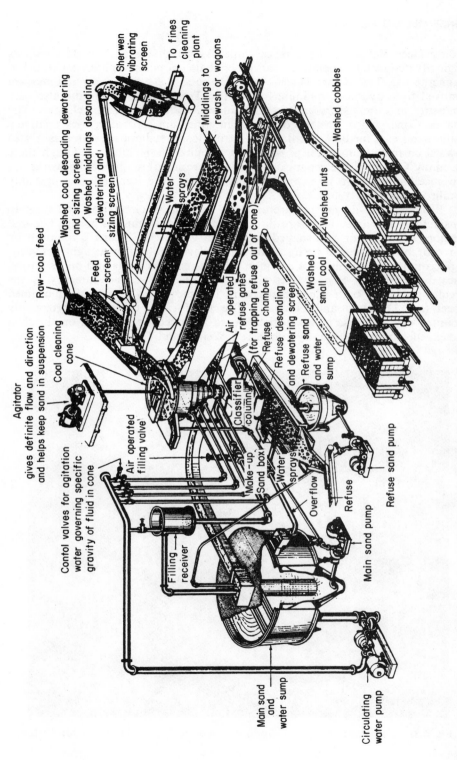

Fig. 6.32 Pictorial layout of the Chance system[27].

DENSE-MEDIUM SEPARATION

In general, the recent trend has been towards shallow baths because of the obvious savings in space and the smaller medium volume required within the circuits. In the early days, deep baths were preferred because medium-density control was much slower in response to attempts by the operators to change it and coals were often relatively easily cleaned.

Many of the earlier deep-bath designs were successfully modified to obtain three-product separators (i.e. low-ash coal, middlings and discard) and although some shallow-bath separators can be modified to obtain three products (e.g. Tromp shallow bath and Wemco drum) it is more common to employ two separate vessels, as shown in Fig. 6.31 which features two Teska baths.

Chance cone A pictorial layout of an original installation of the Chance system is shown in Fig. 6.32.[27] The cone which is the separating vessel is conical in shape with water inlets at various levels to ensure an even distribution throughout its depth. Each water inlet has its own density-control valve. A slow-moving agitator imparts a rotating motion to the medium to enable the floated coal to be carried from the feed point to the overflow weir and hence to the desanding screen. At the back of the cone are two air-operated slide gates, in between which is mounted a refuse chamber for receiving the sinking discards. These gates are automatically controlled for a cycle time of 15–20 s. During this time, the upper gate closes and the lower gate opens discharging the discards. It then, in turn, closes again and the chamber is filled with water before the upper gate opens again to collect more discards.

Both the cleaned coal and refuse products pass to screens for desanding and

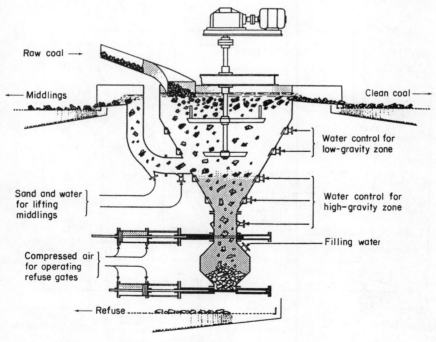

Fig. 6.33 Three-product Chance cone separator.[27]

dewatering and these drainage products are then transported by a launder to the main sand sump. This sump permits the sand to settle and therefore densify, and the settled dense slurry is then pumped back, via the control valves, to the cone water inlets. The main sand sump also acts as a storage reservoir for excess sand to act as a buffer against losses and also to provide for density adjustment. Sand losses in operating this process are usually between 0.5 and 1 kg/t, and because this is a relatively small quantity, topping up is normally only required once per shift. Figure 6.33 shows a three-product version still in use today, which incorporates a middlings column connected to the side of the cone to extract higher ash-content coal 'teetering' in the mid-section of the vessel. Ecart probable moyen (epm) values typical for the two-product separator operating at a separating density of 1.55 with 100 × 9 mm raw coal lie in the range 0.025–0.033. (Chapter 8 deals with efficiency testing of gravity concentrators, for those who are unfamiliar with the above terminology.) The capacity of Chance cones is about 20 t/h/m² of pool area which, for a five-metre diameter unit represents an upper limit of about 350 t/h (practical capacity).

Barvoys separating bath By the late 1950s this separator had become one of the most widely used dense-medium baths for coal cleaning. It was the first commercial 'static' bath, that is to say that the medium remained stable without the need for upward-current water flows. The medium initially used was ground barytés stabilized to some extent by the addition of clays or by a natural clay build-up from the raw coal. Later installations used magnetite. The separator incorporated a middlings tube in most installations, and the middlings product was either sold as a product, crushed and passed to the small-coal circuit, or

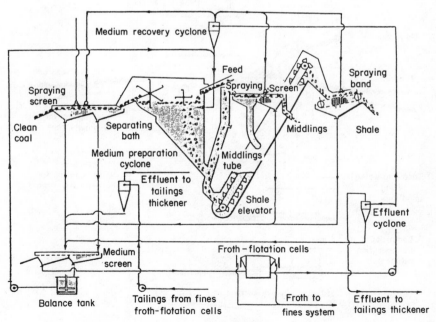

Fig. 6.34 The original Barvoys separating bath.[28]

DENSE-MEDIUM SEPARATION

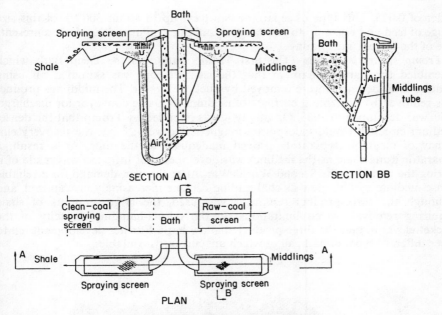

Fig. 6.35 Small-coal Barvoys washer.

added to the rejects. Figure 6.34 shows the original barytes/clay medium system later to be replaced by magnetite.[28] The cleaned coal is paddled over the end of the bath to the drain-and-rinse screen. The discard sinks to the bottom of the vessel into the boot of a bucket elevator to be elevated to the discard drain-and-rinse screen. Medium drains back into the vessel, from the elevator, and that which drains from the initial section of each screen gravitates to the balance tank and is pumped back to the separation vessel. The weir at the top of the middlings tube is at a lower level than that within the shale elevator, which causes an upward current of medium. This in turn raises the separation relative density of the medium in the bottom of the vessel causing mid-range material to be lifted up the middlings tube. The middlings product also passes to a drain-and-rinse screen. Density adjustment for the middlings cut-point is achieved by altering the weir height. This type of unit has been effectively used for cleaning coal sized from 200 mm to as small as 3 mm, but a special small-coal washer was developed for treating 12 × 3 mm material. The unit, as shown in Fig. 6.35, featured different product-removal systems. The clean coal is gently removed from the bath by paddle as with the coarse-coal machine, but the heavier products sinking in the vessel are subjected to an upward current in the middlings tube, created by an air lift which both lifts and separates the intermediate product. Quality regulation is by variation of the strength of this upward current. The final discard may be separated either by a second air-lift tube or by a bucket elevator. Up until the development of the dense-medium cyclone this small-coal Barvoys separating vessel was highly regarded.

Ecart probable moyen (epm) values typical of the coarse-coal Barvoys bath operating at a separating density of 1.55 with 100 × 9 mm raw coal are in the order of 0.025. This type of separator can treat up to about 300 t/h of this size range of feed and reject up to 160 t/h and is about 4 m in width, which represents one of the largest units built.

Tromp separating baths The first Tromp bath was the Deep Bath which resembled the Barvoys bath in that the cleaned coal was skimmed off using paddles and the rejects were removed by bucket elevator. The middlings product was removed by horizontal currents of medium to a side conveyor for discharge. This was developed in 1933. Originally it was intended by Tromp that the dense-medium circuitry would incorporate magnetic recovery,[28] but the recovery efficiency of magnetic separators proved inadequate at the time. As a result, a separator using shale as the medium was developed, but little use was made of it during the period of the Second World War. After the war, demand for a reliable dense-medium system grew as coal-mining became increasingly mechanized, and although the three-product separator was used, the large amounts of shale requiring removal placed limitations on the rejects-handling capacity of the bucket elevator, and the three-product shallow bath. Further development led to three different types of bath, all of which are currently available:

Fig. 6.36 (a) Tromp two-product wheel-type shallow bath (courtesy Birtley). (b) Tromp three-product dense-medium vessel (courtesy McNally).

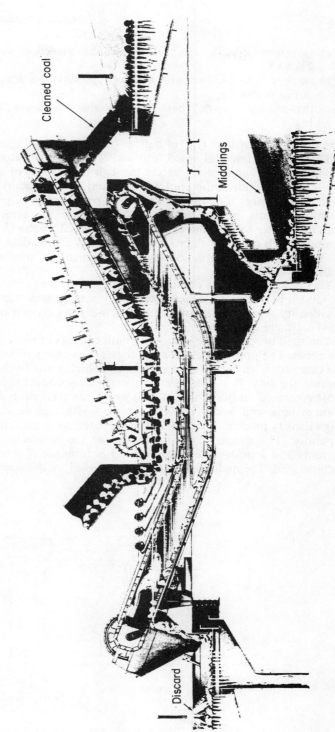

Fig. 6.36(b)

(a) two-product with a wheel type sinks-extraction system, as shown in Fig.6.35;
(b) two- or three-product with a chain type sinks-extraction system, as shown in Fig. 6.36(a, b);
(c) two-product 'static' bath with a sinks-discharge tube, as shown in Fig. 6.37.

The two-product wheel type of bath is capable of treating coal sized 250–5 mm and is a semicyclindrical vessel into which medium is introduced at a predetermined density. The medium enters through the nozzles of a uniform horizontal flow over the width of the pool. Float material is carried by the flow of medium over the discharge weir onto a drain-and-rinse screen. A paddle, located at the discharge points, assists the larger lumps onto the screen. Sinking material descends to the base of the vessel and is removed by the slowly revolving wheel equipped with conveyor flights which gradually lift the discard, allowing medium to drain through grills in the flights. All remaining medium is removed by a drain-and-rinse screen.

This type of washer is obtainable in sizes up to 3 m in width and can treat up to 450 t/h per unit. It has been commonly used in two-stage separations, with a low-density separation followed by a high-density separation. Density ranges of 1.3–1.8 r.d. are attainable.

The shallow, chain type separating bath consists of two superimposed separating zones. The upper zone has a lower relative density of separation which floats the clean coal, while the lower zone has the higher end of the density gradient to capture the sink. It is therefore a natural three-product separating vessel and is most commonly employed as such. Low-density medium is fed to the vessel from a surge tank and is distributed over the width and depth of the upper zone. High-density medium is continuously prepared by recirculation from a magnetic separator. The cleaned coal is removed by the combined action of horizontal currents and a push-plate conveyor and is transported from the vessel over an inclined grid. The middlings product is removed by the combined action of hori-

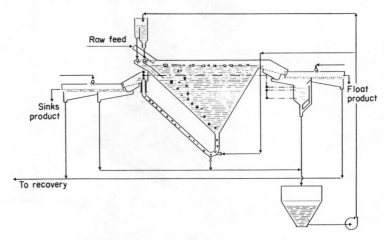

Fig. 6.37 Tromp 'static' bath (courtesy Birtley).

DENSE-MEDIUM SEPARATION

zontal currents and the upper strand of a drag conveyor and is also removed via an inclined grid. The rejects product is raised clear of the bath medium by the lower strand of the drag conveyor.

This type of washer has produced exceptionally high separating efficiencies for a three-product separation and is renowned for its low magnetic losses. It can be used over a wide range of separating densities: 1.3–1.9 r.d. Capacities of up to 450 t/h are possible for a bath of three metres in width.

The Tromp static bath behaves in a similar way to laboratory float–sink apparatus. The depth of medium required to conduct the separation (especially for most coarser coals) is probably less than 0.5 m, and providing the medium remains stable and relatively undisturbed, this thin layer on the surface of a separator can be regarded as the actual static separation zone. This phenomenon led to the development of the shallow-bath separator, but most use the fluid velocity to maintain medium stability and to transport the float product. Tromp designed the static bath specifically to utilize the static separation zone. The continuous renewal of the medium below this static zone and replacement of this by correct-density medium automatically ensures that the medium within the zone is also at the correct density. The floating coal is gently carried away by the steady surface current to an overflow weir. The sinks are withdrawn from the bottom of the vessel and are lifted through a discharge tube by a current of medium.

The medium used for creating the surface flow is taken from an orifice-controlled outlet of a constant-head feed tank, whilst an orifice of the same size controls the outlet of the receiver for the medium drained from the floats.

The head of the medium in this receiver is maintained at the same level as that in the feed tank. This ensures that the volume of medium to the vessel balances that flowing over the weir.

This type of bath is capable of exceptionally high accuracy. Ecart probable moyen values of below 0.01 have been reported for coal sized from an upper limit of 80–3 mm. Bath width varies with the amount of floats requiring removal, but units of up to 6 m in length are operating treating up to 200 t/h.

Table 6.6 gives washing performance for all three types of Tromp separating bath.

Wemco separating baths The cone separator was originally designed for coal applications but has found wide application in the minerals-dressing field as a preconcentrating vessel. Sink fractions are removed by pump or by external or internal air lift (more common for coal shales) as shown in Fig. 6.38(a). Cleaned coal overflows a weir and both cleaned coal and discard pass to drain-and-rinse

TABLE 6.6 Washing Performance of Tromp Separating Baths

Tromp separating bath	Partition density	Expected Epm value		
		150 × 50 mm	50 × 25 mm	25 × 10 mm
Two-product wheel unit	1.4	0.014	0.017	0.025
	1.7	0.022	0.024	0.033
Three-product chair unit	Primary 1.4	0.011	0.015	0.020
	Secondary 1.7	0.022	0.024	0.033
Two-producer static unit	1.4	—	0.012	0.018
	1.7	—	0.020	0.025

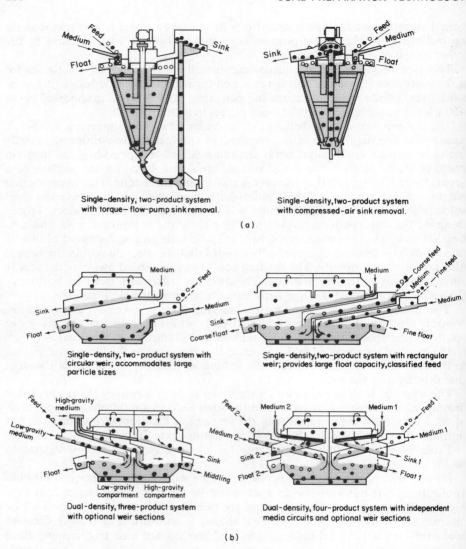

Fig. 6.38 Wemco separating baths: (a) Wemco cones separator types; (b) Wemco drum separator types.

screens for medium recovery by conventional systems. Cones with diameter of up to 6 m and capable of handling up to 400 t/h are available.

The drum separator has in recent years become one of the most widely used dense-medium baths. It is capable of highly efficient sharp separations and is simple in operation and easily maintained. Various types of cleaning system are possible, as shown in Fig. 6.38(b) and all but the four-product system are currently in use for coal cleaning. The raw coal and medium enter at one end of the drum, and clean coal floats and flows over a weir at the discharge end where it

DENSE-MEDIUM SEPARATION

Fig. 6.39 Wemco drum installation (courtesy Amcoal).

passes to a drain-and-rinse screen. The sinks are lifted from the bottom of the medium pool by lifters fixed to the inside of the drum shell. This product drops into a central, suspended trough in which a current of medium flushes the discard onto a drain-and-rinse screen. The three-product arrangement involves another separating compartment, as shown in Fig. 6.38 (b). Floating coal is prevented from being collected by the revolving lifters by means of skirt plates which form a central flow-channel through the bath.

Two-product drums range in diameter from 2 to 4 m and in length from 3.35 to 6 m. Capacities of up to about 400 t/h are possible for the single-overflow unit. Three-product drums also range in diameter from 2 to 4 m but length may be up to 6.5 m and capacities of up to 325 t/h are possible.

Despite the relatively large unit volume of this type of separator the layout inside the plant can be economic in terms of space utilization, as shown in Fig. 6.39.

Table 6.7 gives typical performance guarantees for a Wemco two-product drum cleaning 150 × 10 mm raw coal, based on a pool loading of 20 t/m²/h.

Drewboy and Teska separating baths Both of these baths involve the use of a wheel for sinks removal, but the basic mode of operation is similar to that described for the Tromp and Wemco separators.

The Drewboy separator is a French-designed washer capable of cleaning raw coal sized from 800 to 6 mm at capacities of up to 1000 t/h. The separator vessel widths range from 0.500 to 5 m with corresponding wheel diameter of 2.5–8.0 m.

TABLE 6.7 Washing Performance for a Wemco Drum Separating Bath

Partition density kg/litre	1.4	1.5	1.6	1.7	1.8
Epm	0.044	0.052	0.060	0.068	0.076

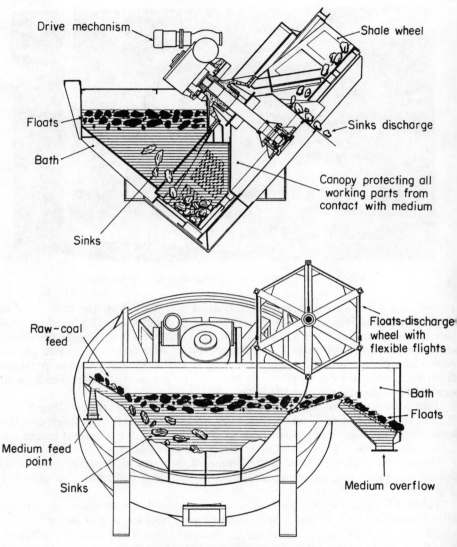

Fig. 6.40 Drewboy separating bath.

Figure 6.40 shows the basic features of the unit. The inclined-wheel unit is most commonly employed for coal cleaning but a vertical-wheel type is also available for treating smaller coal (120 × 6 mm) with capacity of up to 100 t/h.

The Teska separator (Fig. 6.41) is a German-designed washer capable of cleaning raw coal sized from 250 to 5 mm. The largest machines are 3-m wide bath, 6.5-m diameter × 1.5-m wide wheel capable of handling up to 430 t/h of discard or 750 t/h of cleaned coal.

Both the Drewboy and Teska vessels have performance characteristics similar to those given in Table 6.7 for the Wemco drum.

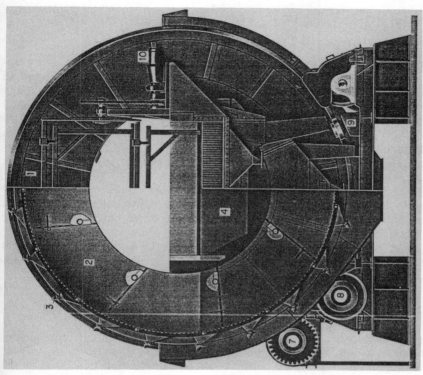
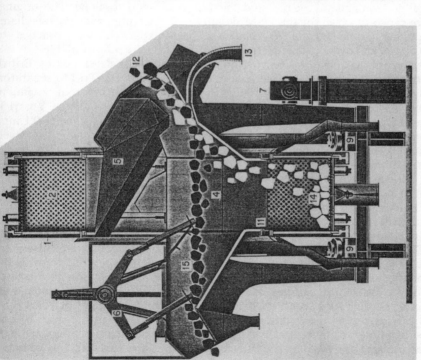

Fig. 6.41 Teska separating bath. 1, Bucket wheel for discharge of sinks; 2, buckets; 3, nozzles; 4, separating compartment; 5, sinks-discharge chute; 6, floats-discharge device; 7, bucket-wheel drive; 8, support; 9, thrust rollers; 10, drive for the floats-discharge device; 11, pneumatically clamped sealing ledge; 12, coal to be washed; 13, dense medium; 14, sinks; 15, floats.

6.6.2 Dense-medium Cyclones

Much has already been said of the vast contribution made by the coal preparation engineers of the Dutch State Mines, but their greatest contribution of all has undoubtedly been the dense-medium cyclone process. The original design manual for the process, first produced in the late 1950s, contained details of all facets of its design and technology and formed the basis for the design of all installed DSM cyclone systems world-wide. By the mid-1980s over 20 companies in 12 different countries were appointed licencees for the Stamicarbon dense medium separation system and over 300 plants had been built incorporating the system in 24 countries.[29]

While many of the design concepts and associated data remain proprietary information of Stamicarbon, several formulae and sizing calculation methods have now been adopted for general coal-preparation usage. Several of these have already been referred to in this chapter.

The DSM work pioneered the development of the familiar cylindriconical cyclone separator for sorting coal from waste, but other forms of cyclone separator have also emerged. All such forms can be categorized as follows:

(a) cylindroconical;
(b) cylindrical.

Category (a) includes the original DSM design and other similar types including McNally cycloids, Krebs cyclone and the Kilborn cyclone. It also includes other developments utilizing the conventional hydrocyclone profile, such as the Chinese electromagnetic cyclone and the Japanese 'swirl' cyclone. The original feed size-range specified for the DSM design of 10×0.5 mm, with cyclone diameter ranging from 200 to 700 mm, has been expanded with practice to 50×0.5 mm, and cyclone diameters of up to 1000 mm are now available.

Category (b) includes the Vorsyl separator designed and developed by British Coal; the Dynawhirlpool, an American development; and the Tri-Flo separator, an Italian-designed two-stage unit. These three units all treat raw coal ranging in size from 25 to 0.5 mm, although the latest-developed version of the Vorsyl is capable of efficient cleaning of coal from as coarse as 50 mm to as fine as 0.1 mm. Barrel diameters of up to 720 mm are available. One other British Coal development has placed another type of separator in this category. This is the LARCODEMS separator (an acronym for LARge COal DEnse Medium separator) which was specifically developed for treating coal sized 100×0.5 mm. It is similar in concept to the Vorsyl, with a 1.2-m diameter barrel and 3.6 m in length, which can treat up to 250 t/h. Instead of being installed vertically it is inclined at 30° to the horizontal.

DSM cyclone separator A schematic drawing of the separator is shown in Fig 6.42. The feed comprising raw coal and medium is introduced at a precise pressure into the tangential inlet. The ensuing flow is rapid and spiralling towards the apex of the unit, and in the core of the cyclone a very fast flow-rate creates centrifugal classification causing shale to move outwards towards the inner wall of the conical shell. As a result, shale is discharged from the spigot or nozzle and coal is carried by the rising internal spiral towards the vortex finder to be discharged from the overflow. Because of the centrifugal acceleration effect, the particles comprising the medium are affected as well as the feed and this causes

DENSE-MEDIUM SEPARATION

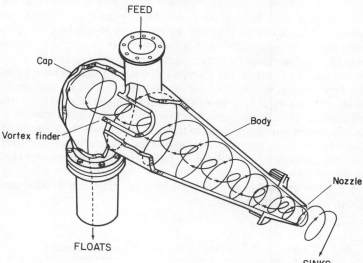

Fig. 6.42 DSM cyclone separator (courtesy Stamicarbon b.v.).

gradual increase in relative density of the medium towards the apex. Therefore, the relative density of separation in a dense-medium cyclone will always be higher than that of the feed. Hence, material having a relative density higher than that of separation moves towards the inner wall of the cyclone and follows a non-tangential path along this wall to be discharged through the apex opening. Material having a lower relative density is forced towards the cyclone centre and

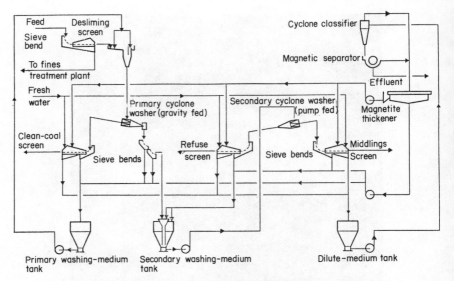

Fig. 6.43 Stamicarbon dense-medium cyclone plant, for coal.

transported to the overflow opening by the strong non-tangential current. Only the very heavy and very light grains pass directly to their respective openings Others are forced to pass into internal recirculation which leads to good separation and results in the characteristic high accuracy of separation achievable with these units. The cyclone may be either pump fed or gravity fed. In the case of the former, raw coal and medium are pumped directly to the cyclone from a mixing box of the type shown in Fig. 6.43 for a secondary stage of cyclone separation. Alternatively, the raw coal may be gravity fed, as in the case of the primary stage in the same figure. To achieve the correct feed pressure for the gravity-fed system DSM has recommended that a head of at least nine times the cyclone diameter, d, be used, i.e. the distance from the overflow level of the mixing box to the inlet of the cyclone should be $9 \times d$. Figure 6.44 and Table 6.8 give commonly used data regarding the gravity-fed cyclone.

Table 6.9 gives typical washing performance data for a 600-mm diameter cyclone cleaning 10×0.6 mm coal.

TABLE 6.8 Dense-medium Cyclone Capacity

Cyclone diameter (mm)	Medium flow (m^3/h)	Product flow		Minimum head (mm)	Recommended head (mm)
		Feed (t/h)	Sinks (t/h)		
200	15	6.5	4.0	1800	2000
350	50	20.0	12.5	3150	3500
500	125	50.0	30.0	4500	5500
600	185	75.0	45.0	5400	6500
700	270	105.0	65.0	6300	7500
750	300	120.0	75.0	6750	9000

DENSE-MEDIUM SEPARATION

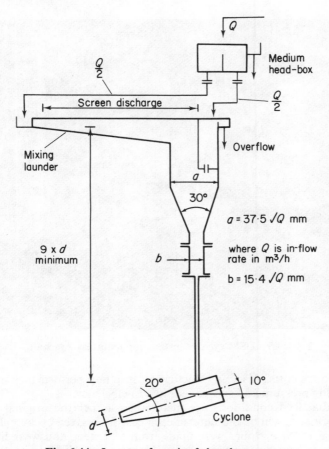

Fig. 6.44 Layout of gravity-fed cyclone system.

Gravity-fed cyclones are most suited to friable raw coal likely to be excessively broken down by pumping. Despite the extra height required in installing these systems, the pressure head, being built in is unaffected by pump performance and remains more or less constant as long as an overflow is maintained at the mixing box. Figure 6.45 shows a typical installation.

Pump-fed cyclones are simpler in terms of plant layout and involve a lower plant profile because the desliming screen need not be installed above the level of the cyclone. Although they are generally cheaper in capital cost, they may prove more expensive to operate due to greater wear and tear on pumps and pipeline

TABLE 6.9 Washing Performance for a 600-mm Diameter Dense-medium Cyclone Separator (10 × 0.5 mm raw coal)

Partition density (kg/litre)	1.4	1.5	1.6	1.7	1.8
Epm	0.075	0.084	0.070	0.097	0.106

Fig. 6.45 DSM cyclone installation (courtesy Amcoal).

and added power costs. Also, any fluctuation in pump performance may lead to lower operating pressure and subsequent loss in efficiency.

Provided that feed conditions are maintained as per the original design the cyclone performance will be the same in each case for a given cyclone diameter.

The original DSM cyclones were made from cast steel, and wear life, particularly at the apex, was quite severe. About 15 years ago Ni-Hard casting became available and component life was significantly increased (five- to ten-fold). Most recently, ceramic materials have been introduced which can be cast into suitable replaceable sections for insertion into the cyclone or else bonded into position as pre-formed tiles, or both. This latter option, utilizing high-alumina ceramic materials, has resulted in a further ten-fold life increase. Cyclone life being prolonged in this manner, the performance characteristics also became more stable and such units can be expected to operate for long periods at peak efficiency. The difference in price between Ni-Hard and the best ceramics is a factor of about 4 to 5. Figure 6.46 shows a cyclone design which features interchangeable Ni-Hard or ceramic components.

The Bretby Vorsyl separator This dense-medium separator, like the cyclone, relies on centrifugal force, but is vertically inclined with a tangential feed-inlet, as shown in Fig. 6.47. It is usually gravity fed from a suitable mixing or feed box located adjacent to the separator. The vortex finder passes up through the body of the separator from its base. As the separation commences, clean coal follows the inward flow, as in the case of the cyclone, and then takes a spiral path up the outside of the vortex finder before passing down the vortex finder to the discharge or the floats drain-and-rinse screen. The shale discard remains at the wall

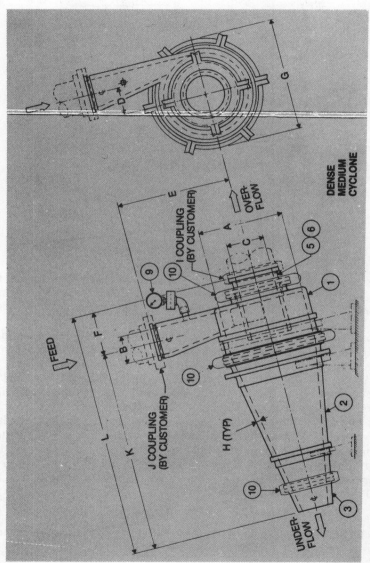

Fig. 6.46 Kilborn dense-medium cyclone with interchangeable parts (courtesy Industrial Equipment Company).

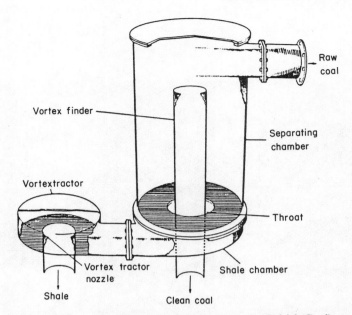

Fig. 6.47 Bretby Vorsyl separator (courtesy British Coal).

of the separator moving downwards, until at the base it is forced to move inwards to the throat before being discharged through the shale chamber and vortextractor. This device permits the shale and medium to be discharged without loss of pressure in the body of the vessel.

This separator was first developed in 1967 by the Mining Research and Development Establishment (MRDE) of British Coal for treating coal sized from 50 to 0.5 mm. The original unit was 610 mm in diameter with a feed capacity of about 75 t/h. About 50 such units were successfully installed by British Coal prior to the development of the 750-mm diameter prototype which was tested at 100 t/h feed. In 1980 it was decided to re-examine the design requirements for both the 610-mm unit and a 720-mm unit,[30] and as a result, more efficient standard units were developed, fabricated from Ni-Hard, with high-alumina ceramic nozzles to ensure maximum wear life. In 1983 the cost of the Bretby II Vorsyl was £1945 for Ni-Hard 2 and £2050 for Ni-Hard 4. The new Bretby II Vorsyl with a 610-mm diameter can treat 80 t/h and the 720-mm can treat 120 t/h. Both can clean coal up to 50 mm in size.

TABLE 6.10 Washing Performance for a 610-mm Diameter Bretby Vorsyl Separator[30]

Size-range treated (mm)	Partition density (kg/litre)						
	1.3	1.4	1.5	1.6	1.7	1.8	1.9
				Epm value			
25 × 12	0.025	0.025	0.030	0.036	0.041	0.046	0.054
12 × 0.5	0.033	0.035	0.039	0.044	0.047	0.050	0.057

DENSE-MEDIUM SEPARATION

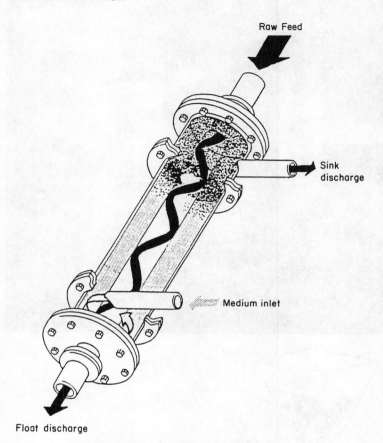

Fig. 6.48 Dynawhirlpool separator.

Table 6.10 gives typical performance figures for 610-mm Vorsyl separator. This type of separator appears to perform best when the medium contains 80–85% magnetics, and the size distribution conforms to the following:

between 105 and 75 μm—less than 4% by mass
between 75 and 50 μm —more than 6% by mass
between 50 and 10 μm —more than 9% by mass
below 10 μm   —less than 5% by mass

Dynawhirlpool and Tri-Flo separators The Dynawhirlpool separator of Minerals Separation Corporation of Arizona is used predominantly in the USA and South Africa for treating coal sized 30 × 0.5 mm. It consists of a cylinder inclined at 25° to the horizontal, as shown in Fig. 6.48. Unlike the cyclone, the supply of dense medium and the raw coal do not enter the separator together. The medium is pumped under a predetermined pressure into the lower outlet. The rotating medium creates a vortex effect throughout the cylinder and exits through either the upper tangential discharge or the lower vortex outlet, i.e. the product outlets. Raw coal entering the upper vortex tube is sluiced into the vessel by a small

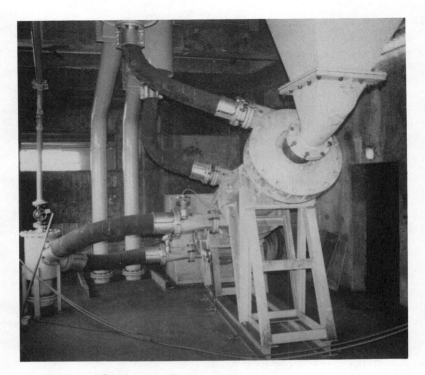
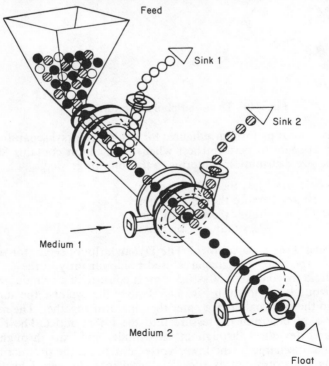

Fig. 6.49 Tri-Flo® separator.

DENSE-MEDIUM SEPARATION

quantity of medium, and a rotating motion initiates the separation. Float material passes down the vortex and does not come into contact with the inner wall of the vessel, which reduces wear potential. The sinks material penetrates the rising medium towards the outer wall of the vessel and is quickly discharged through the sinks-discharge pipe. Because the sinks discharge is close to the feed point of entry, the heaviest and most abrasive material tends to be removed first, thereby minimizing abrasive wear in the shell and discharge outlet.

Nevertheless, high-wear components are cast from Ni-Hard, and ceramic insert units have been tried to maintain high separator efficiency. Like the cyclone the Dynawhirlpool relies upon a certain amount of recirculation within the vessel to refine efficiency. Also, the tangential sinks-discharge outlet is usually connected to a flexible rubber hose, and the height of this hose may be used to adjust back pressure within the vessel to control finely the cut-point.

Dynawhirlpool units with 650-mm diameter are capable of handling up to 100 t/h. Efficiency has been stated as being similar to that reported for the DSM cyclone, but operation is simplified by the direct introduction of deslimed raw coal into the feed pipe of the separator.

The Tri-Flo® separator of Prominco, Genoa shown in Fig. 6.49 is similar in operation to the Dynawhirlpool but incorporates an integral second (re-wash) stage hence increasing efficiency if operated as a two-product separator, or else as a three-product separator[32] producing a lower-ash product from the first take-off. The economics of two-stage cyclone cleaning have been reported as being favourable[33] but for DSM-type cyclones the circuit is relatively complex, as shown earlier in Fig. 6.43. With the Tri-Flo® the circuit is much simplified and the risk of degradation from pumping the coal medium slurry is eliminated.

Both the Dynawhirlpool and Tri-Flo units save building volume by being directly fed with raw coal, as illustrated by Fig. 6.50.[32]

The recently introduced 500-mm diameter Ni-Hard Tri-Flo® unit is able to treat up to 100 t/h. The efficiency of the separator when both float products are

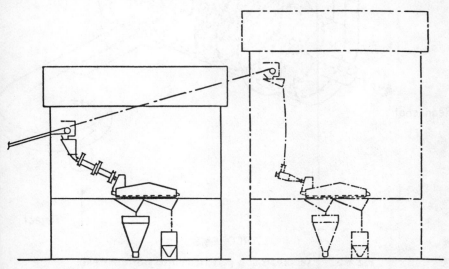

Fig. 6.50 Headroom advantage from using Tri-Flo® separators.[32]

combined exceeds that quoted for dense-medium cyclone separation in a range from 1.5 to 1.8. This is to be expected in view of the fact that the separator is in effect two cyclones in one, performing at very similar separation densities.

Other cyclone separators The *Japanese swirl cyclone* is similar in shape to the standard DSM cyclone, but its installation is inverted, the separator being vertical, with the apex discharge at the top and overflow (floats discharge) at the bottom. The vortex is stabilized by the introduction of an air pipe into the vortex which is open to the atmosphere. The height of this vent may be varied, as can the vortex finder, to act as a controlling feature. The use of coarser magnetite creates little effect on performance but means the medium consumption is lower.[34]

In China[35] recent testing with an *electromagnetic cyclone* has proved encouraging and a commercial unit is currently under trial. A three-phase alternating rotating magnetic field is induced by an electromagnet. This magnet is wrapped around the transition area between the cylindrical and the conical sections of the cyclone body. It creates an artificially high medium density from a feed concentration of the medium of as low as 1.2, and consequently medium abrasiveness is

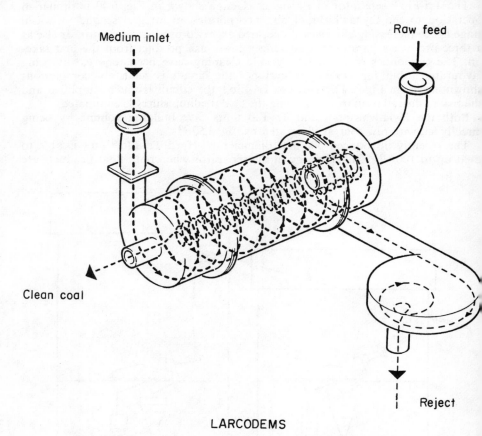

LARCODEMS

Fig. 6.51 LARCODEMS flow pattern (courtesy British Coal).

DENSE-MEDIUM SEPARATION

reduced and magnetite loss is lower. For a 500-mm diameter cyclone, epm values of about 0.05 were obtained with organic efficiencies ranging from 92 to 98% in cleaning 8 × 0.5 mm small coal.

The LARCODEMS unit is a recent development by MRDE Bretby of British Coal. The mode of separation is similar to that of the Vorsyl unit as can be seen from Fig. 6.51. The unit consists of a cylindrical chamber which is mounted at 30° to the horizontal. The feed medium is introduced under pressure either by pumping or by gravity into the chamber via an involute tangential inlet at the base of the chamber. The raw coal is introduced into the separator through an axial inlet at the top end. Clean coal is discharged through an axial outlet at the bottom, and discard reaching the top end of the chamber meets an involute tangential outlet connected to a vortextractor which carefully regulates medium exit rate. This vortextractor had already been successfully used with the Vorsyl unit but was redesigned to have an off-centre outlet to give improved discharge.

A commercial unit installed at Point of Ayr colliery treats 100-mm top-sized raw coal from which the fines have been partially removed by rotating probability screen. The raw coal, deslimed at 0.5 mm, is gravity fed to a 1200-mm LARCODEMS unit (see Fig. 6.52). The medium-control and magnetic-recovery systems are of conventional design and medium is fed to the separator from a headbox located 10.8 m above the inlet to ensure a constant feed pressure. The unit is cast in Ni-Hard 4 with high-chrome cast-iron wear components. The plant treats 250 t/h, of which up to 50% can be rejected. Volumetric capacity is 800 m³/h.

The cost of this plant has been estimated as being below 70% of an equivalent Wemco drum plant capable of only partial cleaning from 100 to 5 mm. Performance of the LARCODEMS has been reported as being equivalent to the tran-

Fig. 6.52 LARCODEMS installation in Britain (courtesy British Coal).

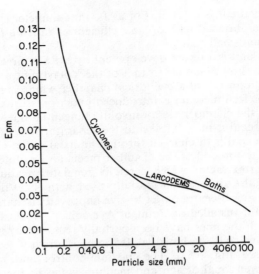

Fig. 6.53 Effect of grain size on the efficiency of dense-medium separators.

sition between dense-medium baths and cyclones, as typified by the graph in Fig. 6.53. Table 6.11 shows results obtained from tests with a prototype unit.

6.7 DENSE-MEDIUM CYCLONE CLEANING OF FINE COAL

Cleaning fine coal in dense-medium cyclones is becoming increasingly attractive for the following reasons:

(a) sharpness of cut compared to other gravity methods and ability to separate oxidized or weathered fine coal;
(b) increasing amount of fines resulting from greater mechanization in mining and handling;
(c) need to eliminate maximum amount of pyritic sulphur;
(d) more-favourable selling price of coal encouraging optimization in recovery;
(e) improved performance of mechanical coal-dewatering equipment.

TABLE 6.11 Washing Performance for a Prototype LARCODEMS Separator[36]

	Size fraction (mm)			
	25 × 0.5	25 × 4	4 × 2	2 × 0.5
Partition density, d_p (kg/litre)	1.436	1.434	1.445	1.490
Epm	0.039	0.035	0.040	0.055
Raw-coal ash (%)	28.42	17.60	33.12	52.98
Clean-coal yield (%)	62.00	74.57	56.46	33.94
Clean coal ash (%)	6.16	6.25	5.54	6.71

DENSE-MEDIUM SEPARATION

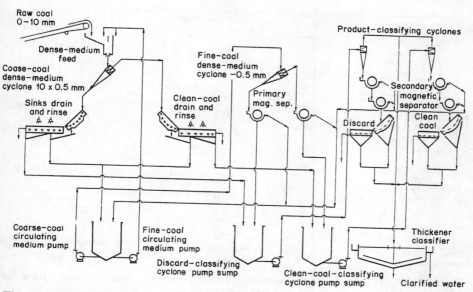

Fig. 6.54 DSM-Stamicarbon/Evence Coppée dense-medium combined circuit for 10 × 0 mm coal (as employed at Tertre and Winterslag, Belgium).

The technology of dense-medium cyclone cleaning of fine coal was developed 25 years ago in Belgium/Netherlands when design work led to the first commercial installation at Tertre. Since then, only limited commercial development has been carried out, despite the high efficiency of the cyclone in cleaning the coal. One of the major problems has been the inefficiency of medium-recovery circuits and the lack of understanding of medium requirements.[37]

New commercial plants have been constructed in recent years in the USA and South Africa with circuits very similar to the original Evence–Coppée/Stamicarbon 1957 design used at Tertre and Winterslag in Belgium.

The first dense-medium cyclone plant in which fine-coal cleaning was attempted was the Tertre plant in Belgium, for which the flowsheet is given in Fig. 6.54. This flowsheet was developed in 1957 by DSM-Stamicarbon and Evence Coppée. It is described[6] as having exhibited very good separation results, but it was another nine years before the next plant—that at Winterslag, also in Belgium—was built. Winterslag was designed with the same flowsheet basis as Tertre and remained in operation, apparently performing as satisfactorily as Tertre, for 17 years.

For Winterslag, it was properly intended that the medium-recovery system be divided as with small-coal circuits, into clean coal and discard sections. In effect, this flowsheet has small- and fine-coal cyclone-cleaning sections but circulating medium is supplied from one sump, thereby sacrificing some control over the separation density of the medium.

It is possible to treat small and fine coal together in a single cyclone, i.e. minus 25 mm, but this cannot provide for the highest potential cleaning efficiency—which might otherwise be accomplished with divided systems—and it requires a higher feed pressure in order to ensure acceptable separation efficiency for the

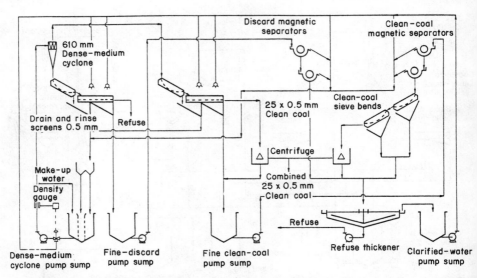

Fig. 6.55 DSM–Stamicarbon/Roberts and Schaefer dense-medium cyclone circuit for 25 × 0 mm (cleaning to zero).

fine-coal fraction. This is achieved by pump-feeding the coal/medium slurry to the cyclone. The biggest single disadvantage of this system is that there is no separate control over the relative density of the two size-fractions, and compromise is required to serve the needs of the fine coal, thereby making some sacrifice in the wear and tear of pumps, pipes and cyclone as a result of the higher pressure. At least three plants have been built in the USA, for which the flowsheet

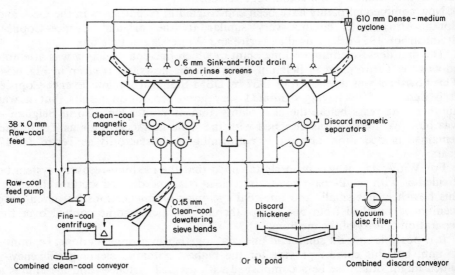

Fig. 6.56 Childress dense-medium cyclone circuit for 38 × 0 mm coal (source Ref. 42).

DENSE-MEDIUM SEPARATION

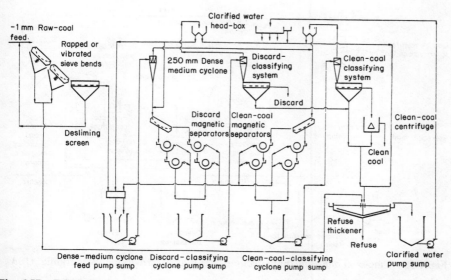

Fig. 6.57 DSM–Stamicarbon dense-medium cyclone circuit for 1 × 0 mm fine coal (after Ref. 6).

given in Fig. 6.55 was developed by DSM–Stamicarbon and Roberts and Schaefer. The performance of these plants has been the subject of a number of papers.[38-40] Another, similar, approach has been reported which was developed by the Childress Corporation and is shown in the flowsheet given in Fig. 6.56.[41,42]

Where gravity sieve-bends were originally used for screening down to 0.15 mm in the Belgium plants, more recent use of rapped or vibrated sieve-bends, or alternatively, Derrick classifying screens, has made it possible to screen down to 0.075 mm without causing blinding or severely limiting the distribution of slurry across the screen.

The circuit currently favoured by DSM–Stamicarbon involves two parallel stages of magnetic separators treating both underflow and overflow on rapped or vibrated sieve-bends cutting at approximately 0.1–0.15 mm, as shown in Fig. 6.57.

This has permitted the extension of the washing range of fine-coal cyclones from 0.5 × 0.15 mm to 0.5 × 0.075 mm, while at the same time accommodating more effective medium recovery.

Two plants in the USA employ Derrick screens for feed classification. Homer City in Pennsylvania, which treats 1 × 0.15 mm coal in dense-medium cyclones, and Marrowbone in West Virginia which treats 0.5 × 0.1 mm coal. At Homer City, medium recovery is initiated by using vibrated sieve-bends, while at Marrowbone, Derrick screens are used, as shown in Fig. 6.58. These plants have been described in numerous technical papers.[43,44]

In South Africa, realizing the potential of the dense-medium process for cleaning fine coal and the need to adapt this process to suit local conditions, the National Centre for Coal Research (NCCR) designed and built a 5 t/h pilot plant which was commissioned in early 1977. South African coals, especially those of

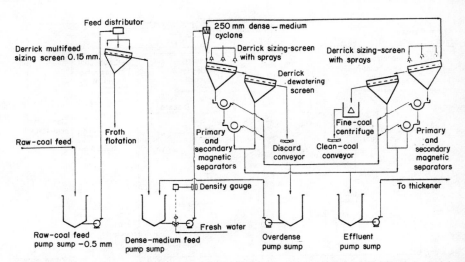

Fig. 6.58 Fine-coal dense-medium cyclone circuit at Marrowbone, W. Virginia, USA.

the Transvaal which have featured in the recent major expansion of South Africa's coal industry, are characterized by extremely difficult washability characteristics. Because of this, separation efficiency must be high in order to maintain the required product quality. The initial work of the NCCR in the early 1970s centred around established fine-coal cleaning systems, including froth flotation, tables and water-only cyclones. Only marginal success was achieved at best and it was decided to build a dense-medium cyclone pilot plant to determine the design and operational requirements of a commercial installation. The NCCR

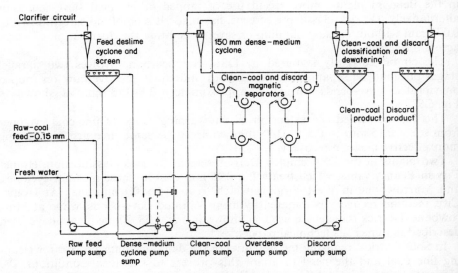

Fig. 6.59 National Centre for Coal Research of South Africa; 5 t/h, fine coal, dense-medium cyclone pilot-plant circuit.

DENSE-MEDIUM SEPARATION

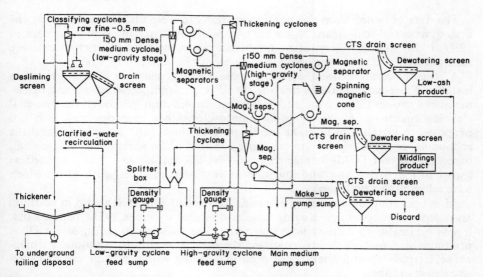

Fig. 6.60 Flowsheet suggested for the Greenside fine-coal, dense-medium cyclone circuit.[51]

circuit, which differs from that originally developed by DSM–Stamicarbon, is shown in Fig. 6.59.

The NCCR pilot plant[45,46] was tested with numerous Witbank area coals and several varieties of magnetite for over a period of approximately two years. This test work culminated in a 70-hour continuous test programme from which detailed test results were obtained.[47]

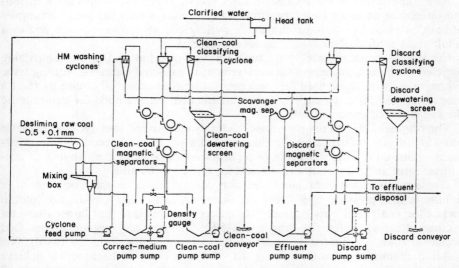

Fig. 6.61 Typical dense-medium cyclone circuit for deslimed 0.5 mm coal (after Lurie, 1978, in Ref. 7).

The data obtained from this continuous test run were utilized in designing the first commercial-scale plant to be built in South Africa. This was completed in 1980 at Greenside Colliery. Some modifications to the pilot-plant circuitry were incorporated into the Greenside flowsheet, as can be seen in Fig. 6.60.

A recently commissioned, 1300 t/h dense-medium plant in South Africa has installed dense-medium baths treating the 75 × 12.5 mm size-fraction, and dense-medium cyclones treating the 12.5 × 0.5 mm size-fraction. Although the fine-coal cleaning circuit employing dense-medium cyclones has been designed, it is not proposed to install it until plant conditions have been fully ascertained, and the economics of installing the circuit can be shown to be acceptable. This circuit, based on current DSM–Stamicarbon and NCCR technology, was proposed as shown in Fig. 6.61. This plant now joins at least two others in South Africa which have adopted this philosophy.

After operating the two early plants at Tertre and Winterslag, and in addition an independent system employing smaller-diameter cyclones, DSM realized that it was preferable to employ finer magnetite. The subsequent work, of the DSM produced size-analysis charts indicating the preferable size-distributions for magnetite, ferrosilicon and mixtures of both in cleaning within certain size and relative density ranges.

Work carried out at the pilot plant of the National Centre for Coal Research (NCCR), South Africa,[48] highlighted the importance of grinding the medium finer in order to achieve the best operational performance of the cyclone. This is shown in the size-distribution chart given in Fig. 6.62. The NCCR stated that for sharp separations, at least 50% of the magnetite should be finer than 10 μm. However, if the magnetite is milled too fine, the losses may become excessive. With magnetite milled to the optimized specification, the losses are controlled at an average of approximately 1 kg/t of coal treated. In the 35 t/h plant built subsequently at Greenside colliery, losses of this order have been reported.

As a consequence of the NCCR findings, most South African operators, including Greenside, incorporate magnetite-milling circuits within the plant. Examples of both batch and continuous magnetite-milling circuits are in evidence and as a result, tight control of medium fineness is exercised.

Control of medium stability by both fineness of grind and by demagnetizing recycled medium is an important requirement in fine-coal cleaning. Settling tests, synonymous with those used for thickener settling-rate control, or else by the F5 index method, discussed earlier, are recommended test methods for determining the in-plant medium condition.

The original DSM development work clearly established that cyclone diameter has a significant influence on the sharpness of separation, especially in the treatment of fine coal. This sharpness of separation can be largely attributed to centrifugal force and is especially significant for fine-coal cleaning.

Dense-medium cyclones employed for coal cleaning usually operate at a minimum feed-pressure of 9 times the diameter, d, in metres of liquid column (m.l.c.) for coal sized plus 0.5 mm. Mengelers[49] points out that larger-diameter cyclones treating wider size-ranges, e.g. 50 × 0.5 mm, may require higher feed pressures than $9 \times d$ m.l.c. in order to perform as well as original DSM 500–600 mm diameter cyclones treating 10 × 0.5 mm. This is because, to achieve similar centrifugal acceleration, a higher feed pressure is necessary. Figure 6.63 was used to illustrate this point.

DENSE-MEDIUM SEPARATION

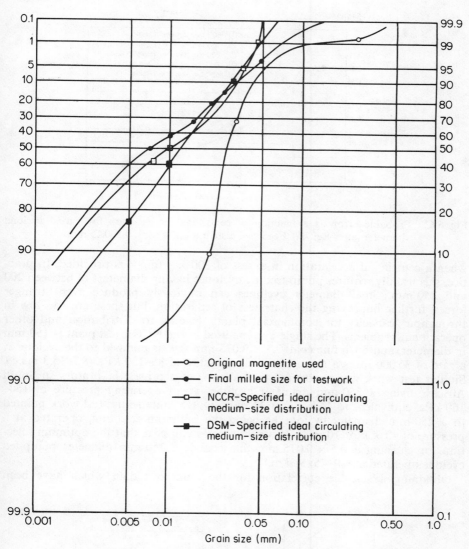

Fig. 6.62 Size-distribution chart for ground magnetite.

In cleaning fine coal the benefits of using smaller-diameter cyclones are clearly evident from Fig. 6.63, confirming work reported by Deurbrouck.[50] Dutch State Mines has long since established that by using a 500-mm cyclone to treat 10 × 0 mm coal, the efficiency obtained for the 0.5 × 1 mm size-fraction will be roughly half that obtained for the 4 × 1 mm size-fraction. It is therefore clearly apparent, that in order to treat efficiently the 0.5 × 0.15 mm fraction, cyclones of smaller diameter, operated at higher feed pressures, will be required.

The pilot-plant test work of both the DSM and the NCCR has confirmed that a suitably sharp separation of the 1 × 0.15 mm size-fraction can be achieved

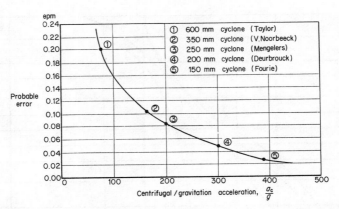

Fig. 6.63 Probable error as a function of centrifugal acceleration for varied cyclone diameter (after Ref. 49). Feed size-distribution of coal: 0.5–0.15 mm.

when a centrifugal acceleration in excess of $200 \times g$ (m/s^2) is provided. In practice, this usually requires pump-feeding cyclones having diameters of between 200 and 250 mm. Small-diameter cyclones can obviously produce even stronger forces, further improving the sharpness of separation, but these are too low in throughput capacity for commercial plants, leading to distribution and other operational problems. The single cyclone used in the NCCR test plant is 150 mm in diameter, and 5 t/h fine coal, 0.5×0.075 mm size, is pumped to the cyclone at a rate of 35 000 litres/h at an operating pressure of 85–150 kPa (8.7–15.3 m.l.c.). Similar test work was conducted by Lathioor at a private laboratory in South Africa[37] using a 250-mm diameter cyclone, at an operating pressure of 150–160 kPa, equivalent to $44.5 \times d$ or 11 m.l.c. Dutch State Mines test work pointed to a 250-mm diameter cyclone as being the optimum selection, operated at a pressure of $40 \times d$ or 10 m.l.c. It would therefore appear that the optimum selection for cleaning a 0.5×0.075 mm fine coal is a 250-mm diameter pump-fed cyclone operated at $(40-45) \times d$ m.l.c.

Obtaining a suitable correlation for the numerous data which have been

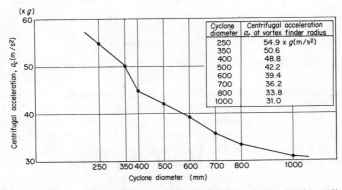

Fig. 6.64 Centrifugal acceleration at the vortex finder of various cyclone diameters (after Ref. 49). Feed pressure used was 9 × cyclone diameter metres liquid column.

reported for pilot- and full-scale, fine-coal dense-medium performance tests is not easy. Mengelers,[49] in discussing the potential pitfalls of being lured towards selecting larger-diameter cyclones for the higher unit capacities that can be achieved, used the graph in Figure 6.64 to demonstrate how potential loss in efficiency can occur.

REFERENCES

1. T. M. Chance. A new method of separating materials of different specific gravities, *Transactions American Institute Mining Engineers* **lix**, 263, 1918.
2. H. Louis. Trough washers, upward-current washers, sand flotation, Chapter 3 in *The Preparation of Coal for the Market*, Methuen, London, 1928.
3. R. H. Richards and C. E. Lock. Coal dressing, Chapter 19 in *Textbook of Ore Dressing*, McGraw Hill, New York, 3rd edn, 1940, p. 522.
4. C. Krijgsman. The Dutch State Mines cyclone-washer, Symposium on Coal Preparation, University of Leeds, November, 1952. Published by the University of Leeds, England.
5. M. G. Driessen. The use of centrifugal force for cleaning fine coal in heavy liquids and suspensions with special reference to the cyclone washer, *Journal Institute Fuel*, Dec. 1945.
6. J. Mengelers and C. Dogge. A new technique for the treatment of 0–1 mm fine coal by means of heavy media cyclones, 8th International Coal Preparation Congress, Donetsk, USSR, May 1979, Paper B1, Congress Proceedings (Ed.) I. S. Blagov.
7. R. A. Lathioor and D. G. Osborne. Dense-medium cyclone cleaning of fine coal, 2nd International Conference on Hydrocyclones, Bath, Sept. 1984 Proceedings (Ed.) L. Svarovsky, BHRA, Cranfield, England.
8. L. Svarovsky. Operating characteristics, Chapter 6 in *Hydrocyclones*, Holt Saunders, Eastbourne, England, 1984.
9. D. V. Keller, Jr. The Otisca process: an anhydrous heavy-liquid separation process for cleaning coal, in *Physical Cleaning of Coal—Present and Developing Methods*, (Ed.) Y. A. Liu, Marcel Dekker, New York, 1982, pp. 35–38.
10. T. Fraser and H. F. Yancey. American experience with the air-sand process of coal cleaning, *Coal Preparation*, **3**, 24–26, 1967.
11. M. Goransson, S. G. Butcher and J. M. Beeckmans. Coal cleaning by counter-current fluidized cascade, AGM Canadian Institute of Mining and Metals (CIMM) Toronto, April 1980.
12. W. R. Chapman and G. F. Eveson. Cleaning coal in liquid carbon dioxide, *Coal Preparation* **3**, 27, 1967.
13. L. Jonker. The development of standard procedures for the evaluation of magnetite for use in heavy-medium separation, Council for Mineral Technology (South Africa), Randburg, South Africa, Report M144, May 1984. ISBN 0 86999 656 8.
14. C. C. Graham. The application of magnetite for coal preparation in Australia, Paper 82–349, 1st International SME of AIME Meeting, Honolulu, Hawaii, September, 1982.
15. C. C. Graham and R. Lamb. *Coal Preparation—Dense-media Rheology: a Review of Measurement and Control*, ACIRL Research Report, P. R. 82–3, April 1982, Australian Coal Industry Research Laboratories, North Ryde, NSW, Australia.
16. J. Morgan and L. G. Parry. *The Nature and Properties of Magnetite*, ACIRL Research Report, P.R. 68–3, 1968, ACIRL, North Ryde, NSW, Australia.
17. L. H. Savage. *The Selection and Treatment of Magnetite for Use in Dense-Medium Coal Preparation*, ACIRL Research Report, P.R. 68–5, 1968, ACIRL, North Ryde, NSW, Australia.

18. C. Bensley. Private communication on magnetite quality in Australia, January 1986.
19. T. G. Hawker. *A Comprehensive Handbook of Permanent Magnetic Wet Drum Separators*, Boxmag-Rapid Company Publication, Birmingham, England, 1975.
20. J. G. Hervol. Swedish discovery: magnetite fines recovery, *Coal Age*, **89**, 64–67, 1984.
21. R. A. Lathioor. Private communication, July 1986.
22. Anon. *Dense-Medium Coal Preparation*, Simonacco Publication, UK, 1962.
23. Anon. *Guide to the Calculation of Cyclone Washeries and Heavy-medium Float and Sink Washeries*, Stamicarbon, Geleen, Netherlands, 1968.
24. H. H. Driessen. *Stamicarbon's Heavy-medium Process for the Separation of Minerals and Coal*, Stamicarbon, Geleen, Netherlands, Memorandum revision, January 1975.
25. E. R. Palowitch and A. W. Deurbrouck. Dense-medium separation, Part 1 of Chapter 9 in *Coal Preparation*, 4th edn, (Ed.) J. W. Leonard, AIME, New York, 1979.
26. A. M. Gaudin. Heavy-fluid separation, Chapter 11 in *Principles of Mineral Dressing*, McGraw Hill, New York, 1939.
27. Anon. The Chance sand flotation system of coal washing, University of Newcastle upon Tyne, *Kings College Mining Society Bulletin* **8**, Bulletin No. 1, Coal Cleaning No. 9, 1958.
28. Anon. *Birtley–Tromp Separating Baths*, Birtley Publication, UK, 1973.
29. Anon. *Mineral Beneficiation Processes and Equipment*, Stamicarbon, Geleen, Netherlands, January 1976.
30. S. R. Shaw. The Vorsyl dense-medium separator: some recent developments, *Mine and Quarry* **13**, 28–31, 1984.
31. Anon. *Vorsyl Separator*, Mining Research and Development Establishment (MRDE) Information, British Coal, February 1979.
32. H. J. Ruff. New Developments in dynamic dense-medium systems, *Mine and Quarry* **13**, 24–28, 1984.
33. J. Abbott, K. J. A. Fawbert, J. Hillmand and G. S. Smith. The economics of secondary separation with recirculation in dense-medium processes, 9th International Coal Preparation Congress, New Delhi, 1982 Congress Proceedings (Ed.) S. K. Bose, paper G.3.
34. R. E. Zimmerman. The Japanese swirl cyclone, *Mining Engineering* **30**, 189–193, 1978.
35. F. Gu and R. Hong. The application and development of hydrocyclones in China's coal preparation plants, 2nd International Conference on Hydrocyclones, Bath, England, publd BHRA, Cranfield, England, 1984, paper G.3.
36. C. L. Shah, S. R. Shaw, E. Strike and W. T. Deeley. A 250 T/H centrifugal dense-medium separator for minus 100 mm raw coal, 10th International Coal Preparation Congress, Edmonton, 1986.
37. D. G. Osborne. Fine-coal cleaning by gravity methods: a review of current practice, *Coal Preparation* **2**, 207–241, 1986.
38. J. Mengelers and J. H. Absil. Cleaning coal to zero in heavy-medium cyclones, *Coal Mining and Processing* May 1967, pp. 62–64.
39. W. S. Taylor, W. L. Chen and D. Chedgy. Feed to zero in heavy-medium cyclones, 107th Annual Meeting, SME, AIME, Denver, Colorado, March 1978.
40. W. S. Taylor and W. L. Chen. Heavy-media cyclone cleaning of 1.5" × 0 raw coal, American Mining Congress, Chicago, May 5, 1980.
41. F. E. Baumgartner. Washing raw coal to zero with heavy-media cyclones, *Mining Congress Journal* 45–50, Oct. 1978.
42. R. E. Zimmerman. Breakthrough in heavy-media cyclone operation, *World Coal*, **3**, 19–21, 1978.
43. N. T. Esposita and S. T. Higgins. Modification of the fine coal circuit of Homer City preparation plant, 111th Annual Meeting, SME, AIME, Dallas, February 1982.
44. E. Skolnik. Heavy-medium cleaning 4-28 mesh coal, *Mining Engineering*, **32**, 1235–1237, 1980.

45. P. J. F. Fourie and T. C. Erasmus. Beneficiation of fine coal in the Republic of South Africa, 3rd Symposium on Coal Preparation, Louisville, October 1977. *Coal Confirmed*, (Ed.) J. F. Boyer, National Coal Association/Bituminous Coal Research, pp. 96–106.
46. D. W. Horsfall. Pilot plant data for fine coal beneficiation plant, Colloquium—*Merits of Pilot Plants and Problems of Scale-up*, South African Institute Mining Metallurgy, March 1978.
47. P. J. F. Fourie. Dense-medium beneficiation of minus 0.5 mm coal in the Republic of South Africa, 5th International Conference on Coal Research, September 1980, Dusseldorf, Paper C5.
48. P. J. F. Fourie, P. J. van der Walt and L. M. Falcon. The beneficiation of fine coal by dense-medium cyclone, *Journal South African Institute Mining Metallurgy*, **80**, 357–361, 1980.
49. J. Mengelers. The influence of cyclone diameter on separating performances of economy, 9th International Coal Preparation Congress, New Delhi, 1982, Congress Proceedings (Ed.) S. K. Bose, Paper B4.
50. A. W. Deurbrouck. Washing fine-size coal in a dense-medium cyclone, USBM Report on Investigations, RI. 7982/1974. Published by US Bureau of Mines, Department of Energy.
51. Anon. Putting a fine wash on it, *Coal, Gold and Base Metals of South Africa*, November 1980, pp. 34–37.

Chapter 7

HYDRAULIC AND PNEUMATIC SEPARATION

7.1 INTRODUCTION

There has always existed a certain amount of confusion between the terms used in coal preparation for improving the quality of raw coal and those used in mineral (or ore) dressing for upgrading ores or concentrating minerals. Hopefully, such confusion can be cleared up by some simple definitions and by the glossary at the end of volume 2.

First of all, it should be remembered that there is a very important difference in the manner in which the need for these processes of separation has arisen.[1] Where metalliferous minerals are concerned, the necessity for purification was indigenous and arose out of the needs of the industry itself. In the case of coal mining, this need was forced upon the industry by extraneous agencies. Therefore, for the former it is coeval with the industry itself and therefore ore dressing can fairly be stated as having been an integral part of metalliferous mining from the outset. Coal cleaning was, on the other hand, brought about long after coal mining was a fully developed industry, in order to supply the exigencies of the users of the coal. Also, it should be remembered that whilst metalliferous mining has existed for countless ages, coal mining is less than nine centuries old. It was therefore to be expected that when the modern need for coal cleaning began to make itself felt in the early part of the present century, the natural tendency would be to adopt the already thoroughly developed and well-tried methods and appliances of ore dressing. Most of these used hydraulic separators. With this adoption came the simultaneous use of the same terminology and hence the ensuing confusion.

In ore dressing, the term *classification* is used to describe the exploitation of the relative-size property in order to effect a separation. Hence, classification will be influenced by other factors, such as shape and relative density, as well as by the fluid, e.g. water or air, containing the grains that are to be separated. The term *concentration* is used to describe the exploitation of the relative-density property in order to effect a separation, whereby the minerals having the highest density can be extracted from those of lower density. In almost all cases the former represents the valuable components, known as the concentrate, and the latter the waste known as gangue or tailing. Gangue is usually a coarse reject, while tailing is fine and often in slurry form. The degree of concentration that can be achieved

HYDRAULIC AND PNEUMATIC SEPARATION

is usually very dependent upon the degree of liberation of the values (i.e. the amount of crushing and grinding necessary to separate them physically from the mined ore). There are several terms that are used to describe the success with which the operation of concentration has been carried out. The first is known as the *ratio of concentration* which is the relative mass of feed to that of concentrate i.e. $K = F/C$. The second term is known as the *recovery*. From the masses of the concentrate and tailing and their assays, that fraction of a certain metal or gangue contained in the feed which is recovered in the concentrate, expressed as a percentage, is the recovery, R, i.e.

$$R = 100 \frac{C_c}{F_f}$$

where C and F are concentrate and feed masses and c and f are metal contents expressed as percentages by mass.

Joint consideration of recovery and ratio of concentration is the traditional method of expressing metallurgical results, at least in connection with metal mining. In coal washing, the latter is so much more important than the former as to have usurped the term. Hence, the washing recovery is the ratio of concentration, expressed as a percentage. This quality is more properly termed the *yield*.[2] This subject is further expanded and discussed in detail in the next chapter.

The reason why the foregoing has been included at the beginning of this chapter and not one of the others is simple. All of the early methods used for mechanical cleaning of coal were the so-called gravity methods employing water as the cleaning medium. Later, use was made of separators using air as the medium. Hence, this chapter deals with both hydraulic and pneumatic separation.

The historical development of coal-cleaning methods, although more recent than that of ore dressing, is nevertheless no less interesting.

In the beginning, use was made of familiar ore-dressing jigs, sluices and troughs, and flowing-film devices such as shaking tables. All of these underwent some form of modification to suit the specific needs of coal cleaning, not the least being the fact that the valuable component was now the lightest contained in the feed. Another important fact was the relatively close densities of values and waste (i.e. 1.4 and 1.6 in some cases) which led to the adoption of some devices used in mineral treatment for classification, as separators in coal-cleaning applications, e.g. troughs and upward-current classifiers and autogenous cyclones.

Once the need for coal cleaning had been firmly established and traditional ore-dressing methods and appliances thoroughly applied, the rate of development of specific coal-cleaning techniques proceeded at a rapid pace. During these early days, much of the effort of development was directed towards hydraulic separation. However, pneumatic separation and froth flotation were regarded as having great future potential, as remarked by Louis in 1928 in one of the first textbooks written on the subject of coal preparation.[1]

The basic types of separators employing water as the separating medium can be categorized as follows:

(a) direct-flow or upward-current separators;
(b) jigs;
(c) flowing-film separators;
(d) autogenous cyclones.

The first three categories represent the traditional forms of coal-cleaning methods used widely until dense-medium separation was commercially developed, i.e. until the late 1940s. The last category appeared more or less simultaneously with the introduction of dense-medium cyclones, as a lower-cost alternative for easier-to-clean coals.

Pneumatic separation reached the height of its popularity in the 1930s, being widely used in Europe and the USA for the preparation of coking coals in particular. Various forms of table, spiral and jig were developed and relatively widely used but their attractiveness rapidly dwindled as a result of the following interrelated factors:

(a) rapid increase in mechanized mining and widespread use of water for dust suppression, rendering the run-of-mine coal both finer in size and wetter;
(b) growing awareness of environmental problems from air-borne coal dust, creating the need for health-hazard legislation;
(c) increasing cost of dry-coal-preparation methods to counteract (a) and (b), i.e. thermal drying of feed coal and capturing airborne dust.

Despite these problems, interest in pneumatic separation methods is still very much apparent and several commercially available separators are in current usage.

7.2 THE MOVEMENT OF SOLIDS IN FLUIDS

The general rules governing the motion of solids in fluids are applicable to both classification and concentration and indeed also apply in solid–liquid separation, as we shall see in Chapter 10. Some discussion of the rules relating to settling of solids in fluids was included in the previous chapter, and much of the basic theory is available in more general texts.[3]

Of paramount importance in coal-cleaning separation employing fluids as the medium, is the internal resistance to motion exerted by the fluid or the solid. Even the most fluid liquid or the most tenuous gas exerts a characteristic resistance, and since this resistance is zero when the solid is at rest with respect to the fluid, but not zero when it is in motion, the value of the resistance, R, is a function of only the velocity, v, i.e.

$$R = f(v) \tag{7.1}$$

If this force, which we term resistance, becomes equal in amplitude and opposite in direction to the resultant of all the other forces acting on a solid in a fluid, the acceleration of the solid will become zero and the velocity will be constant.[2] This terminal or maximum velocity, v_{max}, is of special importance in coal and mineral dressing, as we shall see later. In determining the settling behaviour of solids it is therefore necessary to evaluate v_{max} in terms of the following:

(a) the physical characteristics of the solid, i.e.
- relative density
- size
- shape

HYDRAULIC AND PNEUMATIC SEPARATION

(b) the physical properties of the fluid, i.e.
  - relative density
  - viscosity
  - temperature (and its influence on the above)

(c) the forces that are created by the motion of both the solid and the fluid.

Much of the theory relating to the movement of solids in fluids has initially considered the behaviour of smooth spheres. *Stokes' law* suggests that smooth spheres settling under laminar-flow conditions encounter a resistance to motion of

$$R = 6\pi\mu r v \qquad (7.2)$$

in which μ is the viscosity of the fluid, r is the radius of the sphere and v is the settling velocity, all in SI units.

The Second Law of Motion (i.e. the sum of the forces equals mass times acceleration) applied to the falling sphere takes the form:

$$m\frac{dv}{dt} = mg - m'g - R \qquad (7.3)$$

in which m is the mass of the sphere, m' is the mass of the displaced fluid, v is the settling velocity and t is the time.

If density = mass/volume then by substituting equation (7.2) in equation (7.3) and assuming $dv/dt = 0$ as being the terminal velocity, v_{max}, gives

$$v_{max} = \frac{2}{9}\frac{(d_s - d_1)r^2 g}{\mu} \qquad (7.4)$$

where d_s and d_1 are the relative densities of solid and fluid respectively. This is the form in which Stokes' law is usually written.

Newton's law, which was published almost 200 years before that of Stokes, stated the resistance to the motion of a sphere in a fluid as being

$$R = \frac{\pi}{2} d_1 r^2 v^2$$

which is often written as

$$R = Q\frac{\pi}{2} d_1 r^2 v^2 \qquad (7.5)$$

to include the coefficient of resistance, Q, which makes the equation a more accurate one. In this form, this equation is sometimes called the Rittinger equation after one of its chief protagonists.[4]

Whereas Stokes' law deals with laminar-flow conditions and is applicable for small spheres of up to about 50 μm in size, Rittinger's equation or Newton's law deals with motion under eddying or turbulent resistance, and is valid for spheres larger in diameter than about 5.0 mm.

The maximum or terminal velocity based on Newton's law can be obtained as before and the following equation is obtained

$$v_{max} = \left[\frac{8}{3Q} \times g\left(\frac{d_s - d_1}{d_1}\right) \times r\right]^{1/2} \qquad (7.6)$$

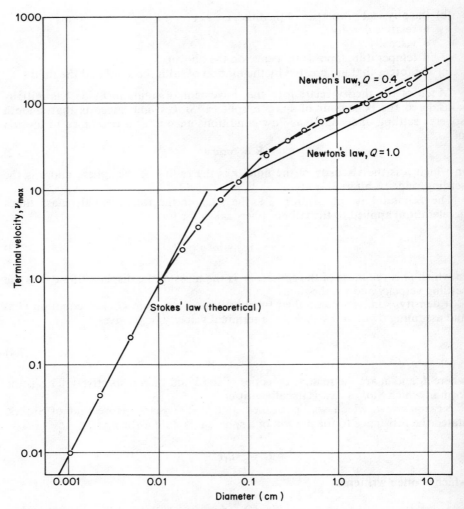

Fig. 7.1 Settling velocities of quartz spheres in water, as determined experimentally, compared with theoretical settling velocities according to Stokes' and Newton's laws.[2]

The important thing to remember about formulae (7.4) and (7.6) is that in laminar flow the terminal velocity varies as the *square* of the solid particle's diameter, and in turbulent flow it varies as the *square root*.

Clearly, then, Stokes' law tends to be most applicable in terms of fine-coal concentration and classification and in thickening, whereas Newton's law is more applicable to coarser-sized solid separations.

Figure 7.1 shows the settling velocities of quartz spheres in water as determined by these two laws. The graph is drawn with $\log r$ and $\log v$ as coordinates, to permit the entire size-range to be presented, and the portions of the curve that fit exactly the Stokes' equation and Newton's equation are straight lines because in one case $\log v$ varies as $2 \log r$, and in the other as $\frac{1}{2} \log r$. The Newton equa-

HYDRAULIC AND PNEUMATIC SEPARATION

tion is drawn for $Q = 0.4$, and the parallel line below the factual curve ($Q = 1$) is that which would be required for the Newton formula to be strictly accurate.

A more suitable way of presenting the same data is to make use of the coefficient of resistance or 'drag coefficient' as it is sometimes called, Q, and the Reynolds number, Re. The Reynolds number is an absolute (dimensionless) number. It was devised by Osborne Reynolds in 1883 and has become indispensible in fluid dynamics. Hence,

$$Q = \frac{8r(d_s - d_1)g}{3v^2 d_1} \quad (7.7)$$

and

$$Re = \frac{2rvd_1}{\mu} \quad (7.8)$$

The quantities permit graphical representation of data for all solids and all fluids, as shown in Fig. 7.2.

More familiar and useful versions of Stokes' and Newton's formulae convert radius to effective particle diameter as follows:

Stokes' equation

$$v_{max} = \frac{(d_s - d_1)d^2 g}{18\mu} \quad (7.9)$$

where d = diameter of the sphere, and $Q = 24/Re$, which applies only in the region where $Re > 0.2$, i.e. in laminar or streamlined flow.

In Chapter 4 we saw how these laws of motion affect the performance of classifiers and learned the importance of operating them with relatively closely sized material. In terms of separation, the same laws apply and are equally critical.

Consider the following generalities in terms of the context of separating coal from other rocks and minerals in a fluid:

(a) two spheres of equal relative density but different size—the larger will fall faster;

(b) two spheres of equal size, but different relative density—the heavier will fall faster;

(c) two rock fragments of the same relative density and the same mean size—the rounder will fall faster;

(d) the terminal velocity in the Stokesian regime is proportional to d^2;

(e) the terminal velocity in the Newtonian regime is proportional to $\sqrt{d}$;

(f) the period of initial acceleration prior to the terminal velocity being achieved under Stokesian conditions is negligibly small;

(g) the period of acceleration prior to v_{max} being achieved in the Newtonian regime is appreciable;

(h) the terminal velocity is reduced by increasing volumetric concentration of solids.

Newton's or Rittinger's equation

$$v_{max} = \left[\frac{(d_s - d_1)3gd}{d_1}\right]^{1/2} \quad (7.10)$$

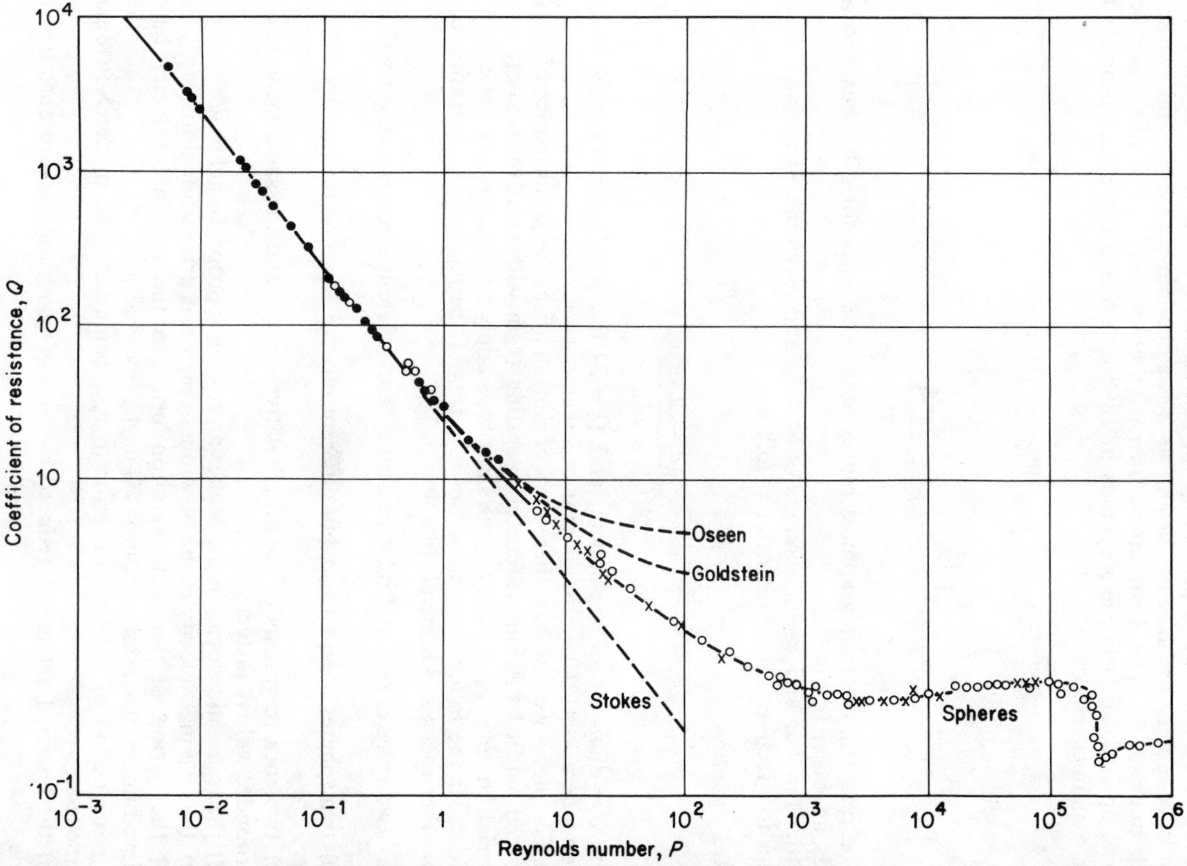

Fig. 7.2. Relationship between drag coefficient and Reynolds number

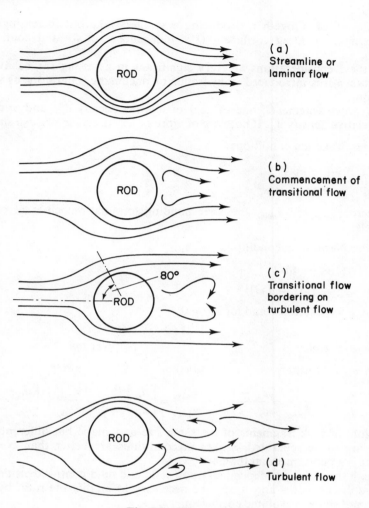

Fig. 7.3 Flow regimes.

There is, therefore, an intermediate range of particle size which corresponds to the range in which a great deal of the coal-preparation classification and separation operations are performed. In this range neither law is truly applicable. In Fig. 7.1, for quartz, this range is between 0.1 and 5 mm and is usually called the transitional regime. All three conditions can be readily seen when a fluid is caused to flow past a stationary horizontal rod, as shown in Fig. 7.3. In the laminar regime, a thin boundary layer exists between the fluid and the surface of the rod that does not separate: case (a).

During the transitional regime, a point of separation of the boundary layer begins to move upstream to reach a point ($Re = 1000$) with contact points of about 80° angle forming a wake larger than the rod diameter. Energy is therefore dissipated in the vortex formed. Finally, in the turbulent regime the separation point is destroyed altogether, producing a smaller wake but increasingly large energy dissipation. Drag coefficient Q is independent of Re and continues to remain so.

All of these considerations are very significant in assessing the behaviour of raw coal when it is introduced to hydraulic or pneumatic separators, as we shall see later on.

Consider two fragments, one of coal of relative density d_{s_c}, and another of shale of relative density d_{s_s}. If both are of more or less the *same size and shape*:

(a) under Stokesian conditions:

$$v_{max_c} \propto (d_{s_c} - d_1)d^2$$

and $v_{max_s} \propto (d_{s_s} - d_1)d^2$

where v_{max_c} and v_{max_s} are corresponding terminal velocities of coal and shale.

(b) under Newtonian conditions:

$$v_{max_c} \propto [(d_{s_c} - d_1)d]^{1/2}$$

and $v_{max_s} \propto [(d_{s_s} - d_1)d]^{1/2}$

and d_1 for water $= 1$ and for air $= 0$

Hence

	Stokes		Newton	
	air	water	air	water
	$\dfrac{v_{max_s}}{v_{max_c}} = \dfrac{d_{s_s}}{d_{s_c}}$	$\dfrac{v_{max_s}}{v_{max_c}} = \dfrac{d_{s_s} - 1}{d_{s_c} - 1}$	$\dfrac{v_{max_s}}{v_{max_c}} = \left[\dfrac{d_{s_s}}{d_{s_c}}\right]^{1/2}$	$\dfrac{v_{max_s}}{v_{max_c}} = \left[\dfrac{d_{s_s} - 1}{d_{s_c} - 1}\right]^{1/2}$

Therefore, if rock fragments of the same size and shape but different relative densities are to be separated, Stokesian conditions are better than Newtonian and water is a better medium than air.

Although the ratios are dependent upon solid and fluid relative densities, shape cannot be disconnected and it may be necessary to modify the ratio by adding some form of shape or volume coefficient.

If the two earlier fragments are different in size but similar in shape, i.e. d_c and d_s are mean diameters for the coal and shale respectively, the earlier relationships become:

(a) Stokesian regime

$$v_{max_c} \propto (d_{s_c} - d_1)d_c^2$$

and

$$v_{max_s} \propto (d_{s_c} - d_1)d_c^2$$

(b) Newtonian regime

$$v_{max_c} \propto [(d_{s_c} - d_1)d_c]^{1/2}$$

and

$$v_{max_s} \propto [(d_{s_s} - d_1)d_s]^{1/2}$$

HYDRAULIC AND PNEUMATIC SEPARATION

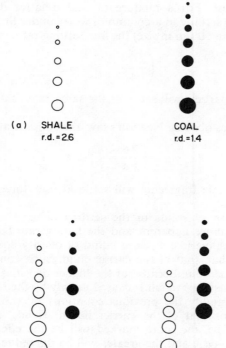

Fig. 7.4 Equisettling conditions: (a) equisettling diameters; (b) free-settling conditions; (c) hindered-settling conditions.

Bearing these in mind and the corresponding ratios obtained between several differently sized coal and shale fragments, there will obviously exist corresponding pairs of equisized coal and shale fragments which could behave the same (identical terminal velocities) in a separator. Figure 7.4 shows this phenomenon pictorially. These equisettling fragments can be defined by a further ratio and hence the larger the value of the so-called 'equisettling ratio', the easier will be any separation.

If the particles are in different flow regimes

$$\frac{d_c}{d_s} = \frac{(d_{s_s} - d_1)^{m_1}}{(d_{s_c} - d_1)^{m_2}}$$

but if they are assumed to be in the same regime then

$$\frac{d_c}{d_s} = \left[\frac{(d_{s_s} - d_1)}{(d_{s_c} - d_1)}\right]^m \tag{7.11}$$

where $m = 1$ for Newtonian regime, and $m = 0.5$ for Stokesian regime.

For low-volume concentrations this is termed the *free-settling ratio*, and where settling occurs in water under Newtonian conditions, the expression has been termed the *concentration criterion*, although this expression is rarely used by coal-preparation engineers. For a mixture of coal (relative density 1.4) and shale (relative density 2.6) settling in a column of water under Stokesian conditions (i.e. all particles finer than 50 μm in size) the free-settling ratio is

$$\left(\frac{2.6 - 1}{1.4 - 1}\right)^{1/2} = 2$$

i.e. the small shale particle will settle at the same rate as a small coal particle of twice its size.

For larger particles obeying Newton's law, the concentration criterion is

$$\left(\frac{2.6 - 1}{1.4 - 1}\right) = 4$$

and therefore the shale fragments will settle at four times the rate of similarly sized coal fragments.

As the proportion of solids in the settling column increases, the effect of crowding becomes more apparent and the falling rate begins to decrease. The system begins to behave like a dense liquid of density equivalent to that of the suspension rather than that of the carrier fluid. Such conditions create what is known as *hindered settling*. Because of the higher densities involved and increasing viscosity, the resistance to fall is caused mainly by the turbulence created. It is now necessary to modify the previous equations by substituting the apparent fluid density, d_a, for that of the carrier liquid. Hence, the lower the relative density of the particle, the more marked will be the effect of reduction of the effective density $(d_s - d_a)$, and the greater will be the reduction in falling velocity. Similarly, the larger the particle, the greater the reduction in rate of fall as pulp density is increased.

Using the previous example for a denser mixture of coal and shale fragments in water creates an apparent fluid density of 1.3, i.e.

$$\left(\frac{2.6 - 1.2}{1.4 - 1.2}\right) = 7$$

so that now, the shale fragments will settle at seven times the rate of similarly sized coal fragments but at much lower velocity than for the free-settling case.

We have seen the ramifications of such benefits in the previous chapter dealing with dense-medium separation where, by selecting a pulp density between that of coal and shale, the velocity of the coal can be rendered negative by causing it to become buoyant.

When the formula for hindered-settling ratio,

$$\frac{d_c}{d_s} = \left(\frac{d_{s_s} - d_a}{d_{s_c} - d_a}\right)^m \qquad (7.12)$$

(Newton) $1 > m > 1/2$ (Stokes)

is used to calculate hindered-settling ratios for shale and bituminous coal, useful tabulations may be obtained for slurries of various concentrations, as shown in Table 7.1.

The relevance of these relationships and the value of the hindered-settling phenomenon will become more evident later on, particularly when jig separation is

TABLE 7.1 Calculated Hindered-settling Ratios for Coal–Shale in Water[2]

%age Shale by mass	m	%age Solids by volume				
		0	10	20	30	40
0	1	5.00	5.45	6.00	6.72	7.67
	0.5	2.24	2.34	2.45	2.59	2.77
25	1	5.00	6.00	7.67	11.00	21.00
	0.5	2.24	2.45	2.77	3.31	4.59
50	1	5.00	6.72	11.00	41.00	∞
	0.5	2.24	2.59	3.31	6.40	∞
75	1	5.00	7.67	21.00	∞	∞
	0.5	2.24	2.77	4.59	∞	∞
100	1	5.00	9.00	∞	∞	∞
	0.5	2.24	3.00	∞	∞	∞

discussed. Many of the early forms of coal-cleaning separator depended a great deal upon the relative density and size differences existing between coal and shale, and the basic principles of separation were carefully adhered to in preparing coal for cleaning by these methods, i.e. prescreening, classification and desliming. Almost all coal-cleaning methods which depend upon relative density differences are based upon use of the hindered-settling condition. As opposed to dense-medium separation, these other processes employ a principle of creating a loose bed of raw coal which is constantly maintained in a mobile or fluid state by water (and occasionally air) flow, together with some appropriate form of motion. This causes the bed to stratify, that is to say denser fragments are encouraged to make their way to the bottom and lighter ones to the top. The first separators to utilize this form of separation were trough or launder types of separator.

7.3 TROUGH AND UPWARD-CURRENT SEPARATORS

Early examples of these types of coal-cleaning devices included the Elliott and Blackett washers which were inclined-trough types, and the Draper washer which was an upward-current type. All three are shown in diagram form in Fig. 7.5 because, despite the fact that they all more or less disappeared with the introduction of more sophisticated versions in the 1920s, they were subsequently resurrected many years later for coal waste-tip reclamation, and several barrel and launder types are still being effectively used today, as can be seen by Fig. 7.6. The reason for their longevity is undoubtedly their inherent simplicity and low cost. However, to be really effective the raw feed must be carefully prepared to comply with the equisettling ratio requirements. This is, therefore, the secret of their success or the cause of their failure.

Washers incorporating combined along-trough flow and upward-current water added a new dimension to coal-cleaning technology in the 1920s, and both efficiency and capacity benefited greatly from the introduction of washers such as the Rheolaveur and later, the Hoyois types of trough washer which were used extensively until the 1950s when either jigs or dense-medium separation gradually eclipsed them. Because, with such types of washer, use is made of both horizontal

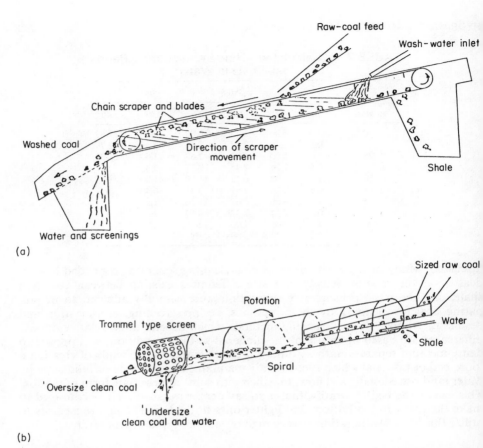

Fig. 7.5 Diagrams showing the principles for (a) the early Elliott trough and (b) the Blackett barrel washers.

Fig. 7.6 Modern-day version of a barrel-type trough washer.

HYDRAULIC AND PNEUMATIC SEPARATION

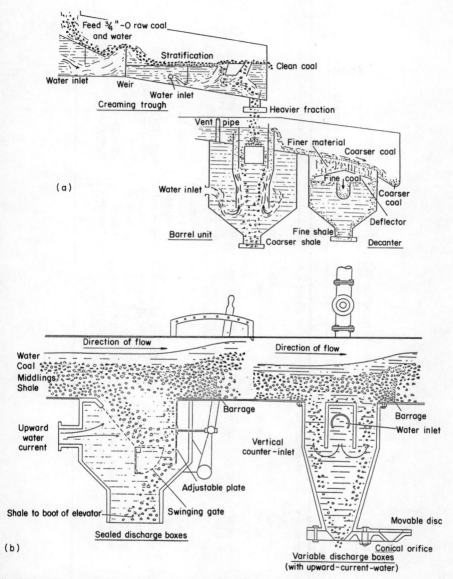

Fig. 7.7 Diagrams of (a) the Hoyois trough, barrel and decanter units and (b) Rheolaveur shale-removal apparatus.

and vertical currents of water, the sorting process is protracted in time and the accuracy of the washer is enhanced. Figure 7.7 gives simplified diagrams of the Hoyois trough, barrel and decanter units and the Rheolaveur shale-removal apparatus. These show clearly how the upward-current water is used to refine the separation occurring as a result of stratification as the coal bed is swept along the washing trough. Various arrangements of troughs and upward-current boxes

TABLE 7.2 Data for Various Types of Trough and Upward-current Washers

Washer	Type	Separation	Raw-coal size (mm)	Approx. dimensions (m)	Shale removal	Remarks
Elliott	inclined trough (1 in 15)	direct flow stratification	75 × 25	0.3 m wide × 18 m long	scraper chain	<12 t/h (introduced c. 1885)
Blackett	inclined barrel (1 in 15)	direct flow/ stratification	75 × 25	1–1.2 m wide × 6–9 m long	inner spiral	<25 t/h (introduced c. 1895)
Draper	vertical tube	upward current	50 × 10	0.5 m dia. tube	star valve and screw conveyor	up to 5 t/h per tube × 10 tubes per battery (patent 1918)
Robinson	stirred conical tank	upward current	20 × 0	3.5 m dia. cone	sliding gates and screw conveyor	up to 50 t/h (introduced c. 1890)
Rheolaveur	inclined trough (2–5%)	direct flow with upward current	100 × 12 12 × 1 –1 mm	>0.9 m wide <20 m long	gates and bucket elevator	up to 100 t/h (introduced c. 1912)
Hoyois	inclined trough	direct flow with upward current	125 × 20 20 × 0	0.75 m wide 3 m long	bucket elevator	up to 100 t/h (introduced c. 1930)

HYDRAULIC AND PNEUMATIC SEPARATION

were utilized, depending upon the size range of the raw coal and the washability characteristics.

These types of washer continued to be used until the 1960s for cleaning small coal (20 mm top size) and occasionally for fine-coal or slurry washing in the case of the Rheolaveur washer.

Table 7.2 gives data for various types of trough and upward-current washers.

7.4 JIGGING SEPARATORS

Jigging is amongst the oldest of mineral concentrating methods, and one of the first methods to be adopted for coal cleaning, about 100 years ago. The simplest jig consists of a framed sieve, hand-held and pulsed up and down in water, to produce stratification of the solids remaining on the sieve. This is a simple and useful technique for demonstrating the exact mechanism employed by modern jig washers capable of treating up to 1000 t/h of raw coal. Figure 7.8 shows the jig family tree. The early coal jigs were those adopted from ore-dressing applications, i.e. piston types like the Harz jig and diaphragm types like the Jeffrey jig. These are characterized by a relatively large stroke and frequent pulsations, but capacity is small. With the development of fixed-sieve, air-pulse jigs, of which the parent was the Baum jig developed in Germany by Fritz Baum and known in Europe since 1892, came the type of washer ideally suited to coal cleaning. Simon–Carves introduced the first coal-cleaning Baum jig into Britain in 1903.[5] The Baum system soon became popular in Britain and was increasingly widely adopted by other coal-producing countries. Up to the period of the Second World War, the original washbox design remained more or less unchanged and coals of up to 50 mm were effectively treated. After the war, a sequence of major

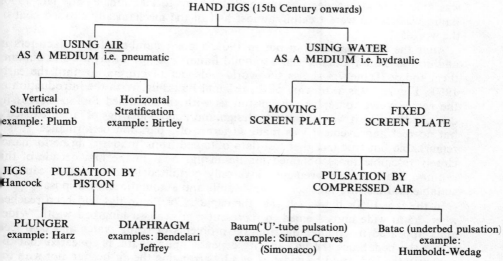

Fig. 7.8 The jig 'family tree'—classification of jigs by type.

TABLE 7.3 List of Baum Jig Manufacturers after Simon–Carves in 1903[6]

In Germany
Schüchtermann and Kremer-Baum, Aktiengesellschaft für Aufbereitung, Dortmund, since 1891
Klöckner-Humboldt-Deutz AG., Cologne, since 1950
Westfalia Dinnendahl Gröppel AG., Bochum, since 1950

In France
Préparation Industrielle des Combustibles, Fontainebleau, since 1937

In Great Britain
Simon–Carves Limited, Stockport, since 1903
The Coppée Company Limited, London, since 1920
Blantyre Engineering Company Limited, Blantyre, since 1922
Norton–Harty Engineering Company Limited, Tipton, since 1925
Nortons (Tividale) Limited, Tipton, since 1925

In Japan
Kobukuro Iron Works Company Limited, Tokyo, since 1917
Imamura Seisakusho Company Limited, Tokyo, since 1927
Nagata Seisakusho Company, Tokyo, since 1933
Sumitomo Machinery Company Limited, Osaka, since 1948
Kawasaki Dockyard Company Limited, Tokyo, since 1950
UBE Kosan Company Limited, Tokyo, since 1950
Mitsubishi Heavy Industries Limited, Tokyo, since 1960

In the United States of America
Link-Belt Company, Chicago, since 1928
Jeffrey Manufacturing Company, Columbus, Ohio, since 1932
McNally Manufacturing Company, Pittsburg, Kansas, since 1932

improvements occurred, with the introduction of automatic pulsion control combined with new and highly efficient rotary air-valves and automatic shale discharge control. As a result, the Baum jig became a more versatile washer and was able to treat a wider size-range feed, i.e. up to 200 mm. By the late 1950s Baum washboxes were treating almost half of the mechanically cleaned coal in the world.

After the basic patents ran out in 1909, a great number of manufacturers in addition to Simon–Carves began to build Baum-type jigs. Many are still building them today. Table 7.3 shows the world-wide manufacturers up until the early 1960s. Figure 7.9 is a drawing of the original Baum jig. With the introduction of the rotary valve to replace the piston as with the modified Simon–Carves jig described earlier, it was possible to design and build wider jigs of larger capacity but no real improvement was made in terms of separation performance. It is a remarkable but true fact that test-data obtained from 'modern' jig performance closely resemble those obtained for Baum jigs built in the first decade of this century. Machine improvements have only permitted increase in unit capacity suitable feed distributions across wider units and automatic shale disposal.

In the late 1950s it was felt that the capacity of Baum jigs had been reached when 2.5-m wide units seemed to represent a critical washing-bed width. Wider units appeared to counter distribution problems of rising water and coal feed across the bed. Since the jig capacity depends upon width, it appeared that the only way the bed could be widened was by arranging the air header, not with the flow but perpendicular to it. This required that the air header be placed beneath

HYDRAULIC AND PNEUMATIC SEPARATION 305

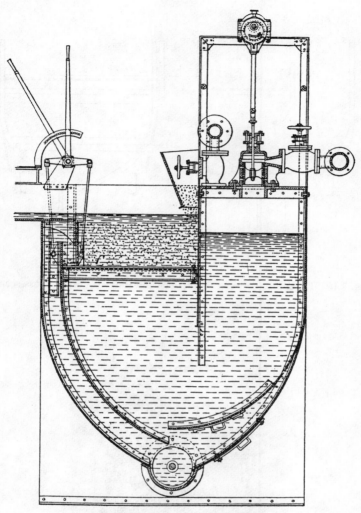

Fig. 7.9 Drawing of the original Baum jig.[7]

the washing bed which would then remove the limitations in width imposed by the familiar U-tube arrangement employed by the Baum jig. Figure 7.10 illustrates this principle. The first jig of this kind was designed and built in Japan in 1958 and was later named the Tacub jig after the original inventor Takakuwa.[8] Following an association with the German company Schuchtermann and Kremer Baum AG which produced a new form of discharge control, the jig was fully commercialized as the Batac jig (Baum–Takakuwa) and many of these machines have since been built and installed all over the world.[6]

With the continued demand for economical cleaning of relatively easily cleaned coals, the jig continued to prosper despite the growing popularity of dense-medium separation. Both types of jig are able effectively to treat raw-coal feeds of

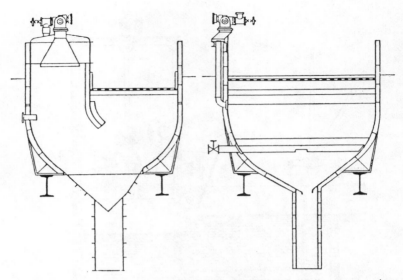

Fig. 7.10 Difference between Baum (left) and Batac (right) jig cross-sections.

Fig. 7.11 Simonacco/British Coal five-metre-wide test Baum jig rig (courtesy Simonacco)

HYDRAULIC AND PNEUMATIC SEPARATION

Fig. 7.12 Modern Baum washers (courtesy Simonacco).

up to 200 mm top size, but both exhibit rapid deterioration in efficiency below 0.5 mm. However, with the introduction of a felspar ragging, which will be described later, even fine coal can be effectively cleaned. Baum jigs of up to 5 m in width are now available for treating 200 mm × 0 (relatively easily cleaned) coal at rates of up to 750 t/h.

The increase in Baum jig capacity came as a result of improvements made in rotary air-valve design and improved hydrodynamic flow conditions produced by washbox profile improvements. The British Coal-preferred washbox design of up to 5 m wide was confirmed by the full-scale test rig shown in Fig. 7.11 as designed by Simonacco and British Coal. Figure 7.12 shows a modern Baum washbox installation in England. Batac jigs of up to 6 m in width are able to treat the same size and type of coal (preferably deslimed at 0.5 mm) at up to 1000 t/h. Figure 7.13 shows a modern Batac jig installation in Germany. The jig is 6 m wide and handles up to 1000 t/h of 10–120 mm coarse coal. Fine-coal jigs of similar design but with a felspar bed are up to 6 m wide and capable of handling up to 600 t/h of 10–0.5 mm fine coal. Both types of high-capacity unit are more or less computer controlled to maintain optimum bed thickness and shale-extraction rate.

We shall look at the various types of jig in more detail later. In the meantime, a review of the basic principles of jigging may prove useful.

7.4.1 Basic Principles of Jigging

The separation of coal from shale is accomplished in a form of fluidized bed created by a pulsing column of water which produces a stratifying effect on the raw coal. This is quite different in its resultant effect from dense-medium separation, as can be seen from Fig. 7.14 which compares jigging with dense-medium

Fig. 7.13 Modern batac jig installation (courtesy K.H.D. Humboldt-Wedag).

separation. As we shall see this stratifying effect results in a definite order of deposition of all fragments contained in the bed. The main purpose of the rising and falling column is to create what is known as 'dilation' or opening up of the bed, and it is the extent to which this dilation may be controlled which governs the effectiveness of the separation.

During the pulsion, or rising part of the cycle, the bed is elevated *en masse*. But as the velocity decreases towards the end of the pulsion stroke, the bed begins to dilate, with the bottom ceasing motion first and the lowermost fragments commencing their descent. This produces an element of freedom of movement for all fragments signalling the commencement of the various principal effects leading to stratification.

In the classical description of the principles of jigging as described by Gaudin,[2] the most influential effects occurring during jigging are, in order of occurrence:

- dilation;
- differential acceleration;
- hindered settling;
- consolidated trickling.

HYDRAULIC AND PNEUMATIC SEPARATION

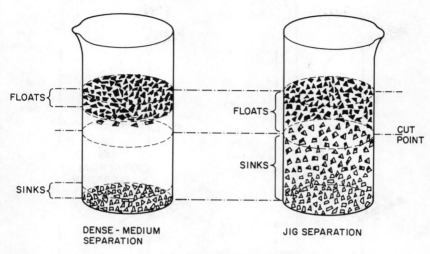

Fig. 7.14 Comparison of separation techniques.

Figure 7.15 illustrates the contribution which each makes towards the eventual stratification. Differential acceleration occurs, immediately following the end of the pulsion stroke. The theory has been advanced that, during this very brief period, the heavier fragments have a greater initial acceleration and velocity than lighter ones.

Think back to the equation of motion (i.e. equation (7.3)):

$$m \frac{dv}{dt} = g(m - m') - R \qquad (7.3)$$

in which v is the velocity of the solid within the fluid, m, m' are the masses of the solid and the equivalent volume of displaced fluid, and R is the fluid resistance.

Initially, R is zero and

$$\frac{dv}{dt} = \left(1 - \frac{d_s}{d_1}\right)g \qquad (7.13)$$

where d_s and d_1 are the relative densities of solid and fluid, respectively.

In a coal (r.d. = 1.4) and shale (r.d. = 2.6) suspension in water, the initial ratio of accelerations would be

$$\frac{\text{shale}}{\text{coal}} = \frac{1.6}{0.4} = 4:1$$

Their initial velocities would then be approximately as 4 is to 1, even although their ultimate speeds may be the same, i.e. if they were equisettling diameters. If the repetition of fall is frequent and the duration of fall short enough, the distance travelled by dissimilar fragments should bear more resemblance to their initial acceleration than to their terminal velocities. Under such conditions stratification would gradually occur as a result of differences in relative density alone.

The hindered-settling effect observed as occurring in jigging is a little different

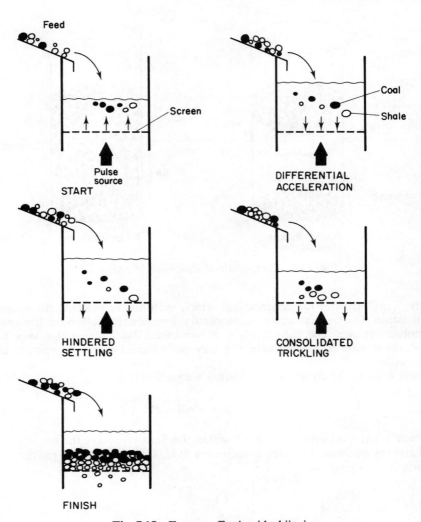

Fig. 7.15 Factors affecting ideal jigging.

from that in sizing classifiers or trough washers. In jigging, the fluid mixture is so thick that it resembles a loosely packed bed of granular solids containing interstitial fluid as opposed to a moving fluid carrying along a large number of suspended solids. In other words, the jig bed lies at the extreme limit of solid–liquid suspension. Such a thick solid–liquid mixture cannot be maintained for any length of time and the amount of rearrangement that can occur is limited. Nevertheless, this high apparent density creates increasingly high settling ratios which change rapidly as the bed approaches the screen and begins to come to rest again. All things being equal, higher settling ratios are attainable during jigging than in other forms of classifier and this effect makes an important contribution in the separating mechanism of jigging.

Very small grains of coal and shale are likely to be the most influenced during

HYDRAULIC AND PNEUMATIC SEPARATION

the pulsion stroke, and during dilation the very smallest may be pushed clear of the bed by the rising fluid. Coarser fragments, being most influenced first of all by differential acceleration and then by hindered settling, eventually bridge against one another and quickly become incapable of further movement. Fine fragments, despite the fact that they may have followed the coarser fragments down into the forming bed, will remain free to move between the interstices of the coarse fragments. Aside from any velocity that may be imparted to these small fragments by the moving fluid, they are bound to settle under the influence of gravity in these interstitial passages. This phenomenon is most pictorially described as consolidation trickling. While this effect is clearly evident in coal-cleaning jigs, it is much more significant in mineral dressing when pulsion–suction jigs, particularly those employing large strokes, are used to treat minerals of high relative density. Also, smaller-size feed material is more commonly treated than in coal-cleaning separations.

The ultimate result of prolonged pulsion and suction components of the jig cycle is a fairly thorough stratification of the bed, the degree of which is largely determined by the following:

machine factors such as:	jig stroke
	jig frequency
	rejects removal system
	water control
coal characteristics such as:	washability
	size distribution
	rejects quantity

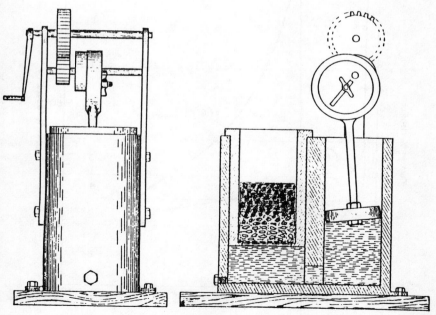

Fig. 7.16 Laboratory jigging test apparatus (source, Ref. 9).

There is no such thing as a perfect separation by jigging even in concept because of the random nature of motion of the materials forming the bed as they pass along the jig unit. Also, the fact that it is necessary to disturb the formed bed to remove the products (coal, middlings and reject) adds further to the need for a truly practical approach to performance assessment.

Referring back to Fig. 7.14 the major difference between dense-medium separation and jig separations is clearly illustrated, and whereas float–sink testing has become the accepted method for obtaining the washability characteristics of a coal there is also merit in considering an alternative jigging 'float–sink' approach.[9] The apparatus used for this form of testing is shown in Fig. 7.16. Although it is rarely used at the present time, it was first introduced and used in the post-World War II era when jig washing predominated in Europe. By removing the successive layers of the formed bed and determining for each, the corresponding mass, relative density and ash content, the washability curve for the coal can be obtained. Figure 7.17 shows typical washability curves obtained using this apparatus and Table 7.4 gives these data in tabulated form. The benefit of this type of approach is that a larger number of densities can be selected and the more difficult density range (i.e. where more near-density material exists) can be more carefully examined. Another benefit is that expensive and often unpleasant organic liquids or toxic salts are not required, and washability data may be readily obtained for small or fine coal 'in the plant'. Such a method also lends itself to automation of float–sink testing, although no such apparatus has yet been reported.

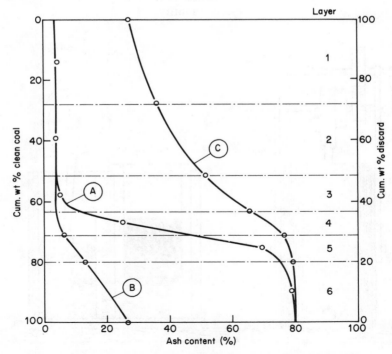

Fig. 7.17 Washability curves obtained from the jigging apparatus.

HYDRAULIC AND PNEUMATIC SEPARATION

TABLE 7.4 Washability data for −3 mm × 0.8 mm Raw Coal (from Laboratory Jig Analysis)

Layer	Direct wt %	Direct ash %	% Wt of ash of total	Cum wt of ash %	Cumulative clean coal wt %	Cumulative clean coal ash %	Sink Wt of ash %	Cumulative discard wt %	Cumulative discard ash %
1	2	3	4	5	6	7	8	9	10
1	27.94	3.8	1.06	1.06	27.94	3.8	25.55	72.06	35.5
2	23.52	3.3	0.78	1.84	51.46	3.6	24.77	48.54	51.0
3	11.39	4.8	0.55	2.39	62.85	3.8	24.22	37.15	65.2
4	7.81	24.9	1.94	4.33	70.66	6.1	22.28	29.34	75.9
5	8.91	68.9	6.14	10.47	79.57	13.2	16.14	20.43	79.0
6	20.43	79.0	16.14	26.61	100.00	26.6	—	—	—
Total	100.00		26.61						

Ash content of original sample (direct determination) 26.4%.

With the stratification phenomenon in mind, consider now the requirements of a practical jig washer. The primary objective is to stratify the raw coal accurately in the time it takes for it to travel from one end of the jig to the other. In order to achieve this objective the jig must meet the following requirements:[10]

1. Accurate stratification must be maintained irrespective of the fluctuations in quantity of feed and irrespective of variations in quantity of shale in the raw coal.
2. The jig must have some inherent and very accurate method of making a clean-cut separation between the cleaned coal and the discard products.
3. The jig must be able to deal with a continuous feed of raw coal and be capable of dealing with the largest possible tonnage.

7.5 THE BAUM JIG

7.5.1 Application

The historical development of the modern Baum jig has already been outlined. In current practice, Baum washing systems have tended to become standardized in many respects. Their use is applicable in those cases where the balance between the economics of coal cleaning and the degree of difficulty in washing the coal to the required specification at a reasonable yield comes down in favour of economics. In other words if the jig can wash the coal efficiency with a yield loss compared to dense medium of, say for example 2–3%, the jig would probably get the job. Baum jigs are usually limited to the treatment of raw coal within the size range of 150 mm to zero with an upper size of not less than 25 mm. Normally, the Baum jig would not be expected to upgrade raw coal finer in size than 0.5 mm. Cut-point densities rarely fall below 1.50 for coal sized above 50 mm and 1.55 for coal sized 25 × 0.5 mm.

7.5.2 Construction

Schematic illustrations of sections through a Baum jig are shown in Fig. 7.18. When viewed in cross-section ((b) in Fig. 7.18) the jig is divided into two com-

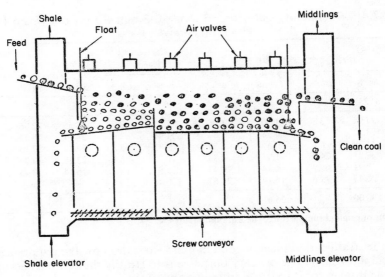

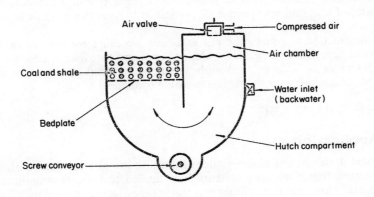

Fig. 7.18 Sections of a typical Baum washbox.

partments lengthwise, one completely sealed from the atmosphere—called the air chamber—and one open section, which receives the material to be separated and accommodates it during the stratification process. The water valve allows admission of 'back water' at a level below that of the bed plate. The longitudinal section shown in the figure illustrates that the box is further divided into several sections or compartments along the direction of flow. This can vary from three to six depending upon the capacity, but five compartments would appear to be the most common. The purpose of this is to provide control over the separation as the material moves along the box, hence each of these sections has its own individual air and water controls. The illustration also shows two main divisions

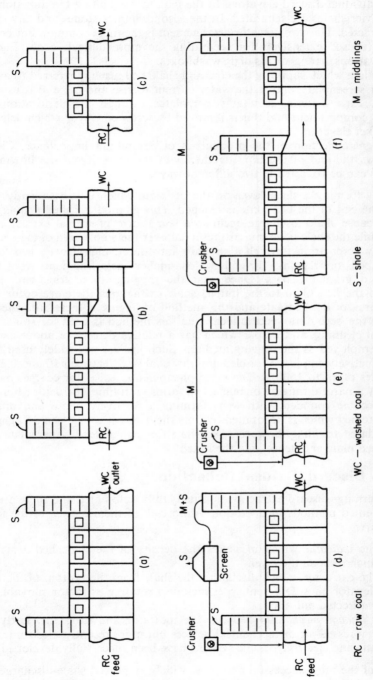

Fig. 7.19 Diagram showing various arrangements of jig washboxes: (a) two shale elevators; (b) three shale elevators; (c) one shale elevator; (d) one shale elevator, one middlings elevator; coarse middlings crushed and re-washed; (e) one shale elevator, one middlings elevator; all middlings crushed and re-washed; (f) two shale elevators, one middlings elevator; all middlings crushed and re-washed.

RC – raw coal   WC – washed coal   S – shale   M – middlings with individual discard elevators. In the first, moving along the direction of flow, the heavier shales are separated. In the second, lighter stones and any middlings are extracted. This two-elevator arrangement is the most common but boxes with three are used extensively for separating the more difficult coals. Figure 7.19 shows various arrangements of jig washboxes.

The plate which supports the coal and shale bed, usually referred to as the bed plate or screenplate, allows the water current to rise and fall and is usually perforated. Fine material inevitably percolates through the perforations to the 'hutch' compartment and this is removed by screw conveyor which delivers it to the bucket elevators.

Efficient collection of the products is of paramount importance. Clean coal overflows the end of the box together with the majority of the flowing water. Discard can be collected in two different ways:

(a) by forming a shale layer on the bedplate which is subsequently removed when the end of the bed plate is reached. This method is called *jigging over* the bed, because shale discharge occurs via some form of special extractor mechanism while the shale bed remains preserved over the whole area of the bedplate.

(b) by percolation through a specially introduced high-density bed forming a 'refuse bed' or 'ragging' through which smaller shale fragments from the raw-coal feed gradually pass. Once past the ragging, these fragments percolate through the bed plate to the hutch. Shale extractions therefore occur over the whole area of the bedplate and this method of jigging is therefore called *jigging through* the bed. As a specific technique this method is used for small- or even fine-coal cleaning, and felspar which has a relative density of about 2.6 is commonly employed as the ragging medium. Such jigs are not widely used but have been effectively employed for cleaning raw coal sized between 10 and 0.25 mm at capacities of up to 150 t/h. The same phenomenon exists for coarse-coal jigging whereby a natural ragging or refuse bed forms with time. Hardstone lumps gradually become rounded, with wear forming a relatively dense and difficult-to-penetrate bed which is negotiated by only the heaviest coarse-shale fragments. By consolidated trickling, however, fine shale fragments steadily percolate through to the bed plate and hence, into the hutch.

7.5.3 Discard (Refuse) Collection

In modern jigs discard extraction is usually fully automated and there are numerous patented methods in operation. Almost all of them consist of the following three parts:

(a) The *indicator* which measures the density of the shale bed, usually by a float, which rests on the layer.

(b) The *extractor*, which discharges the shale from the bottom of the bedplate to the elevator boot. Different types include a rotating extractor, movable gate or specially directed pulsation.

(c) A *linkage mechanism* between the indicator and extractor. Early devices were compressed-air or mechanical devices but more recently, electronic, electropneumatic and electrohydraulic systems have been successfully developed.

One of the most successful and most widely employed shale-discharge control systems ever developed was that first introduced by ACCO (Automatic Coal

HYDRAULIC AND PNEUMATIC SEPARATION

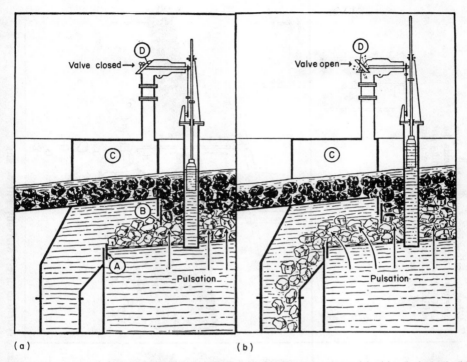

Fig. 7.20 ACCO automatic shale-discharge system: (a) operating with thin shale bed; (b) operating with thick shale bed (courtesy Simonacco.[5]

Cleaning Company, a British subsidiary of PIC of Fontainebleau, France) now Simonacco, in the late 1930s. Figure 7.20 shows the concepts employed. The thicker the layer of shale becomes, the greater the resistance to the water pulsations. This resistance is measured by a float suspended inside a tube with its lower end open to the water. Above the shale discharge gates A and B is an air chamber C with a release valve D. This valve is controlled by the position and movement of the float. With a thin shale bed the float remains in a low position and the valve is closed, trapping air in the chamber. This air acts as a cushion and will gradually harness any water pulsation as the shale thickness gradually begins to increase. During this time no shale is discharged and the bed gradually thickens. When the bed becomes thick the valve opens, releasing the trapped air and permitting pulsation to commence. The shale begins to move once again, being lifted over the gate A at each pulsation to be discharged at maximum capacity. The rate of discharge with this simple and very effective device is therefore infinitely variable from zero to maximum and is automatically controlled in direct proportion to the shale-bed thickness.

Since the float operates only within the steel tube it is protected from the moving bed and can respond accurately under all circumstances. The mechanical linkage used in the ACCO system could be set manually to maintain a bed of any desired thickness. A major reason for the popularity of this system was its simpli-

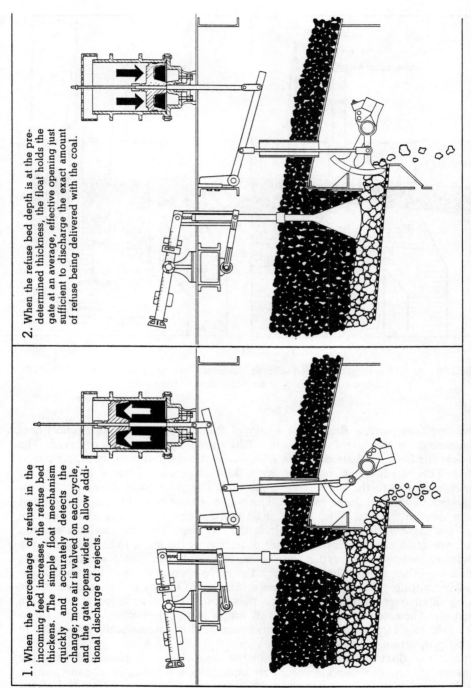

Fig. 7.21 McNally shale-discharge system (courtesy McNally, Pittsburgh).

HYDRAULIC AND PNEUMATIC SEPARATION

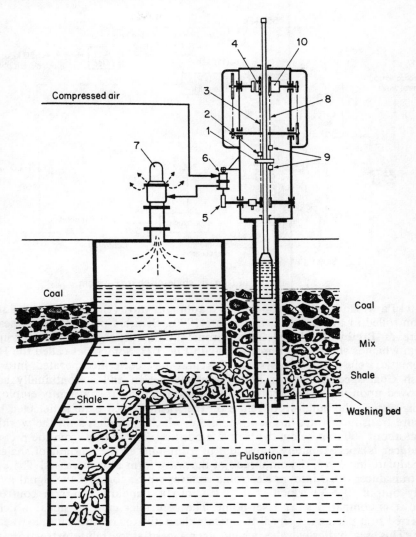

Fig. 7.22 Modern version of the ACCO system of shale discharge (courtesy Simonacco). For explanation of numbers, see text.

city and freedom from any form of direct mechanical extraction, but such systems have been widely and effectively used, as can be seen in Fig. 7.21.

A modern version of the (Simon) ACCO automatic control system is shown in Fig. 7.22. The principle is identical with earlier controllers. The float rises and falls with a stroke in time with the washbox pulsations. With a suitable setting of the striker (1) on the float rod and the adjustable stop (2) on the primary chain (3), it is possible to set a working density level. When the primary chain is moved upwards the top sprocket rotates against a slipping clutch (4), while the bottom sprocket rotates the bottom spindle which rotates a cam (5). The pressure regulator (6) converts the rise or fall motion of the cam to a pressure signal in the range

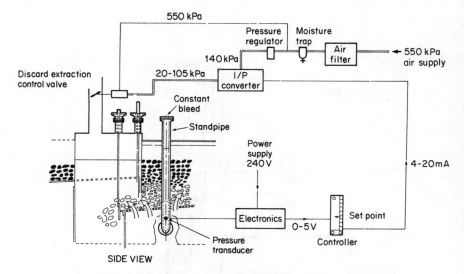

Fig. 7.23 Bretby automatic shale-discharge system.

0–150 kPa which is used to control the diaphragm valve (7) allowing pulsation to be controlled in the chamber. In the late 1970s, British Coal's Mining Research Centre at Bretby developed an electronic automatic shale-discharge control system which is shown diagrammatically in Fig. 7.23. Sometimes called the Hirst Automatic Refuse Control system, this unit has been incorporated into the British Coal specification for Baum washbox design and is continually being improved upon in the numerous installations in which it is currently employed. In this system, the build-up of refuse on the washbox bed plate is detected by a pressure transducer fitted inside a standpipe which is mounted on the washbox side at about 300 mm below the bedplate position. The output from the pressure transducer is approximately sinusoidal in form, and as the amount of shale on the bedplate increases, the amplitude of the wave form also increases. The pressure transducer is fed into an electronic receiver which converts the signal into a steady output (0–5 V) which is then fed into a standard two-term controller where it is compared with a set-point. The controller emits a signal which is converted to a pneumatic signal to operate the activator on the refuse extraction valve. This is a butterfly valve which when closed stops refuse extraction, and when fully open permits maximum refuse extraction.

7.5.4 Air Cycle

The air cycle in a Baum jig consists of four elements, as shown in Fig. 7.24. These are:

(a) *air inlets* during which the air valve opens to the air chamber allowing compressed air to enter thereby creating an upward current;

(b) *expansion*—this is the period between the closure of the compressed-air supply to the air chamber and the subsequent exhaust to atmosphere. During this period, the level of water in the washing compartment served by the valve is higher than in the air chamber, and the dilated bed is momentarily suspended;

HYDRAULIC AND PNEUMATIC SEPARATION

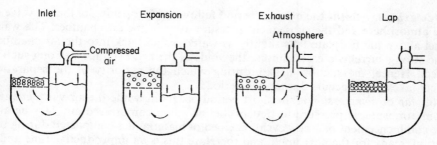

Fig. 7.24 The elements of the Baum air cycle.

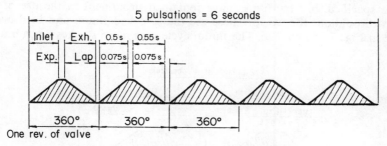

(a) Graphical representation of pulsation versus time.

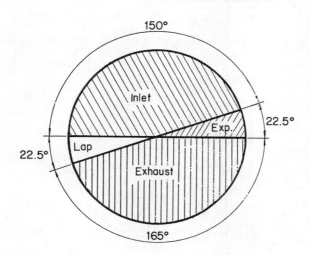

(b) Graphical representation of the division of one cycle.

Speed, 50 r.p.m.

Air pressure, 10–15/kPa

Fig. 7.25 The standard Baum cycle.

(c) *exhaust*—this is the exhaust period following the opening of the air valve to the atmosphere and the subsequent pressure drop in the air chamber. The water level above the bedplate falls slightly as water rushes back into the air chamber. The strong currents created among the voids of the bed produce a strong suction effect. Hence, the suction stroke, during which the majority of the separation effect takes place, occurs during this period;

(d) *lap* or *compression* is the short period occurring while the air valve changes from atmospheric pressure to that of the incoming compressed air. This heralds the commencement of a new cycle. The bed of material lies compacted during this period, ready for the next uplift, and therefore this is an unproductive time and is normally kept to an absolute minimum.

Several types of air cycle are possible since they are dependent upon the variables as specified, but two types have remained prominent in the use of Baum washers throughout the world. These are the so-called Baum and Bird cycles as typified in Figs 7.25 and 7.26. The Baum cycle uses more strokes per minute at a

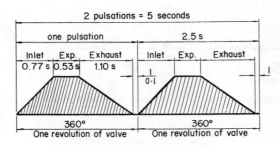

(a) Graphical representation of pulsation versus time.

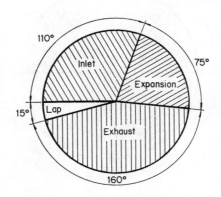

(b) Graphical representation of the division of one cycle.

Speed, 24 r.p.m.
Air pressure, 27·5/ kPa

Fig. 7.26 The 'Bird' cycle.

HYDRAULIC AND PNEUMATIC SEPARATION

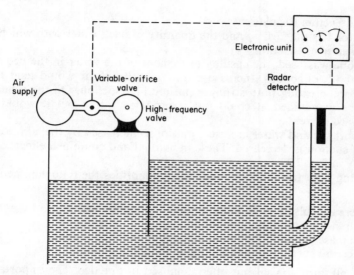

Fig. 7.27 ROSE washbox principle (courtesy Mitchell Cotts).

lower air pressure, while the Bird cycle employs higher air pressure and only about half the number of strokes per minute. Washbox speeds of between 28 and 60 r.p.m. are used for varying types of coal, but despite much controversy as to where the optimum exists, experience appears to have suggested a fairly narrow range of between 40 and 45 r.p.m.

Numerous attempts have been made to tie together in an integrated control system the operation of the air valves and the removal of shale.

One such system is known as the ROSE (Radar Operated Shale Extraction) control unit, which employs short-range radar to measure the height of water in a washbox standpipe as shown in Fig. 7.27. Under the influence of pulsation applied beneath the washing bed, the peak level reached by the water in the standpipe is proportional to the mass of the bed and hence to the shale content. This is similar to other forms of shale extraction controllers, but there are no floats or other kinds of mechanism involved. Specifically designed electronic circuits measure the peak height reached during each pulsation, comparing it with a set-point. Used in conjunction with a large-capacity rotor air-valve with large inlet and outlet ports, the control system is effective even with large-capacity Baum jigs handling raw coal with yields as low as 50%.

7.5.5 Operating Conditions in Baum Jigs

We saw, in Section 7.1, how the operation of the Baum jig is derived from certain basic jigging principles, but there are other, more specific operating conditions which require examination. These include the following:

(a) adjustments to pulsion and suction;
(b) adjustments to the air cycle and air pressure;
(c) adjustments to the jig speed.

Pulsion and Suction Regulation

The operator has the airflow and the quantity of back water with which to regulate the pulsion and suction.

If air alone was used, the ensuing movement of the water in the bed would be similar in both directions. Increase in the quantity of air would tend to elevate the bed more, creating both stronger pulsion and suction. On the other hand, if water alone was added, it could not improve pulsation but it would create a slight upward current.

When both air and water are used, pulsion is increased by the addition of back water, but suction is decreased. These individual and combined effects are shown graphically in Fig. 7.28.

In making adjustments to air and water quantities the following general rules apply:

- to increase pulsation: increase air and/or increase water;
- to increase suction: increase air and/or decrease water;
- to decrease pulsation: decrease air and/or decrease water;
- to decrease suction: decrease air and/or increase water.

which are all fairly logical but often confused in practice. The important choice now is: which one to vary, air or water? The decision depends upon what requires correction, the main objective being to maintain a completely mobile bed with a moderate suction.

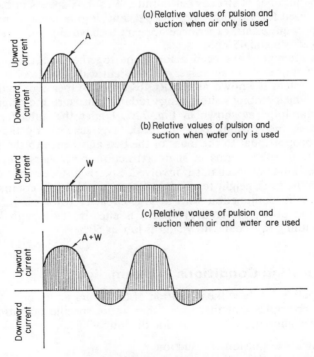

Fig. 7.28 A simplified graphical representation of the effect of air and water on the pulsion and suction of a Baum jig.

HYDRAULIC AND PNEUMATIC SEPARATION

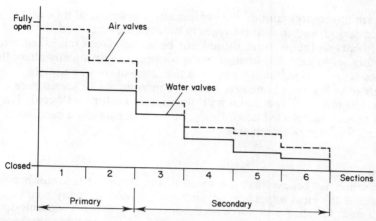

Fig. 7.29 Initial setting of water and air controls for a six-compartment Baum jig.

At the top of the pulsion stroke, the whole bed must be fully dilated, to enable a maximum separation effect to occur. Mobility must be achieved throughout the entire jig bed, and to test for this a common method used is to judge the 'feeling' or compactness of the bed with a long, slim wooden or metal rod. With each pulsion stroke the rod should rise a little, then descend farther with each suction stroke, and with no hard pressure should come to rest on the bed plate. The action of pulsion and suction strokes on the bed of material can best be appreciated by pushing a hand down into the bed and sensing the rush of small grains being pushed upwards, during the pulsion stroke, or downwards during suction. If either effect feels too violent, adjustment will be required to reduce water in the former or air in the latter case. Figure 7.29 shows a commonly used arrangement for air and water settlings. Finer adjustment would eventually give optimized conditions.

The size of the perforations in the bed plate has an important effect on water flow and velocity acting on the bed. Large holes allow greater suction and it is common for the holes to be tapered, with wider diameters facing downwards in order to ensure freedom from blockage. Bed-plate perforations vary from 5 mm to about 30 mm, with the smaller holes, i.e. 5 to about 20 mm being employed for coarse-coal jigging (100 × 0.5 mm) and the larger holes with artificial beds (felspar) for small-coal jigging. It is common for plates of varying aperture to be used as a means of reducing suction.

Air-cycle and Pressure Regulation

In practice there is very little need to vary the air cycle (proportions of inlet, expansion, exhaust and compression). Very few plants deal with such a variety of feeds that frequent adjustment to the air cycle is justified. There are many variations in use and just as many theories supporting them. Once the air cycle has been chosen it is advisable to leave it as a fixed factor. The same applies to the air pressure. The average Baum cycle requires 12–20 kPa pressure and this is supplied by a rotary blower, as a rule. The use of such pressure may be very satisfactory down to a speed of about 30–35 r.p.m. At lower speeds or when the proportion of expansion is increased substantially, the centrifugal types of blowers are unable to maintain the pressure for the length of time required in the expansion period, and so positive types of blowers are employed.

Adjustments to the pressure should not be made, once established, except for maintaining set values. Any change in the air pressure drastically alters the characteristics of separation, taking it beyond the range of routine control.

The air quantity requirement is around 4 m^3/min/m^2 of screen area of mixed feed (e.g. 150 mm × 0) and about half of this amount for small coal. The blower capacity should be selected accordingly, but will normally be capable of supplying up to 7.5 m^3/m^2 of screen area.

Jig Speed

Several authorities believe that the speed of the Baum jig (i.e. number of cycles per minute) is the most important adjustment.[11] However, many boxes have been manufactured, installed and successfully operated without a variable-speed air-valve drive. When mechanical means are available, alterations to the speed of pulsation provide a rapid means for optimization of capacity and/or efficiency. There is therefore a tendency to provide installations with variable-speed drives.

Factors determining the correct speed are:

1. Size distribution of feed. A closely sized fraction can be jigged faster than the original unsized coal and there may be merit in desliming the feed where the fines content of the feed is excessive (i.e. plus 25%).

2. Tonnage feed to box. The general rule is that the higher the throughput the lower the speed should be. Longer strokes tend to be a lot more efficient than short bursts of pulsation, because with short bursts only a small part of the work of dilation of the bed is completed. Many jigs tend to run too fast because this point is not recognized.

3. Depth of the bed. A 'shallow' bed is easier to lift than a deep one, and therefore, a faster speed can be used with 'shallow' beds.

4. Relative density of material. Material of lower density requires reduced speed of jigging because of slower settling velocities. This point is also often overlooked, and speeds that are higher than necessary are employed, with obviously detrimental results.

Feed Size-distribution Effects

In terms of varying size distribution one important point worth bearing in mind is the inherent ability which the jig has for treating various sizes at different densities because of the stratification effect. Figure 7.30 shows the Tromp cut-point against size range as achieved in a modern three-elevator Baum box. As seen, the smaller sizes are separated at considerably higher relative densities. This trend is also shown to be true at every point of separation by the parallel tendencies of the lines representing successive points of separation. For this reason the treatment of presized feed calls for special consideration. In order to effect good separation the jig relies on all sizes to be represented in the washing bed. When some of these are removed prior to washing, the bed will assume different characteristics. It is most desirable to have a certain amount of fines in the feed. These, by filling the voids in the beds, help regulate the effects of suction on the larger sizes and also reduce the quantity of water required, without sacrificing efficiency.

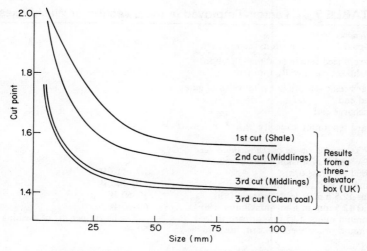

Fig. 7.30 Variation in cut point with size—Baum jig.

Jigs are not normally employed on cleaning coals from which some of the intermediate sizes have been completely removed.

As briefly discussed earlier, the treatment of small coal with a fixed upper size, e.g. 25 mm × 0, or 10 mm × 0, requires the use of artificial bed. Generally felspar is used because of its relative density (2.6) and its favourable cleavage. Operation of a felspar jig differs to some extent from the normal Baum jig mainly because 'jigging through' the bed is practised and the controls therefore work on different considerations.

Difficulties through lack of some sizes are often overcome by recirculation from one of the elevator products, usually middlings, screening out the correct sizes before discarding the unwanted fractions. This practice is always at the expense of a reduction of throughput capacity.

Excessive amounts of certain sizes can be just as embarrassing as insufficient amounts. Every washbox is designed to take a certain range of sizes, and generally the maximum amount of fines in the feed is specified. An excess reduces the efficiency of the separation unless the feed tonnage is correspondingly reduced. Excess fines cause blockage in the bed and sluggish pulsation, i.e. loss of mobility. Readjustment in such cases is very difficult, if not impossible, without correcting in some way for the unbalanced size-distribution. An attempt to retard the extraction rate of the larger discard sizes could constitute a compromise temporary measure. Other factors being the same, the higher the top size of raw feed, the larger will be the tonnage per square metre area that the washbox can treat.

Discards Capacity

There is a maximum proportion of shale in the feed with which any Baum washbox can deal. This relationship is due to the physical limitations to stratify and extract more than a certain amount of discard over a given area; consequently the width of the discharge gates is a limiting factor too.

The various factors connected with capacity and generally used in the process of selecting washboxes for a given duty are tabulated in Table 7.5, together with other useful design data.

TABLE 7.5 Factors Employed in the Selection of Washboxes

(a) *General rating*

Raw-coal feed rate in t/h/m² screen area	15–30
Raw-coal feed rate in (150 mm–0) t/h/m washbox screen width (normal)	100–120
Discard removal rate in t/h/m width of gate	
feed end	30–40
discharge end	15
Discard separation capacity t/h/m² screen area	3–8.5

(b) *The specific rating of the washbox feed*
If W = width of washing area in metres

Normal feed rate (t/h)	100–110 W
Maximum feed rate (t/h)	120–140 W
Minus 12.5 mm in feed—normal (t/h)	50–60 W
Minus 12.5 mm in feed—maximum (t/h)	60–80 W
Maximum minus 12.5 mm discard in feed	3.00–5.00 t/h per m² of washing area
Maximum primary sill discard rate (t/h)	50 W

(c) *The minimum rating of washbox elevators*
(i) Two elevator washboxes

feed end elevator	—90% of total maximum discard/middlings
discharge end elevator	—50% of total maximum discard/middlings

(ii) Three elevator washboxes

feed end elevator	—80% of total maximum discard/middlings
centre elevator	—40% of total maximum discard/middlings
discharge end elevator	—30% of total maximum discard/middlings

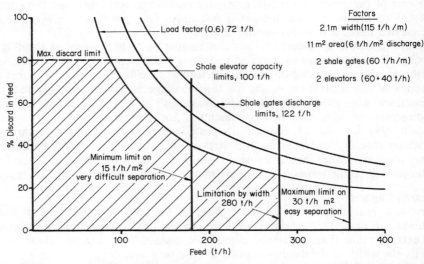

Fig. 7.31 The relationship between discard content of feed and physical limitations of washbox.

The use of these factors is illustrated in Fig. 7.31. Elevator capacity is added to the above limitations and since it is a mechanical limitation it can gradually be adjusted.

Many of the factors included in the table are, of course, arbitrary and the result of many years of observations and experience. Their values are not rigid and depend on the ease of separation, size distribution in the feed and physical limitations of construction.

In Fig. 7.31 the safe capacity is within the shaded area, defined by the load factor and width limit, but if the feed is a difficult coal to wash, or circumstances change, some of the other factors may become limiting.

Exceptionally clean feeds may be difficult to treat efficiently because they contain too small an amount of reject material. In order to achieve separation a heavy bed is a necessity, and lack of it can lead to losses of large coal. If such conditions are temporary, measures designed to retard the discard extraction rate would eliminate unnecessary losses.

Bed Thickness

The density of the bed is a decisive factor in determining the separation. The minimum thickness of the bed is governed by the refuse-gate opening, which must be at least one and a half times the maximum size treated. Also, the sizes treated determine, to a large extent, the total depth of the bed. In the case of larger sizes (with feeds containing +100 mm top size) the bed depth must be around three times the size of the largest lump in order to ensure that these are separated in a water–solids medium and do not just sink onto the bedplate as soon as they enter the box.

Within the limitations of the above considerations, the bed should be kept as shallow as possible, to cater for the prerequisite of quick and efficient mobilization on the pulsion stroke. The correct bed thickness is derived by experimenting with the shale-gate opening which also acts as a retarding device, preventing the bed from sliding out quickly with the pulsation. Similar experimentation is necessary with the discharge-control float. The float automatically maintains the bed thickness but it is only efficient within a designed range of variation in feed composition.

Adjustments to the shale bed are among the routine duties of the operator. Thickness and mobility must be regularly checked and verified by visual observation of the elevator products. Loss of the shale bed must be prevented. The

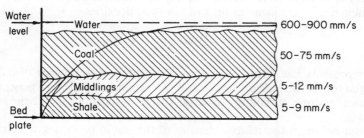

Fig. 7.32 The velocity of the various layers in the Baum jig.

various layers of material have varying retention time in the box, as shown in the diagram in Fig. 7.32.[12]

As a result, a sudden loss of shale bed takes a considerable time to replace. As a washbox of 300 t/h capacity may take as much as half an hour to build up a full shale bed, separation during this period is inefficient. Loss of bed is easily recognized—the elevator product contains coarse coal, together with the normal sizes of shale.

Water Control
Water control in the Baum jig has crucial significance as to the way in which the separator performs. Water is needed at two points: under the bedplate (backwater) and with the incoming raw material (top water). The backwater provides the suction effect and promotes mobility, while the top water thoroughly wets the feed and helps it slide into the box. The proportion of backwater to top water is about 2 : 1 in practice. Excessive top water creates horizontal currents of undesirably high velocity, introducing particle-size and -shape effects into the process of separation.

The total amount of water in circulation varies with the size of the feed and the type of air cycle adopted. Fine coal and unsized raw coal require a great deal less water than preclassified material. Coal containing large amounts of fines may necessitate the use of extra top water in order to prevent 'balling', i.e. conglomeration of small and fine fragments rolling along the top of the bed without being cleaned and contaminating the clean-coal product.

The conventional Baum cycle uses about 1.5–2.5 m^3/h of water per t/h of feed of which about 0.75–1.25 m^3/h are admitted under the bed. These quantities of water are slightly reduced when the washbox is large.

The pure Bird cycle (22 r.p.m.) uses only about half (or less) the quantity of water for a Baum cycle, and relies on higher air pressure and prolonged expansion to provide the pulsion.

The pressure of the water is also very important. It should not be altered, as a complete readjustment of all other settings would then be necessary. A constant level head-tank is the safest source of pressure, working on the overflow all the time.

Feed Solids Concentration
It has often been said that a 15–20% solids concentration in the wash water is beneficial in separation as it creates a dense-medium effect. The ultimate falling velocity in water is a function of size as well as density, and a large piece of coal will attain the same ultimate falling velocity as a smaller piece of shale, when the ratio of their sizes is the same as the ratio of their densities.

$$\frac{d_{\text{shale}}}{d_{\text{coal}}} = \frac{2.5 - 1.0}{1.3 - 1.0} = 5 : 1$$

where d represents size. (This does not apply to the initial stages of fall—which occur at the top of the pulsion stroke and the expansion period in general—when all particles assume a falling velocity strictly according to their respective densities.)

As shown before, if the relative density of the liquid increases, it increases the ratio of the sizes accordingly.

HYDRAULIC AND PNEUMATIC SEPARATION

This means that, theoretically, with dirtier water a wider size range can be separated with the same efficiency or the same size range with increased efficiency.

However, practical experience shows that washing in clean water always produces cleaner end-products. This is contradictory to what has previously been emphasized on theoretical grounds. The explanation lies in the fact that with higher relative densities of the 'medium' the velocity of water required to lift the particles is reduced. So for the same upward velocities of the jigging action there are shale pieces of a larger size range lifted to higher layers of the bed and these contaminate the clean coal. The answer therefore seems to be in an adjustment to the pulsion if these effects are noteworthy.

Use of very dirty wash-water can noticeably affect the ash content of the clean coal by surface adhesion. Installation of efficient water sprays over the clean-coal dewatering screens is strongly recommended in this case.

7.5.6 Efficiency of Washboxes

From the foregoing, the number of variables affecting efficiency may appear to be almost countless and the requirement to achieve reasonable separation, formidable. However, with a systematic approach it is reasonably easy to produce good results. In order to complete the picture of Baum washing, some aspects of efficiency are now discussed and then some typical circuits.

The efficiency of Baum jigs is superior to most of the separators relying on water currents and is therefore only inferior to that of the dense-medium separators.

On mixed sizes, the normal range of e.p.m. values, i.e. 0.04–0.08 compares very favourably with results from tables and water cyclones. When considering that a Baum plant is comparatively inexpensive to construct, the efficiency on coarser sizes makes the Baum jig an attractive alternative to dense-medium separation

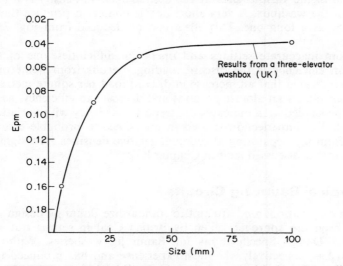

Fig. 7.33 Variation in probable error (e.p.m.) with size—Baum jig.

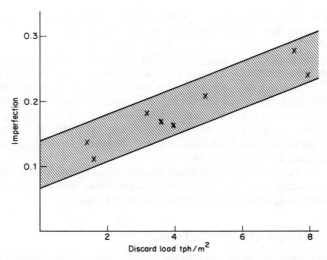

Fig. 7.34 Imperfection versus discard load for 25–10 mm feed in the Baum jig (after Wallace[10]).

providing that the coal is not too difficult to wash or does not require a low density of separation.

On small coal, however, this efficiency is not as comparable with that of dense-medium cyclones. Figure 7.33 further illustrates this point. The results in Fig. 7.34 are from the same washbox at the same time and so provide an accurate comparison. It has been shown that the decrease in efficiency with size accelerates very significantly at around 1/2 mm. Below this size there is virtually no effective cleaning taking place.

Separation of the various sizes at the desired relative density is a function of the length of the washbox. A very short box is unable to produce the same cut and accuracy as a long one. This effect can be deduced from Fig. 7.30 shown previously.

Apart from inaccurate settings and operation difficulties, one of the major influences on efficiency is the discard loading. Results from numerous installations have indicated that an increase in discard load per square metre of screen area per hour effects an almost proportional decrease in efficiency, as shown in Fig. 7.34. The shaded area indicates the trend in efficiency with variations in the discard load. The imperfection is used in preference to probable error as these results are from jigs separating at different relative densities. The significance of this will be further discussed later in Chapter 8.

7.5.7 Typical Baum Jig Circuits

Perhaps the only nationalized attempt to standardize Baum jig design and application has been the approach taken by British Coal in setting out a Code of Practice and Design Specifications for Baum jig washeries. Without doubt British Coal has, collectively, the most experience and has produced the largest number of Baum jig installations anywhere in the world, and therefore its stan-

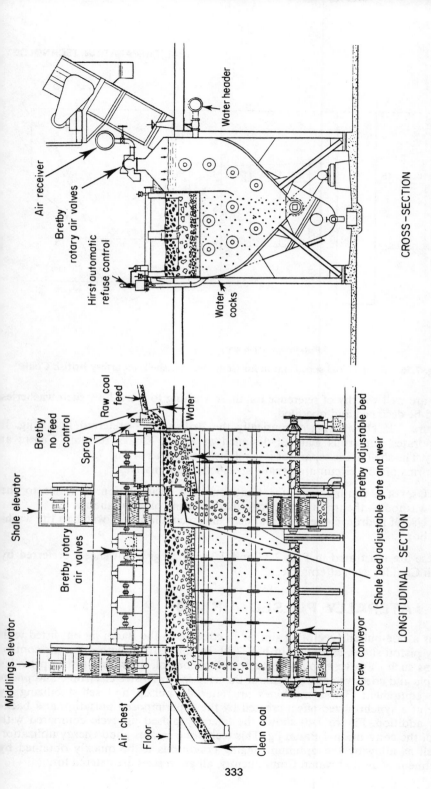

Fig. 7.35 The British Coal-preferred Baum jig.

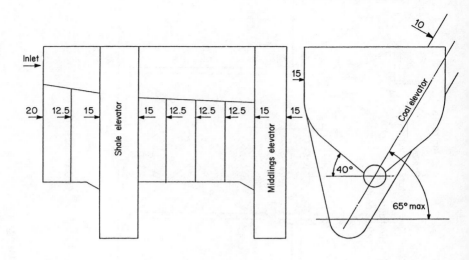

Plate thicknesses in mm

Fig. 7.36 British Coal specification for Baum washer shells (courtesy British Coal).

dards are well worthy of reference for those wishing to learn how such washeries should be designed and operated.

Figure 7.35 shows diagrammatically the British Coal-preferred Baum jig. It incorporates numerous features designed at British Coal's research Centre at Bretby. The shell of the separator is carefully specified as in Fig. 7.36.
Two forms of flow circuit are usually specified as alternatives:

(a) Use of a settling cone for treatment and recirculation of water to the washbox—the traditional circuit used in many older washeries.
(b) Use of hydrocyclones for treatment and recirculation of water to the wash box.

These are illustrated in Figs 7.37 and 7.38 which give flowsheets preferred by British Coal for each alternative.

7.6 LATERALLY PULSED AIR JIG

This is a side-pulsed type of jig similar in concept to the Baum jig but fitted with rotary piston slides, as shown in Fig. 7.39. In comparison with other air-control systems such as valve, flap and eccentric slide control, the rotary piston provides a simple and effective control without the need for expensive electronic and pneumatic equipment. These machines are laterally pulsed and self-stabilizing by virtue of a synchronized effect created by the superimposed pulsation and back water addition. Figure 7.40 shows the laterally pulsed jig cycle compared with that of the conventional Baum jig. This technique creates good energy utilization as well as allowing the optimum jigging conditions to be quickly obtained by adjustment of air and water. Consequently, all grain sizes are catered for.

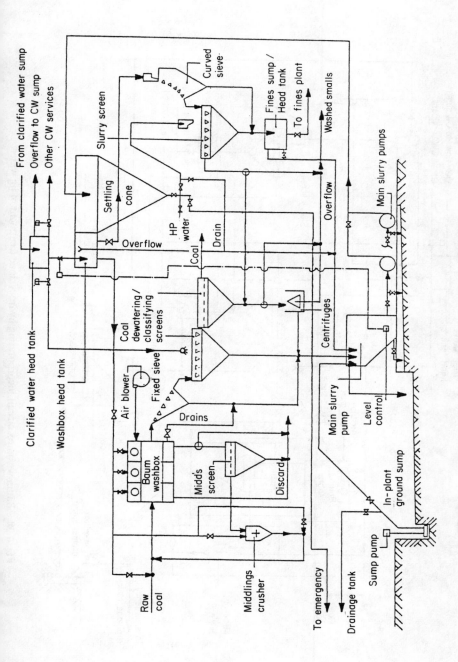

Fig. 7.37 Baum washbox circuit incorporating settling-cone treatment of jig water.

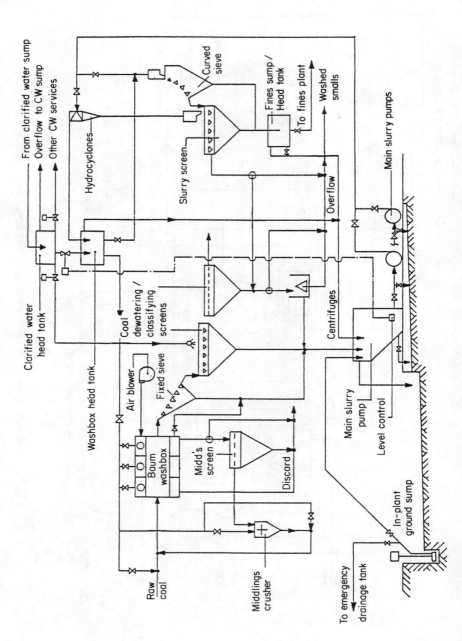

Fig. 7.38 Baum washbox circuit incorporating hydrocyclone treatment of jig water.

HYDRAULIC AND PNEUMATIC SEPARATION

Fig. 7.39 GHH laterally pulsed coarse-grain jig (courtesy MAN–GHH).

This type of jig is available for treating coarse, fine and ultrafine (or slurry). Table 7.6 gives typical operating data for the three types of jig design.

The common design of coarse jig is a single-cut, two-product, twin type, as shown in Fig. 7.39. In this machine coal and reject are discharged using a float-

TABLE 7.6 Laterally Pulsed Jig-design Data

	Grain size		
	Coarse (80–10 mm)	Fine (10–0.5 mm)	Slurry (2–0 mm)
Air requirement ($m^3/m^2/min$)	5.0	3	2×1
Working air pressure (bar)	0.35	0.25	0.15–0.10
Backwater requirement ($m^3/m^2/min$)	0.8	0.5	0.2

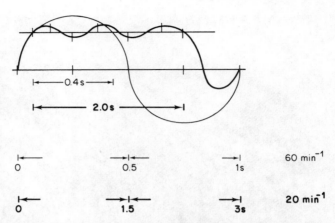

Fig. 7.40 Diagram showing the laterally pulsed jig cycle superimposed onto a normal Baum jig cycle (courtesy MAN–GHH).

activated discharge mechanism, as shown in Fig. 7.41. The rejects fall to bucket elevators in the bottom of the hutch. This type of jig, designed to treat coal sized 150 × 0.5 mm, has two beds each 3 m wide, and is capable of treating each up to 600 t/h. Another version, with two beds each of 3.5 m in width, is capable of treating up to 530 t/h of fine coal sized 10 × 0.5 mm. A third type, representing a recent milestone in fine-coal cleaning, is the slurry jig which is shown in Fig. 7.42 and treats 2 mm × 0 fine-coal slurry.

Further development of the concept of the electronically controlled superimposed air cycle has resulted in a new design of slurry-fed jig. This unit is currently undergoing further development in Germany and is based on the original Cortix air-pulsed felspar jig. It can treat coal sized 2–0.1 mm at up to 10 t/h.[15] The Cortix jig is fed with a slurry of constant solids concentration (controlled by flowmeter and density gauge) and because of their feed-control feature, the quan-

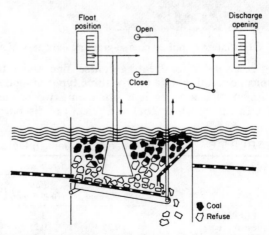

Fig. 7.41 GHH laterally pulsed jig shale-discharge control (courtesy MAN–GHH).

HYDRAULIC AND PNEUMATIC SEPARATION

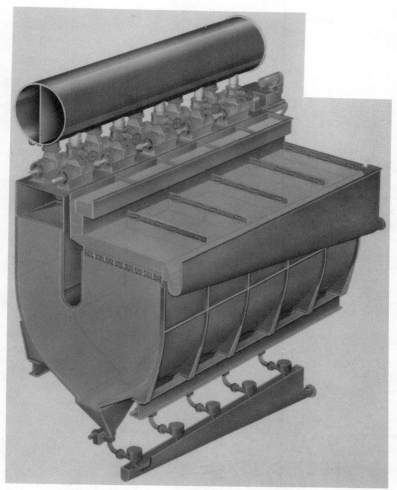

Fig. 7.42 GHH slurry jig (courtesy MAN–GHH).

tity of jig water can be very accurately regulated, in conjunction with the air-cycles, to obtain rapid response to any significant changes occurring in the washability or size distribution of the feed. It is expected to be especially useful in the treatment of coals containing weathered or oxidized material (thus not amenable to flotation) and those coals that contain large amounts of feed pyrite. Figure 7.43 is a diagram of the Cortix jig illustrating the parallel-operational, twin air-pulsation system, a unique feature of this type of jig, based on its laterally pulsed-air, coarse-jig predecessors.

7.6.1 Under-air Jig

Application

The historical development and Japanese origin of this concept of jig has been discussed already. Its emergence in Europe, first as a viable alternative to Baum

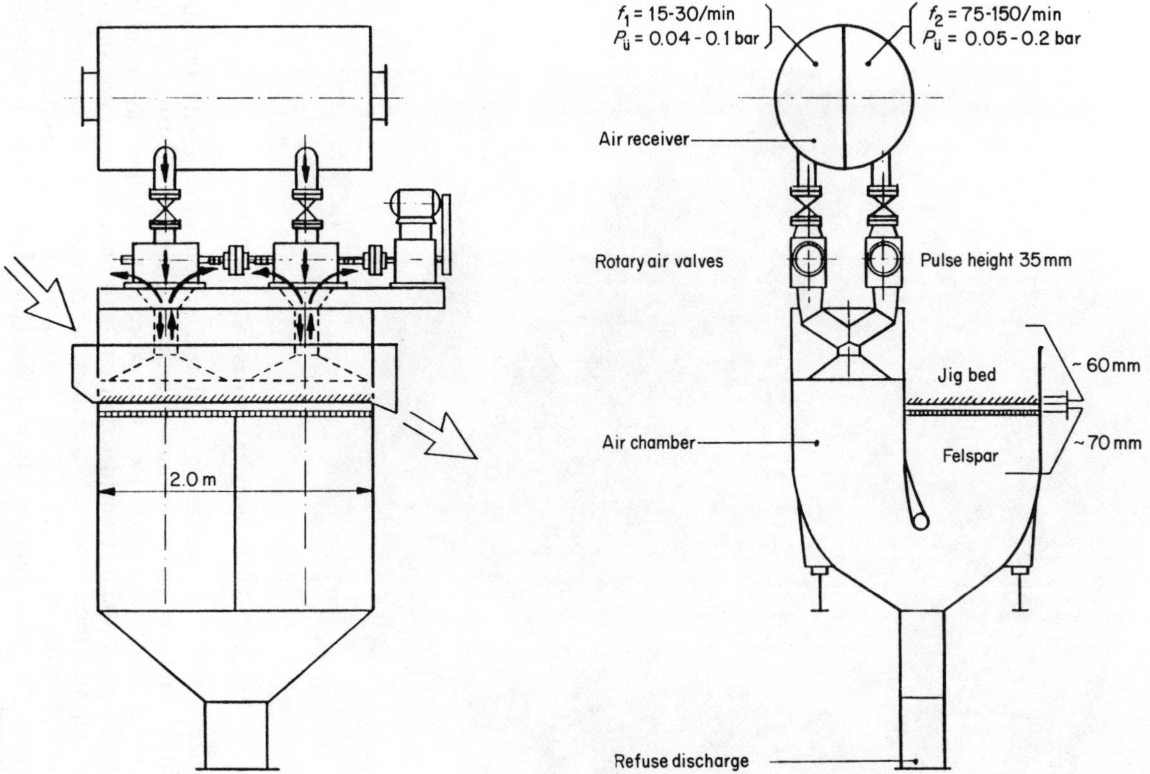

Fig. 7.43 Fine-coal jig with superimposed air cycle.

jigs and subsequently as a means of replacing widely popular but increasingly costly dense-medium systems in Germany and France, has led to numerous successful installations. Humboldt, Fives Cail and SKB currently offer high-capacity, under-air jigs with capacities of up to about 1000 t/h. As we have already seen, in the under-air jig the principle of causing pulsations to occur within the raw-coal bed is the same as with the Baum jig. However, the method of air addition, using a new type of valve system, is considerably different. Most significant is that there is no side chamber, the air chamber running across the width of the jig transverse to the flow of the bed. This form of design tends to led itself to higher-capacity units with performance characteristics similar to those of the smaller-sized ones, as a direct result of the good pulsion distribution provided by the air-valve system. The application of this type of jig is more or less the same as the Baum jig, except that the under-air type occupies less space and in the case of the Batac coarse-coal design is capable of treating larger capacity.

Construction
Figure 7.44 shows a three-compartment, seven-cell, under-air jig of Batac design, for coarse-coal cleaning. Discard is extracted via a discharge device located at the end of the bedplate in each compartment. In the example shown, the air valves used are of a flat disc design which produces a characteristically sharp cut-off of both intake and exhaust movements. These valves are infinitely variable throughout the range of speed and length of stroke ranges afforded by the unit. They are operated from a pre-programmed, electronic, solid-state controller, usually located in the plant control-room. In Fig. 7.44, the air pipes delivering air from the valves to the air chambers cause the pulsations within each cell beneath the coal bed. Each cell has two air chambers mounted beneath the bed. Backwater is admitted at the bottom of each cell, creating exactly the same effect as in a Baum jig. The cleaned-coal product passes over the air at the end of the box. Control of shale is created by a discharge controller, activated by two floats preceding the reject gate opening in each compartment, together with an auxiliary float near to the feed end. These are clearly visible in Fig. 7.13. The floats respond by induction coils which send a signal that is a function of actual bed depth, for control comparison to an adjustment-reference signal (see Fig. 7.45). The feed float, when sensing low bed depth, activates two magnetic switches to close a throttle valve in the air header which prevents any disturbance of the bed. The downstream sensors control the hydraulically operated reject gates, maintaining bed depths and hence controlling the separation cut-point.

Operating Conditions
The capacity of the Batac under-air jig increases with the upper size of the feed and ranges from 18–20 t/h/m^2. Units have been installed ranging in width from 3 to 6 m. The raw coal is first pre-wetted and screened to provide a feed sized between 150 and 10 mm for coarse-coal units jigging mainly over the bed plate, and 10 mm and 0.5 mm for fine-coal units jigging through the bed and utilizing a felspar bed form of discharge. Referring again to Fig. 7.44, which shows the coarse-coal jig cross-section, cells 1, 2 and 5 contain 8-mm round-hole screen-beds, and cells 3, 4 and 6 have larger, 16-mm square-hole openings supporting felspar ragging (shown cross-hatched) which allows fine rejects to pass to the hutch. The slopes of the screen-beds in the first two cells are adjustable, while the others are horizontal. The felspar is laid in small compartments on the bedplate

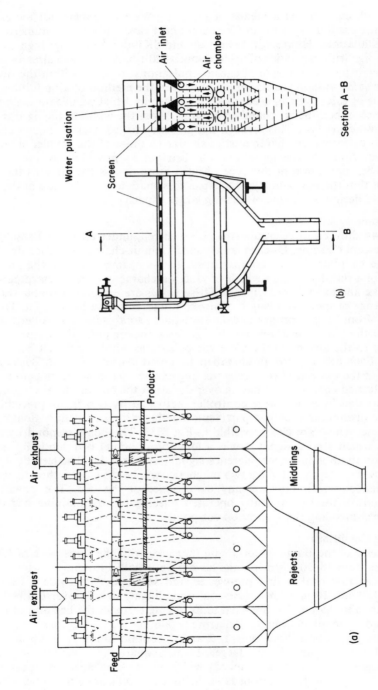

Fig. 7.44 (a) Side-view and (b) end-view cross-sections of the Batac jig (courtesy Humboldt–Wedag, Ref. 13). (a) Jig is designed with a series of multiple air chambers, usually two to a cell, extending under the jig screen for its full width so as to provide uniform air distribution. (b) Batac jig cross-section. High-density material in the coal discharges through the screen plate perforations and at the end of the compartments through shale ejectors.

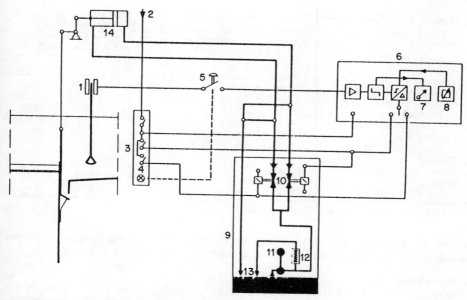

Fig. 7.45 Electro-inductive hydraulic shale-discharge system for the Batac jig (courtesy (Humboldt–Wedag). 1. Inductive coil for level indication; 2. voltage supply; 3. switch combination for manual actuation; 4. measuring-point control; 5. automatic release of measuring points; 6. electronic slide-in control units; 7. indication of actual valve; 8. nominal valve adjustment; 9. pressure oil station; 10. electromagnetic three-way valve; 11. oil pump with electric motor; 12. pressure-regulating valve; 13. oil tank; 14. hydraulic servomotor.

to maintain a uniform distribution. Hence, the two shale ejectors are located immediately preceding each felspar bed, to remove coarse rejects. Only two elevators are employed in a single unit but the secondary rejects (or middlings) can

TABLE 7.7 GHH Jig Efficiency Data[14]

Grain size (mm)	e.p.m.	Imperfection	Separation density (kg/dm^3)
25–10	0.05	0.12	1.39
	0.09	0.09	1.91
10–2	0.06	0.14	1.45
	0.10	0.12	1.89
2–0.5	0.06	0.12	1.49
	0.11	0.10	2.03
Combined	0.06	0.13	1.43
	0.10	0.11	1.92
Feld	Relative density minus 1.5 +1.5 − 1.8 plus 1.8	% Mass 37.5 26.0 36.5	

TABLE 7.8 Guaranteed Performance for a Batac Jig (Coal source: USA Lower Kittanning B seam raw coal crushed to minus 9 mm)[13]

	9 × 0.6 mm mesh			0.6 × 0.15 mm mesh		
	A^a	B	C	A	B	C
dp—Separating density	1.50	1.65	2.00	1.65	2.00	2.20
e.p.m.—Ecart probable moyen	0.075	0.092	0.140	0.182	0.208	0.313
I—Imperfection	0.15	0.14	0.14	0.28	0.26	0.26
OE—Organic efficiency (%)	97.6	99.6	99.8	95.6	98.2	99.6
Theoretical yield (%)	80.2	84.0	90.3	88.7	90.3	94.9
Clean-coal yield (%)	78.3	83.7	90.1	84.8	88.7	94.5
Clean-coal ash (%)	5.72	6.95	6.64	5.86	6.64	9.01
Clean-coal sulphur (%)	1.80	1.92	2.31	1.92	2.04	2.52
Reject ash (%)	53.3	62.8	72.7	48.9	58.0	71.4
Reject sulphur (%)	11.6	14.2	18.6	12.4	15.1	20.6

^{a} A, B and C are three different settings used with the jig.

either be discarded as a reject or recycled with the jig feed (with or without crushing) in much the same manner as described earlier for the Baum jig.

Performance
Performance is reported as being equal to, or better than other types of jig separating at densities of 1.45 or higher. At higher than 1.55 the efficiency determined for some installations has proved better than for dense-medium systems in cleaning coals with relatively difficult washability characteristics. Capital cost of Batac jig installations lies somewhere between that of the Baum jig of comparable capacity and the dense-medium bath. Under normal circumstances, when treating coal of similar size (150 × 10 mm) and washability characteristics, the performance of the Batac jig also lies somewhere between the other two. Table 7.8 gives performance data for the Batac jig as tested with USA coal crushed to minus 9 mm. These data represent the guaranteed level of performance and have been improved upon in numerous large commercial units.

7.6.2 Moving Screen Jig

Reference to this mode of jigging was made in the opening paragraph of this chapter and the category is shown in the jig family tree shown in Fig. 7.8. Although a number of moving-screen jigs were developed in the latter part of the last century and the early part of this one, their biggest drawback was found to be their low mechanical availability. With the emergence of the Baum jig, the interest in moving-screen jigs quickly faded and only one, the Hancock type, remained in common usage by the 1920s. Interest was revived in the early 1980s with the need to deshale (or preconcentrate, as it is referred to in mineral-dressing applications) run-of-mine coal prior to conventional cleaning. The alternatives available for this operation are:

(a) hand picking, which is outmoded, costly in labour and inefficient;
(b) rotary breaker, which will work if the coal is not too hard relative to the stone; and
(c) dense-medium bath, which is costly and likely to be regarded as 'overkill'.

Accordingly, a German company, KHD Humboldt–Wedag, developed a

HYDRAULIC AND PNEUMATIC SEPARATION

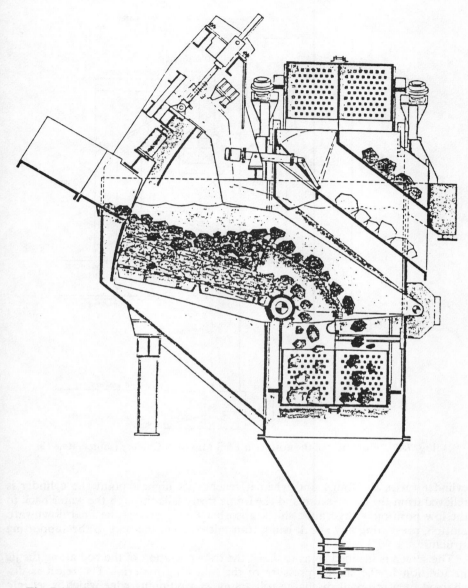

Fig. 7.46 Schematic drawing of the ROMJIG (courtesy K.H.D. Humboldt–Wedag).

movable screen jig referred to as ROMJIG, designed as a single-cut jig for separating large stones from 400 × 40 mm sized raw coal. This unit is shown diagrammatically in Fig. 7.46. The ROMJIG consists of a unilaterally supported screen-frame with a slotted perforation 15 mm wide. The frame is hydraulically moved up and down in a water-filled tank. The screen drive is via a hydraulic cylinder suspended on gimbals which perform the lifting function. The hydraulic

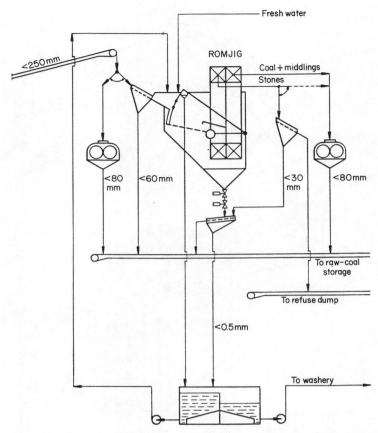

Fig. 7.47 Flowsheet of the ROMJIG at Emil Mayrisch Colliery (source, Ref. 15).

cylinder raises the frame and when it reaches its highest point the cylinder is relieved from the oil pressure and the frame freely falls through the water back to the low position. A hydraulic shock absorber finally arrests the final downward motion, preventing any shock being transmitted from the unit to the supporting structure.

The screen is inclined so as to assist the transportation of the bed along the jig in addition to the sliding pressure of the following material. The reject is discharged by a resonant discharge roll acting as a retaining edge which is set into circular motion by a hydraulic, motorized chaindrive. The speed of the discharge roll depends on the shale-bed thickness. The coal resting on top of the shale bed passes to a weir and is discharged by a resonant bridge. The discharge of both products is effected by a twin-bucket wheel, adequately sized in order to provide good drainage. To avoid the discharge of misplaced material, in particular during start-up, the reject discharge chute is provided with a changeover gate to enable heavy material to be added back with the light. Unlike the pulsed coal jigs, the ROMJIG does not need backwater and only a small quantity of water is withdrawn from the hutch to discharge accumulated fines. This is largely recovered from the 0.5-mm drainage screen to which this material is fed.

The first machine was installed at Emil Mayrisch colliery in Germany in 1985 and has worked very efficiently. The maximum feed rate is 300 t/h. The flowsheet is shown in Fig. 7.47.

7.7 FLOWING-FILM SEPARATORS

The behaviour of coal and shale fragments contained within a flowing stream of water has been discussed previously and methods of separation i.e. launder and trough, have been described. When the water stream becomes so thin as to form a film which may, in some cases, not totally immerse the feed material, some differences occur in the behaviour of the flowing material. These will be described a little later when the basic principles involved are briefly reviewed, but this different form of separation is usually referred to as a flowing-film separation. Its use in mineral processing is arguably more important[16] but during the past few years there has been a significant increase in the popularity of flowing-film concentrating devices for cleaning fine coal. At the forefront of this upsurge has been the new variety of fibre glass, compound trough design of coal-cleaning spirals which has found international recognition for treating fine material in a range from 3 to 0.075 mm. Spirals have been used in the coal industry for a number of years.[17] The recent development in Australia of the new lightweight designs intended for coal use[18–20] has been largely responsible for the renewed interest. When compared with mineral-treatment designs, the features of coal spirals take account of the small relative-density differential required to separate accurately raw coal and provide the following:

(a) a smaller pitch to effectively reduce the velocity of the pulp and increase residence time;
(b) a larger outside diameter and pulp-volume carrying capacity.

It would also appear that, besides the introduction of new lightweight materials of fabrication, the interest in spirals was rekindled by the combined attempts in Australia to use another flowing-film device—the Reichert cone concentrator and the Reichert design of spiral in a dual role of cleaning coal in two or more stages. Although the combination was apparently successful, an all-spiral flowsheet eventually emerged following the design changes described earlier.[19]

Long before the upsurge in interest in spirals occurred, the shaking table had established itself as the only flowing-film coal-cleaning unit to be universally adopted for fine-coal cleaning. Even with the development of fine coal jig separation in the early 1970s, the shaking table remained a very popular choice. In the USA, in particular, tables were the preferred separator for fine-coal cleaning in the eastern coalfields of Appalachia, where they are still widely used, especially for coking-coal treatment. In the USA they remain the preferred alternative to spirals, being able to treat a coarser feed (up to 10 mm top size) but also to treat coal as fine as 0.075 mm.[21] The earliest forms of tables specifically used for coal cleaning were designed to treat coarse size-ranges, and riffle heights of up to 25 mm were occasionally employed for treatment of coal as coarse as 50 mm top size. With the advent of efficient trough-and-jig types of washers the size-range of

Fig. 7.48 Table-treatment section in a USA washing plant (launder lined with basalt blocks) (courtesy Deister Concentrator Company Incorporated).

the tables was gradually reduced to the present upper limit of 10 mm. Even now, the splitting of such feeds into two, or sometimes three, size-ranges is recommended in order to ensure optimal efficiency. The advantages of the table for treating fine coal include: low operating and maintenance costs; high degree of flexibility in operation; and low water requirement, i.e. about 2.5 : 1 ratio of water to solids in the feed. Disadvantages are few if the correct application is selected, but the space occupied by the large number of units usually required is clearly the main disadvantage. Hence, low unit capacity and large space requirement are usually key factors in comparing tables with spirals. To combat the space disadvantages, multideck designs emerged and are now widely employed. Up to four tables are arranged above one another to provide a higher capacity per square metre of plant floor-space occupied by the units. A number of other types of table separators have been developed and a few have been adopted in commercial applications. The most recent innovation is a centrifugal table. This drum-shaped unit is 1.8 m in diameter containing eight arc-shaped decks which are rotated, creating a centrifugal force equivalent to 18–20 times the force of gravity.[22]

Fig. 7.49 Spiral treatment section in a Canadian washing plant (courtesy Luscar-Sterco (1976) Ltd, Canada).

HYDRAULIC AND PNEUMATIC SEPARATION

Fig. 7.50 Novel table tower operating in a German washing plant (courtesy KHD Humboldt–Wedag).

Conventional concentrating tables are able to treat effectively coarser material than coal spirals and have been successfully applied over a feed size-range of 9.5–0.1 mm. Although there are still very few commercial operations where the 0.5–0.1 mm material is tabled separately, it is acknowledged that closer sizing of table feed provides a more effective separation.[21] In the treatment of fine coal in the minus 0.5-mm size range, it is not clear whether the table or spiral is more effective, because of an absence of performance data for either, but it is generally widely regarded that both are very effective in the removal of fine pyrite, and this application is currently a major selection factor. Test work has shown that pyrite finer in size than 74 µm can be effectively separated by either separator. Figures 7.48 and 7.49 show modern table and spiral installations. The table circuits shown treat 6.7 × 0.5 mm raw coal to recover a low-ash coking coal in an American plant. The spiral circuit is used for treating a 1.5 × 0.1 mm raw coal with high clay content in a Canadian plant. Figure 7.50 shows a novel table tower designed and developed in Germany and accommodating 72 tables in 12 levels of six tables each.

7.7.1 Basic Principles of Flowing-film Concentration

Fluid films have a rheological property which may be utilized for the separation of coal, shale and mineral grains. This property is the variation in velocity that occurs at varying depths through the film which ranges, in theory, from zero at the bottom to maximum at a point very close to the top beneath the air friction 'skin' which acts as a slight velocity retardant. This velocity profile is influenced by: the thickness of the film, the viscosity of the fluid and the characteristics of the solids and the surface of the separator. The simplest device for flowing-film concentration consists of an inclined surface or table on which assorted grains are subjected to flowing water. The lighter grains, being most affected, are washed off the table, whereas the heaviest tend to accumulate. This form of separator is

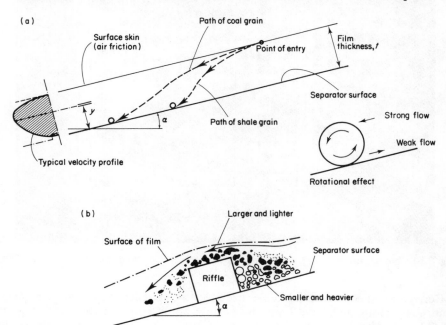

Fig. 7.51 Flowing-film separation on smooth and riffled surfaces. (a) Separation in a flowing film on a smooth inclined surface. (b) Effect of riffle on flowing grains.

HYDRAULIC AND PNEUMATIC SEPARATION

perhaps the oldest type known to man. In an idealized flowing film the velocity gradient is uniform and the flow condition is wholly laminar. If y is the distance within the film from the substratum–film interface to a specific layer, as measured at right angles to the substratum (see Fig. 7.51), the fluid velocity is given by the formula

$$v = \frac{d_1 g \sin \alpha}{\mu}(2t - y)y \qquad (7.14)$$

The volume, Q, of the fluid flowing per unit time and unit width, and the depth of the fluid film, are related. The relation can be obtained by integration between limits 0 and t of the element of fluid volume, dQ

$$dQ = v\, dy = \frac{d_1 g \sin \alpha}{2\mu}(2t - y)y\, dy$$

which gives

$$A = \frac{d_1 g \sin \alpha}{3\mu} \times t^3$$

hence

$$t = \left(\frac{3\mu \times Q}{d_1 g \sin \alpha}\right)^{1/3} \qquad (7.15)$$

which shows that the depth of the flowing film varies as the cube root of the flowrate and inversely as the slope of the surface. While these formulae are rarely used in the design of flowing-film separators, they have been validated by experimental work and have value in the development of separator models for smooth-surface separators, such as the Reichert cone and spirals.

The introduction of riffles on the surface of the table, as featured in the modern shaking table, creates the added separation dimension of stratification, as can be seen in Fig. 7.51(b). The mechanism involved is similar to that previously described for the jig, and the close proximity of each pair of riffles tends to create a form of jigging compartment or trough whereby the pulsations occur as a result of the jigging (or strictly speaking, reciprocating) action of the table. Riffles also create a significant increase in capacity, because of the multilayer separation which occurs, as opposed to the single layer on a smooth surface. Within the riffle troughs, hindered settling and consolidation trickling occur. Because of the loss of motion between successive layers of grains, the stratified material between riffles is acted upon by the lengthwise motion of the deck, to an increasing extent at depth, rather than close to the surface. Since the bottom layer consists of the finest and heaviest grains, these move much faster to the end of the riffles than the largest and lightest grains. Therefore, in respect to size the rate of travel to the discharge end is opposite to that obtained with a perfectly smooth surface. As a result of the differential response of each layer of material to the asymmetrical acceleration imparted to the deck as well as to difference in size, density and shape, the net result in terms of lengthwise travel along the riffles is similar to that shown in Fig. 7.52(a). At the same time the cross-flow of fluid from one riffle to the next also occurs. This is largely the result of the crowding action of the

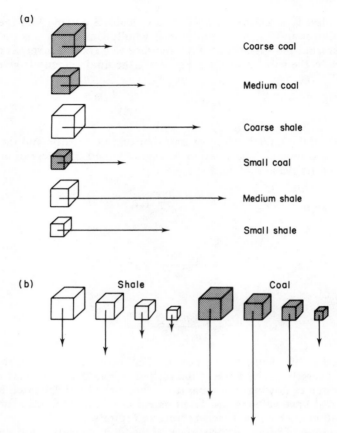

Fig. 7.52 Separation behaviour on a riffled shaking table (after Gaudin, Ref. 2). (a) Vertical stratification within the riffles, and rate of travel. (b) Cross-deck travel across the riffles, and rate of travel.

new feed and partly due to a tapering of the riffle from feed to discharge end. The crowding effect brings the upper stratum within range of the action of the flowing film that moves across the riffles. Lighter and coarser grains are washed downstream and finer and heavier ones move along the line of riffles until the riffle height exposes the coarser grains, causing them to be drawn into the cross-flow. These effects are modified to some extent by eddy current within each trough. Figure 7.52(b) summarizes these effects.

On a riffled table the final segregation is therefore partly according to size and partly according to density, with the fine-heavy grains farthest upstream and the coarse-lightest farthest downstream. This is shown in Fig. 7.53 which is based on the study of shaking tables reported by Richards in 1907.[23]

When the separating surface takes the form of a spiralling trough the force dimension is increased by the centrifugal effect created as the flowing film revolves and descends from the feed to the discharge. Viewed from above the descending ribbon can be seen to spread, the heaviest and coarsest grains remain-

HYDRAULIC AND PNEUMATIC SEPARATION

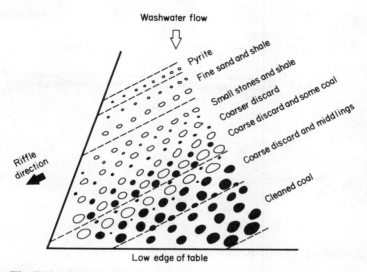

Fig. 7.53 Idealized classification of mixed grains on a shaking table.

ing nearest to the centre and on the flattest part of the trough, whereas the lightest and finest material is caused to ride well up the steep side of the trough. The velocity gradient is influenced by variation in thickness of the fluid film which is also affected by the centrifugal effect, and flow within the fluid takes the 'banana' profile shown in the cross-sectional diagram shown in Fig. 7.54. This creates a constant classifying effect in the sorting zone lying within the flowing liquid, tending to more of the coarser material of all densities travelling farther up the rising surface of the trough.

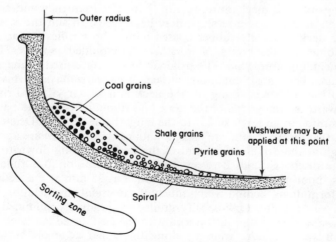

Fig. 7.54 Separation characteristics of spirals.

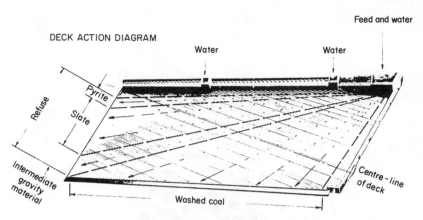

Fig. 7.55 Diagram of James type of shaking table.

7.7.2 Shaking Table Separators

An example of one of the more common forms of coal-cleaning table that is based on the original James sand-table (1907) is shown diagrammatically in Fig. 7.55. The dimensions of the deck are approximately 5 m along the longest edge by 2.5 m in width at right angles to the length. The shape is usually a parallelogram with the coal being discharged along the length of the long, lower edge, the principal slope being across the width at between 10 and 20 mm/m. It is usually infinitely variable between the limits. The deck is supported in an essentially horizontal plane by a pair of steel channels forming a supporting frame and is attached at one corner to a differential-motion drive which imparts a quick-return conveying motion, causing the deck to transport material on the forward stroke and slide beneath it on the return. The original James and the American Deister head-motion systems are the most widely used with modern tables. These produce similar sinusoidal differential-motion curves which are shown, together with drawings of each mechanism, in Fig. 7.56. The frequency and amplitude of stroke as well as the transverse slope may be adjusted to suit the material being treated. The deck is usually rubber coated with rubber riffles bonded into position, but some manufacturers offer other 'all-moulded' surfaces from other materials, including fibre-glass. The original deck surface used in the early tables was linoleum. The most common surface used nowadays is black rubber. Attached to the rubber covering on the deck is a system of solid-rubber riffles tapering toward the refuse end of the table and running in-line and parallel to the conveying motion at an angle of 25° to the longest side. For the Deister diagonal-deck tables, two forms of riffle are used. Standard 'body' riffles are approximately 6–7 mm high at the drive or feed end of the table, tapering down to as little as 2 mm at the discharge end. Between each set of three or four riffles there are often higher 'pool' riffles which may be as much as 25 mm at the feed end. These pool riffles form dams, behind which much of the useful stratification of the feed-bed occurs very rapidly. Low-density grains are gradually skimmed off by the cross-table water-flow, either introduced with the feed, or added later via a wash-bar or wash-water launder running along the upper edge of the table. High-density grains are transported behind the riffles by the differential-motion drive.

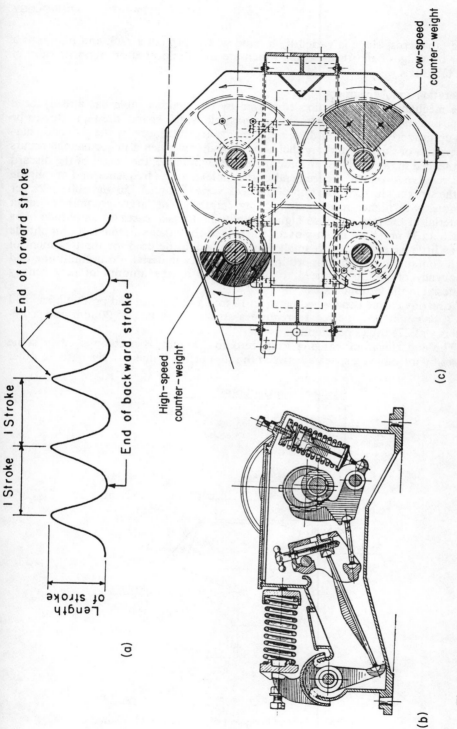

Fig. 7.56 Drawings of James and Deister vibrator units. (a) Differential-motion curve of a shaking table. (b) James simplex vibrator unit. (c) Deister Concenco 88 vibrator unit.

The table cross-slope is varied by a hand wheel through a rack and pinion to a pair of wedges beneath the deck, which are usually located on opposite sides of the table axis.

Operation and Adjustment
The mechanism of separation employed by the shaking table has already been discussed, but several adjustments are possible which enable the separator to be set-up properly.[24] Some of these adjustments are influenced by the rate and characteristics of the feed and are therefore only altered when a major change occurs in the feed material. These include the end elevation (i.e. the height of the discard end relative to that of the feed end of the table) and the frequency and amplitude of the stroke. The reciprocation of the deck varies between 250 and 300 cycles per minute, with higher reciprocation being employed for larger amounts of reject material in the feed and also higher amplitude. In such cases an amplitude of as much as 35 mm may be employed, whereas at the other extreme is an amplitude of as little as 10 mm. Such small amplitude would be used for the treatment of coal containing a high proportion of near-density material. Both amplitude and frequency of the stroke need to be decreased as the amount of near-density material in the feed increases.

A readily cleaned coal sized between 10 mm and 0.5 mm and producing a yield of 70% might be expected to require a stroke amplitude of 20 mm and a frequency of 275 cycles per minute.

The end elevation, ranging from zero to 300 mm, is accomplished in some cases by introducing spacers between the supporting frame and the base. In other

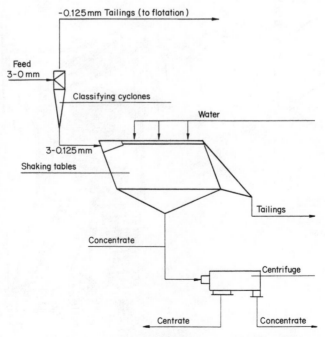

Fig. 7.57 Typical flowsheet of a shaking table plant.

HYDRAULIC AND PNEUMATIC SEPARATION

cases a lever system is employed. In general, higher end elevations would be used for coarser coal. The commonly used 10 mm top-size would only require between 75 and 100 mm. Once set, this slope is rarely changed.

The cross-slope and amount of washing water may both be readily adjusted to suit the feed. Cross-slopes of between 2 and 3° are common, and 5° is normally regarded as the maximum. The amount of water added with the feed is usually 2 : 1 as a ratio of water to solids by weight. About 90% of the total water used reports to the clean coal, and of the total water used, including that introduced with the feed, approximately 25% is wash water.

Both size and density of the materials contained in the feed significantly affect the separation, and where large amounts of near-density material prevail it is advisable to classify the feed into a narrow size-range to ensure the highest efficiency. Sieve bends, Vor-siv and high-frequency sizing screens are commonly used as well as classifying hydrocyclones, as shown in Fig. 7.57.

Under normal loading conditions a single table is able to treat between 9 and 12 t/h of 10-mm top-size coal feed. Capacity is mainly dictated by the discard rate which is between 3 and 5 t/h, depending on the design of table. The feed rate would obviously be reduced if large quantities of middlings were present in the feed or if the top size of the feed decreased. Table capacity will be increased if the upper size of the feed is increased or if the feed size-range is reduced by classification of the feed. Shaking tables are most commonly supplied in multiple-deck form and units comprising two, three and four tables (see Fig. 7.58) are available with differing combinations of drive mechanism.

A recently designed German preparation plant includes a novel table tower that accommodates 12 levels with six tables on each level in a compact design that affords good operator access, without utilizing a large amount of plant area.[22] This was shown earlier in Fig. 7.50, but Fig. 7.59 shows the layout on

Fig. 7.58 Two-deck table unit which may be assembled to form a four-deck unit (courtesy Deister Concentrator Company Incorporated).

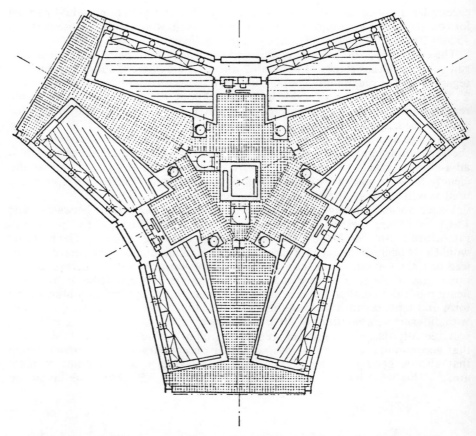

Fig. 7.59 Six-table level of German design (courtesy K.H.D. Humboldt–Wedag).

each deck. Another recent innovation has been the introduction of a process controller which makes end adjustments possible to every table from a single remote station.[25]

A centrifugal table has been developed in China[26] which is claimed to give effective cleaning and removal of pyrite from a fine-coal feed at a unit capacity of up to 15 t/h. The drum-shaped unit is 1.8 m in diameter, containing eight individual, arch-shaped decks with 25 m^2 combined area. The assembly is rotated at 100–150 r.p.m. creating a centrifugal force equivalent to 18–20 × g.

Efficiency of Shaking Tables

In a comprehensive evaluation of table performance carried out by the USBM[24] in the early 1960s, five tabling circuits all located in different plants were carefully studied. Table 7.9 summarizes the performance data obtained for treating coal sized between 9.5 and 0.75 mm. The five plants sampled all showed reasonable and typical efficiency levels for the Deister diagonal-deck table covering a separation density range of 1.52–1.69.

HYDRAULIC AND PNEUMATIC SEPARATION

TABLE 7.9 Summary of Performance Data from Five American Shaking Table Installations[24]

	Plant				
	A	B	C	D	E
Ash (%)					
feed	10.8	10.8	10.2	9.8	20.1
clean coal	3.7	4.7	5.4	5.7	4.6
refuse	61.8	50.6	59.1	65.4	75.6
Actual recovery (%)	87.8	86.8	91.2	93.2	78.2
Theoretical recovery (%)	89.4	89.2	92.1	93.8	80.3
Efficiency (%)	98.2	97.4	99.0	99.3	97.3
Ash error (%)	0.4	0.6	0.2	0.2	1.0
Yield error (%)	1.6	2.4	0.9	0.6	2.1
Float in refuse (% product)	14.8	20.4	10.2	7.4	7.8
Sink in clean coal (% product)	1.1	2.1	0.6	1.1	2.0
Total misplaced material (% feed)	2.8	4.5	1.5	1.5	3.3
Near density ±0.10 material (% feed)	5.0	7.8	1.3	1.1	2.4
Relative density of separation	1.52	1.52	1.69	1.64	1.64
Probable error, relative density	0.074	0.080	0.092	0.115	0.105
Imperfection	0.142	0.154	0.133	0.181	0.164
Error area (cm^2)	52	63	67	73	72
Distribution % to washed coal: relative density fraction:					
float–1.30	99.5	99.1	99.5	99.9	99.2
1.30–1.35	98.2	96.4	99.4	99.5	98.3
1.35–1.40	95.1	90.2	99.2	98.7	97.0
1.40–1.45	85.9	80.4	98.8	95.6	93.7
1.45–1.50	66.2	63.9	96.1	88.0	86.7
1.50–1.60	39.5	41.5	87.6	66.3	70.8
1.60–1.70	16.2	22.3	64.7	47.3	48.7
1.70–1.80	6.3	17.4	34.0	26.2	23.4
1.80–sink	3.0	8.7	5.5	4.8	6.1

Table 7.10 gives similar data obtained for treating 3 × 0.125 mm fine coal with Humboldt twin shaking tables at a German plant. Both sets of data appear to be in general agreement with one another when it is considered that the feed size-range for the German coal is narrower than that of the American coal.

7.7.3 Spiral Separators

Spiral separators have been employed in mineral dressing since 1943 when the Humphrey's unit was introduced. For many years following the introduction of this unit its name was synonymous with spiral separation. However, subsequent development of spirals by a number of different manufacturers has led to much wider use in mineral-dressing applications, and eventually their adoption for coal cleaning. The early spirals tended to be used only for comparatively easy separations, and in consequence had relatively simple and insensitive profiles. They were constructed from cast iron and tended to be excessively heavy in view of their relatively limited unit capacity.[27] The new breed of spirals are made from relatively lightweight materials, initially rubber-lined fibre glass, and more recently fibre-glass spray coated with polyurethane. The increasing use of spirals has led to further developments involving detailed attention to trough profiles, split-

TABLE 7.10 Summary of Performance Data from a Shaking-table Installation in W. Germany[22]

Size distribution of feed			Products (+0.125 mm)		
Size (mm)	% Mass	% Ash		% Yield	% Ash
+ 3.0	14.6	15.38	Concentrate	76.5	6.67
3.0 − 1.0	38.8	17.66	Middlings	10.1	42.25
1.0 − 0.5	15.4	21.22	Tailings	13.4	77.21
0.5 − 0.315	5.8	20.43			
0.315 − 0.125	6.0	21.20			
− 0.125	19.4	22.12			
	100.0	19.11		100.0	19.72

Sharpness of separation (+0.125 mm)

	Low cut			High cut		
Size (mm)	Density	e.p.m.	Imperfection	Density	e.p.m.	Imperfection
+ 3.0	1.48	0.043	0.09	2.13	0.146	0.13
3.0 − 1.0	1.5	0.058	0.116	2.06	0.146	0.14
1.0 − 0.5	1.5	0.066	0.13	1.91	0.111	0.12
0.5 − 0.315	1.46	0.055	0.12	1.87	0.136	0.16
0.315 − 0.125	1.48	0.107	0.22	1.995	0.179	0.18
3.0 − 0.125 (analysed)	1.49	0.062	0.130	1.98	0.135	0.14

Washing plant: Fürst Leopold
Feed: Fine coal seam H_2
Solids content: 376 g/litre
Feed rate per table: 3.2 t/h

ter design and, for certain applications, the elimination of the need for washwater.

The concept of cleaning fine coal using spirals was first considered shortly after the separator was first introduced. Testwork at the time clearly established that the spiral could perform comparably well with shaking tables when treating relatively fine feeds (minus 3 mm). However, the economics of treating fine coal by spirals, particularly in view of the large space and floor-loading requirements for a relatively small number of units, negated the single benefit of the low energy cost of running spirals. The development of spirals specifically for cleaning coal fines is, therefore, a relatively recent event. Current benefits of the newly designed spirals are as follows:

  (a) lightness;
  (b) equivalent efficiency to tables for 3 × 0.075 mm feed size-range;
  (c) no drives, pump fed via distributor;
  (d) trough profile to suit coal characteristics;
  (e) low cost;
  (f) enhanced capacity by selection of twin- or triple-start units.

The treatment of coal fines with spirals utilizes fairly small differences in relative density compared to mineral-dressing applications. Also, the bulk relative density of the spiral feed is generally much lower for coal spirals, at about 1.5 r.d. as compared with mineral slurries which usually exceed 2.8 r.d. These differences are reflected in the current design of spirals specifically intended for coal, i.e.

HYDRAULIC AND PNEUMATIC SEPARATION

Fig. 7.60 Reichert Mark 10 twin-start coal spirals (courtesy Mineral Deposits Limited).

(i) a lower pitch than for mineral separation; the pitch reduction creates lower velocity and increases the pulp stability as well as residence time;
(ii) a larger outside diameter (up to 1 m) and pulp volume-flow capacity permitting larger solids-handling capacity.

A good example of a commonly used 'coal' spiral is the Reichert–Wright Mark 10 spiral, as shown in Fig. 7.60. This separator does not have a wash-water requirement and is manufactured in a twin-start, six-turn version which has the effect of doubling capacity per unit floor area of the plant. Reichert coal spirals have one set of adjustable splitters at the bottom end of the trough. For twin-start units the splitters are gauged to enable simultaneous and equivalent adjustment to splitters located on both volutes on the one column. Product streams from the two spiral volutes in twin spirals are combined in the product collection box as designed. This reduces cost and eases collection launder design. In the Reichert coal spiral, the trough diameter is approximately 700 mm and frame height is 2380 mm. The original Humphrey spiral has also been modified for coal (and light mineral) separation and is fabricated from lightweight materials. It is similar in physical size to the Reichert but has retained the wash-water channel.[17] Another coal-cleaning spiral, the Wyong design, which like the Reichert originated in Australia, is wider at up to 1000 mm and therefore capable of higher unit capacities. A further advantage of the so-called 'plastic' spirals is that they can be fabricated in various colours and therefore, used in different designs for different treatment circuits, remain easily identifiable.

The popularity of the newly designed coal spirals is demonstrated by the following facts. From 1983 to 1985, 15 coal-spiral circuits were installed in New South Wales and Queensland in Australia, six circuits in South Africa and one circuit each in Canada and the USA. The trend continues. Prior to these installations, none had existed in Australia or South Africa. The most common application for these circuits has been between dense-medium cyclones and froth flotation,[28] as shown in Fig. 7.61(a).

Operation and Adjustment

The performance of spirals in treating coal fines is dependent on the characteristics of the feed (i.e. size distribution and washability) and also operating parameters including feedrates, feed-solids concentration and splitter positioning. In some models, wash-water regulation is also a factor.

The common upper limit of feed size is about 3 mm, but depends upon washability. The lower limit is about 75 µm, and therefore the spiral feed should preferably be deslimed prior to treatment.

The nominal design capacity is 2–3 t/h dry solids per start depending upon trough diameter (700–1000 mm) and discard content. Test data have demonstrated that feedrate is in fact the most significant independent parameter governing spiral performance.[17] This means that although the spiral may appear to accept a feedrate of more than the design capacity the performance may deteriorate rapidly. It is therefore the discard handling capacity which really determines feedrate. For Reichert Mark 10 spirals up to 1.5 t/h of discard can be treated without detrimental effect until over 4.5 t/h of feed is encountered or 5 m³/h of pulp. Pulp densities of up to 45% solids in the feed and between 30 and 60% solids in the discard (by weight) are specified as normal. Feed density has been found to be of relative insignificance in spiral performance providing the other stated criteria are met. Feed density depends more specifically on the nature of the feed solids (size distribution and ash content) and are nominally within a range of 20–40% solids by weight.

The splitter positions largely determine the ash content of the final product and constant fine tuning is necessary to obtain optimized ash content and yield values. Figure 7.62 shows the mass and ash distributions occurring across the bottom trough profile of a Reichert Mark 10 spiral, in the treatment of raw coal with 35–40% ash content. These curves clearly illustrate the significance of splitter positioning and yield and ash content.[18]

HYDRAULIC AND PNEUMATIC SEPARATION

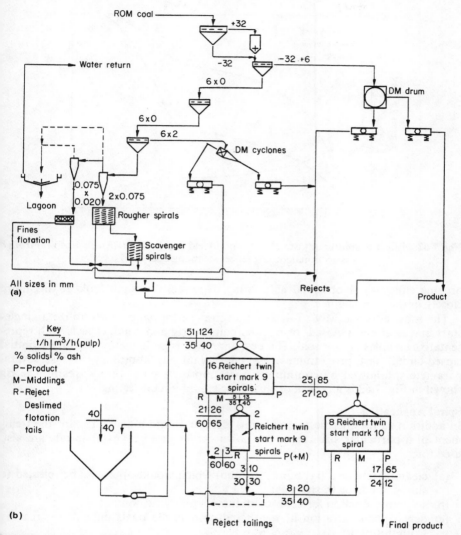

Fig. 7.61 Typical flowsheets incorporating coal-cleaning spirals (after Richards, Ref. 18): (a) complex flowsheet incorporating spirals; (b) typical flowsheet for scavenging 40 t/h coking-coal flotation tailings (nominally $-0.5 + 0.075$ mm).

Efficiency of Spirals

Partition curves after Tromp (see Chapter 8) have been determined for modern coal spirals under varying test conditions and for various types of feed.

Recent performance data from Australia[27] is for Reichert Mark 9 or Mark 10 spirals cleaning 3×0.075 mm raw coal. This is summarized in Table 7.11. A reduction in organic efficiency when treating poorer quality raw coal is to be anticipated and in general, some form of correlation between Epm values and the

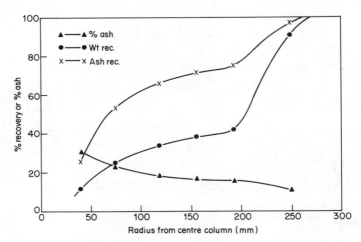

Fig. 7.62 Effect of splitter adjustment on spiral yield and product. (Mark 10 spiral trough cross-sectional performance; steaming coal +75 μm.)

organic efficiency is observed as existing. However, we shall look at this more closely in the next chapter.

The most effective way to determine spiral performance is to utilize a single-start unit in either a closed- or open-circuit test rig and conduct tests with representative samples of raw coal. The best approach is perhaps to do both, i.e. initial closed-circuit test programme in a pilot plant or laboratory to test several parameters, followed by continuous operation on a plant bleed-stream. The rig shown in Fig. 7.63 is a typical one used for closed-circuit testing.

Spiral Applications

In addition to the specific role of being 'designed' into a new plant for the treatment of a particular size range, as shown earlier in Fig. 7.61(a) spirals are also used for:

(a) cleaning fine coal (i.e. minus 0.5 mm) which would otherwise be rejected to slurry ponds or waste piles;
(b) sulphur reduction for coals containing high pyrite content;
(c) density concentration of weathered or oxidized coals difficult to treat by froth flotation;
(d) treatment of fine coal containing high clay content.

Many recent applications have been retrofitted into plants either to replace

TABLE 7.11 Reichert Spiral Performance (+75 μm)

Spiral type	Coal type	Coal product			Coal + middlings		
		Organic efficiency	Ash error	e.p.m.	Organic efficiency	Aah error	e.p.m.
Mark 9	steaming	92.5	2.8	0.15	95.7	2.1	0.10
Mark 10	coking	96.7	1.4	0.16	98.5	1.0	0.11
	steaming	95.1	1.8	0.14	93.4	2.9	0.12

HYDRAULIC AND PNEUMATIC SEPARATION

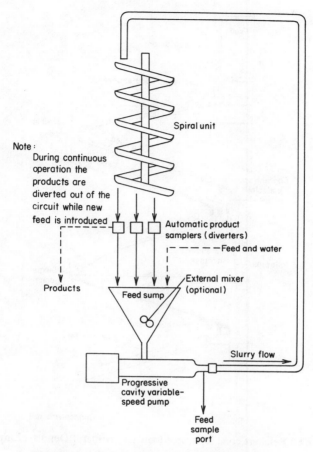

Fig. 7.63 Typical closed-circuit spiral test facility.

other, less successful circuits (water-only cyclone replacement is common) or to treat hitherto untreated coal (as in South African plants). Also, many spiral applications have been found in scavenging of reject or waste streams and slurry ponds or fine-coal dumps. Figure 7.61(b) shows a typical scavenging circuit for treating 0.5 × 0.075 mm coking-coal flotation tailings, of which numerous examples now exist.

7.8 REICHERT CONE SEPARATOR

This separator is primarily intended for high-capacity treatment, normally in the range 65–90 t/h with feed density of between 35 and 45% solids by weight for coal separations. Feed size-range is from 3 to 0.050 mm, or finer if fine pyrite is to be removed. The unit was designed specifically for sand treatment in Australia, where it has since found many other applications, including coal. However,

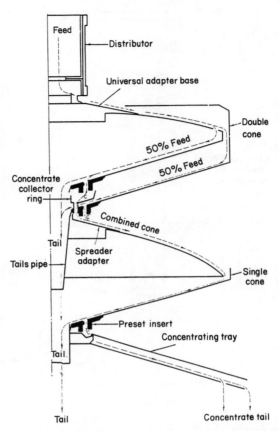

Fig. 7.64 Reichert cone separator (courtesy Mineral Deposits Limited).

despite its undoubted capability to upgrade coal, it qualifies at best as a preconcentrator or rougher, ahead of spirals which perform the final cleaning. Hence, in a two-stage cleaning circuit the cone may be regarded as a candidate for the initial stage. Figure 7.64 shows a cross-sectional drawing of the Reichert unit which comprises several cone sections stacked vertically so as to provide stages of upgrading. The cones are made of fibre glass and are mounted in circular frames over 6 m high. Each cone is 2 m in diameter and has no moving parts. The cones deposit the denser material into external collection trays which also serve to upgrade further the material by acting as a sort of pinched sluice concentrator. The feed slurry is distributed evenly around the periphery of the cone and flows downwards towards the centre. The feed is therefore classified as per the flowing-film concept described earlier. Denser grains separate to the bottom of the film and some of this material is removed by an annular slot in the bottom of the cone. The fine coal is carried over the slot to the next cone and eventually to a central discharge port.

HYDRAULIC AND PNEUMATIC SEPARATION

7.9 WATER-ONLY CYCLONES

The principles of separation of the hydrocyclone have already been discussed in Chapter 4, and the dense-medium cyclone was dealt with in the previous chapter, Chapter 6.

The water-only cyclone has, for many years, been widely applied in the treatment of fine coal sized below 0.5 mm. The principle attractions have been twofold: reduction of pyrite from high-sulphur raw coal; and effective cleaning of fine coal with below 10% near-density material. Cleaning efficiency decreases sharply with increasing fineness in size and for this reason the cyclone is rarely effective

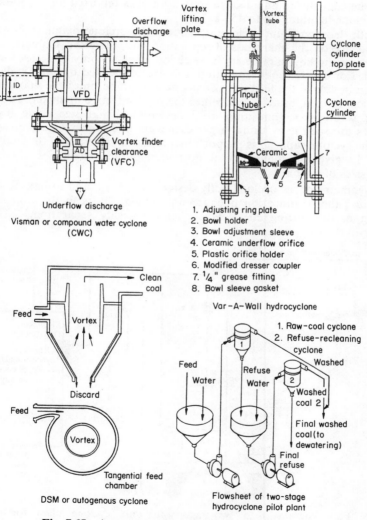

Fig. 7.65 Autogenous (water-only) cyclone designs.

below 0.10 mm in grain size.[29] A variety of typical water-only cyclone designs are given in Fig. 7.65. They are all characterized by a wide cone angle, $\geq 60°$, and are operated to suppress the classification phenomenon by a stratification process which occurs on the conical wall of the separator. By careful adjustment of vortex-finder clearance and apex orifice, it is usually possible to achieve an acceptable separation. More stringent variables include feed pressure and inlet diameter or even the size of cyclone used, but all applications are extremely sensitive to feed variations, and consistency in flowrate, solids concentration and size-consists is most important.

One other design, an air-sparged hydrocyclone,[30] combines flotation and centrifugal separation. By adding methyl isobutyl carbinol (MIBC) to a 15% by weight fine-coal slurry, an effective separation was reported for a 150-mm cylindrical cyclone handling up to 450 kg/h.

As with dense-medium cyclone operation, the cyclone diameter has a significant influence on the sharpness of separation and, therefore, diameter and pressure become the key parameters in selection of water-only cyclones for treating fine coal. For coarser-coal separation, the size of the feed opening is determined by the grain size of the feed (i.e. usually at least three times the maximum grain size) which, in turn, usually determines the most effective diameter. For fine-coal cleaning, where maximum grain size is less critical than size range, higher unit capacities are sought by using larger diameters and higher feed pressures. Normally, the cyclone capacity is proportional to the square root of the feed pressure. Table 7.12 gives an indication of normal minimum capacities obtained with various sizes of cyclones.

The geometry of the cyclone body design has been found to have a significant effect upon the stratification phenomenon which occurs in the conical section, resulting in the sorting effect which causes density separation.[31] The major

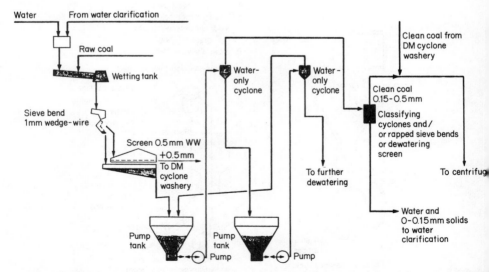

Fig. 7.66 Flow-diagram of a two-stage water-only cyclone plant for coal (after Stamicarbon).

TABLE 7.12 Capacity and Feed-pressure Data for Various Water-only Cyclone Diameters

Diameter (mm)	Pulp flow Normal: (min) m^2/h	Solids flow Normal: (min) (t/h)	Pressure head[a] Normal: (min) metres liquid column
100	8 (3.5)	1.3 (0.6)	5 (1.0)
200	27.5 (13)	4.7 (2.2)	9 (2.0)
350	81.5 (46)	13.3 (7.5)	10 (3.2)
500	171.5 (105)	27.8 (17.0)	12 (4.5)

[a] Normal pressures used (pump-fed cyclones) are between 5 and 12 mlc which creates a substantial increase in unit capacity.

purpose of the vortex finder, which is usually equal to or close to the cyclone diameter in length, is to control migration of misplaced grains. In addition, the introduction of spacer pieces into the cylindrical section of the cyclone, thereby making the overall cyclone length greater, leads to the achievement of greater capacity without loss in performance. This is also applied as a means of attaining a lower relative-density cut-point, especially in the treatment of fine coal sized below 1 mm in maximum grain size.

For the same feed and relative density of separation, the Tromp curve of a water-only cyclone will show a probable error value about three times as high as the dense-medium cyclone treating the same coal. In order to improve upon this performance, it is normal for two stages of cyclone to be used, with up to 20% of the feed solids being recirculated to the first stage. A circuit of the type shown in Fig. 7.65 is recommended by the Dutch State Mines (DSM), the original inventors of this type of washing cyclone. The significance of this combination is shown in Table 7.13 which gives computer-predicted data for other various fine-coal cleaning units, for reference purposes.[32]

The relative performance of each of these separators is clearly shown in Fig. 7.67 which is a graph showing the various partition curves obtained with each separator.

TABLE 7.13 Computer Performance Prediction for Fine-coal Cleaning Units

Separator	Dense-medium cyclone	Table	Spiral	Single-stage water-only cyclone	Two stage water-only cyclone recirculated
Probable error	0.0599	0.1117	0.1233	0.2239	0.2108
Cut-point 1st stage	1.675	1.658	1.649	1.505	1.521
2nd stage	—	—	—	—	1.550
Float–sink yield	87.26	87.26	87.26	87.26	87.26
% Ash—clean coal	11.00	11.00	11.00	11.00	11.00
%Ash—reject	58.34	54.31	51.63	27.49	40.87
Plant yield	85.66	84.32	83.28	58.78	77.26
Plant rejects	14.36	15.70	16.74	41.24	22.74
Organic efficiency	98.16	96.62	95.44	67.36	88.54

Raw-coal feed in each case was 3 × 0.15 mm with 17.8% ash content.

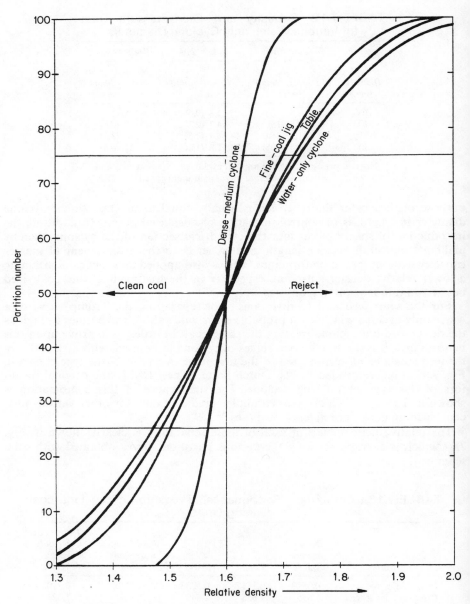

Fig. 7.67 Tromp performance curve of a dense-medium cyclone, fine-coal jig, table and water-only cyclone treating 6 × 0.5 mm final coal.

An interesting application for the water-only cyclone reported some years ago[33] is its use as a clean-coal scalper preceding dense-medium cyclones. By reducing the amount of coal to be treated by the more expensive dense-medium circuit it is possible to utilize fully the capabilities of each type of washer while at

HYDRAULIC AND PNEUMATIC SEPARATION

the same time making considerable savings in cost. There are, however, numerous coals which do not lend themselves to the low-density separations required by the water-only cyclones, and little is gained from such a partnership. Computer simulation can demonstrate the potential benefits accurately, and for any likely combination of raw-coal blend.

7.10 SLURRY DISTRIBUTORS

With the exception of jigs, all other forms of hydraulic separator in current widespread use are relatively small in capacity. Consequently, they require some form of distributor to divide the feed into several streams in order to feed each separator. The distributor, therefore, has a vital role and may greatly influence the performance of the subsequent separators if its own performance is inconsistent in any way.

Figure 7.68 shows a simple classification of distributor types, but there are numerous varieties in common use, many of which are of proprietary design and commonly associated with a particular type of separator. Although several alternatives are likely to be worthy of consideration, it is often difficult to decide which to select and it is worthwhile to review carefully specific requirements. The following factors need to be taken into account:

(a) the type of washing unit to be served;
(b) the degree of division required, e.g. six streams;
(c) the solids concentration and size-distribution of the feed;
(d) the effects of poor distribution on the washing unit;
(e) the possibility of future expansion;
(f) the need for an operational stand-by washing unit;
(g) operational advantages and disadvantages.

All of these factors should be considered in addition to relative costs, wear and tear and ease of maintenance.

In considering the separator, capacity is important. Remember, we have seen early in this chapter that tables are able to treat up to 12 t/h per unit, spirals can treat up to 3 t/h, and water-only cyclones can treat up to 30 t/h. Distributors

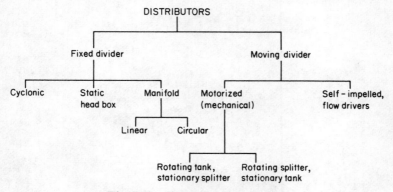

Fig. 7.68 Slurry distributor types.

Fig. 7.69 Concenco revolving feed distributor commonly used for table-feed distribution (courtesy Deister Concentrator Company Incorporated).

Fig. 7.70 Static-head box slurry-distributor commonly used with spirals.

HYDRAULIC AND PNEUMATIC SEPARATION

Fig. 7.71 Water-only cyclone cyclopac (courtesy Quintette Coal Limited).

commonly used with these three types of separator are required to feed up to 8 tables, or 32 twin-turn spirals or 10 water-only cyclones, i.e. approximately 100 t/h, 200 t/h and 300 t/h respectively.

Figure 7.69 shows a motorized distributor commonly used for table-feed distribution of coal sized 9.5 mm × 0; Figure 7.70 shows a static-head distributor found to be very effective for spiral-feed distribution; and Fig. 7.71 shows a cyclopac distributor employing a circular manifold type. These illustrations tend to demonstrate an inconsistency in distributor selection which would appear to have evolved with usage of each type in conjunction with a particular type of separator. It is probably an area of application worthy of closer investigation, but many common-sense applications of simple, low-cost distributors have proved themselves to be worthy alternatives to the more expensive motorized types. The converse has also occurred frequently, whereby fixed distributors have been very successfully replaced by one of their moving counterparts. Coal-preparation engineers therefore bear the burden of responsibility for attention to such details during plant design and in looking for ways and means of improving plant performance.

7.11 PNEUMATIC (OR DRY) SEPARATION

Although wet methods of separation are inherently more efficient, dry-cleaning methods employing air as the separating medium did find wide acceptance for treating coal between 1930 and 1960. Wet methods create two particular problems which promoted interest in dry methods. These are:

(a) they add moisture to the cleaned coal, thereby reducing quality, i.e. adding back water after having removed ash;

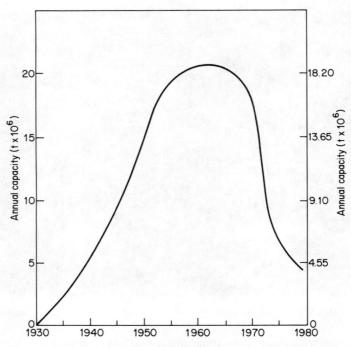

Fig. 7.72 Dry-cleaning plant capacity in the USA.

(b) dewatering and drying costs are the most rapidly increasing treatment costs in coal preparation.

Coal cleaning using dry methods was widely practised in the United Kingdom and the United States up until 1960, as can be seen in Fig. 7.72 and Table 7.14. Almost all of these separators were of pneumatic type. The reasons for the demise are associated with the inherent inefficiencies of the separators and also the general deterioration that has taken place in mined coal quality. Another factor has been the environmental problems encountered as a result of dust emission, and the associated cost of eliminating it. Pneumatic separators reached a peak in their usage in the immediate post-World War II era when mining was mainly underground by hand-got methods. With the advent of mechanized mining and the ensuing dilution by out-of-seam material and increased quantity of fines, the

TABLE 7.14 Dry Cleaning Plants in the United Kingdom

Year	No. of dry cleaning plants	% Dry cleaned of total cleaned
1929	75	8.1
1947	166	13.6
1956	93	6.1
1970	12	<1%
1980	4	<1%

HYDRAULIC AND PNEUMATIC SEPARATION

use of dry separators rapidly declined. In the pre-mechanization period, the average surface moisture content of run-of-mine coal was between 3 and 4% whereas now it is over 6%.[34] Also, the amount of fines sized below 0.5 mm increased from below 5% to over 25%. The combined effect of both on the efficiency of dry screening was dramatic. Because of the fact that pneumatic separation depends greatly upon obtaining a closely sized feed containing a minimum amount of fines, the inefficiency in sizing the feed rendered dry separators almost ineffective. This led to their subsequent decline, and nowadays less than 3×10^6 t of coal per year are treated by pneumatic separation.

7.11.1 Basic Principles of Pneumatic Separation

Pneumatic-separation methods take advantage of differences in relative density to effect a separation between coal and shale and are therefore very similar to hydraulic methods. We have already looked at how this behaviour is characterized in the second section of this chapter. Pneumatic methods fall into three categories:

(a) pneumatic tables;
(b) pneumatic jigs;
(c) pneumatic dense-medium separation.

In the last type, an air/magnetite or air/sand suspension is utilized as described in the final section of the previous chapter. The remainder of this chapter deals with the other two categories.

The earlier concepts of equisettling ratio and hindered settling ratio are most important in determining feed-preparation requirements, although the role which the latter plays in air separations is far less significant than with water separations. Differential acceleration, on the other hand, is a most important phenomenon. In both air jigs and tables a stratification of the bed will quickly occur as a result of an oscillating motion imparted to the bed materials by either air pulsation or deck vibration. The time available for effective separation to occur before critical settling ratios are reached is much less than with water separation, and therefore the remaining separating phenomenon is that of differential acceleration. In theory, for coal and shale particles, if the smallest diameter is one-third of that of the largest diameter the mixture can be separated in an air medium. Theoretical considerations also show that in a tabling or a jigging separator, the time of free fall should be minimized. In practice these phenomena have proved more difficult to define.[34, 36]

7.11.2 Air Tables

Pneumatic tables can be divided into two groups:

(a) riffled tables;
(b) smooth tables.

(a) *Riffled tables* The riffled table is similar in construction and operation to the wet concentrating table. The deck, generally horizontal or slightly tilted, is covered with a fine-mesh screen cloth through which a constant-velocity upwards-air-current flows. The deck is often compartmentalized so that differing air velocities can be applied across the table.

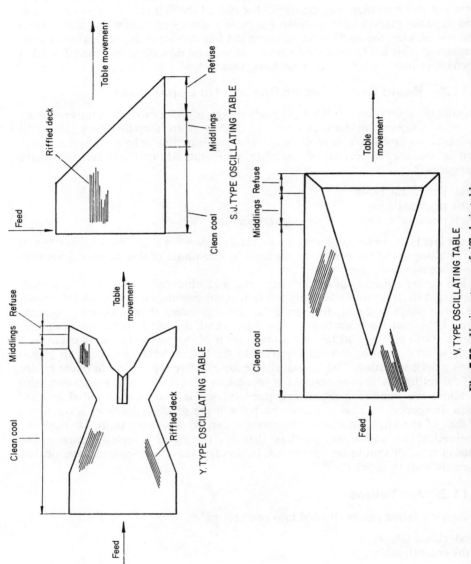

Fig. 7.73 Various types of riffled air table.

HYDRAULIC AND PNEUMATIC SEPARATION

Mobility of the material is achieved by applying longitudinal oscillations to the entire deck. The feed is applied at one end of the table where it immediately encounters the upwards flow of air. The shaking motion of the deck and the upwards-current air enable a stratification of the bed such that the heavier grains settle to the bottom and the lighter grains rise to the top.

The heavier grains have their movement retarded and redirected by the riffles on the deck and travel in the direction of the deck vibrations. The lighter grains, forming the upper strata, move by gravity across the bed. Clean coal, middlings and rejects may be collected at the discharge edge of the deck with the use of diverting blades.

Some of the various deck-and-riffle configurations made available by manufacturers are shown in Fig. 7.73. Performance and operating data for these three types of table are shown in Table 7.15.

(b) *Smooth tables* In the second group of pneumatic tables, the rectangular deck is inclined at about 10° to the horizontal. Longitudinal vibrations are again imparted to the deck which is unriffled and generally manufactured from

TABLE 7.15 Riffled Pneumatic Tables—Operating Data[34]

Designation	SJ Table
Type of table	Riffle, constant-velocity air
No. of products	3
Approximate date of introduction	1922
Locations	North America
Vibration rate	320 r.p.m.
Stroke	10 mm
Standard size	1500 mm wide × 2100 mm long
Capacity	15 t/h table for 3.2 × 1.6 mm feed
Capable size range	2 : 1
Epm	0.15–0.25
Designation	Vee table
Type of table	Riffle, constant-velocity air
No. of products	3
Approximate date of introduction	1930
Locations	United Kingdom
Vibration rate	Approximately 300 r.p.m.
Stroke	10 mm
Standard size	1500 mm × 3000 mm
Capacity	20 t/h table for 6.4 × 1.6 mm feed
Capable size range	4 : 1
Epm	0.15–0.25
Designation	Y table
Type of table	Riffle, constant-velocity air
No. of products	3
Approximate date of introduction	1930
Locations	North America
Vibration rate	330 r.p.m.
Stroke	10 mm
Air pressure under bed	19 mm w.g.
Standard size	1500 mm × 3000 mm
Capacity	20 t/h table or 13.3 t/h/m width
Capable size range	4 : 1
Epm	0.15–0.25

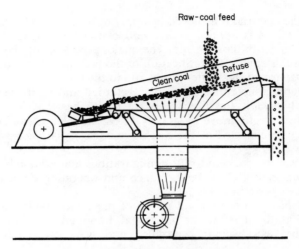

Fig. 7.74 The Birtley contraflow separator.

punched plate. A centrifugal fan delivers air at constant velocity and pressure to the underside of the deck.

Raw coal is fed across the width of the deck at approximately the centre point. Stratification of the bed takes place due to the upwards-current air. The heavier particles sink through the bed, come into contact with the deck and are moved up the deck due to the eccentric reciprocating motion. The lighter coal particles tend to float to the top of the bed and are pushed down the deck by the new feed material. These separators are simple two-product units and thus do not produce a middlings fraction.

The Birtley Contraflow separator is typical of this type of machine (see Fig. 7.74 and Table 7.16).

The FMC Dry Table is included in this category since it has received some recent attention and was evaluated at a coal mine with feed rates of up to 10 t/h.

The separator is similar in principle to an air table, but in this case the device is simply shaken without the introduction of upwards-current air (see Fig. 7.75). The table is wedge shaped so that the light particles fall off the deck first and the

TABLE 7.16 Smooth Pneumatic Tables—Operating Data

Designation	Contraflow separator
Type of table	Smooth deck, constant velocity air
No. of products	2
Approximate date of introduction	1940
Locations	Europe
Vibration rate	500 r.p.m.
Stroke	
Standard size	2400 mm long × 70–2400 mm wide
Capacity	up to 32 t/h for 50 × 25 mm feed; 2400 mm wide deck (13.3 t/h/m width); up to 16 t/h for 3.2 × 116 mm feed; 2400 mm wide deck (6.7 t/h/m width)
Capable size range	2 : 1
Epm	0.12–0.22

HYDRAULIC AND PNEUMATIC SEPARATION

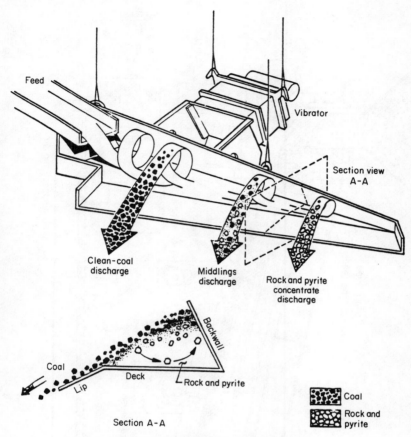

Fig. 7.75 FMC vibrating table.

heaviest particles stay on and are eventually ejected off the end.

The results for all of the tables are very similar, requiring a dry-sized feed with the size ratios varying from 2 : 1 up to 4 : 1. (Note: these figures were supplied by the manufacturers and it is probable that the mean value of 3 : 1 is more realistic.)

The capacities of the machines are approximately 13 t/h/m width for large coal (50 × 16 mm) and 7 t/h/m width for small coal (10 × 3.2 mm). It is unlikely that any of the tables perform an effective separation at particle diameters below 3.2 mm.

There are still a few constant-velocity air-table plants in operation in Europe; and all are of the contraflow type.

Although the acceptability of all types of pneumatic separators has declined jigs. In fact, a survey of pneumatic separators in North America in 1966 revealed that all used pulsating air as the separating medium. The reason for this decline is that air tables are less efficient than air jigs. They also require a closer-sized feed and are susceptible to change in surface moisture-content of the feed.

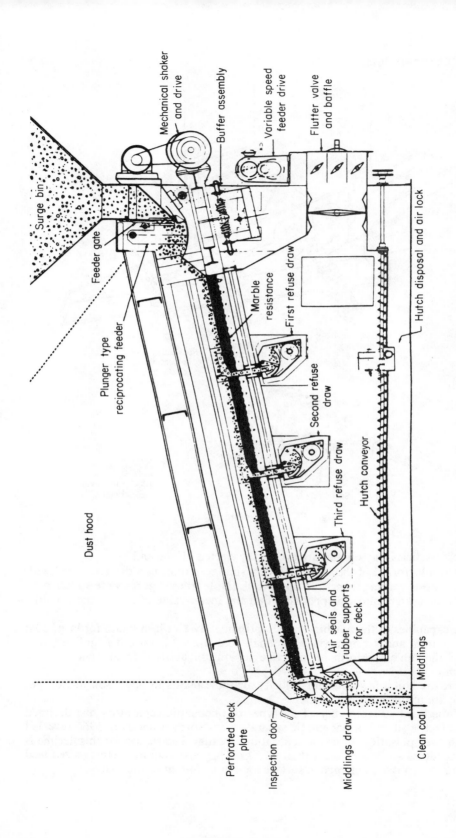

Fig. 7.76 Roberts and Schaeffer stump air jig (courtesy Roberts and Schaeffer Company).

HYDRAULIC AND PNEUMATIC SEPARATION

7.11.3 Pneumatic Jigs

The operating principles of the pneumatic jig are very similar to those employed in the familiar Baum-type jigs, except that air is used as the separating medium instead of water. The principles of operation will be described using the most common North American air jig, namely the Stump Air-Flow jig shown in Fig. 7.76.

(a) *Pulsed air jig* The raw feed enters the machine at the upper end of an inclined, oscillating, porous deck mounted over an air space called a plenum chamber. Air is pulsated through the plenum chamber by means of a rotating butterfly damper. The deck is fastened in place to partition plates which divide the deck into compartments spaced every 110 mm along the length of the deck and extended across the full width. These compartments are filled with ceramic 'marbles', retained in place by slotted plates, which slide over each other to give an adjustment from zero opening to the full open area of the slots. The marbles help diffuse the air over the entire bed, and thus eliminate short-circuiting regardless of variations in the depth of the material bed.

Stratification of the bed occurs as a result of the air pulsations and the oscillations imparted to the deck.

In a recent study[35] by the US Department of Energy, performance data were given for a series of tests of a Super Airflow jig. The results are summarized in Table 7.17.

The heavier grains migrate through the bed and eventually pass along the surface of the deck until they encounter slots or draws which extend across the full width of the deck. The reject grains pass through these draws. There are normally four under a conventional deck 2400 mm wide and 2700 mm long. The first three usually collect discard material and the fourth collects middlings, which can be recirculated. The deck is totally enclosed by means of a dust hood. The dust-laden air is pulled through a cyclone dust collector and cloth filter. A flowsheet of a typical plant is shown in Fig. 7.77, and typical specifications for this and other air jigs are shown in Table 7.18.

TABLE 7.17 Summary of Performance Data for Super Airflow Jig (Performed by US Department of Energy, April, 1979)

Size of table	2400 mm wide × 2740 mm long			
Feed rate (t/h)	100	150	150	90
t/h/m²	13.8	20.7	20.7	12.4
Feed surface moisture (%)	2.8	2.1	2.1	4.9
Feed top size (mm)	50	50	50	25
Feed bottom size (mm)	0	0	0	0
Percentage minus 0.5 mm in feed	23.9	5.3	5.0	6.6
Feed ash (%)	26.0	14.4	16.2	15.8
Clean-coal ash (%)	18.7	9.1	9.1	9.0
Refuse ash (%)	67.6	47.2	40.7	29.9
Yield (%) (weight)	85.1	86.1	77.4	67.4
Float in refuse[a]	62.7	31.2	34.6	57.7
Sink in clean coal[a]	5.2	3.2	5.0	6.9
Percentage near-density material	8.1	2.0	3.8	6.1
Effective r.d. of separation	2.67	1.99	1.78	1.56
Epm	—	0.334	0.296	0.312

[a] %age of product.

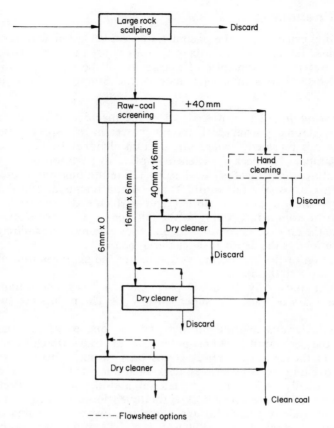

Fig. 7.77 Typical dry-cleaning plant flowsheet (courtesy Roberts and Schaeffer Company).

(b) *Rotating air jig* Symonds[34] developed a laboratory-scale rotating air jig. This was essentially a conical-shaped basket with perforated sides through which pulsed air was passed. Performance data for the machine gave e.p.m. values as low as 0.086 at surface moisture values of 8.2%. The feed material was 6.4 × 1.6 mm raw coal. The data also showed that the separator was best suited to act as deshaler operating at cut-points in excess of 1.80 r.d. No commercial separator of this type is known to exist.

There are many air jigs still in operation in the United States and a few in Europe. The reasons for the demise of air jigs are the same as those described earlier for tables, namely inability to treat adequately raw coals with elevated surface moisture, fines and rejects contents.

It has been estimated that air jigs produced approximately 5.8×10^6 t of clean coal in the United States in 1976. The current figure is approximately 4×10^6 t (this is less than 1% of the total coal cleaned per annum in the USA).

The main suppliers of air jigs are:

HYDRAULIC AND PNEUMATIC SEPARATION

TABLE 7.18 Specifications and Operating Data Air Jigs[a]

A.	Jig type	Pulsed air
	Jig designation	Super airflow
	Manufacturer	Roberts and Schaeffer
	Top size treated	50 mm
	Minimum size treated	0.5 mm
	Oscillation rate of deck	600 strokes/min
	Throw of deck	6.3 mm
	Air quantity	60.9 m^3/min/m^2 at 0.74 kPa for 1.7 mm coal
		274 m^3/min/m^2 at 1.49 kPa for 38 mm coal
	Size of table	2400 mm wide × 2740 mm long
	Size of deck holes	1.83 mm
	Air pulsations	190/min
	Jig capacity	100 t/h for 2400 mm wide, 15 t/h/m^2
B.	Jig type	Pulsed air
	Jig designation	Humboldt–Wedag
	Top size of particles treated	50 mm
	Minimum size of particles treated	0.5 mm
	Jig capacity	21.5 t/h/m^2 for 25 mm coal
		5.4 t/h/m^2 for 10 mm coal

[a] These are about half the values of a Baum jig operation.

Roberts and Schaeffer Company, Chicago, Illinois, USA;
The Hycaloader Company, Lake Providence, Louisiana, USA;
Humboldt/Wedag, W. Germany.

In some of the current applications the dry cleaner is used as a deshaling device

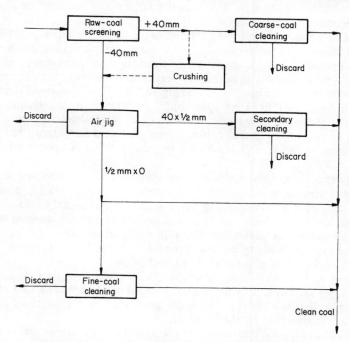

Fig. 7.78 Typical flowsheet with air jig as deshaling device.

prior to conventional wet cleaning (such as a dense-medium separator). Figure 7.78 shows a modified flowsheet.

When assessing these results it should be remembered that the feed to these units is either 50 mm × 0, or 25 mm × 0 material (i.e. the feed material was not presized before it was fed to the air jig). In addition, it is generally recognized that the top size of feed to an air jig should be 20 mm. Coarser particles require more air, and disruption of the bed is often encountered with the result that excessive misplaced material is found in the smaller (minus 6 mm) sizes. However, these results do serve as a basis for judging the performance of comparable or new pneumatic separators.[36]

The data also show that pneumatic jigs (and tables) are basically inefficient units and are effectively 'deshalers'. That is to say they are best suited to performing a separation at a high cut-point (+1.80 r.d.) in order that the shale and rock can be removed from the raw coal. Usually, little coal is lost in the refuse, but the 'clean coal' does contain high proportions of misplaced reject material.

However, the jig has outlasted all other dry-cleaning systems because it has high throughput characteristics, is more efficient than the other commercial processes and is not quite so seriously affected by damper feeds (6% surface moisture feeds, however, still cause serious problems).

7.11.4 Dust and Noise Control

Both dust and noise are of concern in the application of dry-cleaning methods.

Modern dust-extraction systems are far more efficient than some of the older shaker type baghouse systems. They require far less maintenance and as such are likely to be better looked after.

Air jigs and air tables employing the use of air as part of the separating process employ baghouses for the removal for the entrained particles. The development of modern synthetic fibres and the use of continuous bag-cleaning methods have considerably reduced the size of the dust-extraction systems, enabling air-to-cloth ratios to be increased from around 2 : 1 up to 6 or 7 : 1. Entrained dust in an air stream can be highly abrasive. The older systems required significant maintenance just to keep pace with hood and ducting replacements.

The use of rubber liners and rubber sheets in metal frames can have the dual effect of reducing noise and rendering the ducts less susceptible to wear than steel.

In any preparation plant, noise and the exposure of the operators to noise, has become more and more important. In dry-cleaning plants, noise is generated primarily by falling material impinging on metal surfaces, from the interaction of the mechanical equipment with the material and from the air-moving equipment. Although dry-cleaning plants have traditionally been classified as excessively noisy, some innovations can be incorporated to keep the noise levels well below the 90 dBA specified by noise abatement authorities.

Significant reductions in noise levels (up to 30 dBA) have been achieved by the use of energy-absorbing polymer sheets bonded to dust hoods and chutes. The significance in the bond is that the noise generated by the material striking the metal surface is absorbed by the polymer.

The rubber lining of chutes and the incorporation of rubber decks on screens have added significantly to noise reduction. These applications could also be adapted to the plate decks on an air jig.

Silencers or sound mufflers have received considerable attention and have reduced the noise levels of blowers. Isolation of such equipment either in independent sound-absorbing rooms or external adjacent rooms has considerably enhanced the plant environment.

The environmental aspects arising from dust and noise emissions are discussed in more detail in Chapter 19.

REFERENCES

1. H. Louis. *The Preparation of Coal for the Markets*, Methuen, London, 1928.
2. A. M. Gaudin. *Principles of Mineral Dressing*, McGraw-Hill, New York, 1939.
3. J. M. Coulson and J. F. Richardson (original authors). *Chemical Engineering*, Pergamon, Oxford, 1982 (revised edn by J. H. Harker and D. A. Allen).
4. P. R. von Rittinger. *Lehrbuch der Aufbereitungs Kunde*, Ernst und Korn, Berlin, 1867.
5. Anon. *Baum Jig Coal Preparation Plant*, Simon–Carves Publication, 1956.
6. F. Brinkmann and H. Hellwig. The Baum jig, *Gluckauf* **100**, 1249–54, 1964.
7. H. Louis. Chapter 7 (Jigs) in *The Dressing of Minerals*, Arnold, London, 1909, p. 19.
8. T. Takakuwa and M. Matsumura. Suggestions for the improvement of the air-pulsated jig, Paper A118, and International Coal Preparation Congress, Essen, 1954.
9. Anon. A preliminary study of washability characteristics, *King's College Mining Society Bulletin* **8**, Bulletin 1, Coal Cleaning No. 16.
10. W. W. Wallace. The modern Baum jig, *Colliery Engineering*, October 1953 and May 1954.
11. A. A. Hirst. Jig washers in theory and practice, Symposium of Coal Preparation, University of Leeds, June 1952.
12. G. J. Sanders. Private communication, 1981.
13. R. E. Zimmerman. Batac jig—a new, improved Baum-type jig for cleaning coarse and fine sizes of coal, *Mining Congress Journal*, May 1974, 43–49.
14. H. Breuer. Preparation plants and jigs, Type GHH, *Technical Notes* by Man Gutehoffnungshutte GmbH, 1985.
15. A. Bahr, H. Ludke, W. Mehrhoff and P. Wilczynski. Recent developments and research in the separating of fine coal, 2 mm, with the RAG. 9th International Coal Preparation Congress, New Delhi, 1982, Paper C6, (Ed.) S. K. Bose.
16. F. B. Michell and D. G. Osborne. Gravity concentration in modern mineral processing, *Chemistry and Industry* **58**, Jan. 1975.
17. J. Alexis. Cleaning coal and refuse fines with Humphreys spiral concentrator, *Mining Engineering* **132**, 8, 1224–1227, 1980.
18. R. G. Richards. Spiral concentrators for fine coal treatment, Queensland Coal Preparation Society, May 1983.
19. E. G. Kelly, D. J. Spottiswood, D. E. Spiller and C. N. Robinson. Performance of modern spirals cleaning fine coal, SME–AIME Annual Meeting, Los Angeles, February 1984.
20. D. T. Hornsby, S. J. Watson and C. J. Clarkson. Fine-coal cleaning by spiral and water washing cyclone, 2nd Australian Coal Preparation Conference, Rockhampton, October 1983.
21. C. H. Tiernon. Concentrating tables for fine-coal cleaning, *Mining Engineering* **32**, 8, 1228–1230, 1980.
22. S. Heintges. Using shaking tables for the preparation of coal fines or any coarse slurries, South African Coal Processing Society, Johannesburg, February 1981.
23. R. H. Richards. The Wilfley table, *Transactions American Institute Mining Engineers*. Part 1, **38**, 556–580, 1907 and Part 2, **39**, 303–315, 1908.

24. A. W. Deurbrouck and E. R. Palowitch. Performance characteristics of coal-washing equipment—concentrating tables, United States Bureau of Mines, Department of Energy (USBM DOE). Report of Investigation R16239, 1963.
25. J. Christopherson. Deister tables, Private communication, 1985.
26. F. Jiang, Y. S. Yang and Z. S. Liang. Study on application of centrifugal table for fine coal concentration, IX International Coal Preparation Congress, New Delhi, 1982, Paper G5, (Ed.) S. K. Bose.
27. R. G. Richards, J. L. Hunter and A. B. Holland Batt. Spiral concentrators to fine coal treatment, *Coal Preparation*, 1, 2, 207–229, 1985.
28. D. Jackson. Fine-coal spirals, *Coal Age*, **89**, 66–69, 1984.
29. E. A. Draeger and J. W. Collins. Efficient use of water-only cyclones, *Mining Engineering* **32**, 8, 1215–1217, 1980.
30. J. D. Miller and M. C. Van Camp. Fine-coal cleaning with an air-sparged hydrocyclone, AIChE Meeting, Houston, April 1981, Paper 62b.
31. L. Svarovsky. *Hydrocyclones*, Holt Saunders, Eastbourne, 1984, Chapter 9.
32. R. A. Lathioor, Kilborn Engineering (BC) Ltd, Vancouver, Canada. Separator efficiency prediction, Personal communication, March 1986.
33. M. F. Goodrich. Use of water-only cyclones as clean-coal scalpers preceding heavy-media cyclones, SME AIME Fall Meeting. St Louis, Missouri, October 1977.
34. D. F. Symonds. A review of dry cleaning process, Report Coal Mining Research Company, Devon, Alberta, Canada, 81/21-T, March 1981, NorWest Resources.
35. R. P. Killmeyer and A. W. Deurbrouck. Performance characteristics of coal-washing equipment: air tables, USBM DOE Report of Investigations. PMTC-6 (79), 1979.
36. S. G. Butcher. Coal Mining Research Company, Devon, Alberta, Canada. Personal communication, June 1987.

Chapter 8

EFFICIENCY TESTING OF GRAVITY CONCENTRATORS

8.1 EFFICIENCY DEFINITIONS

The only reasonable way of assessing or evaluating the efficiency of a coal-cleaning system which utilizes gravity as the means of separation is to determine the sharpness of separation achieved. In doing so, the method employed must not be influenced in any way by the various physical characteristics of the coal, i.e. washability, size distribution, density, etc.

Most coal-cleaning methods are unaffected by high- and low-density fragments relative to the density of separation, these fragments reporting to their appropriate separation products without difficulty. As the density of fragments approaches the density of separation, however, the separator is more severely tested and its inherent ability to provide sharpness in separating the coal becomes more and more significant.

In testing this phenomenon with a given coal, that is, obtaining a measure of the amount (mass) of misplaced material for various density increments covering the range of fragments in the sample, a distribution curve may be obtained which illustrates the performance of the separator under specific operating conditions. In some respects, this distribution curve may be considered as being the 'fingerprint' of that particular separator in that it represents the inherent capability of that one unit. Despite the manufacturing consistency that is now possible in producing machinery of any kind, it is and always will be impossible to produce physically identical machines having identical performance capabilities. In addition, uneven wear and tear and incidental damage will create further differences. The distribution curves shown in Fig. 8.1 illustrate this; they represent three Ni-Hard dense-medium cyclones operating in parallel streams in one plant. In terms of geometry and feed conditions, their performance should have been the same. The fourth curve is for a dense-medium cyclone of the same size and geometry, but lined with ceramic material and installed at the same time as the other three. Hence, the four 'fingerprints' tell their own story.

Despite the occasional use of other forms of efficiency measurement, the predominantly used calculations are based on results obtained from float–sink testing of the products of separation and comparison with the results obtained from a standard float–sink test on the feed to the separator.

Basic separation efficiencies can then be determined by comparison of these

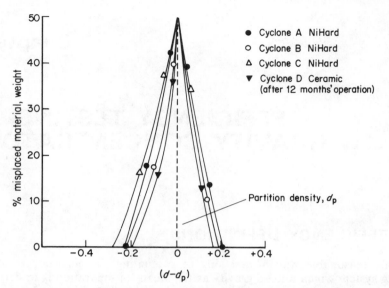

Fig. 8.1 Probability curves for dense-medium cyclone separators.

washability data. Among the most widely used are those of:

Fraser and Yancey (1923):[1]

Organic efficiency, E_o

$$= \frac{\text{actual yield of clean coal obtained} \times 100}{\text{float–sink test yield corresponding to the same ash content}}$$

Hancock (1947):[2]

Process efficiency, E_p = actual yield clean coal

$$\frac{\% \text{ float material in clean coal}}{\text{float–sink test floats in feed coal at clean-coal ash}}$$

$$- \frac{\% \text{ sink material in clean coal}}{\text{float–sink test sinks in feed coal}}$$

The basis for further manipulation of these formulae is usually derived from the simple relationships:

$$\text{Yield, } Y = \frac{\text{clean coal, } C}{\text{feed coal, } F} \times 100$$

$$100 - Y = \frac{\text{discard, } D}{\text{feed coal, } F} \times 100$$

where F, C and D are masses,

So if $F = D + C$, and f, c and d are the corresponding ash-content values, then $F_f = D_d + C_c$

EFFICIENCY TESTING OF GRAVITY CONCENTRATORS

Yield may be derived directly from ash values using:

$$Y = \left(\frac{d-f}{d-c}\right) \times 100$$

and

$$100 - Y = \left(\frac{f-c}{d-c}\right) \times 100$$

Other forms of separation efficiency are occasionally used, including those which are derived on the basis of recovery. Such a form of separation efficiency is defined as the difference between the recovery of light materials (those which are correctly placed) and heavy materials (those which are incorrectly placed) in the clean-coal product. Alternatively, they may be related to the recovery of combustible material in the clean coal.[3] These forms of efficiency are most commonly evaluated in qualitative terms and the parameters of ash content and calorific value are most frequently used.

$$\text{Separation efficiency, } E_s = \frac{\text{actual efficiency}}{\text{theoretical efficiency}} \frac{E_A}{E_T}$$

$$\text{and } E_A = \left(\frac{d-f}{d-c}\right)\left(\frac{100-c}{100-f} - \frac{c}{f}\right) \times 100$$

$$E_T = \left(\frac{d_1 - f_1}{d_1 - c_1}\right)\left(\frac{100 - c_1}{100 - f} - \frac{c_1}{f}\right) \times 100$$

where c_1 and d_1 are determined by float–sink analysis: i.e. c_1 is %age average ash-content of floats in the feed; d_1 is %age average ash-content of sinks in the feed (as opposed to those actually obtained, i.e. c and d) from which:

$$E_s = \frac{Y_c}{Y_{c_1}} \left(\frac{f-c}{f-c_1}\right) \times 100$$

The other common forms of efficiency are those determined by accounting for the errors in the system where the errors are then deduced to represent the inefficiency of the separator. Most of these forms of efficiency are derived from graphical methods.[4,5]

One of the most common[6] is:

ash error = (% of clean-coal ash) − (% float-coal ash at the corresponding yield value)

$$= C - C_Y$$

yield error = (% of clean coal recovered) − (% float coal at the corresponding product-coal ash-content)

$$= Y - Y_C$$

Although much has been published regarding the selection of the most meaningful methods for deriving the efficiency of gravity separators, the approach most commonly adopted is that contained in the current international (ISO) standard[7] which requires that for the standard expression of efficiency, the following four formulae should be used in conjunction:

(a) separation density expressed as
  (i) partition density; and/or
  (ii) equal errors density;
(b) total of correctly placed material at the separation density, expressed as a percentage of the reconstituted feed and where required the misplaced material in each product;
(c) écart probable moyen (e.p.m.) and imperfection (I);
(d) ash error or organic efficiency.

8.2 THEORY OF PARTITION (DISTRIBUTION) CURVES

The partition curve can be explained in simple terms as a histogram of the distribution of groups of coal of various density ranges in the cleaned coal or the discard products.[8] The partition curve approach was first proposed by Tromp in 1938.[9] In Tromp's curve, the percentage recoveries at each density fraction are plotted against the mean value of each of these densities to obtain a curve of the type shown in Fig. 8.2. In an ideal separation, the curve runs parallel to the abscissa at the density of separation. Increasing amount of misplacement of material causes a greater deviation from this axis and the head and tail of each curve tends to become characteristic of a particular type of separator. The previous Fig. 7.67 shows several curves for different separators ranging from a dense-medium bath, which is the most efficient, to the trough washer which is the least.

The partition curve may also be referred to as the Tromp curve, for obvious reasons, or other names such as distribution curve, error curve and grade recovery curve.

8.3 THE PARTITION CURVE

The following description was first included in the *King's College Mining Society (University of Newcastle upon Tyne) Bulletin*[10] over 20 years ago. It is reproduced with permission from the University of Newcastle upon Tyne.

In any separation, misplacement occurs, thus producing washing errors, the value of these errors being related directly to the amount of near-gravity material at the gravity of separation. Tromp,[9] in a study of jig washing, observed that the displacement of migratory particles was, with certain modifications, a normal frequency curve of Gaussian distribution, and from this observation the partition curve in the form of an ogive was evolved.

Since the construction of the curve involves the elementary calculation of partition coefficients (partition or distribution factors), Table 8.1 shows the results for a two-product separation although it must be appreciated that those for a three-product separation can be similarly calculated.[7]

EFFICIENCY TESTING OF GRAVITY CONCENTRATORS

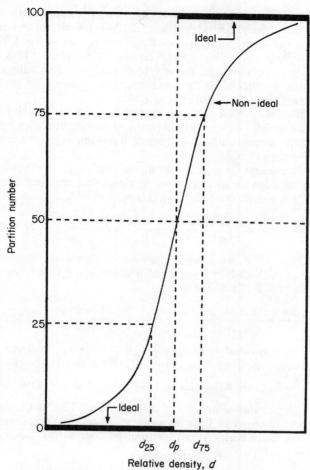

Fig. 8.2 Ideal and non-ideal Tromp partition curve.

TABLE 8.1 Washability Results of a Two-product Separation

Relative density	% Weight		% of Feed		Calculated feed	Mean r.d.	Partition coefficient
	clean coal	reject	clean coal	reject			
1	2	3	4	5	6	7	8
−1.30	55.52	0.27	39.03	0.08	39.11	—	0.2
1.30–1.40	28.99	0.40	20.38	0.12	20.50	1.35	0.6
1.40–1.50	6.34	0.67	4.46	0.20	4.66	1.45	4.3
1.50–1.60	2.99	0.98	2.10	0.29	2.39	1.55	12.1
1.60–1.70	1.34	1.55	0.94	0.46	1.40	1.65	32.9
1.70–1.80	0.67	2.89	0.47	0.86	1.33	1.75	64.7
1.80–1.90	0.20	2.56	0.14	0.76	0.90	1.85	84.4
1.90–2.00	0.15	2.73	0.11	0.81	0.92	1.95	88.0
2.00–	3.80	87.95	2.67	26.12	28.79	—	90.7
	100.00	100.00	70.30	29.70	100.00		

Column 1—Relative density fractions The limits result directly from the test, except that with the very lightest fraction, the lower limit, and in the case of the heaviest fraction the upper limit, are not usually determined, e.g. 1.60–1.70 r.d.

Column 2—Weight percentage of the clean-coal fractions This is a series of direct calculations of percentage, the weight of each fraction being referred to the summation of the weights of all the clean-coal fractions, e.g. wt percentage of the 1.60–1.70 r.d. fraction of the clean coal = 1.34%.

Column 3—Weight percentage of the reject fractions As for the previous column, this is a series of direct calculations of percentage with reference to the summation of the weights of all the reject fractions, e.g. wt percentage of the 1.60–1.70 r.d. fraction of reject = 1.55%.

Column 4—Percentage of feed reporting to clean-coal Direct calculations of the percentage of each density fraction of clean coal referred to a basis of the original feed, e.g. for the 1.60–1.70 r.d. fraction:

$$\frac{\text{column 2} \times \text{\% yield of clean coal}}{100} = \frac{1.34 \times 70.30}{100} = 0.94\%$$

Column 5—Percent of feed reporting to reject Similarly, direct calculations of the percentage of each density fraction of reject referred to a basis of the original feed, e.g. for the 1.60–1.70 r.d. fraction:

$$\frac{\text{column 3} \times \text{\% yield of reject}}{100} = \frac{1.55 \times 29.70}{100} = 0.46\%$$

Column 6—Reconstituted percentage weights of the r.d. fractions in the original feed e.g. percentage weight of the 1.60–1.70 r.d. fraction in the raw feed:

$$\text{Column 4} + \text{column 5} = 0.94 + 0.46 = 1.40\%$$

Column 7—Mean relative density of the fractions This is the average relative density of each of the intermediate fractions, e.g. 1.60–1.70 r.d., mean 1.65.

Column 8—Partition coefficients (partition or distribution factors) Percentage of each r.d. fraction reporting to reject, e.g.

$$\text{partition coefficient of 1.65 r.d. fraction} = \frac{\text{column 5}}{\text{column 6}} \times 100$$

$$= \frac{0.46}{1.40} \times 100 = 32.9\%$$

The partition curve (curve A, Fig. 8.3) is constructed by plotting mean r.d. (column 7) as abscissa against ordinate of partition coefficient (column 8).

The partition curve may also be plotted by turning it upon itself above the 50% value and plotting the higher values in the opposite direction (curve B, Fig. 8.3). When this curve is plotted with a standard ordinate scale of 1 cm = 2% (partition coefficient) and an abscissa of 1 cm = 0.1 (relative density), the area in cm^2 enclosed by the curve is known as the error area.

The partition density, d_p, is the relative density corresponding to a partition coefficient of 50, i.e. 1.7 (Fig. 8.3). This is the relative density at which an infinitesimally small increment of raw feed is equally divided between clean coal and reject.

EFFICIENCY TESTING OF GRAVITY CONCENTRATORS

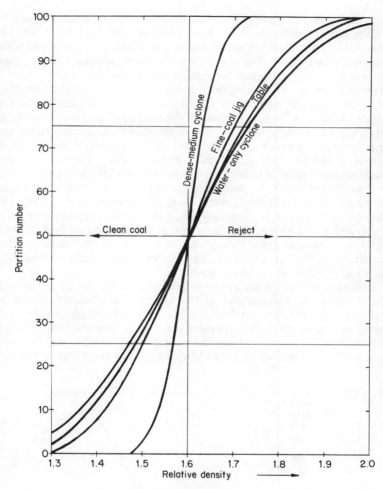

Fig. 8.3 Partition curve.

The probable error, e.p.m., is given by

$$\frac{d_{75} - d_{25}}{2} = 0.105$$

This is usually known as the écart probable moyen and the value gives an indication of the quantitative errors inherent in the separating process at a given separation density.

The imperfection, I, is given by e.p.m./$(d_p - 1)$ for wet processes. The value e.p.m./d_p is sometimes used for dry processes. The definition for dry processes is not given in ISO, but it appears logical that the expression e.p.m./d_p should be used. In this example, it is equal to $0.105/(1.7 - 1) = 0.150$.

Strictly speaking, the partition curve is not independent of the raw-feed characteristics, and examination of a number of curves shows that the 'tails' (upper and lower extremities of the curves) vary not only with different separators, but can also be influenced by different coals. It appears that the friability of the raw coal has a considerable influence on the shape of these 'tails'. As far as the separators are concerned, it is evident that the nearer one approaches a true float-and-sink type separation, the more regular the ends of the curves become.

Nevertheless, for most separations, the centre portion of the curve approximates to a straight line and this is obviously the most reliable portion since it corresponds to the high-frequency zone of the normal Gaussian curve. In general, the d_{25} and d_{75} values define this portion of the curve; consequently the probable error, e.p.m. (the 'semi-interquartile' of the cumulative curve of normal distribution), is independent of raw-feed characteristics, whereas the error area which embraces the 'tails' is influenced by the nature of the raw feed.

Error area suffers from the disadvantage that high-density separations often include low-partition coefficients even at high densities, and a very extensive range of float—sink liquids is necessary. Probable error increases with increase in d_p for a given size of feed and also increases as the size of raw feed diminishes. It is observed to be proportional to $(d_p - 1)$ for wet processes,[11] hence introducing the term 'imperfection' as a measure of efficiency and which, as emphasized by Cerchar, is independent of partition density.

Even although the partition curve is not cumulative by its derivation, it is identified by some as the integral of the normal distribution on the basis of general shape. However, Tromp[12] argues that it is only by plotting $(d - d_p)^{0.6}$ as a parameter on the abscissa that 'frequency curves of migrated material' prove to be identical with Gaussian curves, indicating limit velocity as the governing parameter for the probability of errors (Fig. 8.4). Many partition curves do in fact

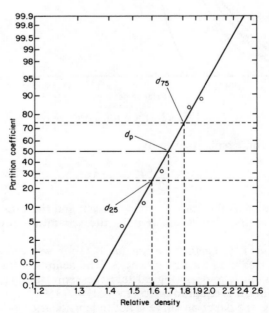

Fig. 8.4 Frequency curve of migrated material.

EFFICIENCY TESTING OF GRAVITY CONCENTRATORS

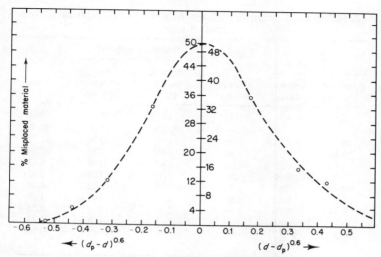

Fig. 8.5 Anamorphosed partition curve.

conform to the normal shape. As one would expect, tests on dense-medium separators, in which the effect of shape and size of fragments is minimized, give approximately symmetrical partition curves. In some cases,[13] it is necessary to plot the logarithm of the apparent relative density $(d_p - 1)$ as the abscissa to obtain an approximately normal distribution.

For accurate error area measurements, it is advantageous to determine the value at which the curve cuts the x-coordinate and in many cases, as mentioned previously, this value is not known when the coefficients do not reach the 100% range. By extrapolation of the anamorphosed partition curve, which is constructed by determining the best straight line through the points in the linear portion of the curve, this value can be obtained (Fig. 8.5). This anamorphosis is produced by plotting the partition coefficients on a probability scale and relative density on a $\log (d_p - 1)$ scale. The highest partition coefficient plotted in Fig. 8.3 is 88.0 at a mean density of 1.95, but the anamorphosed curve (Fig. 8.5) shows 99.9% of the feed reports to reject at 2.4 r.d. This extrapolation is shown in the 'tails' in Fig. 8.3. Using this technique, the error area becomes a function of the separation and not of the raw coal.

The results obtained from the data shown are as follows:

partition gravity 1.70 r.d.
probable error 0.105
imperfection 0.150
error area 70.6 cm^2

8.4 OTHER PERFORMANCE TEST DATA

The data obtained for the calculation of partition coefficients as shown in Table 8.1 are usually expanded to include qualitative results (normally ash content, but occasionally calorific value as well), and the tabulation is extended to include

TABLE 8.2 Correctly Placed Material (100% minus misplaced material)

| % of feed | | | Misplaced material | | |
clean coal (as 4)	reject (as 5)	Relative density	sinks in cleaned coal ↑Σ 9	floats in reject ↓Σ 10	Total (12 + 13)
9	10	11	12	13	14
		—	70.30	0.00	70.30
39.03	0.08	1.30	31.27	0.08	31.35
20.38	0.12	1.40	10.89	0.20	11.09
4.46	0.20	1.50	6.43	0.40	6.83
2.10	0.29	1.60	4.33	0.69	5.02
0.94	0.46	1.70	3.39	1.15	4.54
0.47	0.86	1.80	2.92	2.01	4.93
0.14	0.76	1.90	2.78	2.77	5.55
0.11	0.81	2.00	2.67	3.58	6.25
2.67	26.12	—	0.00	29.70	29.70
70.30	29.70				

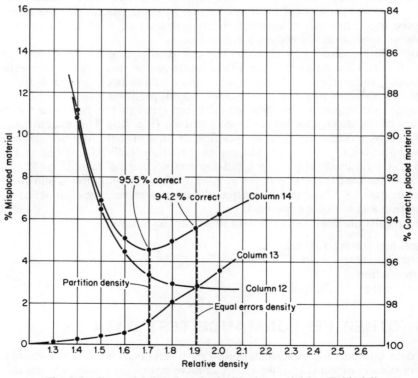

Fig. 8.6 Correctly placed, and misplaced material (see Table 8.2).

EFFICIENCY TESTING OF GRAVITY CONCENTRATORS

calculation of the following:

(a) correctly placed and misplaced material;
(b) reconstituted feed to the separator.

The form which such computations usually take is similar to that specified in the ISO standard[7] and will then only depend upon the type of separation, i.e. two- or three-product. We shall discuss the ramifications of three-product separation later but, first, let us consider the two-product separation reported in Table 8.1. This table has been expanded to include calculation of misplaced material in Table 8.2. The form of computation is included in the table and is self-explanatory. The final three columns in this table are the key ones in that they define the amounts of misplaced material caused by the separator. When shown graphically, as in Fig. 8.6, the intersection of the two sets of misplaced material coincides with what is known as the equal-errors density or Wolf cut-point. This in turn coincides with the 'low-point' in the total misplaced material curve, i.e.

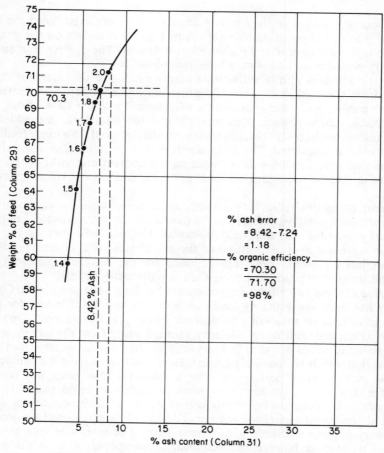

Fig. 8.7 Ash error and organic efficiency (see Table 8.4).

TABLE 8.3 Ash Content of Products and Reconstituted Feed

Relative density fractions (as 1)	Cleaned coal				Reject			Reconstituted feed	
	wt % of feed (as 4)	Ash %	Quantity of ash		wt % of feed (as 5)	Ash %	Quantity of ash	wt % of feed (16 + 19)	quantity of ash (18 + 21)
15	16	17	18	19	20		21	22	23
−1.30	39.03	1.40	54.64	0.08	1.60		0.13	39.11	54.77
1.34–1.40	20.38	7.60	154.89	0.12	7.80		0.94	20.50	155.83
1.40–1.50	4.46	15.90	70.91	0.20	18.40		3.68	4.66	74.59
1.50–1.60	2.10	28.60	60.06	0.29	30.10		8.73	2.39	68.79
1.60–1.70	0.94	36.60	34.40	0.46	37.30		17.16	1.40	51.56
1.70–1.80	0.47	41.60	19.55	0.86	43.90		37.75	1.33	57.30
1.80–1.90	0.14	46.40	6.50	0.76	52.00		39.52	0.90	46.02
1.90–2.00	0.11	57.80	6.36	0.81	59.20		47.95	0.92	54.31
2.00–	2.67	69.20	184.76	26.12	71.90		1878.03	28.79	2062.79
	70.30	8.42	592.07	29.70	67.21		1996.14	100.00	2625.96

when the curve is closest to the abscissa or relative-density axis. When the partition density from the partition curve given in Fig. 8.4 is shown on the graph in Fig. 8.7, it is found to occur at a higher relative density. The difference in amount of correctly placed material between these two points is 1.3%.

By the introduction of qualitative data corresponding to the quantitative data obtained from the washability tests, a further parameter is added to provide an additional means of assessing separation efficiency. Table 8.3 shows how, when ash analytical values of each float product and the final sinks are added, the comprehensive washability characteristics of the original raw-coal feed to a separator may be calculated. This is demonstrated by the tabulations in Table 8.3 and follows on into Table 8.4. By comparing ash content and yield in the graph shown in Fig. 8.7, the important values of ash error and organic efficiency may be obtained.

Ash error is the difference between the actual percentage ash of the product and that shown by the washability curve (based on the reconstituted feed) to be theoretically possible for the yield obtained. Organic efficiency is the ratio (normally expressed as a percentage) of the actual yield of a desired product to the theoretically possible yield (again, based on the reconstituted feed), both actual and theoretical products having the same percentage of ash.

The value of these two efficiency parameters has long been postulated. Organic efficiency has been suggested by many as being an important means for establishing a basis for guarantees of separator (or washery) performance.[14] One of the problems associated with using efficiency values derived from the partition curve alone is that the values are usually derived from the most favourable part of the curve (i.e. that which is invariably linear). Hence, values derived for the purpose of setting guarantees of performance for a separator may often not represent much of a challenge to the separator when required to treat a relatively easily cleaned coal. The situation frequently occurs where, at a low density of separation, the sharpness of separation (as recorded by a low e.p.m.) will be good, yet the recovery organic efficiency will be very poor. The reverse is true at a high density of separation. In practice, rectilinear anamorphosis of a partition curve frequently shows that the curve is in fact distorted (only slightly evident in Fig.

TABLE 8.4 Reconstituted Feed

Relative density fractions (as 1)	Reconstituted feed				Cumulative floats				Cumulative sinks		
	wt % (as 22)	quantity of ash (as 23)	ash % (26 ÷ 25)	Relative density	wt % ↓Σ 25	quantity of ash ↓Σ 26	ash % 30 ÷ 29	wt % ↑Σ 25	quantity of ash ↑Σ 26	ash % 33 ÷ 32
24	25	26	27	28	29	30	31	32	33	34
−1.30	39.11	54.77	1.40	1.30	39.11	54.77	1.40	100.00	2625.96	26.26
1.30–1.40	20.50	155.83	7.60	1.40	59.61	210.60	3.53	60.89	2571.19	42.23
1.40–1.50	4.66	74.59	16.01	1.50	64.27	285.19	4.44	40.39	2415.36	59.80
1.50–1.60	2.39	68.79	28.78	1.60	66.66	353.98	5.31	35.73	2340.77	65.51
1.60–1.70	1.40	51.56	36.83	1.70	68.06	405.54	5.96	33.34	2271.98	68.15
1.70–1.80	1.33	57.30	43.08	1.80	69.39	462.84	6.67	31.94	2220.42	69.52
1.80–1.90	0.90	46.02	51.13	1.90	70.29	508.86	7.24	30.61	2163.12	70.67
1.90–2.00	0.92	54.31	59.03	2.00	71.21	563.17	7.91	29.71	2117.10	71.26
2.00–	28.79	2062.79	71.65		100.00	2625.96	26.26	28.79	2062.79	71.65
	100.00	2625.96	26.26							

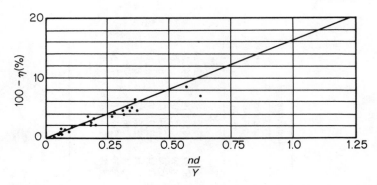

Fig. 8.8 Graph showing relationship between loss of organic efficiency and nd/Y for South African coal.[14]

8.5). Such distortion may have an important bearing on the organic efficiency of the separation without causing much alarm in terms of e.p.m. value. The use of e.p.m. alone is therefore inadvisable. Other parameters such as misplaced material, ash error, etc., do provide some protection for the purchaser of a new separator. However, in deriving such values from coal samples taken at the tendering stage, there is often a risk that the actual raw coal requiring treatment may not have been properly represented. One way of overcoming such problems is to incorporate some form of relationship between partition and washability data into the form of guarantee. One example is as follows:

$$oe = 100 - \left(C \times \frac{nd}{Y} \right)$$

where $oe(n)$ = organic efficiency; nd = percentage of near-density material within plus or minus epm of the cut-point (partition density); C = constant; Y = float-sink (theoretical) yield of coal at the same cut-point.

The graph shown in Fig. 8.8 shows the relationship between loss of organic efficiency, i.e. $(100 - oe)$, and nd/Y obtained for South African coal. These results were obtained from dense-medium baths, dense-medium cyclonic separators, Chance sand cones and Baum jigs, with coals ranging in size from 180 mm to 0.5 mm. The e.p.m. values varied from 0.003 to 0.11, nd varied from 1 to 73% and yield varied from 17 to 90%. Washed-coal products with 5–15% ash were obtained.

8.5 CONDUCTING AN EFFICIENCY TEST

In order to carry out a plant-performance test on a gravity separator, a considerable amount of accurate data must be taken. The plant must be operating under stable conditions, and during the period of the test the average yield must be determined. This can be done by weightometer readings, if available, or by the ash balance method or by both. Samples of clean coal and discard must be taken over the full period of the test and increment weight and frequency must adhere to correct sampling procedures. The clean coal and discard are subjected to float-

and-sink analysis across a full range of densities, usually 1.30–1.80. The density intervals should be as close as possible to give proper definition to the curve. With difficult coals, e.g. a high percentage of near-density material at the cut-point, the interval should be 0.02. South African and Indian coals, in particular, require this sort of detailed analysis.

Other data that should be recorded during a performance test are the feed rate, the instrument recorded density and the actual density of the medium.

Prior to commencing with the compilation of test data, it is recommended that some form of standard tabulation sheet be adopted. The recommended one would be that obtainable from the ISO or British Standards, *Methods for Expression and Presentation of Results of Coal Cleaning Tests*. These are set out in an easily reproducible format that can be readily adapted to enable the results of two- or three-product separators to be recorded.

In carrying out performance testing of three-product separation, it is useful to consider the separator as being a combination of two distinct two-product separations (i.e. a low-density and a high-density cut). It will not matter whether these two stages are in fact carried out in different separating vessels or in different parts of the same vessel.

The diagram in Fig. 8.9 illustrates different combinations possible for the two stages using the symbols F (feed), C (cleaned coal), R (refuse) and M (middlings, or intermediate product). Diagrams (a) and (b) are typical arrangements for two-stage dense-medium separations, the only difference being that the low-density cut occurs first in (a) and second in (b). Diagram (c) represents a three-product Baum jig or Wemco, three-product drum type of separator. Middlings may be

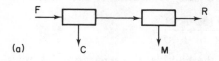

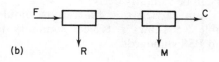

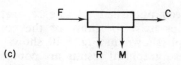

Fig. 8.9 Various combinations of three-product separators: (a) low-density separation first; (b) high-density separation first; (c) three products from one separator; F, feed; C, coal; M, middlings; R, reject.

collected as a separate product or recirculated and, providing that any recirculating middlings are included in the reconstituted feed, F, the balance is unaffected.

The efficiency of a three-product separation may be calculated in two different ways:

(a) by regarding the separation as occurring in two distinct stages, each with its own feed;
(b) by regarding the separations as being comprehensively achieved in one unit, the feed for which would then be the reconstituted raw coal.

Bearing the selected approach in mind, it is fairly simple to derive an appropriate formula to represent the separation. For example:

For diagram (a)

Method (1) low-density separation is

$$\frac{100(M+R)}{(C+M+R)}$$

and the high-density separation is

$$\frac{100R}{(M+R)}$$

Whereas

Method (2) low-density separation is unchanged

$$\frac{100(M+R)}{(C+M+R)}$$

but the high-density separation becomes

$$\frac{100R}{(C+M+R)}$$

The other two options can be similarly derived.

When the test has been carried out, the results should be summarized in tabular form and Table 8.5 shows a typical summary table similar to that included in the ISO and BSS standards.

8.6 RELATIONSHIP BETWEEN ECART PROBABLE MOYEN AND GRAIN SIZE

Considerable data have been published relating separating efficiency to coal grain-size.[18-20] Relationships exist for most of the common types of coal-cleaning unit and the graph shown in Fig. 8.10 shows curves obtained from Canadian washeries.[21] These graphical results are compared with similar ones obtained from plants in the USA in Table 8.6. All of these data serve to demonstrate that there is a distinct size-range over which each of the individual separators is most effective. This is perhaps more clearly shown when grain size is

TABLE 8.5 Performance Test Summary Table

Reference:

Test details		Date of test			Name of plant			
		Tonnages			Results summary			
			Size analysed	Total feed to plant				
			tonnes / % mass	tonnes / % mass				
Upper size of coal analysed (mm)		Raw coal			(a) single cut-point two-product separation products = cleaned coal reject			
Upper size of coal to plant (mm)		Products cleaned coal middling reject			(b) two cut-points three-product separation products = cleaned coal middlings reject			
Type of cleaning unit					(1) high-density cut: cleaned coal + middlings reject			
Rated capacity (t/h)		Plant feed rate (t/h)			(2) low-density cut: cleaned coal/middlings + reject			
Coal seams treated								
Details of blend type A (%) type B (%) type C (%)					Separation	Separating density		Correctly placed material (%)
						(1)	(2)	(1) (2)
Actual test period (min)					Partition			
					Equal errors			
Total stoppage time (min)					e.p.m.			
					Imperfection			
Net test period (min)					Ash error		%	
					Organic efficiency	%		

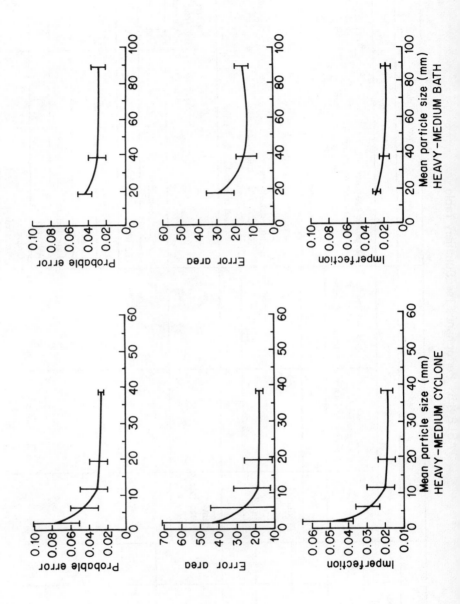

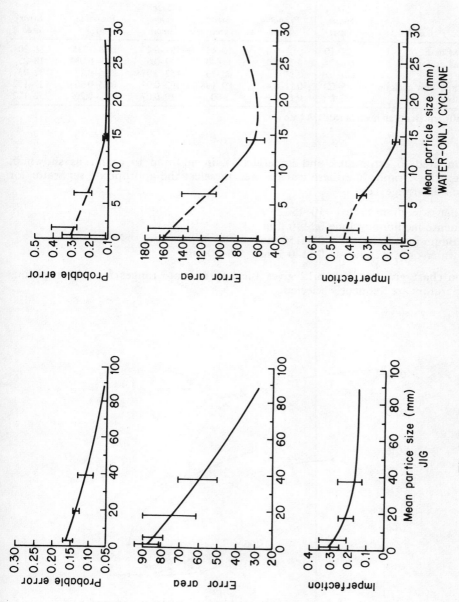

Fig. 8.10 Curves for various separators of e.p.m. versus mean grain size.

TABLE 8.6 Comparison Between Canadian and USA Plant Performance

	Canadian			USA		
Unit	Size range (mm)	Probable error	Error area	Size range (mm)	Probable error	Error area
DM bath	127–10	0.02–0.07[a]	10–45	127–6.4	0.02–0.035	16–30
DM cyclone	50.8–0.6	0.02–0.05	13–40	19–0.6	0.028–0.034	18–25
Jig	127–0.6	0.11	68–75	152–0.07	0.05–0.17	32–93
Water-only cyclone	19–0.1	0.11–0.41	89–196	102–0.07	0.25–0.30	127–152
Table	3.2–0.1	0.16	100	10–0.07	0.07–0.20	52–122

[a] 0.07 represents an overloaded DM bath

compared to error area and imperfection, in addition to e.p.m., as shown in Fig. 8.11. Graphs like these can be used to select the appropriate separator for each size-range, i.e.

dense-medium bath	10–150 mm
dense-medium cyclone	0.5–10 mm
Baum jig	0.5–150 mm
water-only cyclone	0.15–0.5 mm; 0.5–10 mm

The chart given in Fig. 8.12 gives the approximate ranges for which various separators are frequently selected.

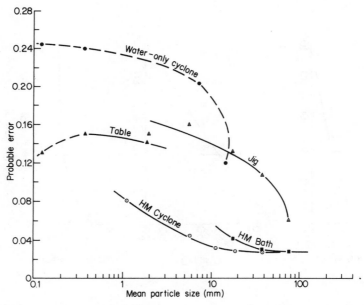

Fig. 8.11 Mean grain size versus e.p.m. error area and imperfection for four types of coal-cleaning unit.

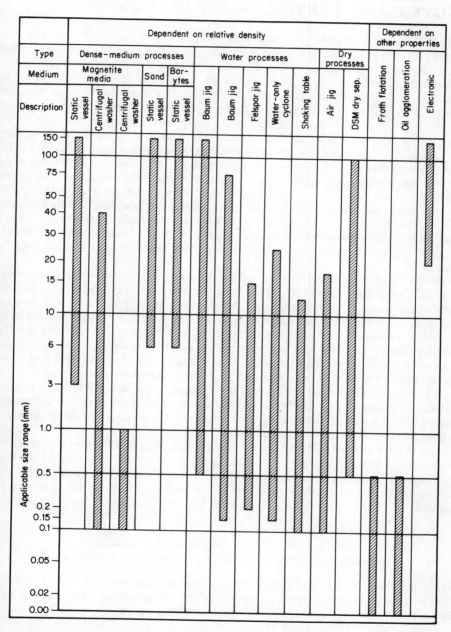

Fig. 8.12 Optimum size range for various coal-cleaning units.

8.7 RELATIONSHIP BETWEEN ECART PROBABLE MOYEN AND DENSITY OF SEPARATION

It is well known that the e.p.m. decreases with density of separation, e.g. at a low density of separation a very sharp separation is achieved. This relationship for any particular separator is rarely defined, e.g. for a dense-medium cyclone the relationship between e.p.m. and density of separation is e.p.m. $= f_1 \times f_2 \times f_3 (0.027 \times d_p - 0.01)$, where f_1, f_2 and f_3 are factors for coal size, cyclone size and contractor's guarantee factors, and d_p is the separation or partition density.

Much of the early work of defining relationships between e.p.m. and separation density was carried out by the Dutch State Mines in evaluating the performance of separators developed by its own research and development activities. For many years, this type of information was carefully protected and restricted in its use by the developers to include only licence holders of the technology. In recent years, however, many new separator manufacturers have produced similar information, and research institutions and universities have advanced the mathematical approaches involved to produce complex mathematical models for separators and other coal-preparation equipment, paving the way to complete plant modelling.[15-17] Many aspects of this type of work are covered later in Chapters 20 and 21.

Reverting to the relationship between e.p.m. and separation density: the most commonly used approach for generating a partition curve typical of a particular type of separator is to plot empirical data on a large graph and draw a smooth curve through these points. Additional points are then taken from the curve to obtain a 'characteristic' partition curve. This master curve can then be used to predict clean-coal yield and ash content or calorific value by reading off the average partition coefficient for each relative-density fraction and applying these values to the washability curves for the raw coal in each size-fraction tested.

When such calculations are carried out manually, the time involved is usually extremely lengthy, but if mathematical representations of the partition curves are derived for the various alternative washing units, it is an easy matter to program a computer to perform these calculations.

The master curve typified by empirical results, obtained from many different tests with a particular type of washing unit, provides the required number of points to generate a polynomial capable of representing the partition curve. The master curve points are first taken from the graph and tabulated as shown in Table 8.7. This curve is dependent upon the following values:

(a) the partition or separation density;
(b) the upper and lower size limits of the sample size range; and
(c) the equipment characteristics for the separator under consideration.

These are used to calculate the e.p.m. of the separation by means of a formula similar to that given in the first paragraph of this section, from which the particular partition curve is derived from the 'characteristic' curve.

Some typical formulae in common usage are:

(a) *dense-medium bath*

    e.p.m. $= f_1 f_2 (0.047 d_p - 0.05)$

EFFICIENCY TESTING OF GRAVITY CONCENTRATORS

TABLE 8.7 Example of a computer data sheet for input data for coal-cleaning-unit simulation

DATA SHEET FOR WACO

1. Data File Content: db or adb

FROM GRAPH		LAB RESULTS				
Float Specific Gravity	Percent Cumulative Weight	Float Specific Gravity	Percent Ash	Additional Parameters		
1.300						
1.325						
1.350						
1.375						
1.400						
1.450						
1.500						
1.550						
1.600						
1.650						
1.700						
1.750						
1.800						
1.900						
2.000						
2.100						
2.200						
2.400						
2.600						
2.800						

2. Responses During Program Execution
   a) Data File Name:
   b) Number of Sets of Laboratory Data:
   c) Description:
   d) Number of Additional Parameters:
   e) Percent Ash Targets for which Theoretical Yields are Required (Last Value = 0.0)
   f) Results File Name:

where f_1 is a feed-coal size factor ranging from about 0.5 for coarse coal to 1.4 for small coal; and f_2 is a guarantee factor for which a typical value might be 1.1–1.2.

(b) *dense-medium cyclonic separators*

e.p.m. $= f_1 f_2 f_3 (0.027 d_p - 0.01)$

where f_1 is a feed-coal size factor calculated for mean sizes ranging from 0.5 mm to 10 mm, giving values ranging from about 2 to 0.75 respectively; f_2 is an equipment factor depending upon unit size (cyclone diameter) and type—typical values 1.1 to 1.2; and f_3 is a guarantee factor, typically 1.1 to 1.2.

(c) *Baum jig*

e.p.m. $= 0.78 f_1 f_2 f_3 [d_p(d_p - 1) + 0.01]$

where f_1 is an equipment factor dependent upon feed-size composition. For feed-size mean values ranging from 1 mm to 100 mm, values ranging from about 3 to 0.5 respectively are used; f_2 is a size-correction factor. For feed-size mean values of 1–100 mm, values ranging from about 1 to 2 are used; f_3 is a guarantee factor typically 1.1 to 1.2.

(d) *water-only cyclonic separators*

e.p.m. $= f_1 f_2 (0.33 d_p - 0.31)$

where f_1 is an equipment factor dependent upon feed-size composition—and diameter, e.g. 100 mm diameter treating 0.5×0.05 mm coal, $f_1 = 0.8$; 500 mm diameter treating 10×1 mm coal, $f_1 = 0.9$; f_2 is a guarantee factor typically 1.1 to 1.2.

8.8 USE OF DENSITY TRACERS

Since the early 1960s, density tracers of one type or another have been used for pilot-plant studies with coal- and mineral-separation equipment. They were used to test mineral jigs,[22] water-only cyclones,[23,24] and dense-medium cyclones.[25] Three types of tracers have been used:

(a) radioactive type;
(b) magnetic type; and
(c) density type.

None of the early work which extended into the mid-1970s was carried out with tracers which were capable of providing accurate separating efficiency determination. It is fair to say that their use was primarily intended to obtain an indication of separator behaviour rather than accurate performance evaluation, although the work carried out in South African diamond plants to study the separating efficiencies of dense-medium cyclones might be considered a possible exception. This work pioneered the more recent development of tracers with closely spaced relative-density increments.[26] Jig and water-only cyclone coal separations are usually adequately defined by tracers widely spaced on the

TABLE 8.8 Properties of JKMRC[a] Density Tracers[27]

Shape	cubic
Nominal sizes	2–100 mm on each edge
Tolerance on edge length	±10%
Nominal relative density	1.20–2.00 in 0.01 increments
Relative density tolerance	±0.005

[a] Julius Kruttschnitt Mineral Research Centre.

relative-density scale and, for diamond separations, the major requirement is to check for misplacement of high-density diamondiferous material.

Density tracers for use in dense-medium separators, which are able to provide very accurate near-density separations with small incremental variation in separating density (± 0.005 r.d.), must be manufactured to much more rigorous specifications. Those in current use are available in 0.01 r.d. intervals and have been found to be suitable for producing precise partition curves. Table 8.8 shows the properties of Australian-manufactured tracers which have been found to be unaffected by continuous immersion in water and sufficiently abrasion and impact resistant to withstand repeated use. They are made of plastic and accurately dosed with heavy-metal salts to obtain the required density. They are brightly coloured and readily seen and, when added to the separator feed, they can be readily collected by hand from the product and refuse streams and sorted into appropriate r.d. increments on the basis of colour coding or other markings.[27] After counting, the partition number of each r.d. is calculated and the Tromp partition curve drawn. Partition numbers are calculated on the basis of those actually recovered, and lost tracers are not included in the calculation. For each relative-density fraction:

$$\text{partition number} = \frac{\text{number of tracers recovered from reject}}{\text{total number of tracers recovered}} \times 100$$

Density tracers allow a quick and accurate estimation of circuit performance to be made. A typical test requires four or five people for about one hour and the results are almost immediately available for use.

TABLE 8.9 Data Obtained for a Typical[a] Tracer Test[27]

Relative density	Number added	Number of tracers				
		Number retrieved		Lost	Retained in cyclone	Partition number
		product	reject			
1.39	30	30	0	0	0	0
1.40	30	30	0	0	0	0
1.41	30	28	2	0	0	7
1.42	30	25	5	0	0	17
1.43	30	7	19	0	4	73
1.44	30	4	25	0	1	86
1.45	30	0	30	0	0	100
1.46	30	0	30	0	0	100
1.47	30	0	30	0	0	100

[a] Size 33 mm tracers; 0.5–45 mm coal feed; test duration 60 minutes.

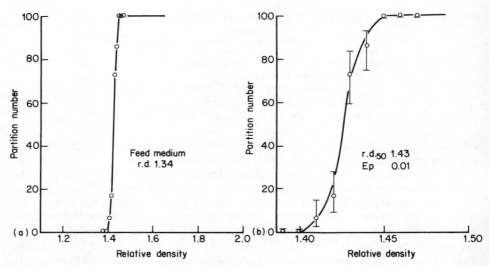

Fig. 8.13 Dense-medium cyclone partition curve on conventional and expanded scales. (a) Normal partition curve. (b) Partition curve drawn to an expanded relative-density scale.

Table 8.9 gives data from a typical tracer test and Fig. 8.13 shows the cyclone partition curve drawn on a conventional and an expanded r.d. scale. Figure 8.14 shows the partition curves before and after a worn cone component was replaced, demonstrating the value of tracers in determining unit wear and other similar major causes of efficiency loss.

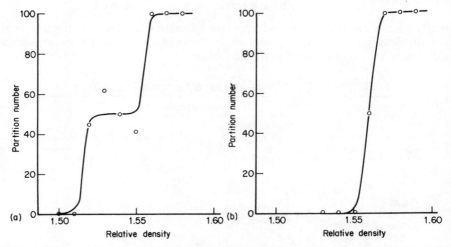

Fig. 8.14 Partition curve for dense-medium cyclone before and after worn cyclone component is replaced. (a) Partition curve for a module with one severely worn cyclone. (b) Partition curve for the same module after replacement of both cyclone bodies.

REFERENCES

1. T. Fraser and H. F. Yancey. Interpretation of results of coal washing tests. *Transactions AIME* **69**, 447–469, 1923.
2. R. T. Hancock. Calculating efficiencies from washing figures, *Colliery Engineering* 140–142, 156, 191–193 and 237–239, 1947.
3. F. F. Peng, A. D. Walters, M. R. Geer and J. W. Leonard. Evaluation and prediction of optimum cleaning results, Chapter 18 in *Coal Preparation*, (Ed. J. W. Leonard), AIME, 1979.
4. H. Heidenreich. Scientific and economic investigation of coal washing, *Gluckauf* **65**, 237–239, 949–956 and 991–997, 1947.
5. L. Valentik. Evaluation of washery performance, *Transactions SME/AIME*, **244**, 3, 344–353, 1969.
6. A. Grounds and L. W. Needham. The performance and efficiency of British Coal preparation plants, Paper G4, 1st International Coal Preparation Congress, Paris 1950, pp. 760–775.
7. ISO Standard. *Coal Cleaning Tests: Expression and Presentation of Results*, ISO923, 1975-12-01.
8. A. Terra. Theory of washing coal, *Revue Industrie Minérale*, **425**, 383–403, 1938.
9. K. F. Tromp. Neue Wege für Beurteilung der Aufbereitung von Steinkohlen (New methods of computing the washability of coals), *Gluckauf* **37**, February 6 and 13, 1937, 125–131, 151–156. Excerpts in *Colliery Guardian* **154**, May 21, 28, 1937, 955–959, 1009.
10. H. Y. Robinson. The partition curve, *King's College Mining Society Bulletin*, **9**, 5, Coal Cleaning Series No. 25.
11. C. Krigsman and J. N. J. Leeman. The application of cyclone washers in coal washeries, 4th International Coal Preparation Congress, Harrogate, 1962.
12. K. F. Tromp. Notion of probabilities in coal washing—some notes on frequency curves of migrated material, Proceedings 1st International Coal Preparation Congress, Paris 1950.
13. P. Belugon and J. Ulmo. Representation of washing results, Proceedings 1st International Coal Preparation Congress, Paris 1950.
14. P. J. van der Walt, C. D. Alexander, P. J. F. Fourie and S. F. Streicher. Organic efficiency as a basis for guarantees of washery performance, Proceedings 6th International Coal Preparation Congress, Paris 1973.
15. K. J. Reid, Lu Maixi and Zhang Shenggui. Simulating coal preparation distribution curves, *Coal Preparation* **1**, 231–249, 1985.
16. B. S. Gottfried, P. T. Luckie and J. W. Tierney. Computer simulation of coal-preparation plants, Report no. DOE/PC/30144-T7, US Department of Energy, December 1982. Available from National Technical Information Service, US Department of Commerce, Springfield, VA 22161, USA.
17. R. X. Rong and G. J. Lyman. Computational techniques for coal washery optimization-parallel gravity and flotation separation, *Coal Preparation* **2**, 51–67, 1985.
18. R. A. Lathioor, Kilborn Engineering (BC) Ltd, Vancouver, Canada, personal communication regarding computer simulation of coal-cleaning units, June 1967.
19. D. S. Davies, H. H. Driesen and R. M. Oliver. Advances in hydrocyclone heavy-media separation technology from ores, Proceedings 6th International Mineral Processing Congress, Cannes 1963.
20. A. W. Deurbrouck and J. Hardy. Performance characteristics of coal washing equipment, United States Bureau of Mines, Department of Energy, Report of Investigation No. 7673, 1972.
21. M. W. Mikhail, J. L. Picard and O. E. Humeniuk. Performance evaluation of gravity separators in Canadian washeries, Paper 4, Canadian Institute of Mining and Metallurgy, 2nd Technical Conference on Western Canadian Coals, Edmonton, 1982.

22. H. Y. Robinson. The application of radioactive tracer technique to a study of the fundamentals of jig washing, *King's College Mining Society Bulletin* **9**, 1, Coal Cleaning No. 21, 1964.
23. J. Visman. The cleaning of highly friable coals by water cyclones, Proceedings 4th International Coal Preparation Congress, Harrogate 1962, pp. 155–163.
24. W. E. Foreman and L. D. Amundson. A magnetic tracer technique to determine partition coefficients for gravity separations, Proceedings 2nd Symposium in Coal Preparation, National Coal Association/Bituminous Coal Research Inc. Coal Conference, October 1976, Kentucky, (Ed.) J. F. Boyer, published by NCA/BCR Inc., pp. 127–131.
25. T. J. Napier Munn. A survey of methods for the prediction and evaluation of gravity concentration processes, De Beers Diamond Research Laboratory, Report No. E8/3.01, May 1985.
26. J. J. Davis, C. J. Wood and G. J. Lyman. Density tracers give option to float–sink tests, *Coal Age* **38**, 60–62, 1986.
27. J. J. Davis, C. J. Wood and G. J. Lyman. The use of density tracers for the determination of dense-medium cyclone partition characteristics, *Coal Preparation* **2**, 107–125, 1985.

Chapter 9

FLOTATION, AGGLOMERATION AND SELECTIVE FLOCCULATION

9.1 INTRODUCTION

Flotation is a physicochemical process that depends upon the selective attachment to air bubbles of some rock species and the simultaneous wetting by water of others. While mineral flotation is widely understood and current practice is frequently sophisticated, coal flotation has been neglected and is still a relatively crude process. Flotation is applicable to treating coal finer than approximately one millimetre in top size and is most commonly used for the treatment of metallurgical coking coals that can earn a sufficiently high selling price to cover the cost of cleaning the coal.

Agglomeration, or bulk oil flotation, as it was first referred to, is a chemical process whereby the coal surfaces, being attracted to oil, become coated with a potential bridging liquid which leads to subsequent agglomeration of coated particles caused by some form of agitating mechanism. Agglomeration tends to be more applicable to finer particles (minus 0.5 mm) and does not appear to offer the same potential for selectivity as flotation.

Selective flocculation is often possible for very fine particles (minus 0.15 mm), but it is a process still very much under development. By careful chemical conditioning, polymers can be utilized to adsorb selectively onto the surface of either coal or non-coal particles causing them to form floccules which may then be effectively classified. Recently, combination processes of selective flocculation and flotation have been shown to recover effectively very fine coal.

In practice, coal flotation is by far the most widely used method of beneficiating fine bituminous coal, especially coal with coking properties. The flowsheet for this process is a familiar one and one that is not usually very flexible. This limitation still tends to cause 'bottlenecks' in many plant designs. Many specific examples exist of coal-flotation equipment, and some of these are reviewed later in this chapter.

Agglomeration, although a proven process, has not been widely utilized because of the cost of oil and the fact that the oil cannot be economically recovered and re-used. Recently, however, oil recovery processes have been developed which improve the cost effectiveness of the process. It is nevertheless a proven and potentially viable process and is generally acknowledged as being a strong future prospect for treating very fine coal slurries, in special applications such as pipeline transportation.

Economic and technical factors have restricted the development and use of selective flocculation, but work is progressing to develop this process for specific applications. One such application, believed to have large future potential, is in the selective separation of pyrite from high-sulphur-containing coals. Another is in the treatment of coals containing large quantities of colloidal clay minerals.

This chapter concentrates in particular on the flotation process because it is recognized as being the long-term 'work-horse' in the cleaning of fine coals. It is undoubtedly a major avenue of research work in every country which has a significant coal industry, and the main thrust appears to be in developing a more scientific understanding of the process and finding the ways and means of controlling all aspects of process operations.

Much has been published about oil agglomeration but, at the present time, the volume of reported research work appears out of proportion with the commercial application of the process.

Finally, selective flocculation, despite the obvious appeal in recovering ultrafine coal particles, is the most problematic prospect of all in terms of commercialization. Not only does it depend upon sophisticated chemical interactions, but the eventual floccules obtained require subsequent dewatering, with resultant good solids recovery and competitive moisture, in order to become commercially viable. In this regard, it would seem probable that for selective flocculation to become more viable and widely applicable, it will need to be applied in conjunction with either flotation or agglomeration methods, thereby enhancing the performance and range of both.

9.2 PHYSICOCHEMICAL PROCESS FUNDAMENTALS

9.2.1 Flotation

Before commencing with a description of the process fundamentals involved with flotation it is essential to discuss some of the properties of coal that are of special importance to the process.

The coalification process, commencing with the formation of peat and proceeding through various definitive phases or ranks of coal, occurs as a result of definite chemical structural changes in the coal. In the transition of brown coal (sub-bituminous) to bituminous coal, the chemical structure of the coal changes as a result of the gradual elimination of polar groups such as hydroxyl and carboxyl. Synonymous with this change is a reduction in the inherent and oxygen contents and a marked increase in carbon, and as a result, the coal substance becomes less hydrophilic. Hydrophobicity increases significantly with rank increase, with bituminous coal reaching a peak at almost 90% carbon content. Above this, a slight decrease in hydrophobicity occurs as the carbon atoms form a structure with increasing order in three dimensions.

More details of the classification of coal types is included earlier in the book in Chapter 2. The discussion included in this section serves to highlight two things:

(a) The rate of flotation of the coal increases with increasing hydrophobicity and this phenomenon is to some extent rank related.

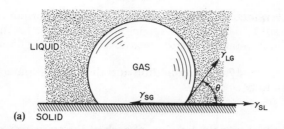

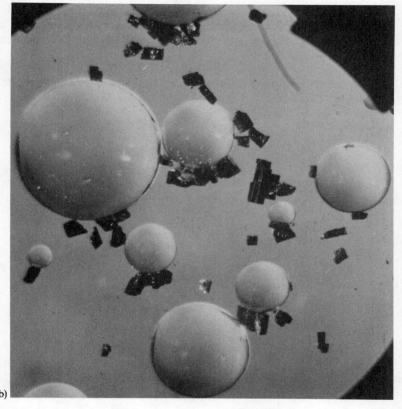

Fig. 9.1 (a) Equilibrium contact angle. (b) Photomicrograph showing contact of coal particles with air bubbles (× 10 approx).

(b) Ease of floatability is related to the function of the total carbon contained in basic structural elements or macrolithotypes, i.e. in increasing order: fusain, durain, clarain and vitrain.

For flotation to occur, an air bubble must become attached to a solid surface of a particle immersed in water. The extent to which this interaction occurs is characterized by the contact angle. Substances with native floatability, called hydrophobicity, of which coal is a good example, exhibit large contact angles even when submersed in pure water.

Bubble–particle contact therefore stabilizes to an equilibrium contact between air bubble and solid. The contact angle, θ, is the angle formed between the liquid–gas and liquid–solid interfaces measured through the liquid as shown in Fig. 9.1.

The general thermodynamic condition for this phase contact is defined by Young's equation to be:

$$\gamma_{SG} = \gamma_{SL} + \gamma_{LG} \cos \theta \tag{9.1}$$

where γ_{SG}, γ_{SL} and γ_{LG} are the tensions of each of the solid–gas, solid–liquid and liquid–gas interfaces.

The change in free energy accompanying the replacement of unit area of the solid–liquid interface by the solid–gas interface is defined by Dupré's equation:

$$\Delta G = \gamma_{SG} - (\gamma_{SL} + \gamma_{LG}) \tag{9.2}$$

which, when combined with equation (9.1), provides an expression for the free-energy changes, and becomes

$$\Delta G = \gamma_{LG}(\cos \theta - 1) \tag{9.3}$$

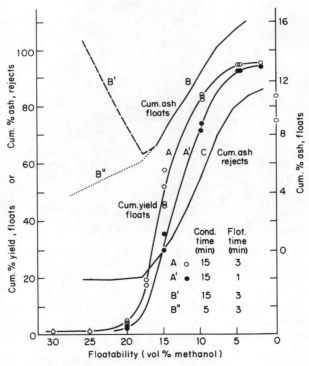

Fig. 9.2 Cumulative flotation–washability relationships for the $(-210 + 149)$ μm size-fraction of coal sample in aqueous methanol solutions using P/S cell without a bubble deflector.

FLOTATION, AGGLOMERATION AND SELECTIVE FLOCCULATION 419

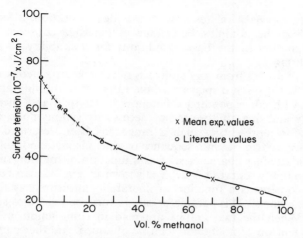

Fig. 9.3 Static surface tension of aqueous methanol solutions.

This equation, often called the Young–Dupré equation, means that if a definite contact angle is obtained, there will be a free-energy decrease once the coal particle becomes attached.

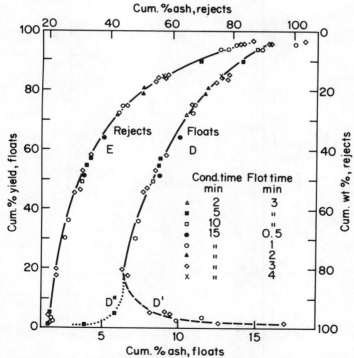

Fig. 9.4 Cumulative yield versus cumulative ash data of floats and rejects for $(-210 + 149)$ μm size-fraction of coal sample, no bubble deflector.

Although this approach relates only to the ideal condition, it forms the basis for understanding the complex behaviour of heterogeneous coal grains, and allows an explanation of the basic conditions for floatability to be defined in mathematical terms.

Recognizing that coal from any source is a very heterogeneous material, it is not surprising that it often displays a whole range of wettability and floatability characteristics, with corresponding differences in physical and chemical properties. Wettability and floatability are not necessarily synonymous, and an interesting method for representing such a range has been described.[2] Small-scale flotation tests are carried out in a sequence of aqueous methanol solutions. These tests produce floatability characteristics as a function of solution surface tension, for which 'washability' curves of the type shown in Fig. 9.2 can be obtained for yield and ash content as a function of floatability measured as the volumetric proportion of methanol. This parameter can, in turn, be related to surface tension (see Fig. 9.3). When the two graphs are read in conjunction, a value for the critical surface tension of floatability can be obtained. Another curve, analogous to the densimetric washability curve (described in Chapter 5), can be drawn as shown in Fig. 9.4 which is termed the flotation–washability curve. While such an approach may have only limited practical application, because it requires to be firmly standardized as a method, it does serve as a useful method for quantifying the critical contact angle for specific types of coal.

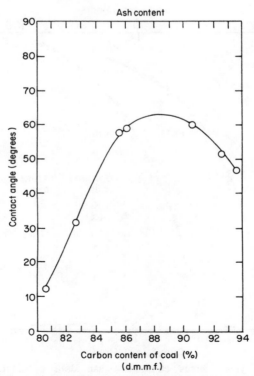

Fig. 9.5 Variation of contact angle with carbon and ash content of coal.

The dependence of coal floatability on rank is shown clearly in Fig. 9.5. This graph plots equilibrium contact angle against carbon content of a coal as derived from experiments.[3] As mentioned earlier, the order of floatability may change as the amount of mineral matter or as content associated with the various types and sources of coal changes. This is also illustrated in Fig. 9.5, lending further credibility to the washability approach mentioned earlier, since it is the coal with the lowest ash content which usually floats first.

The hydrophobic character of a specific source of coal is, as we have seen, influenced by rank (or specifically its hydrocarbon skeleton and oxygen content), and the amount of ash content of the coal. Hence, coals of low rank and those coals which are weathered or oxidized tend to be more hydrophilic due to the larger oxygen content.

The complexities of coal grain and bubble attachment, i.e. the basis for the flotation process, appear to involve several forces:[4]

- coulombic forces arising from the electrical double layer;
- dispersive forces involving London–van der Waals attractive forces;

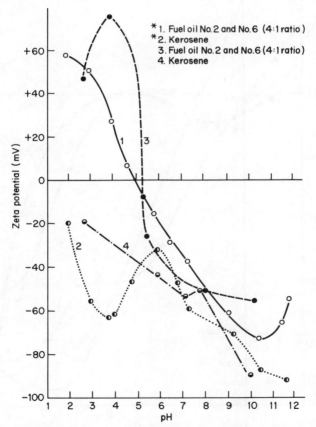

Fig. 9.6 Zeta potential–pH of kerosene- or fuel oil-in-water emulsion droplets. (After Wen and Sun[7]).

- structural forces which occur from specific effects of the solid on the water molecule through hydrogen bonding or hydration.

These forces, when combined, will specifically determine the degree of hydrophobicity of the surface and consequently the rate and equilibrium position for bubble attachment.

In conventional (mineral) flotation, water-soluble collectors are introduced which adsorb on the mineral rendering its surface hydrophobic. Coal flotation is achieved by a slightly different mechanism, often referred to as emulsion flotation. In this process, an apolar, water-insoluble oil hydrocarbon is used as a collector and a water-soluble surfactant is employed to promote the froth. Dispersion of the collector in this process is therefore of paramount importance in achieving good flotation efficiency.

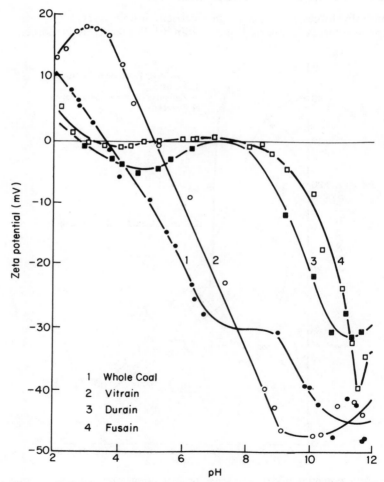

Fig. 9.7 Variation of bituminous whole-coal and lithotype zeta potential with pH.[8]

FLOTATION, AGGLOMERATION AND SELECTIVE FLOCCULATION 423

The thorough emulsification of water-insoluble collectors, by means of high-shear mixers or ultrasonics, results in a vastly more effective agent.[5] It has been shown that there is an optimum dispersion of the emulsion corresponding to a droplet diameter of about 5 μm for paraffin or kerosene.[6] These droplets adhere to the coal and thereby make it more hydrophobic. For example, the contact angle at a vitrain surface was increased from 57° to 82° by addition of emulsified fuel oil at a concentration of 250 mg/litre.

Other work has shown that by measuring zeta potential, kerosene droplets carry a negative electrical charge in water for practically the entire pH range and should therefore be inferior to the fuel oil collectors which become positively charged from pH 5 (see Fig. 9.6). This is of special interest when the zeta potential–pH curves for various coals are studied (see Fig. 9.7).[7]

A similar graph shown in Fig. 9.8 shows the variation of bituminous coal lithotypes in terms of zeta potential and pH.[8] As pointed out by Aplan,[9] the variation between the different types of bituminous coal is often as great as the variation between the microlithotypes of one coal, but at all pH values above about 5, coal is usually either negatively charged or neutral.

9.2.2 Agglomeration

Wet agglomeration processes are those in which the size enlargement occurs among particles suspended in a liquid phase. There are a number of ways in

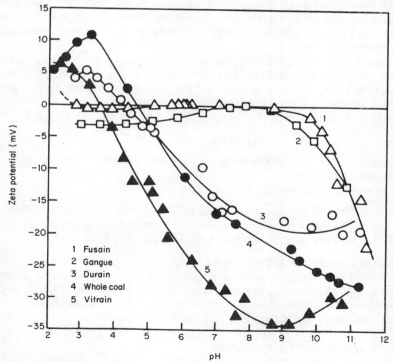

Fig. 9.8 Electrokinetic potentials for macrolithotypes in anthracite. (Courtesy F. F. Aplan).

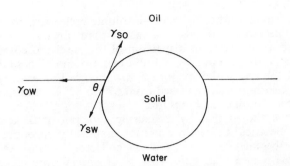

Fig. 9.9 Forces acting on a solid particle at an oil–water interface.

which this can occur, and perhaps the simplest is by the addition of electrolytes which cause a reduction in zeta potential, with the resultant formation of coagulates. The next order of process is the substitution of polymer flocculants for electrolytes, to create the collection of larger quantities of particles into floccules which can in some circumstances become quite massive in size (up to 5 mm). In the third order of process, that is spherical agglomeration, finely divided solids are treated with a so-called 'bridging liquid' which preferentially wets the solid and is immiscible with the suspensoid liquid. On agitation, the bridging liquid becomes distributed over the exposed surfaces of the dispersed solid particles, and upon particle collision the bridging liquid forms junctions. The bonding force between the particles is therefore governed by the interfacial tension of the liquids involved.[10]

The mechanism of particle absorption by the bridging liquid (usually a heavy, inert oil) may be simply explained by the diagram in Fig. 9.9. The position of the solid particle at the interface between oil and water is, as stated earlier, governed by the relative values of interfacial tensions.

Therefore at equilibrium:

$$\gamma_{so} = \gamma_{sw} + \gamma_{ow} \cos \theta$$

where γ_{so}, γ_{sw} and γ_{ow} are the surface-force energies at the solid–oil, solid–water and oil–water interfaces.

The following conditions may then occur:

(a) If $\theta < 90°$, the particle will tend to be drawn into the aqueous phase.
(b) If $\theta = 90°$, the particle will remain at the interface.
(c) If $\theta > 90°$, the particle will be drawn into the oil phase.

Spherical agglomeration of coal occurs when condition (c) is satisfied for the coal particles, and selectivity is assured if condition (a) occurs for any other solid constituents. Problems usually occur when the sulphide mineral pyrite is present because the fresh pyrite surface is often as responsive as coal to the oil phase, and selective agglomeration is therefore more difficult to achieve.

9.2.3 Selective Flocculation

Before beginning to discuss the subject of selective separation, it will be useful to review briefly the various conditions which can occur in an aqueous suspension

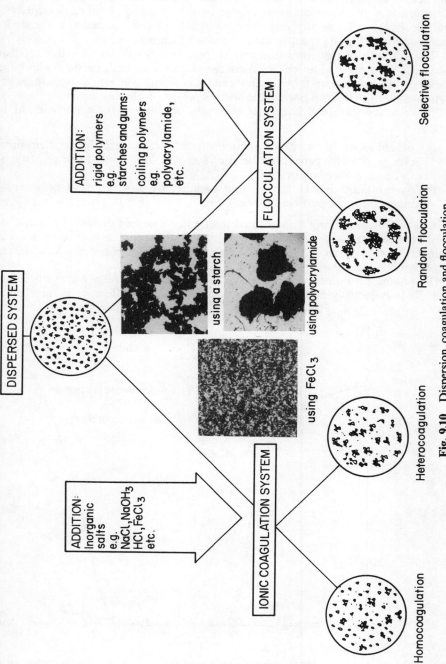

Fig. 9.10 Dispersion, coagulation and flocculation.

of rock or mineral particles. Three significant conditions can occur as a result of adjustment of the chemical environment by the addition of either electrolytes or polyelectrolytes. These are: dispersion, coagulation and flocculation. Figure 9.10 summarizes pictorially the alternative results which may occur.

The stability of dispersed aqueous suspensions is due to the mild repulsion occurring as a result of an electrical charge surrounding each particle. This condition may often be induced by adding a suitable electrolyte which encourages such repulsion for the various particulate species present in the suspension.

Polyelectrolytes are believed to assist the settling of these particles in two ways:

(a) by reduction in the zeta potential of the particles and hence their surface charge, thereby permitting the London–van der Waals force of attraction to become effective; or
(b) by bridging, that is, the binding together by adsorption of polymer segments onto the particle surface.

Coagulation, resulting from mutual interaction between dissolved organic salts and particle surfaces, may induce favourable adsorption conditions for certain mineral or rock types, while not affecting others. Hence, a potential for selective separation is created.

This same phenomenon is also true for certain types of polymer flocculants, usually polyelectrolytes. In such cases, it is invariably a difference in the chemical

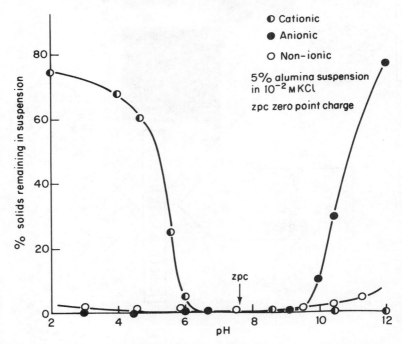

Fig. 9.11 Suspended solids remaining after flocculating Al_2O_3 suspensions with varying polymer charge.

constitution of an effective flocculant which allows adsorption and not a function of molecular weight, polymer type or chain length.

Experimental work[11] has clearly shown that adsorption of a polyelectrolyte can be almost completely inhibited by adjustment of surface charge on the particle. A good example is shown in Fig. 9.11 using alumina and three polyelectrolytes of different charge types. Adjustments in pH cause a transition of the alumina charge from positive to negative.

Polymer dosage has been observed as increasing with mineral-matter content of coal in non-selective flocculation,[12] but dosage in selective treatment is often influenced by other reagents added to disperse the unwanted component, usually shale or pyrite.

Selective flocculation of coal–shale mixtures appears to have been achieved in many cases by inhibition of flocculant adsorption on the shale rather than by promotion on the coal surface. This approach is really one of selective dispersion, since it is caused by electrostatic repulsion due to the presence of pre-absorbed modifying ions on the shale surface and also by the removal of free hydroxyl from the surface of clay minerals or alkali treatment.[13]

An example of such a system is the use of a mildly anionic polyacrylate to flocculate bituminous coal, together with the addition of the inorganic dispersant sodium hexametaphosphate for the treatment of the clay minerals in an alkaline slurry.[14]

9.3 FLOTATION FUNDAMENTALS

9.3.1 Flotation Methods

All flotation systems require the generation of a swarm of gas bubbles to levitate the solid particles. These gas bubbles can be formed by three distinctive methods:

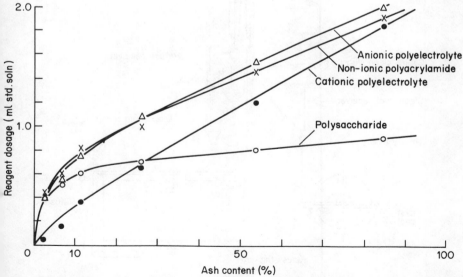

Fig. 9.12 Optimum flocculant dosages for four polymers with each coal–shale sample.

(a) By mechanical means, or *dispersed flotation*: bubbles of about 1–5 mm are formed by an impeller, or by passing or drawing air through a porous plate or nozzle.

(b) By nucleation of gas from solution: this can be achieved either by applying a vacuum to the suspension—*vacuum flotation*; or by saturating water with air under pressure—*dissolved air flotation*; or by air supersaturation into a flowing stream of suspension—*microflotation*. Mean bubble sizes for each of these three methods range from 100, 75 and 50 µm respectively.

(c) By electrolysis of the aqueous phase, or *electroflotation*: this method requires sufficient conductivity to create ultrafine bubbles of below 50 µm in very quiescent conditions.

The mean bubble size attained by each method is very much influenced by the addition of surfactants.

All three methods have been applied to coal separation, but dispersed flotation is by far the most widely applied, and electroflotation is still a purely experimental approach. The latter two methods find application in solid–liquid separation where the objective becomes the removal of *all* solids from the liquid phase. As

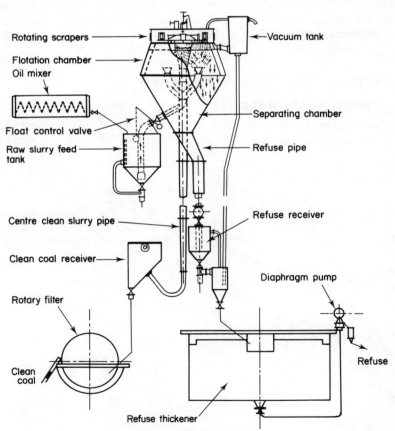

Fig. 9.13 Coppée–Elmore vacuum flotation machine.

FLOTATION, AGGLOMERATION AND SELECTIVE FLOCCULATION

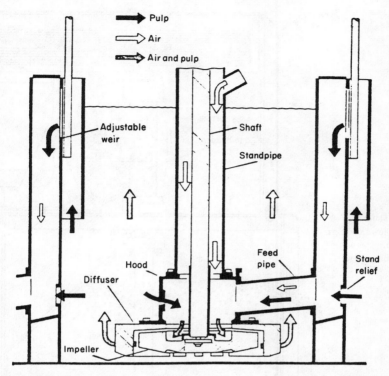

Fig. 9.14 A cross-section of a Denver Sub-A, a typical cell-to-cell machine.

such, feed suspensions are usually very much more dilute than for coal-and-mineral separation 'froth' flotation.

9.3.2 Flotation Machines

It is perhaps worthy of mention that at one time the Coppée–Elmore vacuum (or bulk) flotation machine was widely used in Britain for coal flotation. In this machine, froth was formed by air released by vacuum (80–90 Pa). This machine represents the only non-dispersed air method which has been relatively widely commercially applied in coal cleaning. It is shown diagrammatically in Fig. 9.13.

Dispersed machines can be subdivided into four classes. Much has been written about the various types of machine which belong to each of these classes:[15–18]

(a) *Mechanical*, which are the most common type, comprising a vertical, mechanically driven hollow shaft fitted with an impeller. The shaft and impeller are located centrally in a cubical tank which is filled with slurry. Air is forced or drawn through the shaft to be dispersed throughout the pulp by the impeller.

Two subdivisions of mechanical machines exist:

(i) pulp flow or 'cell-to-cell' machines comprising three to eight cells in a line, each separated by weirs in 'open trough' or 'open flow' fashion (see Fig. 9.14);

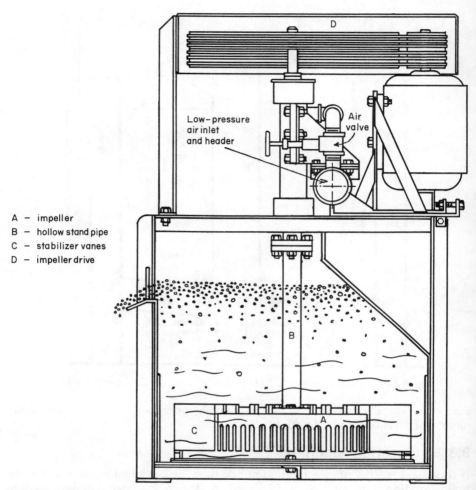

A — impeller
B — hollow stand pipe
C — stabilizer vanes
D — impeller drive

Fig. 9.15 Agitair coal flotation cell—example of an aerated type of cell (courtesy Galigher Incorporated).

(ii) aeration machines that have air pumped to them by a blower; sub-aeration machines use the impeller design to induce larger quantities of air (see Fig. 9.15).

(b) *Pneumatic* Machines in this class have no impeller and rely upon compressed air to agitate and/or aerate the pulp (see Fig. 9.16).

(c) *Froth separators* These machines employ a feed system located centrally above the tank and feed is introduced onto the froth bed, minimizing the transport distance of the concentrate (see Fig. 9.17).

(d) *Column* Flotation occurs in a countercurrent flow of air bubbles and slurry within a fairly tall, quiescent columnar tank (see Fig. 9.18).

Mechanical machines In recent years, a design trend of increasing volumetric

FLOTATION, AGGLOMERATION AND SELECTIVE FLOCCULATION

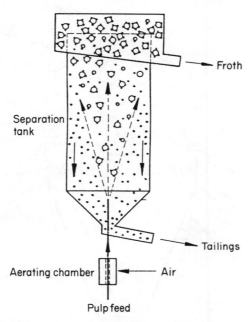

Fig. 9.16 Pneumatic flotation principle.

capacity in individual machines has emerged with the result that most current installations employ machines of between 8.50 and 14.20 m^3. There are, however, a number of larger machines (28.30 m^3) available, although many plant designers regard the 14.2 m^3 machines as being optimum[16] when taking into account performance and capital and operational costs.

Table 9.1 shows the principal design characteristics of many of the more widely employed coal-flotation machines. Figure 9.19 shows cell volume and power consumption comparisons for several popular machines. Although emphasis has been placed upon enlargement of individual units, other innovations have appeared despite the fact that little change in conceptual design has emerged. Modern machines tend to be simpler in construction, and therefore easier to maintain. Modern materials such as cast and sprayed polyurethanes, AR (wear-resistant) plate, polypropylene, ceramics and epoxy compounds are utilized to provide protection from wear and abrasion, corrosion and chemical attack.

The original mechanical coal-flotation machines were applied to freely floating coals and tended to be of simple, open-trough design. Generally, cell volumes of 2.8 m^3 were the most popular in an open-trough arrangement of five to seven cells, i.e. 14–22.5 m^3 volume.

The cell-to-cell arrangement became employed when coal of a more difficult nature was encountered, and from these applications it was generally realized that better control in selectivity and higher potential recovery can often be obtained as a trade-off for slightly higher operational costs. As a result, the cell-to-cell configuration has now become widely employed.

Few of the many varieties of cell tank shapes and impeller mechanisms of coal-flotation machines can be claimed as designs specifically developed for coal

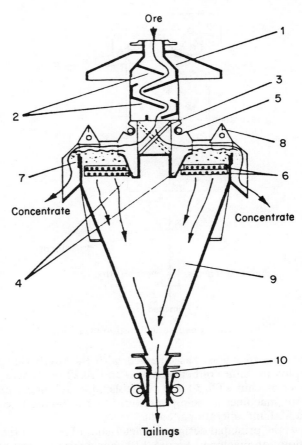

Fig. 9.17 ØNC-16 Froth separator: 1, feed tank; 2, sloping baffles; 3, divider; 4, preliminary aeration channel; 5, nozzles; 6, aerators; 7, overflow lip; 8, froth paddle; 9, pyramid-shaped tank; 10, valve.[16]

applications. However, many have been modified to suit the hydrophobic nature of coal and the consequent ease with which it floats when compared to many minerals. Some of these designs are compared in Figs 9.20 and 9.21. Of these examples, Wedag, Denver Sub-Aeration and Unifloc can be readily arranged in a cell-to-cell configuration, whereas the remainder require an additional feed-box arrangement to divide each cell or else some form of customized divider. The Denver Sub-A coal-flotation machine serving as a good example of a cell-to-cell design (see Fig. 9.14) comprises: an adjustable weir with a feed pipe, which transports slurry from the weir to the pump impeller; a self-aerating impeller and large-capacity froth paddles, to accommodate the large concentrate yields characteristic of coal flotation. Most specific coal designs incorporate self-aerating impellers, being cognisant of the hydrophobicity of most coal sources for which flotation can be applied. In selective flotation applications, and in particular reverse flotation involving floating pyrite from a depressed coal concentrate, air injection may be likely to prove useful.

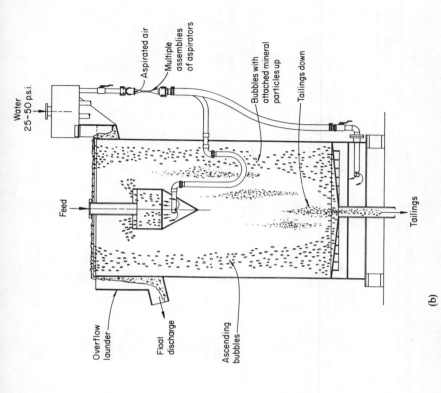

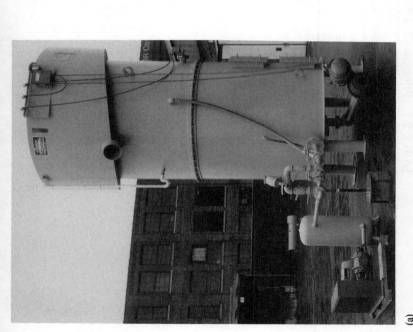

Fig. 9.18 Flotaire flotation column: (a) 2.4 m diameter example; (b) diagram to show principle (courtesy, Deister Concentrator Company Incorporated.)

TABLE 9.1 Principal Characteristics of Mechanical Flotation Machines

Trade name	Model No.	Tank internal dimensions							Blower needed	Specific air flow (m³/min)/m³ pulp	Impeller power		Impeller size		
		Length L (m)	Width W (m)	Depth D (m)	Pulp volume (m³)	Surface area (m²)	Specific area (m²/m³)	Specific length (m/m³)			Motor (kW)	Specific consumpt. (kW/m)	Diam. Ø (m)	Aspect ratio 2Ø/(L+W)
						OPEN FLOW								
Denver	100	1.57	1.57	1.21	2.8	2.46	0.88	0.56	Yes	1.2	7.5	2.1	0.61	0.39
D-R	180	1.82	1.82	1.62	5.1	3.31	0.65	0.36	Yes	0.9	11	1.7	0.61	0.34
Coal	300	2.23	2.23	1.82	8.5	4.79	0.58	0.26	Yes	0.8	18.5	1.5	0.69	0.31
	500	2.69	2.69	2.01	14.2	7.24	0.51	0.19	Yes	0.6	22	1.2	0.84	0.31
						CELL-TO-CELL								
Denver	100	1.52	1.52	1.22	2.8	3.91	1.40	0.54	No	—	11	3.1	0.61	0.40
Sub-A	200	1.83	1.83	1.59	5.7	4.72	0.83	0.32	No	—	15	2.1	0.69	0.38
Coal	300	2.10	2.10	1.89	8.5	6.83	0.81	0.25	No	—	18.5	1.7	0.69	0.33
	400	2.30	2.30	2.12	11.3	8.18	0.72	0.20	No	—	30	2.1	0.84	0.37
	500	2.70	2.70	1.98	14.2	10.8	0.76	0.19	No	—	37	2.1	0.84	0.31
Agitair	90C × 300	2.29	2.29	1.73	8.5	5.24	0.62	0.27	Yes	0.6	15	1.4	0.69	0.30
Coal	102C × 500	2.74	2.74	2.01	14.2	7.51	0.53	0.19	Yes	0.5	18	1.0	0.76	0.28
	144C × 1000	3.58	3.30	2.13	28.3	11.8	0.42	0.13	Yes	0.4	30	0.8	0.84	0.24
Humboldt	3	—	—	1.20	3	3.53	1.18	—	No		15	4.0	0.58	—
Wedag	5	—	—	1.50	5	5.51	1.10	—	No		18	3.0	0.63	—
	8	2.20	3.80	1.20	8	8.36	1.05	0.28	No		30	3.0	0.75	0.25
	10	2.37	4.10	1.30	10	9.72	0.97	0.24	No		30	2.4	0.75	0.23
	12	2.60	4.30	1.40	12	11.2	0.93	0.22	No		37	2.5	0.85	0.25

	Model													
Krupp	TR 5000	1.82	2.77	—	1.60	5	5.04	1.00	0.32	No	0.3–1.3	15	2.4	0.43	—
OK	1.5R	—	—	—	—	1.5	—	—	—	Yes	0.3–1.0	5.5	1.0–2.7	0.50	0.53
	3R	1.52	1.52	—	1.21	3	2.31	0.77	0.51	Yes	0.5–1.3	7.5	1.0–2.0	0.63	0.29
	8R	2.29	2.29	—	1.88	8	5.24	0.66	0.29	Yes	0.5–1.1	15	1.0–1.6	0.75	0.27
	16R	2.95	2.69	—	2.46	16	7.94	0.50	0.18	Yes	0.5–1.1	30	0.9–1.4	0.75	—
	16U	—	—	—	—	16	—	—	—	Yes	0.3–0.8	30	0.9–1.4	0.90	0.25
	38U	3.49	3.59	—	3.23	38	12.5	0.33	0.092	Yes	—	55	0.9–1.2	—	—
Sala	AS2-1	1.76	0.88	—	0.77	1.2	1.55	1.29	1.47	Yes	1.3	4	2.3	—	0.44
	AS2-3	2.47	1.24	—	0.87	2.7	3.06	1.13	0.91	Yes	0.9	11	2.9	0.55	—
	AS2-4	2.47	1.24	—	1.32	4.0	3.06	0.77	0.62	Yes	0.6	11	1.9	—	0.44
	AS2-6	3.48	1.74	—	1.06	6.6	6.06	0.92	0.53	Yes	0.8	22	2.3	0.77	—
	AS2-9	3.48	1.74	—	1.51	9.1	6.06	0.67	0.38	Yes	0.6	22	1.7	—	0.45
	AS2-12	4.26	2.13	—	1.28	12	9.07	0.76	0.36	Yes	0.8	37	2.2	0.95	—
	AS2-15	4.26	2.13	—	1.63	15	9.07	0.60	0.28	Yes	0.8	—	—	1.10	0.45
	AS2-18	4.92	2.46	—	1.44	18	12.1	0.67	0.27	Yes	—	44/60	1.7/2.3	1.10	0.45
	AS2-22	4.92	2.46	—	1.79	22	12.1	0.55	0.22	Yes	1.0	44/60	1.4/1.9	—	0.45
	AS2-28	5.62	2.46	—	1.79	28	13.8	0.49	0.20	Yes	0.8	44/60	1.1/1.5	1.10	0.45
	AS4-36	4.92	4.92	—	1.44	36	24.2	0.67	0.14	Yes	0.6	88/120	1.7/2.3	1.10	—
	AS4-4	4.92	4.92	—	1.79	44	24.2	0.55	0.11	Yes	1.0	88/120	1.4/1.9	1.10	0.45
Wemco	44	1.12	1.12	—	0.51	0.57	1.25	2.19	1.96	No	0.8	3.75	3.9	0.22	0.20
1+1	56	1.42	1.42	—	0.61	1.1	2.02	1.84	1.29	No	1.0	5.5	3.8	0.28	0.20
	66	1.52	1.68	—	0.69	1.7	2.55	1.50	0.89	No	1.0	7.5	3.5	0.32	0.20
	66D	1.52	1.68	—	1.19	2.8	2.55	0.91	0.54	No	1.0	7.5	2.1	0.33	—
	84	1.60	2.13	—	1.35	4.2	3.41	0.81	0.38	No	0.9	11	2.1	0.41	0.22
	120	2.29	3.05	—	1.35	8.5	6.98	0.82	0.27	No	0.8	22	2.1	0.56	0.21
	144	2.74	3.66	—	1.60	14.2	10.0	0.70	0.19	No	0.8	30	1.7	0.66	0.21
	164	3.02	4.17	—	2.36	28.3	12.6	0.45	0.11	No	0.7	45/55	1.3/1.6	0.76	0.21

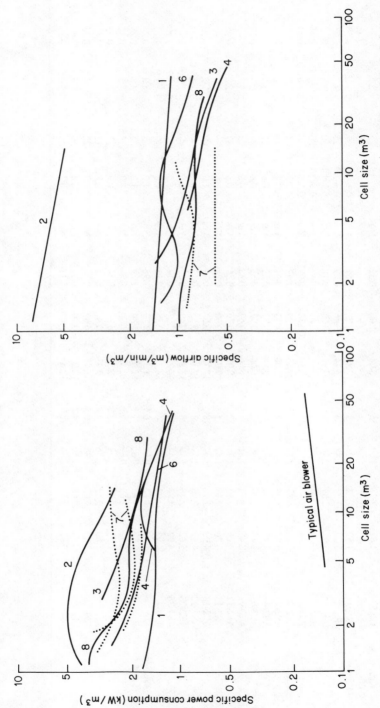

Fig. 9.19 Selected characteristics of some mechanical flotation machines: 1, Aker; 2, Booth; 3, Denver D-R; 4, Agitair; 6, OK; 7, Sala (two tank depths); 8, Wemco.

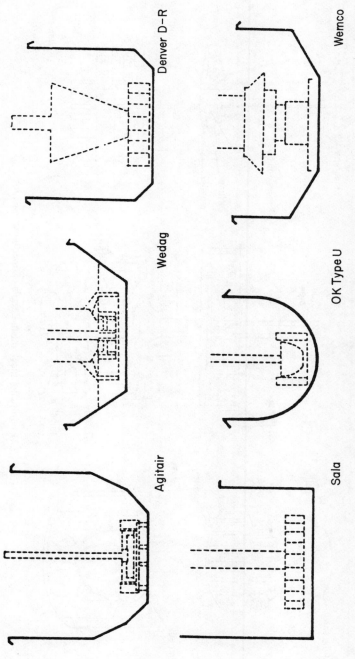

Fig. 9.20 Cell tank profiles of open-flow machines.

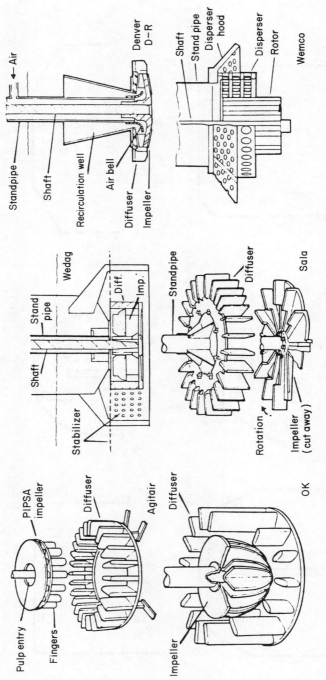

Fig. 9.21 Impeller mechanism details of open-flow machines.

FLOTATION, AGGLOMERATION AND SELECTIVE FLOCCULATION

Pneumatic machines Very few pneumatic machines are currently used for coal flotation. One recent development from Germany, the Schauenburg machine, has been effectively used for coal flotation. It comprises an aerating chamber (see Fig. 9.16) in which the air bubbles are created by porous pipes. The aerated slurry enters a separation tank with a conical base and several similar units can be connected in parallel to expand total throughput capacity. As much as 1000 m^3/h can be treated in this manner. Advantages of this type of flotation system are:

- lower capital cost mainly due to low space requirements;
- no moving parts, with consequent low energy cost;
- easy regulation of air- and slurry-flow with good amenability for measuring and control of operation;
- effective flotation of wide size-range, 0.3 mm to 0.01 μm.

Feed-solids concentration of 30–50 g/litre is a common range for coal, but this can be exceeded to 60 g/litre. A similar design of flotation cell is in use in the USSR, but coal applications are few. A different type of machine called the Heyl and Patterson (H and P) Cyclo-Cell has been used in the USA. This unit, shown in Fig. 9.22, comprises an elongated trough similar in shape to that used in mechanical machines. Instead of mechanical impellers, this cell has submerged vortex chambers which agitate the slurry and introduce air. The vortex chambers create cyclonic motion of the slurry, and low-pressure air is introduced at the centre of the turbulent zone creating shearing of the incoming air into a multitude of fine bubbles. Air requirement is small, and pumps capable of providing 100–200 kPa, (1–2 bar) pressure are used. The same advantages apply to this design, as mentioned previously for the conical tank, and neither type can be arranged in a cell-to-cell configuration.

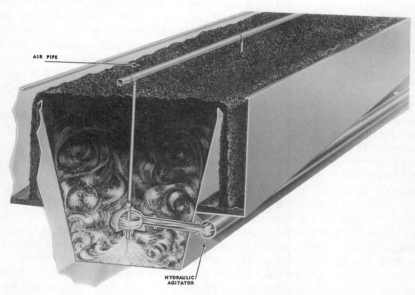

Fig. 9.22 H and P high-energy Cyclo-Cell (courtesy Heyl and Patterson Incorporated, Pittsburgh).

Froth separators Froth separators are still of major interest in ore treatment, but some coal use has been reported in the USSR.[25] Potential for application lies in the treatment of very coarse coal slurries (from 1 mm to 50 µm), possibly those containing finely divided clays. The principle of froth separation is to concentrate and discharge conditioned-feed slurry within a froth blanket, thereby entraining hydrophobic grains within the bed. This is achieved partly by a filtering process within the froth itself. The success or failure of the process depends upon the utilization of the potentially enormous air–water interfaces existing within the froth, and the residence time required for concentrates to be captured by the froth. The absence of turbulence, especially in the froth zone, is therefore a prerequisite and a key feature in the design of the separator and in its range of application. The benefits in terms of coal treatment are that larger particles can be captured and high treatment rates are possible—both well in excess of mechanical machines. Power costs and air consumption are also lower. One design (see Fig. 9.17) will treat up to 50 t/h of slurry at 50–70% solids by weight. Air is introduced through aeration pipes at 115 kPa in volumes of up to 2 m^3/min.

Column flotation The original column-flotation separators were designed in Canada for the treatment of fine-ore suspensions. Commercial columns of up to 2000 mm diameter are capable of treating large tonnages of up to 100 t/h of sulphide ore.

Coal-flotation columns tend to rely on the formation of a froth bed on the surface of the column and a washing phenomenon caused by interaction of the rising bubble mass with newly introduced slurry insurges through the pulp.

Porous air diffusers composed of hydrophobic substances such as sintered metal overcome the tendency to become plugged, and in some cases air is actually introduced as a fine dispersion in water, as in the case of the Flotaire cell (shown earlier in Fig. 9.18) which comprises a vertical cylindrical tank about 4.6 m in height and 2–3.7 m in diameter. Air is introduced with water beneath a constriction deck located at the base of the cell and passes through the constriction deck which uniformly distributes the air. Tailings are continuously discharged from the bottom of the cell via a central discharge pipe which is partially restricted by a suspended plug-valve supported from an adjusting cable. Concentrates overflow a peripheral launder at the top.

In contrast with the previously described columns, the Leeds or Norton–Hardy Flotation Column, shown in Fig. 9.23, was specifically invented and designed by Dell[21] at Leeds University to provide effective countercurrent froth cleaning. In the current designs, the machine incorporates a mechanism with a special impeller design. This is a horizontal disc, located close to the floor of the rectangular tank, with upwardly pointing vertical fingers located around the periphery. Air is pumped to the impeller zone via two pipes. The impeller is surrounded by baffle vanes fitted into the base of the tank which serve to generate the bubbles. In between the impeller zone and the froth base are a number of horizontal barriers which separate the stages of cleaning. These barriers act like non-return valves, their role being to collect froth beneath, becoming more and more bouyant until the bubbles escape to beneath the next barrier. Each escape causes the bubbles to pass through a downflow of wash-water which step by step removes entrained waste. Besides collecting a buoyant bubble mass beneath the barriers, the lower bars also tend to collect any surges of down-flowing grains on

FLOTATION, AGGLOMERATION AND SELECTIVE FLOCCULATION

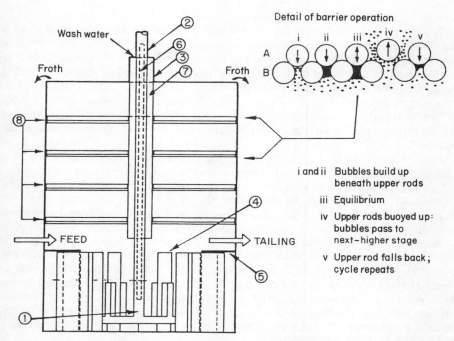

Fig. 9.23 Schematic illustration of a single Norton–Leeds column cell: 1, Impeller; 2, shaft; 3, partition box; 4, baffle vanes; 5, corner baffle plates; 6, dual air line; 7, wash water chamber; 8, barrier levels—four shown.

their upper surface. When the bar is no longer able to support the mass of settled solids, it moves downwards, allowing the sediment to be gradually removed by the wash-water flow. The pulp volume below the lowest barrier is deliberately minimized to reduce the total deposit of the cell and, because of this, capacities are similar to mechanical cells of similar floor-area size.

9.3.3 Reagents and Selectivity

In most cases, flotation 'reagents' are added in advance of the flotation cells, often in a conditioning tank containing an agitator. The conditioning time, as determined by the volume of the tank in relation to pulp flow rate, is usually determined by test work. Occasionally, further reagent addition may be necessary following initial flotation, and extra dosage is then usually added directly into the cells.

Unlike mineral flotation, most coals are readily floatable and reagent systems are often simple by comparison. It is worth bearing in mind that in the case of coal, it is quite common for over 80% of the feed to be floated off as a concentrate, whereas for minerals it is quite rare for as much as 50% of the feed to be floated. Hence, coal flotation tends to be a non-selective bulk flotation system involving only two main reagent functions: frothers and collectors.

Frothers facilitate stable froth and regulation of bubble size in dispersed-air systems. For coal, they are usually heteropolar, and coal generally responds well to aliphatic alcohols, terpinol and cresols.

Collectors create contact between coal and air bubbles by the formation of a thin coating over the particle surface, rendering it hydrophobic. Some frothers also act as promoters in that they assist with emulsification of the oily collector (methyl isobutyl carbinol is a good example) and therefore frother/collector agents are often utilized if effective in this way.

As mentioned earlier in Section 9.2.1 of this chapter, an apolar, water-soluble oil hydrocarbon is often employed as the collector, and thorough dispersion of this oily substance is an important requirement in ensuring maximum effectiveness.

Some reagents include an emulsifying agent, but such components are also often added independently. Other reagent functions, i.e. depressants, pH regulators and dispersants, tend to have a much more secondary role than in mineral flotation. It is possible to depress either the coal or the associated waste or gangue, although little commercial application of flotation involving depression is made. One significant development has been the depression of coal in a second-stage flotation operation involving the removal of pyrite as a froth concentrate, and this is dealt with later in Section 9.6 of this chapter. Pulp pH control may also be regarded as a form of depressing coal because yield tends to reduce sharply for most bituminous coal at both high and low values, but most pH adjustment is to cater for acidity of the feed slurry and is more concerned with corrosion protection than with feed conditioning. Slime contamination, caused by hydration of shales into clay colloidal suspensions, is often effectively controlled by adding a depressant.

Later on, in Chapter 12, a more detailed account of froth-flotation reagents is given. In most cases, froth-flotation separation needs to be carefully optimized to achieve good selectivity and high yield. Many coals are so hydrophobic that very high flotation rates are possible and selectivity can be sacrificed, partly due to hydrophobicity of most coal microlithotypes, and also entrainment. With such a free-frothing substance, entrainment is the major source of contamination, but others do exist, e.g. slurry turbulence at the froth–liquid interface and slime coating of coal particles.

Selectivity is improved by the following:

(a) deep froth layer;
(b) reduced turbulence at the interface;
(c) froth washing, which is created in semistable froths, especially those which coalesce;
(d) desliming of the feed;
(e) feed dilution; and
(f) conditioning with gangue dispersant.

Froth depth is usually variable on mechanical cells and can be maintained at up to 400 mm. Obviously, froth is greatest at the initial cells, where the froth is also most stable, and least in the final cells, where coalescence is common.

Slurry turbulence is often a characteristic of the machine in treating a particular type of slurry, but for coal flotation it is less of a problem than for other applications.

Most fine coal is cleaned in mechanical machines. The European industry has tended to develop a preference for cell-to-cell machines over open-flow, because

FLOTATION, AGGLOMERATION AND SELECTIVE FLOCCULATION

of the common occurrence of hydrating shale with the coal, which breaks down and can impair flotation unless good selectivity combined with good hydrodynamic cell conditions can be obtained. The relatively small group of flotation cells, which are designed for the special needs of coal cleaning, have been demonstrated as providing better in-cell circulation and more even distribution of potential concentrate material over the entire bank of cells. This reduces the tendency to overfloat in the first and underfloat in the last cell, and is attributed as being the major reason for better selectivity.

Only the Leeds cell design provides for froth-washing by virtue of its layers of baffles.[19] Some mechanical cell designs cause a recirculation effect of pulp close to the impeller zone, but little evidence exists that this feature creates improved selectivity for coal treatment.

9.4 TESTING PROCEDURES AND FACTORS AFFECTING FLOTATION

9.4.1 Apparatus

Numerous types of laboratory apparatus are available for flotation testing, from contact-angle measurement devices and small unit cells such as the Hallimond tube, for precise analytical testing, to laboratory flotation-cells ranging in capacity from 250 g to 2000 g. The latter comprise an impeller–diffuser system of similar design to that used in commercial mechanical designs, inside a stainless-steel or glass vessel, acting as the cell.

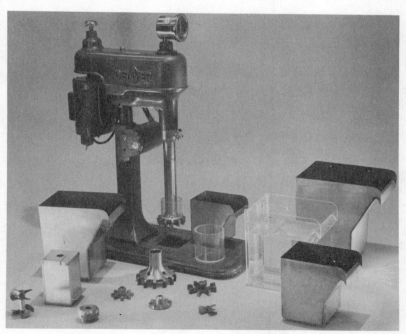

Fig. 9.24 Laboratory mechanical flotation machine.

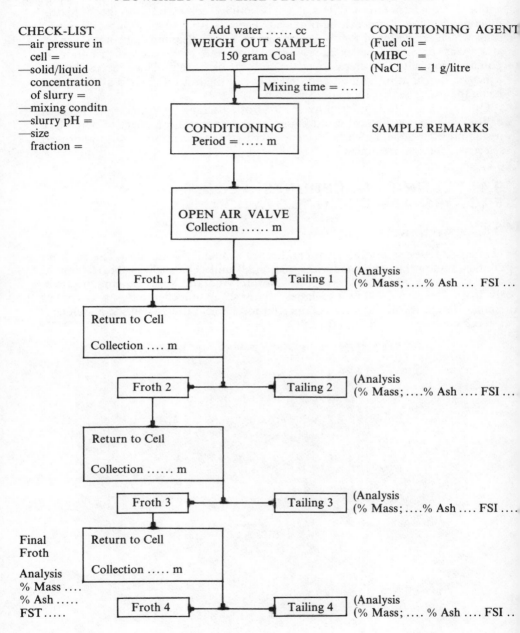

Fig. 9.25 Flowsheet for a time-recovery type of flotation test.

FLOTATION, AGGLOMERATION AND SELECTIVE FLOCCULATION 445

FLOWSHEET: TIME/RECOVERY FLOTATION TEST

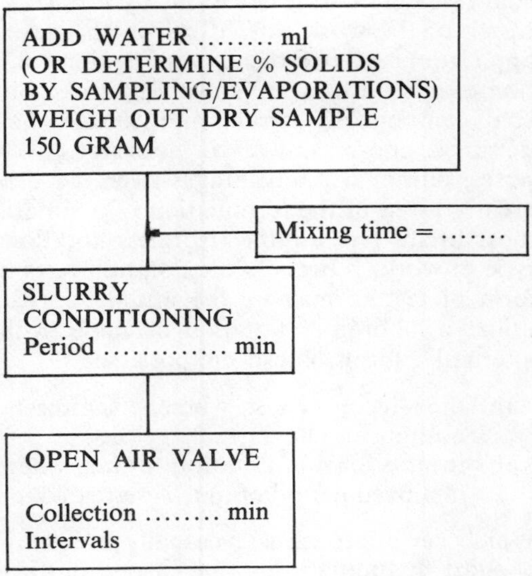

CK LIST

ressure = ...

y
entration = ...

on = ...

IDITIONING AGENTS

oil = ...

C = ...

rs = ...

LYSIS

loats (individual
ample):
 —moisture
 —ash
 —total sulphur
 —gross calorific value
Tailings:
 —moisture
 —ash
 —total sulphur
 —gross calorific value

ADD WATER .... ml
(OR DETERMINE % SOLIDS
BY SAMPLING/EVAPORATIONS)
WEIGH OUT DRY SAMPLE
150 GRAM

Mixing time = ...

SLURRY
CONDITIONING
Period .... min

OPEN AIR VALVE

Collection .... min
Intervals

TABLE OF RESULTS

No.	Fractional %				Cumulative %			
	Mass	Ash	Sulphur	Time (min)	Mass	Ash	Sulphur	Time (min)
1								
2								
3								
4								
5								
6								
7								
8								
9								
10								

MARKS

Fig. 9.26 Flowsheet for a release or reverse-flotation type of flotation test.

The use of contact-angle and Hallimond tube devices is usually confined to coal research work carried out by surface- or colloid-science research specialists. Much of the work carried out in the development of new reagents for specific applications utilizes this type of equipment, but other work investigating separation of microlithotypes is usually carried out with flotation columns. The bright-coal components, vitrain and clarain, have the strongest coking properties; durain is poorer and fusain does not coke at all; whereas for combustion purposes, durain has advantages over the others. Basic small-scale test work can greatly assist in the formulation of a suitable reagent system to separate out the most useful types. Other frothless-flotation columns are also available for this type of work.[20] Bench-scale flotation tests usually follow, but in many cases this form of testing may be the initial approach taken. Most in-plant testing will utilize a laboratory flotation machine of the type shown in Fig. 9.24, and two alternative forms of tests are possible:

(a) time–recovery test, whereby samples are taken at fixed time-intervals until nothing else floats;
(b) reverse form of flotation testing, whereby each flotation product recovered in a fixed period of time is retreated in another step.[21]

Whichever approach is eventually selected, it is absolutely essential that a series of short preliminary tests be carried out in order to establish the correct conditions for the standardized test. Before indicating what form these would normally take, reference is made to the flowsheets in Figs 9.25 and 9.26 which outline the general format for each test approach.

Also, it is perhaps worthy of note that several reports of recent work carried out[26] by the Australian Coal Industry Research Laboratories (ACIRL) suggest that any form of batch test conducted with a laboratory flotation cell can still lead to a fairly substantial discrepancy between laboratory and plant results. It is therefore recommended that simple, timed flotation-rate tests or eventually, tests that in some way simulate the envisaged plant circuits, should always be correlated with a continuous laboratory- or pilot-scale test to obtain the closest assessment of plant performance. A 3.5-litre cell was used by ACIRL in a circuit that included a thickener, clarified-water sump, feed system and reagent addition/conditioner system. This approach has been in use since 1974, and several reports are available which describe its accuracy for several Australian coals.[27] In the UK, the so-called 'Leeds cell' has been used with some success to achieve continuous conditions on a laboratory scale, but only limited success is reported for coal applications.

Use of the data from the reverse-flotation alternative requires the application of an additional correlating factor to compensate for the idealized successive-flotation effect and also the extrapolation of the desired ash and corresponding yield. In most cases, the yield obtained is a further exaggeration, above that obtained by the successive form of testing. Some users report a greater bias resulting from the use of this type of test when compared with continuous testing, and even more markedly, when compared to the results from plant operations. Nevertheless, because of the reproducibility of the method, the reverse flotation approach is commonly utilized apparently despite mixed results being reported. These differences could well be the result of inconsistency in test procedures used.[28]

For the most uniform *pulp* conditions, the following general rules should be followed for whichever method is selected:

(i) The pulp mixing time should not be less than 5 min, and water used to make up the slurry should be checked for pH 7–8, or an otherwise stated value, ± 0.5. Mixing time, when selected, should remain unchanged.

(ii) Slurry conditioning (after mixing) following the primary addition of reagents should not be less than 5 min and should remain unchanged when a suitable time has been established.

(iii) Sodium chloride may, if required, be added at a concentration of 1 g/litre to stabilize the ionic concentration. Tap water used in the test work can be expected to vary significantly in total dissolved solids (TDS) value, and if high TDS level or wide variation is to be expected, stabilization is advisable.

For the most uniform *test* conditions, the following general rules should be followed:

(i) The coal used for the test should be screened to pass a standard test sieve, usually 100 mesh. (In some circumstances, additional test work may be required for the size-range 28–100 mesh. In such cases, the coal used should *all* lie within this size-range prior to mixing.)

(ii) Exactly prescribed proportions of coal sample and water should be used to ensure pulp density. Accuracy should be: mass, ± 0.01 g, volume ± 0.5 cm^3 (to include the added reagent solutions).

(iii) Aeration conditions should be clearly prescribed. Normally, no forced air is employed and the valve to the standpipe is opened fully to maximize the impeller air-intake capabilities of the cell. However, in some cases, for example for freely floating coal, it is necessary to reduce the air intake by partially closing the valve in order to slow down the flotation recovery rate. When this is necessary it should be clearly specified in the test report, e.g. valve half-open, etc.

(iv) Extreme care should be taken to ensure that, at the prescribed impeller speed, no pulp spills over the froth lip. This is an indication of overfilling the test cell and should be corrected by suitable reduction of pulp level in the cell.

(v) Froth removal using a blade should only be applied to coax the froth over the cell lip. Persistent removal close to the froth–pulp interface reduces stability and also creates a tendency for non-float material to be removed. It should therefore be avoided at all costs. Normally, no coaxing is required during the initial one-minute period because the froth is generated quickly and remains stable for a sufficiently long period to ensure safe carriage from the cell. Later, when coalescence occurs and a consequent loss in stability arises, the blade should be used to coax gently the froth to the lip and over it. *No pulp* should be scalped over the froth lip by the blade.

(vi) When an initial concentrate is refloated, the pulp should be re-agitated and thus mixed for a fixed period of time (normally in excess of one minute). If more reagent is added, the further conditioning period (including the mixing time) should not be less than five minutes. This ensures reproducibility in technique and is not intended to simulate operating procedure.

For the most reliable *analytical* results, the following general rules should be followed:

(i) Time intervals used in the time–recovery test should be regarded as flexible, and as even a distribution of froth product as possible should be aimed for. During the first 30 s a large recovery is normally obtained and the increment interval should initially be short, to be increased gradually as the test proceeds. As a guide, a more or less equal volume of froth should be aimed for in each increment recovered during the first three minutes of each test.

(ii) All froth and tailings samples should first be filtered to provide moist cake samples which may then be air dried in an evaporating oven with a thermostat controller. The most common filter used for such work is the pressure chamber into which fine filter paper is fitted to ensure almost total recovery of solids. As an alternative, the Buchner funnel–vacuum apparatus may be used, but with certain slurries, particularly tailings, the filtration rate is slow unless flocculant is used. This may, in some cases, be inadvisable for coal-float products because of possible detrimental effects on the coal coking properties. Hotplate drying is inadvisable unless unavoidable because of the high contact temperature often incurred. The heating arrangements employed should be thermostatically controllable and operated at 105 °C. It should be remembered that fine coal, because of the surface area involved, is more sensitive to heating than coarser coal.

(iii) Test results should be tabulated and plotted on a graph similar to that shown in Fig. 9.27.

Other types of laboratory-scale flotation cell are available, e.g. 138-mm Leeds laboratory column as shown in Fig. 9.28(a), and several other columns similar to that shown in Fig. 9.28(b).[22]

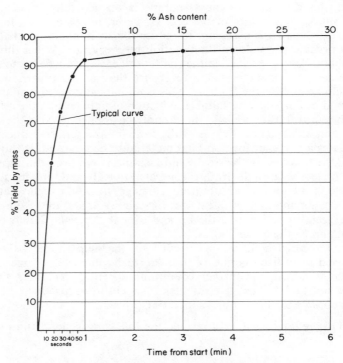

Fig. 9.27 Flotation test—yield/time/ash graph.

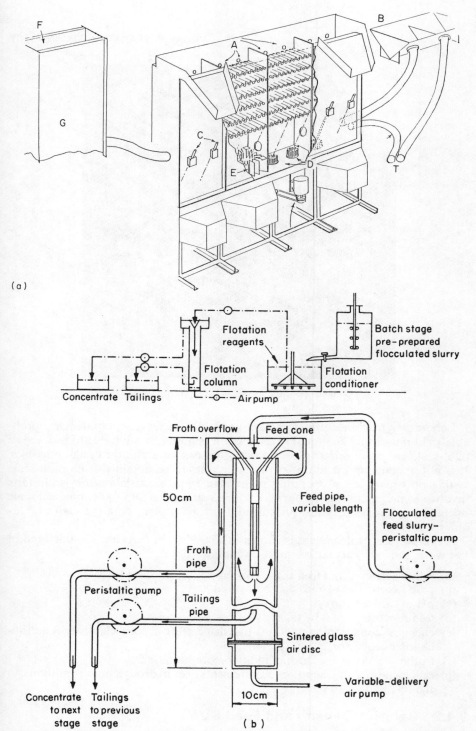

Fig. 9.28 (a) Laboratory Leeds flotation column. A, Reflux water inlets; B, tailings overflow box; C, air inlets; D, impellers; E, stabilizers; F, feed inlet; G, conditioner; T, tailings outlet pipes. (b) Example of laboratory flotation column.

Fig. 9.29 Pilot-scale mechanical flotation unit.

Laboratory machines usually give a good idea of reagent requirement, probable yield ranges and coal-quality data, but it is normally advisable to proceed to pilot-scale testing in order to obtain data that can be useful for design purposes. It is not uncommon for more favourable results to be obtained from pilot-scale testing on-stream in an operating plant than from batch laboratory-scale tests involving small quantities of slurry. For this reason, scale-up to commercial scale operation is a more practical proposition with pilot-plant data than with laboratory test data.

A typical pilot-scale system is shown in Fig. 9.29. In carrying out this kind of test work, the following factors must be considered:

(a) relative density and feed size-consist;
(b) pulp density (feed-solids concentration);
(c) pH of the feed slurry;
(d) feed rate (range of flow rates);
(e) chemical conditioning to adjust pH, disperse or depress clays, etc., emulsification of collector;
(f) frother-collector dosage and addition/conditioning;
(g) cell agitation and aeration requirements, i.e. hydrodynamic conditions in the cell.

9.4.2 Relative Density and Feed Size

The floatability of a coal slurry depends on coal rank and ash content as well as upon its relative density and size-consist. The latter two properties must be determined in order to ascertain the floatability over a definite period of time. In batch

TABLE 9.2 Flotation Rate of Coal as a Function of Grain Size (after Brown and Smith[23])

Particle size BSS mesh	Flotation rate constant (min^{-1})
+30	0.03
30 × 44	0.08
44 × 60	0.14
60 × 85	0.23
85 × 120	0.30
120 × 200	0.45
−200	1.70
Overall constant for first minute	0.31

Continuous flotation test: sub-A type cell; frother—cresylic acid; coal—Winter seam, West Yorkshire, 85% carbon, dry, ash-free basis, 10.3% solids concentration

testing carried out in the laboratory, the finest-sized coal tends to be floated off first. In order to obtain the coarsest coal grains, it is often necessary to accept some gangue minerals. From such tests, it has been determined that flotation rate at any instant in time is directly proportional to the concentration of the solids remaining in the pulp.[23] For the range of particles sized from 30–200 mesh, there is a significant increase in the flotation rate of over 50-fold, as shown in Table 9.2. The normal practical range of coal flotation is 500–100 μm, but in numerous cases, material up to one millimetre is contained in the feed and some coal is recovered of below 50 μm in size. Miller[24] has indicated that floatability is proportional to the surface area and inversely proportional to particle mass within this size-range.

9.4.3 Pulp Density

In coal flotation, it must be borne in mind that the majority of the feed is recovered as a concentrate and the optimum feed-solids concentration is lower than for ore treatment, being usually 10–15% by weight. However, pulp-density control in a coal-preparation plant is often difficult to maintain consistently and actual ranges may be as wide as 5–20% by weight. Bearing in mind also, the foregoing comments about solids density and particle size, it is probably wise to think in terms of an optimum value of about 10% solids by volume unless the feed contains high clay content in which case a more dilute feed will probably result in optimized selectivity.

9.4.4 Pulp pH and Chemical Conditioning

Because coal is reasonably hydrophobic over wide extremes in pH, some coals can be effectively frothed in both acidic and alkaline slurries. Others can only be effectively collected within a narrow pH range about a definite optimum value, close to the neutral point. At this point, i.e. pH 7, coal appears to have a slightly negative charge. A slight increase in acidity of the pulp creates hydrogen adsorption, reducing the surface charge to zero. At this point the coal is at its most hydrophobic. In a more acid pulp, the coal surface becomes positively charged and hydrophobicity falls. In an alkaline pulp, adsorption of hydroxyl ions creates a stronger negative charge which also reduces hydrophobicity. The reasons why some coals are more sensitive to such adjustments than others rests with differences in the proportions of the microlithotypes and the freshness or freedom from weathering or oxidation of the coal.

Change in pH also influences reagent performance, dispersion or coagulation of non-coal components such as clays, and in some cases chemical precipitation of dissolved salts.

9.4.5 Cell Agitation and Aeration

Experience has shown that pulp agitation and aeration are both very important factors in achieving good froth flotation of coal. Too low an impeller speed gives insufficient mixing, poor emulsification and dispersion of the reagents and inadequate bubble–particle collision. On the other hand, excessively high speeds can destroy air dispersion, eliminate bubble–particle contact and produce an unstable froth. An optimum exists between these limits, above which flotation rate ceases to increase and wastage of power occurs. To identify such an optimum condition usually requires the implementation of a test programme, ideally with the full-scale flotation unit, although useful information may be obtained from pilot-plant tests.

Controlled test work with a subaeration flotation machine has suggested an optimum air-bubble content of about 30% by volume with a mean bubble-size of just less than one millimetre, but in a commercial cell it is difficult to confirm whether or not this is a typical condition.[29]

For the Elmore–Coppée vacuum–flotation cell, a higher optimum air-bubble volume, closer to 50% by volume, is achievable, but the mean size of the bubble is smaller at 0.25 mm. The value of the air-to-solids ratio required in order to float fine coal by whatever technique selected, can only be determined by continuous pilot-plant test work involving a realistic range of hydraulic loadings. In dissolved-air systems such as the Elmore machine or solid–liquid separation methods involving an air saturator, this relationship is far more critical and must be carefully determined. Henry's law states that the theoretical mass of air available at atmospheric pressure (i.e. available for flotation) is proportional to the saturation pressure at which the air is dissolved:

$$M_a = C_s P$$

where M_a is mass of air dissolved in water in mg/litre, C_s is solubility of air in water at STP, in mg/m^3, and P is pressure above atmospheric in kPa.

At the saturation pressures involved in dissolved-air systems (250–500 kPa), the values of M_a range from about 30 to 75 mg air per litre.[30] In dispersed-air systems, this quantity of air increases for pneumatic machines and for mechanical cells.

9.5 WEATHERING AND OXIDATION EFFECTS

Coals of bituminous or lower rank undergo atmospheric oxidation which begins as soon as the coal is mined and proceeds during transportation and storage. In some cases, prolonged storage can lead to detrimental effects in terms of the effectiveness of froth flotation. The degree of oxidation has been shown to be very dependent upon temperature and the time of exposure[31] and these two factors most influence froth flotation.

Not all oxidation effects have occurred after mining. Many coal deposits contain coal which is substantially oxidized before mining. Such an effect usually occurs as a result of the close proximity to the surface where the coal is overlain by disturbed or pervious overburden material. Once outcropping or shallow coal has been removed, the amount of oxidation diminishes rapidly. In some cases, large quantities of coal may be affected, especially in areas where geological disturbance has been substantial, or where percolation of water from pervious overlying strata has occurred over long periods of time. Good examples of such deposits are found in the Rocky Mountain low-volatile coking-coal deposits in Canada and in the Bowen Basin in Queensland, Australia.

High-quality coking-coal deposits contain large quantities of coal affected by oxidation, and coking properties have been seriously impaired by this effect. Free-swelling indices normally in the range 6–8 are reduced to 1–2, and fluidities are also severely affected. Oxidization effects on the fine-coal component are generally more pronounced than for coarser coal, and even if flotation can be promoted by chemical methods the subsequent inclusion of oxidized coal is often detrimental to the overall cleaned-coal quality. Consequently, much of this coal is wasted or added to thermal coal products.

Figure 9.30 shows photomicrographs of finely polished, thin sections of oxidized and unoxidized coal.[32] In the USA, a test has been developed[33] which uses Safranin-O dye to stain oxidized fragments of coal to facilitate the quantitative assessment of the degree of oxidation of the coal using microscopic techniques. The analysis obtained can be further used to related oxidation effects with those on coking or thermal properties.[34]

Whichever type of formation mode has occurred to cause oxidation of the coal, the effect is characterized by the formation of acidic groups at the coal surface. This always results in a reduction of the hydrophobicity and floatability of the coal, which can be restored to some extent in the laboratory by the dissolution of the oxidized layer using caustic solution or by reduction using benzidine in benzoyl alcohol.[35] In commercial flotation circuits, cruder methods such as scrubbing have been attempted with some success, but highly friable coals are not amenable to this approach.

Reactivity towards oxygen varies considerably with coal rank. The low-rank coals are very amenable to oxidation and rapidly lose their floatability, but there is a general decrease in reactivity as the carbon content increases. Cationic collectors such as alkyl amines have been suggested[31] and others such as polyglycol ether-based reagents are reported to be effective for highly oxidized coal,[36] especially when used in conjunction with a so-called 'conditioning' agent which improves the hydrophobicity of the oxidized coal surface.

The behaviour of weathered coal during the flotation process has been the subject of numerous studies and an attempt has been made to develop a flotation model for the coal surface.[37] The following observations can be taken as a basis for such a model:

(a) Weathered coal which has become hydrophilic can be made to float at pH 2 when deslimed and/or scrubbed.

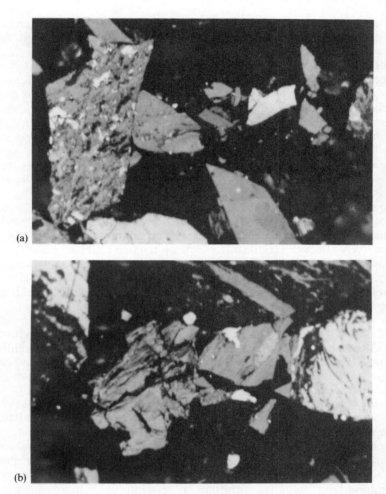

Fig. 9.30 Photomicrographs showing oxidized and unoxidized coal: (a) stained and cracked vitrinite grain, oil immersion × 325; (b) marginally green-stained vitrinite grain, oil immersion × 325. (Courtesy David Pearson and Associates.)

(b) While treatment with dilute hydrochloric acid (HCl) removes Ca, Fe and some Al from the surface, it does not impart or restore natural hydrophobicity to the deslimed coal surface. A 10% hydrofluoric acid (HF) solution removes Si from the surface and does restore hydrophobicity appreciably.

(c) Decarboxylation of deslimed coal imparts partial hydrophobicity.[38]

9.6 PYRITE REDUCTION

In recent years, much emphasis in coal-flotation research has been placed on reducing the sulphur content of coal. Of the two types of sulphur in coal, inorganic and organic, only the inorganic, represented mainly by the mineral pyrite, may be separated from coal. The floatability of coal-pyrite is different from that of ore-pyrite and its separation from coal presents a difficult problem.[39] Although much research has been devoted to determining ways and means of achieving an effective method, little success has so far been reported in commercial applications. A two-stage reverse flotation process has been developed in which the first stage is conventional flotation to obtain coal and pyrite in the froth product. This product, reconditioned with dextrin which depresses coal in the second stage, is then floated with xanthates, leading to the removal of pyrite as the second froth product.[40] The two-stage process is based on the selective adsorption of dextrin on hydrophobic particles through surface interactions. Another reagent, carboxymethyl cellulose, has also been tested for the same purpose with some success. The earlier compounds, used with mixed success and depending much upon the surface exposure of the pyrite, involved the use of hydrolysed metal ions derived from $FeCl_3$, $AlCl_3$, $CrCl_3$ and $CuSO_4$.[42] After development and optimization of the coal-pyrite flotation process in the laboratory, efforts were directed to scaling up to pilot-plant level.[41] A plant was designed and constructed with one eventual flowsheet, as shown in Fig. 9.31 The first stage was done in two parallel banks of four Wemco Fagergren cells each,

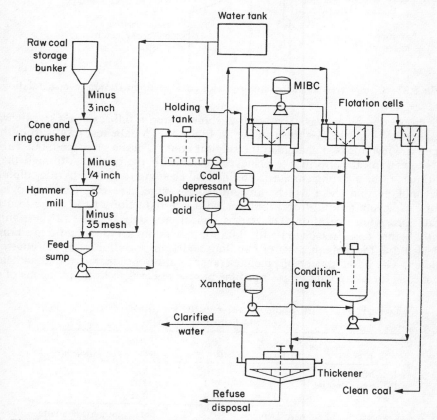

Fig. 9.31 Pilot-plant flowsheet for coal-desulphurization reverse flotation circuit.

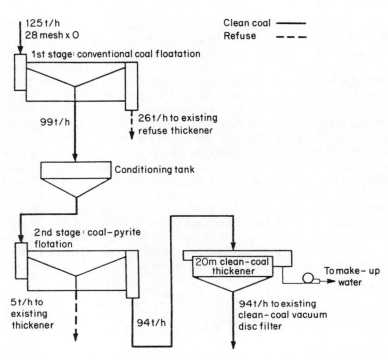

Fig. 9.32 Commercial flowsheet employed for coal-desulphurization reverse flotation.

and the second stage was a two-cell machine. Later, a full-scale test circuit was developed,[43] with the flowsheet shown in Fig. 9.32. To date, test results from this circuit are few due to a change in the sulphur content of the plant feed, but Table 9.3 shows typical performance achieved so far with the circuit. Although there continues to be a strong emphasis on fine-coal cleaning to reduce pyritic sulphur, it is unlikely that there will be any immediate large-scale emergence of plants built exclusively around froth flotation, with size reduction of the entire feed to create maximum liberation. This seems logical, not only because of the significantly higher costs associated with flotation, including grinding and dewatering, but also due to the high costs of handling and transportation of an all-fine-coal product. The development of pipeline transportation, originally viewed as being the means for overcoming this impediment, now appears to bring problems of its

TABLE 9.3 Full-scale Coal–Pyrite Flotation Process Test Results (see Fig. 9.33)

Product	Weight (%)	Ash content (%)	Pyritic sulphur (%)	Total sulphur (%)
Underflow	93	19.0	0.64	1.05
Froth concentrate	7	9.7	3.07	3.49
Feed	100	18.3	0.81	1.22

Reagents: 0.40 kg/t Aero Depressant 633; 0.35 kg/t Potassium amyl xanthate; H_2SO_4 added to pH 6.7

own which negate, to some extent, the originally anticipated benefit of lowering transportation costs.

9.7 FLOTATION CIRCUITS AND PRACTICE

As mentioned earlier in discussion of mechanical dispersed air-flotation machines, the most common flotation circuit employed in coal cleaning is an open-trough cell bank catering for a residence time of between 3 and 6 min. The development trend has been towards the use of larger-capacity cells thereby

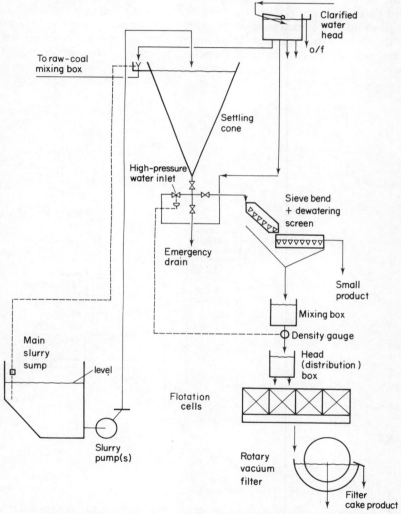

Fig. 9.33 Flotation feed-concentration control using settling tower.

eliminating numerous parallel cell-banks and ensuring better feed distribution and much more uniform feed conditions. It is still not common for conditioning to precede the flow into the cells, this facility often being required for coals of inconsistent floatability, such as weathered or oxidized coals. Occasionally, some attempt is made to provide for regulation of the feed-solids concentration. In Britain, for example, especially in plants utilizing Baum-jig cleaning of the coarse coal, the fines are pre-thickened in a settling tower or thickener and then diluted to the required feed concentration for flotation by addition of overflow water from the tower. A control circuit involving a density gauge, as shown in Fig. 9.33, is commonly employed. Other similar circuits employ hydrocyclones which serve to provide a thickened (underflow) feed to the flotation circuit and also effective desliming of clay-mineral content of the feed, as shown in Fig. 9.34.

Although single-stage flotation is regarded as being the conventional method of coal cleaning, there are a number of alternative circuits in use employing more than a single cell-bank for each feed-stream. One approach, becoming increasingly common in the treatment of metallurgical coking coal, is the use of two

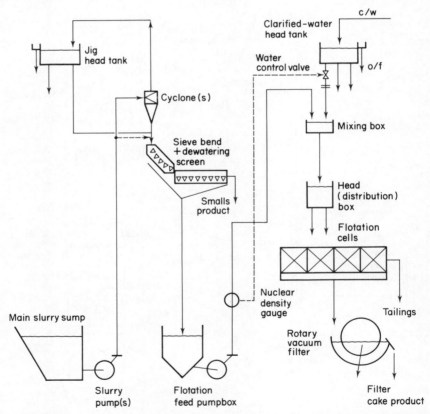

Fig. 9.34 Flotation circuit employing desliming hydrocyclone to prepare flotation cell feed.

FLOTATION, AGGLOMERATION AND SELECTIVE FLOCCULATION

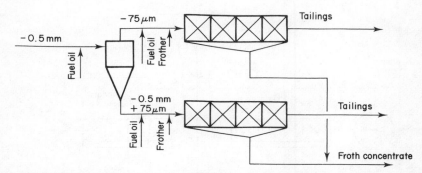

Fig. 9.35 Twin single-stage flotation treatment of a coarse and fine coal, following pre-classification using a hydrocyclone.

cell-banks for treating a preclassified feed, i.e. coarse- and fine-coal feeds (see Fig. 9.35).[44]

A hydrocyclone classifies the flotation feed into two differently sized feed-streams, often 0.6×0.15 mm and 0.15 mm $\times 0$, which are then conditioned with the appropriate reagent dosages prior to treatment in different cell-banks. This pretreatment step has been shown to result in an optimized fine-coal yield and reagent consumption condition when compared with single-stage 0.6 mm $\times$ 0 treatment. Test work, including pilot-plant tests, will usually establish the viability of this option.

In some cases, there may be benefit in using a two-stage circuit employing a second cell-bank for recleaning the concentrate obtained from the first stage. Again, this circuit has found its most frequent application in the treatment of high-quality metallurgical coking coals, especially those in which clay contamination of the feed has been a problem. In some cases, very hydrophobic coal with a strong tendency to respond rapidly to flotation, i.e. high rate of recovery, has also been shown to benefit from a recleaning stage. Figure 9.36 shows a typical two-stage flotation circuit and Fig. 9.37 shows an alternative employing a desliming cyclone for preclassifying in the secondary stage. This circuit is of special value in the treatment of flotation feeds with high clay contamination. In one such circuit, density control in the second cell-bank was controlled by bleeding

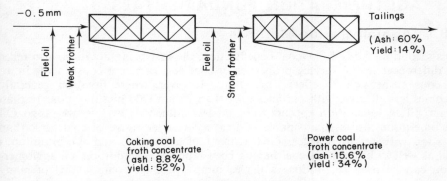

Fig. 9.36 Two-stage flotation circuit.

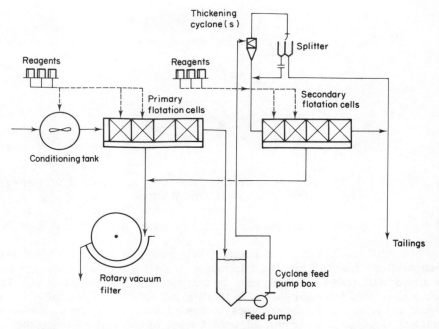

Fig. 9.37 Two-stage flotation circuit with preclassification of the primary tailings.

back a regulated amount of cyclone overflow product. This also served to introduce a stabilizing effect on the 0.5 × 0.05 mm feed to the second cell-bank.

Three-product separation involving a two-stage flotation circuit has also been operated effectively in Australia. The primary cell-bank was used to obtain a low-ash-content coking-coal by using 'starvation' reagent dosages. A thermal coal product of 15–16% ash content was then obtained by reconditioning the primary tailings with additional reagents and floating off a second product from the secondary cell-bank.[44]

9.8 AGGLOMERATION FUNDAMENTALS

9.8.1 Agglomeration Methods

This second method of fine-coal beneficiation, the first being flotation, also relies on differences in the surface properties of coal and non-coal constituents in an aqueous suspension to effect separation. However, where flotation generally becomes less effective with particles finer than 75 μm (200 mesh) or coarser than 1 mm, oil agglomeration appears to be far less influenced by this size-range. Oil agglomeration also seems capable of treating a wider range of coal rank than flotation with reported successful treatment of anthracitic and sub-bituminous coal as well as bituminous coal having been reported. In addition, oil agglomeration can be used to produce a dense, lumpy product compounded of many particles of coal. Each agglomerate may contain coal fragments varying in size from as coarse as 2 mm to ultrafine particles of a few micrometres in size, and will possess the inherent strength to remain intact during transportation. Because of the lipophilic (oil loving) and hydrophobic (water hating) contrast of the coal surface, the formation of the agglomerate tends towards maximum evacuation of water from within its structure. This phenomenon has arguably as much significance in the application of the oil agglomeration to fine-coal cleaning as the ability to eliminate non-coal by differential-size separation.

Fine-coal fragments in an aqueous suspension can be agglomerated in a number of different ways. The simplest method is a peri- or orthokinetic flocculation occurring as a result of the addition of electrolytes to reduce the surface change on the surface of the coal allowing London–van der Waals forces of attraction to become predominant. In this method, only very small particles are affected and relatively small composites result. An advancement of this method is the use of polymeric flocculants to form much larger composites, often called floccules, by a combined mechanism of ion adsorption and polymer bridging. This is the third aspect—or selective flocculation—dealt with in this chapter. The third method involves the addition of a second, immiscible liquid—oil—to wet preferentially the coal fragments, causing adhesion by capillary interfacial forces. In many instances, orthokinetic flocculation and also polymer flocculation may be advantageous in assisting oil agglomeration, provided that their application can be assured as being selective.[55]

Perhaps the first commercial oil agglomeration application was the Trent process.[47] This process consisted of a mixing vessel which agitated a mixture of finely ground coal and water to which oil had been introduced at about 25–30% by mass of the coal contained in the slurry. After about 10–15 min of agitation, agglomeration had proceeded to the extent where the coal agglomerates could be readily separated from the unaffected shale by a screening unit. More recently, the Convertol process was introduced, which uses high-speed mills to create the same effect. The coal agglomerates are removed from the slurry by a centrifugal filter unit.[48]

Perhaps the most successful avenue in terms of process efficiency combined with oil usage has been that of spherical agglomeration. Both of the previous processes involved the use of large quantities of oil, creating poor economic potential for the technique. Spherical agglomeration processes are those which treat finely divided solids in aqueous suspension with a so-called bridging liquid—oil. On agitation of the suspension, the oil covers the lipophilic surface of the coal, and small agglomerates result from contact between the coated particles. The rate of growth of these initially formed agglomerates depends upon the amount of oil, the degree and type of agitation, and on the size distribution of the coal, as well as on surface conditioning of the coal. With adequate oil and ideal formation conditions, the unconsolidated, initially formed flocs become progressively larger, densified spheres, eventually turning into pasty lumps in which the solids are essentially dispersed in the bridging liquid. Where up to 20% of the pore volume is occupied, bridging liquid is in a pendular state, being formed as lenses between contacting particles. This produces a flocculated product with an open structure which can be easily disintegrated again. With continued agitation, the pore volume becomes continually filled, with capillary action drawing particles closer and closer together, producing spherically shaped agglomerates. At this juncture, the bridging liquid is in the funicular state. Finally, when the pores

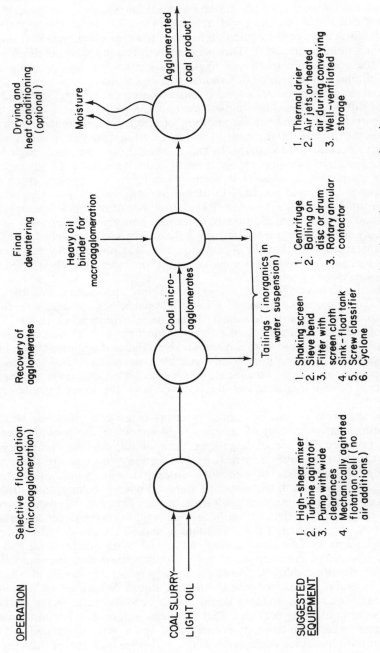

Fig. 9.38 Coal agglomeration flowsheet indicating alternative steps and equipment.

FLOTATION, AGGLOMERATION AND SELECTIVE FLOCCULATION 463

are almost completely filled with bridging liquid, with water exclusion almost total, the capillary state is reached and the spherical agglomerate has achieved its peak of strength and sphericity.[56] Spherical agglomeration has been mainly developed as a process for coal treatment by the National Research Council of Canada (NRCC).[46,56,57] Related recent developments also include the LoMAg process in the USA,[49] the Shell Pelletizing Separation process in Holland,[50,51] the BHP/BPA IPTACC system in Australia[53,54] and work done by the Indian Central Fuel Research Institute[58] and Polish researchers.[59]

9.8.2 Spherical Agglomeration Systems

There are a number of different flowsheets containing a variety of types of equipment that have emerged during the course of development of the various methods of spherical oil agglomeration. The process flowsheet will depend upon the size and type of agglomerators required, which in turn is dependent upon the type and quantity of oil used, the method of agitation and the required moisture content of the final product. Figure 9.38 represents a generalized flow-diagram with suggested equipment for coal agglomeration.[60] The sequence of operations includes selective flocculation or microagglomeration, agglomeration recovery with rejection of non-coal material, size enlargement and, if required, final treatment of the agglomerates into the final product form, e.g. pellets.

Although two stages of agglomeration are shown, the small agglomerates produced in the first stage may be adequate for many applications such as when the product is to be transported in a slurry pipeline as for the Integrated Pipeline Transportation and Coal Cleaning (IPTACC) system which results in further agglomerate growth during the transportation phase.[61]

Where larger and denser agglomerates are required to facilitate easier storage and handling, the heavy binding oils used to form larger agglomerates do not perform well for further ash rejection and the second stage is therefore devoted more to product preparation than to cleaning.

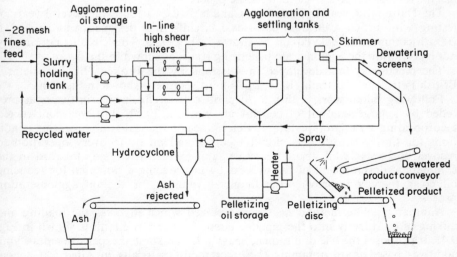

Fig. 9.39 Flowsheet of 200 t/h NRCC spherical agglomeration circuit.[61]

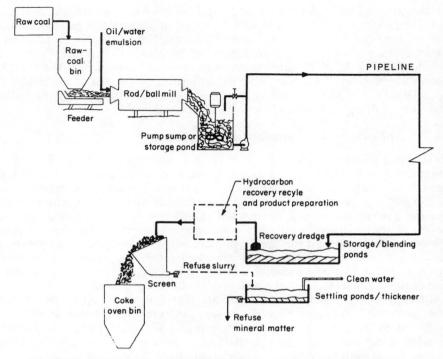

Fig. 9.40 BHP/BPA integrated pipeline and coal-cleaning system flowsheet.

Oil agglomeration displaces moisture very effectively, and for larger-sized agglomerates it is not normally necessary to employ thermal dewatering unless it has the further purpose of hardening the pellets.

The National Research Council of Canada (NRCC), following several years of testing with bench-scale and pilot-scale circuits, has designed and studied a 200 t/h commercial-scale circuit as shown in Fig. 9.39. This circuit may or may not incorporate the agglomerate pelletizing stage, as shown in the figure.

The pipeline system developed in Australia by the Broken Hill Proprietary/ British Petroleum Australia (BHP/BPA) joint venture is shown in Fig. 9.40.

Following initial work with a 5 t/h pilot-scale circuit, which in turn was preceded by a large amount of bench-scale work, a 30 t/h unit was constructed incorporating a 1.6-km pipe-loop to determine commercial-scale operating conditions for the system.[61] This circuit incorporated an oil-recovery operation to recover and recycle oil where necessary, and represented a step forward in the commercial development of the process by introducing a potential for reducing oil costs thereby improving the financial viability for the oil agglomeration process.

Another pipeline method has been reported which involves the loading and mixing of oil directly into the pipeline circuit.[62] The test circuit is shown in Fig. 9.41. It proposes the use of bitumen, heavy oils and some inexpensive emulsifying additives based upon obtaining these commodities locally in Alberta, Canada, and applying them to the treatment of the enormous sub-bituminous and lignitic

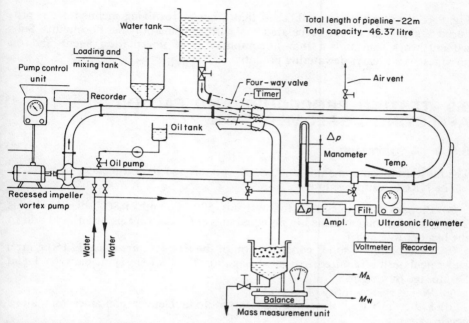

Fig. 9.41 Schematic diagram of equipment used in Alberta Research Council pipeline loop reactor.

coal reserves in the Canadian plains. The ensuing fuel would then be used in coal-fired power plants. Finally, a Polish agglomeration system concentrates upon dewatering as well as de-ashing of the coal as an on-line dewatering system for introduction in conventional coal-preparation plants.[55] This unit is shown in

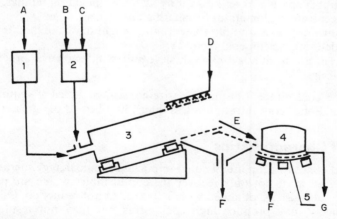

Fig. 9.42 Drawing of the Polish coal-fines dewatering agglomerator. 1, Slurry tank; 2, flocculant tank; 3, dehydrator; 4, pressing drum; 5, belt conveyor; A, coal slurry; B, flocculant; C, water; D, compressed air; E, agglomerated product; F, clear water; G, final solid product.

Fig. 9.42. Test work had established that a gently revolving inclined drum produced agglomerates from flocculated coal slurries using polymer flocculants. Subsequent work has shown that a combination of flocculation agent and oil provides even greater dewatering potential.

9.9 TESTING PROCEDURE AND FACTORS AFFECTING AGGLOMERATION

9.9.1 Apparatus

The most useful equipment is a standard, variable-speed food-blender.

The NRCC test procedure is:

(a) Prepare the reagents as follows: mix 500–1000 ml of 80% diesel oil plus 20% No. 6 fuel oil emulsified in an equal mass of water (in this order) and add to the blender to emulsify.

(b) Determine the solids concentration of the feed mixture (ideal is 150 g/litre) and calculate the required reagent dosage for 50 ml of slurry at between 1 and 5% dosage by weight.

(c) Weigh this quantity of reagent.

(d) Either mix-feed slurry or add mixed to a blender container and allow gentle agitation.

(e) Test pH and adjust to alkaline by adding quantities of sodium carbonate.

(f) Add reagent and allow to mix for one minute.

(g) To commence the test, increase blender speed to maximum and maintain for 30 s.

(h) Reduce to 60% full speed for a further 30 s.

(i) Remove sample and transfer to 100 mesh (Tyler) test sieve (spray-washing the blender container to remove all solids).

(j) Use rubber spatula to remove gently all water including tailings, and transfer coal-concentrate agglomerates by rotating basket centrifuge.

(k) Remove wet cake (and filter paper) and weight dry cake (and remove filter paper) and determine clean-coal recovery.

(l) Add centrate to minus 100 mesh sieve slurry, filter or evaporate to obtain weight of tailings.

Slurry is fine coal minus 100 mesh in size-consist, deslimed if required at 200 mesh or finer. Solids concentration is less than 150 g/litre at pH greater than 7.0.

9.9.2 Oil Characteristics

Oil type and concentration are of equal importance in coal agglomeration. To be effective, the oil must not only recover the combustibles which are present, but the product must also be obtained with a controlled ash reduction. Oils of low to medium density may produce high coal yields but may not result in strong enough agglomerates to facilitate commercial recovery. High-viscosity oils, on the other hand, present problems of emulsification and dispersion in the slurry and may not be selective enough to achieve the required degree of recovery. Pre-dispersion appears to be the key to most effective oil agglomerators and if

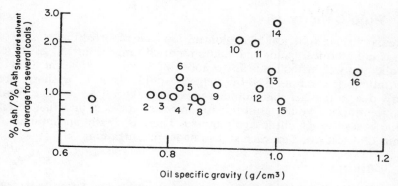

Fig. 9.43 Ash content of agglomerates using various oils as bridging agents. 1, Hexane; 2, coal-tar raffinate; 3, Stoddard solvent; 4, kerosene; 5, diesel fuel oil; 6, No. 2 fuel oil; 7, paraffin O-119; 8, paraffin O-120; 9, BTX and coal-tar light oil; 10, bunker C; 11, coal-tar (30–60%) in xylene; 12, catalytic fractionator bottoms; 13, petroleum resin in light oil; 14, light coal tar; 15, No. 6 fuel oil; 16, middle heavy creosote.

this is assured, it may be concluded that there is nothing inherent in the coal-surface wetting properties of most oils (lipophilicity) to prevent their use as bridging liquids for bituminous coal.

Figure 9.43 gives details on bridging oils which may be used effectively for coal.[45] This figure plots the ratio of ash levels as obtained for various coal samples using different oils to ash levels obtained with Stoddard solvent for the same coals. Stoddard solvent is a highly paraffinic oil known to perform well as an agglomerating reagent. A distinct change in characteristics occurs above and below 0.9 g/cm³. Those oils with densities below 0.9 g/cm³ appear to give good ash rejection.

9.9.3 Coal Characteristics

The extent to which a coal can be cleaned is determined mainly by the dissemination of mineral matter throughout its structure. This is not a function of coal rank, and indeed many sub-bituminous coals have low ash contents and many bituminous coals have high ash contents. As for mineral dressing, size reduction is the key to liberation but it is still not common for coal to be ground or pulverized to fine size to achieve liberation prior to cleaning. Much recent research has been devoted to fine-coal cleaning, or deep cleaning as it is often called, and such concepts have generally involved wet or dry milling prior to beneficiation.

Sub-bituminous and lignitic coal, characterized by higher oxygen and moisture content, tend to be quite hydrophilic in nature, and in order to agglomerate them many oils are required. Apparently, the nitrogen, oxygen and sulphur functional groups contained in heavy oils are able to absorb onto the coal surface.

Gondwana coals of the Southern Hemisphere, characterized by higher inertinite content, also behave differently to Northern Hemisphere bituminous coals. Hence Indian, South African and in particular Australian researchers have more success with medium–heavy oils than with light oils.

9.9.4 Pulp Density

Most researchers in spherical oil agglomeration have reported that slurry concentration is not a critical factor and values reported range from 150 g/litre (equals 13% solids by weight for coal) up to about 25% solids by weight. Low slurry concentrations (i.e. minus 5% by weight) would result in a sharp drop in recovery, due to the fact that intimate mixing would not be possible. Similarly, high concentrations would result in entrapment of non-coal material, despite the probable overall lower oil dosage required. There is clearly an optimum, to be determined by test work, for each system under investigation.

9.9.5 Mixing

Agitation time has been shown to increase as oil density and viscosity increase. In general, the mixing time required to form resilient coal agglomerates decreases as the agitation is intensified. In the test blender described earlier, the impeller tip velocity of 8–12 mm/s was reported for a residence time of 3–5 min. Power

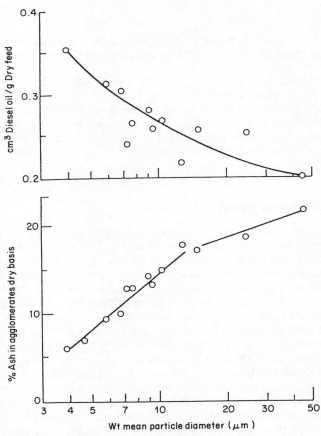

Fig. 9.44 Effect of particle size on ash content and oil dosage.

FLOTATION, AGGLOMERATION AND SELECTIVE FLOCCULATION

input for spherical agglomeration has been established as lying between 10 and 40 kW/m^3 of mixer volume.[46] Some reduction in power required may be obtained by pre-emulsification of the agglomerating agents. Attrition of the solids in the pulp may also prove beneficial for some coals.

9.9.6 Particle Size-Consist

Figure 9.44 shows the effect of particle size on ash content and oil requirement in recovery of coal from flotation tailings. As discussed in Section 9.3.3 of this chapter, liberation is an important criterion in maximizing recovery, but over-milling the coal will result in wastage of reagent. Milling studies are therefore of paramount importance in determining operating requirements for a particular coal source when studying ultrafine coal cleaning.

9.9.7 Moisture Content

The amount and type of oil used for agglomeration is the single most important factor in determining the amount of moisture reduction. Figure 9.45[46] is a generalized graph which simplifies the relationship between oil concentration, and moisture content of and size of agglomerates, as the formation process is

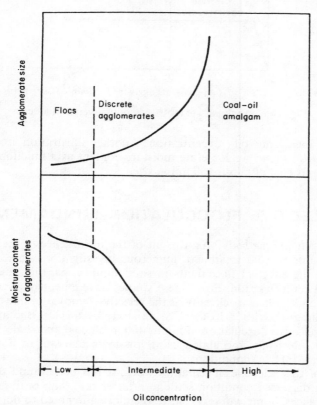

Fig. 9.45 The effect of oil concentration on agglomerate size and moisture content.

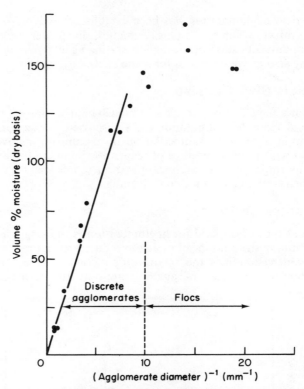

Fig. 9.46 Agglomerate moisture content versus diameter.

extended by increasing oil concentration. Another illustration from the same source (Fig. 9.46) shows agglomerate moisture-content as a function of diameter for a bituminous coal of minus 150 μm in size.

9.10 SELECTIVE FLOCCULATION FUNDAMENTALS

Selective flocculation methods As a result of the upsurgence in the use of polymeric flocculants in coal treatment, numerous investigations have been carried out to study the effect of flocculants in solid–liquid separation, i.e. thickening, filtration and centrifugation. From these studies have emerged several that have gone further and looked specifically at the selective removal of fine-coal particles from coal-tailings slurries.[62] In the USA, a study by the US Bureau of Mines[63] investigated selective flocculation of minus 400 mesh coal and shale. In this work, fine coal was selectively flocculated using low-molecular-weight non-ionic polymers in alkaline (pH 11) conditions.

In much of this selective flocculation work, it has been found that non-coal particles in a dispersed condition settle at a faster rate than coal, mainly due to density differences. Some workers[64] have therefore attempted to flocculate selectively the non-coal constituents, leaving the coal dispersed. However, because of

FLOTATION, AGGLOMERATION AND SELECTIVE FLOCCULATION

the very specific nature of separation required for each non-coal substance, i.e. clays, silica, pyrite, etc., the approach most likely to be applicable in commercial practice is that of coal flocculation.

Certain polymeric flocculants, especially non-ionic and anionic ones, have been shown to be very strongly adsorbed onto coals. The flocculation of coal by polyethylene oxide has been found to be good over the pH range 2–10, even in the presence of large quantities of other particulate with which hydrogen bonding could have occurred.[65] Since the hydrophobic character of coal is related to the amount of exposed OH^- group (i.e. on the surface of the coal) and, as this is a function of rank, coals of higher rank have less wettable surfaces because more carboxyl groups are replaced by aromatic groups. Oxidation of the coal surface effectively reverses this trend. These kinds of inconsistencies in the surface characteristics have tended to make selective separation very inconsistent when testing

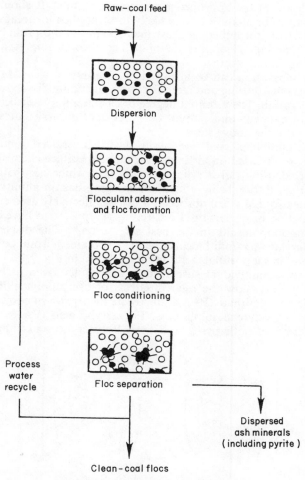

Fig. 9.47 Selective-flocculation concepts.

with plant slurries as opposed to carefully prepared and 'clean' suspensions in laboratory research work.

In concept, the selective flocculation of coal has been described as involving four stages:[66]

(a) total dispersion, possibly created by addition of chemical dispersant additives for non-coal and pyrite components;
(b) addition of polymeric flocculant to create selective flocculation of coal particles;
(c) selective flocculation resulting from conditioning the slurry; and
(d) separation of flocculated coal.

Figure 9.47 demonstrates these conceptual steps. The first step, dispersion, is achieved in a tank fitted with a slow-stirring impeller. The volume of the tank provides for the necessary residence time to achieve a semistable suspension following the addition of the appropriate dispersing agent(s). It should be remembered that selective flocculation is only likely to be applied for treating very finely sized material. Most researchers suggest that 100 μm is the probable upper size limit for coal cleaning using this method. The second step is achieved in a separate tank, again containing a slowly revolving agitator. Floc formation and the subsequent separation can be achieved in a variety of ways. Methods so far attempted are gravity settling and froth flotation, including dissolved-air flotation and oil agglomeration. The required degree of cleaning may not be achieved in a single stage of treatment and it may be necessary to carry out several stages involving all four of the above steps.

Selective flocculation of coal and pyrite has also received some attention.[67] Successful separation based on distinct differences in surface-chemistry properties has been reported using polyacrylic polymers into which xanthate groups had been incorporated. Pyrite, being a sulphate mineral, has an affinity for xanthate (dithiocarbonate), so that when absorbed it can be effectively used either to create selective flocculation or dispersion. Research work appears to have demonstrated that the latter option results in the best separation.[68] The dispersing reagent, sodium poly(acrylate-acrylodithiocarbonate), is obtained from polyacrylic acid and xanthate and has a postulated structure as shown in Fig. 9.48.

When tested with individual suspensions, the results given in Fig. 9.49 were obtained. The graphs show the relative effects of using polyacrylic acid (PAA) and polyacrylic acid xanthate (PAAX) for pyrite dispersion in conjunction with a coal flocculant, polystyrene sulphonate. Polyacrylic acid (PAA) was found to inhibit flocculation of both pyrite and coal. Polyacrylic acid xanthate (PAAX)

$$\left[\left(\begin{array}{c} CH_2-CH \\ | \\ C=O \\ | \\ ONa \end{array} \right)_{n-d} \left(\begin{array}{c} CH_2-CH \\ | \\ C=O \\ | \\ O-H-O \\ | \\ C=S \\ | \\ SNa \end{array} \right)_d \right]_n$$

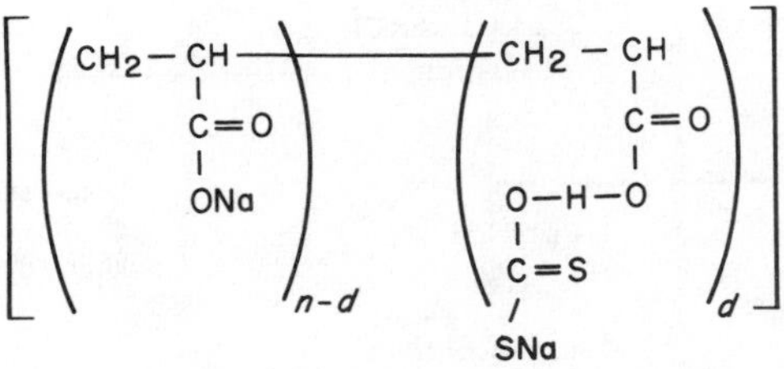

Fig. 9.48 Postulated structure of pyrite; dispersant, polyacrylic acid xanthate (PAAX).

FLOTATION, AGGLOMERATION AND SELECTIVE FLOCCULATION

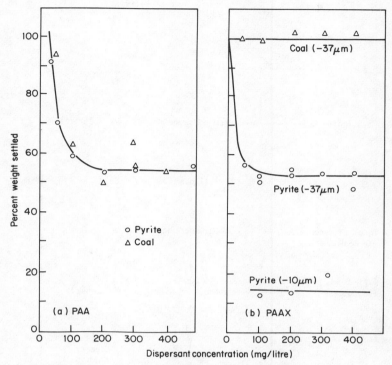

Fig. 9.49 Effect of polyacrylic dispersants on pyrite flocculation.

seemed to influence only the pyrite. A proposed flowsheet for selective flocculation of coal from pyrite is shown in Fig. 9.50.

9.11 CONCLUDING COMMENTS

This chapter has dealt with three important coal-cleaning methods, all of which may be regarded as being in an evolutionary stage of development. Although coal flotation has been used for many years, it is only comparatively recently that research endeavours into selective coal cleaning have been effectively translated into commercial practice. Flotation still does not seriously compete with gravity cleaning of readily treatable coals and is therefore viewed as a means for treating more difficult-to-clean, fine coal in the minus 0.5 mm size-range, especially in the treatment of coal with high pyrite content.

Oil agglomeration is still viewed as being of dubious economic potential except perhaps in one area, that of ultrafine coal cleaning to obtain coal–water mixtures. However, oil cost is still the key factor in the application of the process, and research is being directed towards, ways and means of recovering oil to reduce this cost.

Selective flocculation appears to be emerging as an effective means of treating ultrafine (minus 100 μm) coal slurries or of preparing such slurries for subsequent

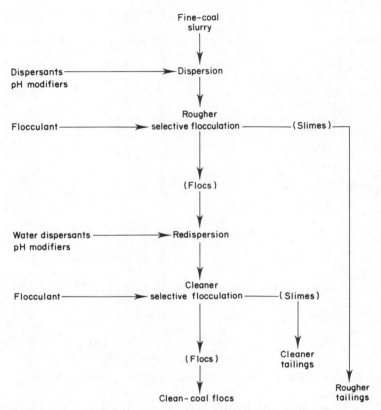

Fig. 9.50 Flowsheet for selective flocculation of coal from shale and pyrites.

treatment for either flotation or agglomeration, using flocculants or oil or a combination of both.

REFERENCES

1. D. J. Brown. *Coal Flotation in Froth Flotation*, 50th Anniversary volume, (Ed.) D. W. Fuestenau, AIME, New York, 1962, pp. 518–538.
2. D. T. Hornsby and J. Leja. A technique for evaluating floatability of coal fines, using methanol solutions, *Coal Preparation*, **1**, 1, 1–19, 1984.
3. R. M. Horsley and H. G. Smith. Principles of coal flotation, *Fuel*, **30**, 54–63, 1951.
4. J. S. Laskowski and J. D. Miller. New reagent in coal flotation, Conference on Reagents in the Minerals Industry, Rome, September 1984. Transactions IMM.
5. S. C. Sun, L. Y. Tu and E. Ackerman. Minimal flotation with ultrasonically emulsified collecting agents, *Mining Engineering*, **7**, 650–660, July 7, 1955.
6. V. I. Klassen, I. N. Plaskin and N. S. Vlasova. 3rd International Coal Preparation Congress, Brussels-Liège, June 1958, Paper E5.
7. W. W. Wen and S. C. Sun. An electrokinetic study on the oil flotation of oxidized coal, *Transactions AIME* **262**, 174–180, 1977.
8. J. A. L. Campbell and S. C. Sun. Bituminous coal electrokinetics, *Transactions AIME* **247**, 111–114, 1970.

9. F. F. Aplan. Coal flotation, in *Flotation*, A. M. Gaudin Memorial Volume, AIME, New York, 1976, Chapter 45, pp. 1235–1264.
10. I. E. Puddington and B. D. Sparks. Spherical agglomeration processes, *Minerals Science Engineering* 7, 3, 282–287, 1975.
11. R. W. Dexter and D. G. Osborne. Principles of selective flocculation, *Journal Camborne School Mines* 73, 34–41, 1973.
12. D. G. Osborne. Flocculant behavior with coal–shale slurries, *International Journal of Mineral Processing* 1, 243–260, 1974.
13. A. D. Read and G. T. Hollick. Selective flocculation techniques for recovery of fine particles, *Mineral Science Engineering* 8, 3, 202–213, 1976.
14. R. Hucko. Beneficiation of coal by selective flocculation—a laboratory study, USBM Report on Investigations, No. 8234, 1977.
15. E. N. Wharton and G. A. Mason. The practice of coal flotation, in *Minerals Engineering Handbook, Mine and Quarry*, File No. FF/04/1.
16. P. Young. Flotation machines, *Mining Magazine* January 1982, 35–59.
17. W. Irwin. Coal flotation practice, SME-AIME Fall Meeting, Minneapolis, October 1980.
18. C. C. Harris. Flotation machines, in *Flotation*, A. M. Gaudin Memorial Volume, AIME, New York, 753–815, 1976.
19. C. C. Dell and B. W. Jenkins. Leeds flotation column, Paper J3, 7th International Coal Preparation, Sydney, Australia, 1976.
20. A. C. Partridge and G. W. Smith. Small sample flotation testing: a new cell, *Transactions Institution Mining Metallurgy* Section C, 80, C199–200, 1971.
21. C. C. Dell. An improved release analysis procedure for determining coal washability, *Journal Institute Fuel* 37, 149–150, 1964.
22. D. G. Osborne. Recovering of slimes by a combination of selective flocculation and flotation, *Transactions Institute Mining Metallurgy* Section C, 87, C189–C194, 1978.
23. D. G. Brown and H. G. Smith. The flotation of coal as a rate process, *Transactions Institute Mining Engineering* 113, 1001–1020, 1953–54.
24. F. G. Miller. Reduction of sulphur in minus 28 mesh bituminous coal, *Transactions AIME*, 229, 7–14, 1964.
25. V. A. Malinovskii, N. V. Matveenko, O. M. Knaus, Y. P. Uvarov, N. N. Teterina and N. N. Boiko. Technology of froth separation and its industrial application, Transactions 10th International Mineral Processing Congress, London, 1973, (Ed.) M. J. Jones, pubd Institution of Mining and Metallurgy, London, 1974, Paper 43, pp. 717–727.
26. A. J. Le Page and F. Pollard. Methods for providing reliable data for coal preparation plant design. 7th International Coal Preparation Congress, Sydney, Australia, 1976. Congress Proceedings, Paper D2.
27. A. J. Le Page. Laboratory froth flotation of coal on a continuous basis, Australian Coal Industries Research Laboratories (ACIRL), Report No. L.R. 75–4, June 1975.
28. R. G. Burdon. Some factors influencing the rate of flotation of coal, Proceedings 4th International Coal Preparation Congress, Harrogate, England, 1962, Paper D6, pp. 273–285.
29. V. I. Klassen and V. A. Mokrousor. *An Introduction to the Theory of Flotation*, Translated J. Leja and G. W. Poling, Butterworth, London, 1963.
30. R. J. Gochin. Flotation, Chapter 19 in *Solid–Liquid Separation*, (Ed.) L. Svarovsky, Butterworth, Guildford, 1981, 2nd edn, pp. 503–524.
31. S. C. Sun. Effects of oxidation of coals on their flotation properties, *Mining Engineering* 199, 1954, 396–401, 1954.
32. D. E. Pearson. The quality of Western Canadian coking coal, *Canadian Institute of Mining and Metallurgy Bulletin* 73, 47–53, 1980.
33. R. J. Gray. Safranin stain technique for determining the extent of oxidation of bituminous coals, Technical Report, US Steel Corporation, Pittsburg, USA, 1972.

34. D. E. Pearson, of Victoria, B.C., Canada, personal communication regarding testing of oxidized coals, June 1985.
35. I. N. Plaksin, (Ed.) *Physicochemical Principles of the Action of Apolar Collectors in Flotation of Ores and Coals*, (Russian), Moscow, Izd. Nauka, 1965.
36. K. H. Nimerick and B. E. Scott. New methods of oxidized coal flotation, American Mining Congress, Chicago, May 1980.
37. B. Yavur and J. Leja. Correlation of zeta-potential and floatability of weathered coal, SME-AIME Annual Meeting, Dallas, February 1982.
38. M. H. Hubacher. Determination of carboxyl group in aromatic acids, *Analytical Chemistry* **21**, 8, 945–950, 1949.
39. J. Laskowski, M. R. Bustin, K. S. Moon and L. L. Sirois. Desulphurizing flotation of Eastern Canadian high-sulphur coal, International Conference on Processing and Utilization of High-sulphur Coals, October 1985, (Ed.) Y. A. Attia, pp. 247–266, Elsevier, Amsterdam/New York.
40. K. J. Miller and A. W. Deurbrouck. Froth flotation to desulphurize coal, International Conference on Processing and Utilization of High-sulphur Coals, October 1985, (Ed.) Y. A. Attia, pp. 255–291, Elsevier, Amsterdam/New York.
41. K. J. Miller. *Flotation of Pyrite from Coal: Pilot-plant Studies*, Report on Investigations No. 7822, Bureau of Mines, US Department of the Interior, Washington, DC, 1973.
42. A. F. Baker and K. J. Miller. *Hydrolyzed Metal Ions as Pyrite Depressants in Coal Flotation: a Laboratory Study*, Report of Investigations No. 7518, Bureau of Mines, US Department of the Interior, Washington, DC, 1971.
43. S. R. Taylor, K. J. Miller, R. E. Hucko and A. W. Deurbrouck. New methods of coal desulphurization, 8th International Coal Preparation Congress, Donetsk, USSR, May 1979, (Ed.) Congress Proceedings, I. Blagov, pubd Soviet Organizing Committee.
44. S. K. Mishra. Trends in fine coal flotation in Australia, *Mining Engineering* **37**, 1121–1125, 1985.
45. C. E. Capes. Agglomeration, in *Coal Preparation*, (Ed. J. W. Leonard), 4th edn, AIME, New York, 1979, pp. 10–105 to 10–116.
46. C. E. Capes and R. J. Germain. Selective oil agglomeration in fine coal beneficiation in physical cleaning of coal, (Ed. Y. A. Lui), Marcel Dekker, New York, 1982, Chapter 6, pp. 293–351.
47. G. St J. Perrott and S. P. Kinney. The use of oil in cleaning coal, *Chemical and Metallurgical Engineering* **25**, 182, 1921.
48. K. Lemke. The cleaning and dewatering of slurries by the convertol process, 2nd International Coal Preparation Congress, Essen, Germany, September 1954.
49. L. Messer. LoMAg Technology, low moisture agglomerates and recycled water from coal slurries, American Minechem Corporation, Coraopolis, PA, USA.
50. R. B. Brown, H. H. Brookman and C. G. Haupt. A continuous process for agglomeration and separation, *Proceedings Institute Briquetting and Agglomeration (USA)*, **II**, 61, 1969.
51. E. Verschuur and G. R. Davis. The Shell Pelletizing separator: key to a novel process for dewatering and de-ashing slurries of coal fines, 7th International Coal Preparation Congress, Sydney, Australia, May, 1976, Paper H1.
52. W. Blankmeister. Optimized dewatering below 10 millimetres, 7th International Coal Preparation Congress, Sydney, Australia, May 1976, Paper H4.
53. L. W. Armstrong, A. R. Swanson and S. K. Nichol. Selective agglomeration of fine-coal refuse, *BHP Technical Bulletin* **22**, 1, 37–41, 1978.
54. G. R. Rigby and T. G. Callcott. A system for the transportation, cleaning and recovery of Australian coking coals, 5th International Conference on the Hydraulic Transport of Solids in Pipes, British Hydromechanical Research Association (BHRA, Cranfield), May 1978, Hanover, W. Germany, Paper E5, pp. 65–82.

55. J. Szczypa, R. Sprycha and W. Janusz. Agglomeration of flocculated fines—a competitive method for coal slimes dewatering, *Coal Preparation* **1**, 2, 251–257, 1985.
56. I. E. Puddington and B. D. Sparks. Spherical agglomeration process, *Mineral Science Engineering*, **7**, 3, 282–287, 1975.
57. C. E. Capes, A. E. McIlhinney, R. E. McKeever and L. Messner. Application of spherical agglomeration to coal preparation, 7th International Coal Preparation Congress, Sydney, Australia, May 1976, Paper H2.
58. G. C. Sarkar, B. B. Konar and S. Sakha. Demineralization of coals by oil agglomeration technique, 7th International Coal Preparation Congress, Sydney, Australia, May 1976, Paper H3.
59. J. Szczypa, J. Neczaj-Hruzewicz, W. Janusz and R. Sprycha. New techniques for coal fines dewatering, Proceedings International Symposium Fines Particles Processing, AIME, New York, *Transactions* **2**, 1676–1686, 1980.
60. C. E. Capes, A. E. McIlhinney and A. F. Sinianni. Agglomeration from liquid suspension: research and applications, in *Agglomeration 77* (Ed. K. V. S. Sastry), AIME, New York, 1977, pp. 910–930.
61. G. R. Rigby, A. D. Thomas, C. U. Jones and D. E. Mainwaring. Slurry pipeline studies on the BHP-BPA, 39 t/h demonstration plant, 8th International Conference on the Hydraulic Transport of Solids in Pipes, British Hydromechanical Research Association (BHRA, Cranfield), August, 1982, Johannesburg, RSA.
62. D. G. Osborne. Flocculant Behavior with coal–shale slurries, *International Journal Mineral Processing* **1**, 243–260, 1974.
63. R. E. Hucko. *Beneficiation of Coal Fines by Selective Flocculation*, US Bureau of Mines, Report of Investigations No. 8234, 1977.
64. Z. Blaschke. Beneficiation of coal fines by selective flocculation, 7th International Coal Preparation Congress, Sydney, Australia, 1976, pp. 1–11.
65. R. J. Gochin, M. Lekili and H. L. Shergold. The mechanism of flocculation of coal particles by polyethylene oxide, *Coal Preparation* **2**, 19–33, 1985.
66. A. D. Read and G. T. Hollick. Selective flocculation techniques for recovery of fine particles, *Minerals Science Engineering* **8**, 202–213, 1976.
67. Y. A. Attia. Cleaning and desulphurization of coal suspensions by selective flocculation, 1st International Conference on High-sulphur Coals, Columbus, Ohio, 1985, Elsevier.
68. Y. A. Attia and P. W. Fuerstenau. Feasibility of cleaning high-sulphur coal fines by selective flocculation, 14th International Mineral Processing Congress, Toronto, Canada, 1982. Congress Proceedings pubd by Canadian Institute of Mining and Metallurgy, Sherbrooke, Quebec, Canada, 1982.

Chapter 10

SOLID–LIQUID SEPARATION

10.1 INTRODUCTION

This chapter covers all solid–liquid separation, other than simple drainage. In other words, all forms of equipment that create a significant reduction in liquid to provide a predominantly solid product are discussed. The resultant, predominantly liquid product is usually of lesser interest in coal-preparation applications. These various types of equipment can be either stationary (fixed) or moving in their mode of separation, as will become evident later in the chapter when the most common examples of the currently used equipment are described.

Solid–liquid separation can be conveniently divided into the following distinct categories:

(a) sedimentation; clarification and thickening—with relation to fine solids, normally sized below 1 mm, often in semistable suspension;
(b) filtration; screening and centrifugation—with relation to a wider size-range of solids, often sized below about 25 mm in the case of the latter two.

Sedimentation is the collective term describing the gravity separation of the fine solids, usually under quiescent conditions, resulting in the formation of a sedimentary layer of solids and a relatively clear supernatant liquid.

Clarification is the special situation where solids, in dilute suspensions (below 1–2%) are caused to settle out to produce a clear overflow.

Thickening, on the other hand, is the special situation whereby solids in more concentrated slurries are caused to settle out to produce a thickened underflow, of specific concentration, and a clear overflow. The extent to which each is taken depends upon the relative stringency in quality required for each of the two phases.

Screening, specifically to dewater a slurry, is widely practised in coal preparation. Either the solids are truly captured by arrestment by a fine-screen surface—almost a draining process, or else, and more commonly, they are caused to form a cake which entrains the finer solids that would not normally be arrested by the apertures of the screen surface.

Most screens are moving (mechanical) types, often employing high vibrational forces, but fixed, flat inclined sieves and sieve bends are also common.

Filtration is the entrapment of fine solids into a cake against a porous filtering media, resulting from the action of some form of prime force. This force may be other than or in addition to gravity. Vacuum and pressure filters are both common in coal preparation.

Centrifugation is the collection of solids resulting from the action of centrifugal force. Because of the high forces involved, centrifugal dewatering is usually the most effective mechanical method. It is also often the most costly, and for this reason other forms may be preferred despite the lesser degree of dewatering obtained. Centrifugal dewatering machines are available which can collectively dewater a wide range of sizes, normally of 25 mm upper limit. These include vibrating basket or basket types; pusher and peeler types; solid bowls and screen bowls and several others. Chemical methods of dewatering and solid–liquid separation include oil agglomeration and flotation methods (as described in Chapter 9). The chemical agents employed for such methods are discussed later in Chapter 12 which deals in detail with solid–liquid separation, in particular in terms of thickening and filtration, where chemical agents are almost invariably responsible for ensuring highly effective and consistent solid–liquid separation.

A special application area is in tailings or effluent slurry disposal. In coal preparation, where economic considerations are usually very significant, the method selected for tailings disposal can often have critical impact on the profitability of a mine. Environmental considerations and the restrictions imposed by legislation are factors which are becoming increasingly felt in financial terms. Environmental aspects are fully dealt with in Chapter 19 but the approach to selection of the most effective combination of solid–liquid separation equipment is one which is dealt with in this chapter.

10.2 PRINCIPLES OF SEDIMENTATION

The gravity-settling behaviour of discrete particles through a homogeneous fluid, based essentially on modified versions of the original Stokes' law, has featured in the studies of many different solid–liquid systems. The earliest work tended to simplify matters by considering the settling behaviour of uniformly shaped solids, usually spheres of differing diameters, for a wide range of concentrations. From these humble beginnings has evolved a reasonably sound knowledge of the sedimentation characteristics of many particulate types and systems, including coal and its associated rocks and minerals. It is, nevertheless, still true to say that much of the design data which are employed in designing present-day thickening and clarifying equipment have been derived empirically.[1]

However, it is possible to define initially in mathematical terms fundamental relationships governing the main parameters involved in gravity settling, and later tying them together into a specific relationship by the introduction of empirically derived constants.

Coe and Clevenger[2] suggested two regimes of sedimentation. In one, particles are assumed to settle without any mechanical support from the particles beneath, their mass being borne solely by interaction with fluid. They called this behaviour 'free settling' and assumed that under such conditions settling rate, u, will vary only with the solids concentration, c.

In their other regime, the individual particles are sufficiently close to each other to make frequent contact, and at least some supporting force is transmitted through interparticle contact. Such behaviour results in each discrete particle

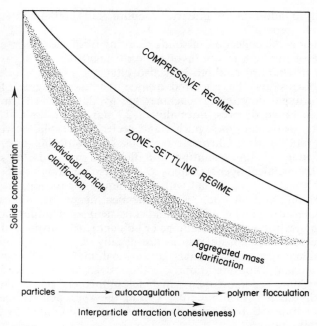

Fig. 10.1 Solids concentration versus interparticle attraction (after Fitch, Ref. 3).

exerting a collective compressive effect on those beneath, and hence the regime is named compression. As long as the particles are subsiding, some of their mass is borne hydrodynamically, but when subsidence is complete, this component disappears altogether. Coe and Clevenger recognized that the simplistic behaviour of uniform and discrete particles went only a little way towards explaining the settling characteristics of coagulating or flocculating minute rock and mineral fragments in a complex suspension of the type usually encountered in mineral treatment.

They recognized that these fragments may settle in any one of several different manners, governed by their concentration and the relative tendency for them to cohere.

Four different types are suggested by Fitch,[3] as shown in Fig. 10.1. This diagram shows how the increasing solids concentration of a suspension exploits the relative tendency of the various solid particles to cohere.

Clarification occurs when the solids concentration is very low and the particles are therefore far apart from each other and free to settle individually.[4] A limited number of collisions is inevitable because of the difference in particle size, and therefore some coagulation or flocculation will occur. Owing to this, there are two settling modes—particulate and aggregate. Clarification behaviour is easily recognizable in a settling test because slower settling masses string out behind faster ones and the supernatant fluid does not materialize as a clear phase. At increasing solids concentration, the individual settling masses crowd more closely together and thereby become increasingly influenced by one another. Those masses which have any tendency to cohere, tend, as a result of this close contact, to link to form an almost plastic structure and the combined solids mass usually settles as one forming a clear interface between the pulp and the supernatant. This is called the 'zone-settling regime' and free-settling conditions prevail. When settling tests are performed in clear glass cylinders this phenomenon can be clearly distinguished from clarification because the interface moves downwards as opposed to upwards in the case of clarification. Understandably, the boundary between pulp and supernatant is more dilute than farther below, but as settling proceeds the structure becomes firmer and even the larger masses are supported. Such an increase in concentration will continue until the pulp structure becomes so firm that it develops compressive strength. Each hypothetical layer is able to provide mechanical support for all overlying layers, and subsidence is retarded by this support as propagated from the base of the containing vessel. This solids stress occurs in what is called the compression regime. As Fitch points out, all regimes are rarely present simultaneously in a thickening unit but it is useful to assume so when developing a hypothetical thickener model of the type shown in Fig. 10.2.[5]

The feed slurry, being denser than the clarified water in the upper zone of the model, spreads out as a feed layer. Initially it is contained by a feed well, but later it is free to disperse over the full area of the vessel. Simultaneously, liquid destined to join the clarified zone is displaced upwards by the settling particulate masses. In doing so, there is an attendant tendency for very fine particles to accompany the liberated fluid and move upward to the overflow. There is still the possibility that they will flocculate with other particles and rejoin the downward flow.

In the extreme circumstance that the vessel is being fed with excessive solids,

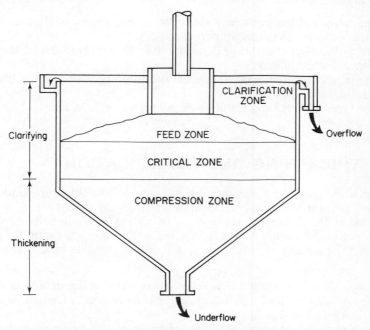

Fig. 10.2 Zones in a continuous thickener (after Fitch, Ref. 5).

such that it cannot thicken and discharge them at the required rate, a critical zone will gradually develop. This will continue to grow until the appropriate rate of feed solids is restored. In practical terms, the ability of a thickener vessel to avoid the formation of a significant critical zone becomes a function of both the design capacity and maintaining a steady feed close to specified rate and solids concentration.

Once zone settling has become established in the lower reaches of the feed zone, a compression zone occurs almost simultaneously, and in descending towards the underflow, the solid mass becomes more and more compacted or thickened.

In considering these various zones it must be clearly borne in mind that this model is a dynamic one, and that at all times during normal operation there is a balanced flow of solids- and liquid-rich phases. In some cases, for highly flocculated pulps for example, this balance can prove to be a delicate one, and the formation of the critical zone becomes an ever-present threat. Alternatively, the solids-handling capacity of the thickener may not be catered for by the amount of feed solids presented to the vessel and an increasingly dilute underflow will be the result.

Each zone, for a given type of feed, will demand some pool area. The effective area required will therefore be controlled and determined by the largest of these demands. Some zones will demand depth also and the amount provided must therefore be equal to the sum of all individual amounts for each zone. In commercial thickeners short-circuiting, turbulence, unequal flocculant distribution, etc. will call for further allowances to be made.

In summary, then, the most important factors affecting the rate and mode of settling are as follows:

(a) the nature of the particles—size distribution, shape, relative density, mineralogical and chemical properties, etc.;
(b) the proportion of solids to liquid forming the suspension, and the resultant concentration effects;
(c) the type of pretreatment—chemical conditioning, flocculation, heating and cooling, etc.;
(d) the type of containing vessel—size, shape, wall effects, etc.

10.3 THICKENING AND CLARIFICATION

In order to design a thickener we must know what is required to handle a specified flow, prevent unwanted solids from passing into the overflow, and deliver underflow at the desired concentration. In simple terms it boils down to two simple questions: how much area is required, and how much depth?

As we have already discussed, there are several possible modes of sedimentation:

• *Clarification* in which solids settle either individually or are collected into separate floccules, each of which then settles at its own characteristic settling rate, closely relating to Stokes' law;
• *Zone settling* in which particles cohere into a structure such that all in a

SOLID–LIQUID SEPARATION

given neighbourhood subside at the same rate, but the structure does not lend mechanical support;
- *Compression* in which the structure is capable of mechanical support.

10.3.1 Clarification

Clarification zone requirements are commonly expressed in terms of overflow rate, v_o, and t, detention time. They relate to pool area and depth so that:

$$v_o = \frac{Q}{A}$$

$$t = \frac{V}{Q} = \frac{Ah}{Q} = \frac{h}{v_o}$$

where A is the settling area of the clarifier in m², h is the depth of the clarifier in m, V is the clarification-zone volume in m³, and Q is the volume of overflow per unit time in m³/h. The pool area, A, is either the horizontal plane area or the area projected onto a horizontal plane where inclined-plate (e.g. lamella type) clarifiers are considered.

10.3.2 Thickening

In the volume beneath the feed zone of a continuous thickener, the slurry is in the process of thickening. This thickening process embraces any critical-zone settling which may occur and compression. In the combined thickening zone, therefore, flows are considered to be acting vertically downwards and uniformly distributed in the lateral plane. It is therefore usual practice for design purposes to consider all flows as being one-dimensional and in steady state.

In the original Coe and Clevenger treatise the downwards velocity of solids at any point is the sum of two components:

(a) settling velocity, v, of the solids with respect to the pulp as a whole;
(b) a transport velocity, v_u, of the pulp as a whole resulting from underflow removal.

The resultant mass rate of solids flow per unit of area, G, or solids flux as it is usually called, is:

$$G = C(v + v_u)$$

and

$$v_u = \frac{Q_u}{A}$$

where Q_u is volume flow rate of underflow in m³/h, C is solids concentration at one point in kg/m³, A is area of the thickener vessel in m², and v is settling velocity of solids in m/h.

Therefore

$$G = C\left(v + \frac{Q_u}{A}\right)$$

By eliminating Q_u/A, the classical Coe and Clevenger equation is obtained:

$$G = \frac{v}{\dfrac{1}{C} - \dfrac{1}{C_u}} \text{ t/h/m}^2$$

where C_u is underflow concentration in kg/m³.

For many years this equation provided the basis for determining thickener areas required for a given slurry. By carrying out numerous jar-settling tests at differing slurry concentrations, values of u and C could be introduced into the equation covering the entire range of concentrations, from that of the dilute feed, C_f, to that of the eventual desired underflow, C_u. From the results so obtained, thickener area was based upon the maximum value obtained. Usually a factor of safety was added to this. Empirical data eventually showed that very often underestimation of area requirement occurred, resulting from overestimation of solids flux.

Coe and Clevenger assumed that in the zone-settling regime (see Fig. 10.2) the settling rate, v, would be a function of concentration only. If this was true, it would be expected to have the same value in a batch test as in a full-size thickener. When values of v and C are derived from several batch tests, and values of G are calculated for each pair of values at any fixed value of C_u, the resultant graph of G against C would look like that given in Fig. 10.3.

This graph highlights an important empirical fact. There is a maximum for G at some concentration, C_a, and a minimum, G_c, at some higher concentration, C_c. Thickener flux is limited to the lowest value of G observed between the feed concentration, C_f, and that of the underflow, C_u. If, as in most practical cases, the flux obtainable at the feed concentration exceeds that of G_c and the under-

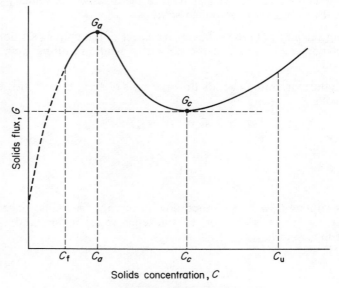

Fig. 10.3 Solids flux versus concentration.

SOLID–LIQUID SEPARATION

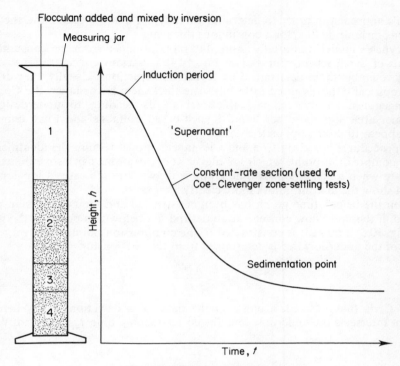

Fig. 10.4 Batch-settling test.

flow concentration, C_u, exceeds C_c, the maximum flux that can pass to the underflow will be G_c, i.e. that corresponding to C_c. At higher solids loadings than this, a critical zone of concentration, C_c, will form in the thickener. When this occurs, solids, being unable to pass through as fast as they are supplied, add to the zone causing it to fill up the thickener and overflow. For this reason zone-settling imposes an area demand on thickener design.

The converse is also true when feed-solids flux decreases below the same critical level and the mass balance is again lost. However, since the zone now dwindles and eventually disappears altogether, no provision of depth in thickener design is necessary to facilitate zone-settling. The critical concentration is the only zone-settling one that can cause solids accumulation in the thickener and as such will exist only if solids feed equals or exceeds the critical zone-settling capacity.

In a batch test, settling commences with a uniform initial concentration of solids (see Fig. 10.4). The concentration in the compression zone (3) must range between that of the original concentration in zone (2) and that of the final condition in zone (4). If the solids handling capacity per unit of area is lowest at some intermediate concentration, a zone of such concentration must begin forming, due to the fact that the rate at which solids enter this zone will be lower than that at which they leave. The mathematician, Kynch[6] showed how to determine the concentrations and fluxes in these zones by geometrical constructions on the transitional sections of a single-batch settling curve. Talmage and Fitch[7] used

Kynch's approach in order to determine directly, for a single-batch test, the critical zone, or limiting flux, for a continuous thickener.

If Kynch's model was totally valid, flux plots obtained from the constant-rate sections of batch-settling curves (see Fig. 10.4) at various concentrations should closely resemble those determined by Kynch constructions. Usually, they do not, and empirical data obtained from full-scale thickeners have shown the Coe and Clevenger test to overestimate thickener fluxes, leading to underdesign of thickener area. On the other hand, Kynch-based Talmage and Fitch constructions appear to do the opposite.

All procedures based on Coe and Clevenger's model require the identification of the compression point, which for higher concentrations can prove difficult to ascertain when a height-versus-time graph is drawn. In such cases a log–log plot should show the discontinuities more clearly.

When the height–time graph has been constructed and the compression point located, if the underflow concentration desired lies below this point on the graph (see Fig. 10.5) a tangent is constructed to the compression point. The level on the graph of the underflow line is determined from the relationship

$$h_u = \frac{h_0 C_0}{C_u}$$

where C_u is that selected, usually on the basis of a detection test. Where the tangent intersects the underflow line, the corresponding time, t_{u_1}, is noted. Where

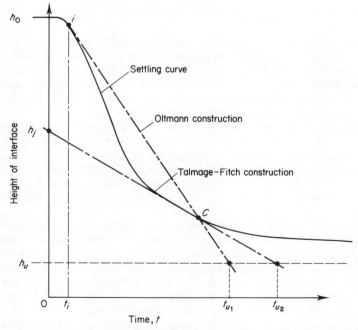

Fig. 10.5. Batch-settling test with Talmage–Fitch and Oltmann constructions.

the settling curve has a distinct induction period, the induction time estimated from the graph is deducted from t_{u_1} to give the adjusted settling time, t_u, i.e. t_{u_1} minus t_i equals t_u.

Corresponding thickener flux is then obtained from

$$G_c = \frac{C_0 h_0}{F t_{u_1}}$$

where C_0 is the solids concentration of the initial pulp; h_0 is the height; F is an empirical constant, usually a 20% allowance, i.e. f becomes 1.2; G_c is usually expressed in $t/m^2/h$ of solids loading; and $1/G_c$ is therefore the critical thickener unit area in $m^2/t/h$.

As mentioned earlier, the approach suggested by Talmage and Fitch can, on occasion, lead to overdesign of a thickener, and an alternative to this graphical construction, developed by one of Fitch's colleagues, Oltmann, has been found to result in a more accurate estimation of thickener area. This has apparently been particularly applicable to coal and coal-tailing slurries.[5]

For this method, the form of construction shown in Fig. 10.5 is used. If there is an induction period, the linear part of the curve is extended back to its intercept, as shown, to point i. When points i and c, the compression points, are connected to intercept with the underflow line, u is formed as shown. The adjusted settling time then becomes t_{u_2} minus t_i equals t_{u_1} or the time to achieve the true underflow concentration. Likewise h_0 is adjusted to h_{0_1} to compensate similarly. Corresponding critical thickener flux, G_c, by this method becomes

$$G_c = \frac{C_0 h_{0_1}}{F t_{u_1}} \; t/h/m^2$$

where F again is usually taken as 1.2, and unit area $= 1/G_c$ as before.

Despite the fact that it is desirable to design a thickener from first principles, so to speak, it is often the case that thickeners are designed on the basis of empirical data accumulated from other, previous applications with similar material.

The most common approach is to consider the calculated solid and liquid flow rates in terms of design loadings, which naturally enough have built-in safety factors. As will be explained later, three categories of thickener exist to which this empirical approach may be applied:

- Conventional thickeners, which do not rely on polymer flocculants for enhanced rates of settlement. Large residence times are therefore a characteristic of this type of thickener.
- High-rate thickeners, for which such flocculants are applied in moderation, but large storage capacity is still a feature of the design.
- High-capacity thickeners, which are totally dependent upon polymer flocculants in order to function efficiently.

The common loadings used to obtain basic thickener sizing are as follows:

	Solids: Design Loading ($m^2/t/h$)	Hydraulic: Design Loading ($m^3 h/m^2$)
1. Conventional	10	0.5
2. High-rate	5	1.0
3. High-capacity	1	4.0

10.4 THICKENING AND CLARIFICATION EQUIPMENT

Circular thickeners are usually regarded as the normal form used in coal-preparation applications, but a number of rectangular basins have been used and occasionally a lamella type may also be selected.

10.4.1 Conventional Thickeners

The conventional circular thickener comprises a circular tank, into which is fitted a mechanism consisting of a drive head, shaft, or cage and rotating arms. The last has the dual purpose of conveying sedimented solids to a central point of discharge and creating a channeling effect which releases further water from the sediment, thereby creating a dense but manageable underflow slurry.[8]

While the design of this mechanism is of the utmost importance, the cost of the thickener tank often greatly exceeds that of the mechanism. For this reason, careful consideration must be given to the overall design and configuration in order to minimize costs. The tank, which can represent more than half of the total cost of the thickener, may be constructed of steel, concrete or both. Steel tanks usually prove to be most economical for thickeners up to 25 m in diameter, but larger ones than this are nevertheless still quite common. The bottom of the tank can be flat while the mechanism arms are sloped towards the central discharge area. With this design, settled solids bed in to form a sloping false floor. Some formed-steel floors are also encountered but the extra cost is usually significant. Concrete sides and bases become more commonly used in the larger sizes (50–100 m) but combinations of concrete floors with steel sides or concrete sides and sand-filled bases become more evident in the largest-sized units (>100 m). The very largest of all (>300 m), unusual in coal applications, are essentially excavated basins lined with an impermeable membrane, while the sides are simply developed spoil berms through which overflow pipes are fitted.[1]

The method of supporting the mechanism depends on the tank diameter. In small sizes, the drive head is supported overhead on a superstructure or bridge spanning the tank with the rake arms being fitted to the central drive shaft. Above approximately 25 m, a stationary centre column of either steel or concrete is used to support the drive head and mechanism. In this type, the rake arms are attached to a rotating drive cage which is in turn bolted to the main drive gear. Both of these types are shown in Fig. 10.6 and these represent the most common types of coal-preparation plant thickeners.

A variation of the centre-column type is the caisson type, which is constructed by enlarging the central pier in order to install the underflow discharge pumps. This very substantial form of construction is only employed for the largest-diameter types of thickener, often in excess of 200 m, and the thickener vessel is usually an earthen basin as described earlier.

Feed-well designs have altered very little in conventional thickener designs although one or two new systems have been successfully introduced which provide for high utilization of polymeric flocculants and therefore ensure higher handling capacities.

The rake arm construction must be adequately strong to withstand torques well in excess of normal loading conditions. Many factors influence the design,

SOLID–LIQUID SEPARATION

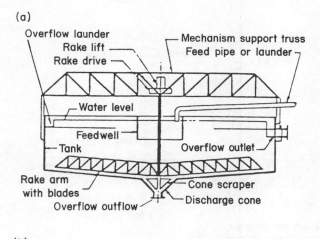

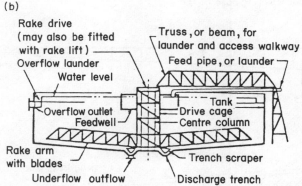

Fig. 10.6 (a) Bridge- and (b) column-support types of thickener.

especially those relating to the properties of the underflow sediment. There are many different types of construction, varying from a single rigid member with blades bolted directly on (up to 15 m maximum diameter) to many truss types of design for the larger tanks. It is often advisable to eliminate large structural members in motion close to the settled sediment, especially in the larger units. In such cases a form of extension arm can be used. These are often called 'thixo' arms since they were developed for and are now almost always used with sediments having thixotropic tendencies. Pipe construction of the arms can further improve flow conditions and for this reason some fairly large-diameter units (100–200 m) are fabricated using round steel piping. Figure 10.7 shows examples of raking arm construction.

In most conventional thickeners, the rake arms are connected to the centre shaft or cage, but there are other alternatives to the rigid form of connection which are in common usage for specific forms of application. One of these alternatives is the cable support type. This consists of a hinged rake arm (see Fig. 10.8) which is fastened to the base of the centre column. The hinge is designed to provide simultaneous vertical and horizontal movements of the rake arm, which

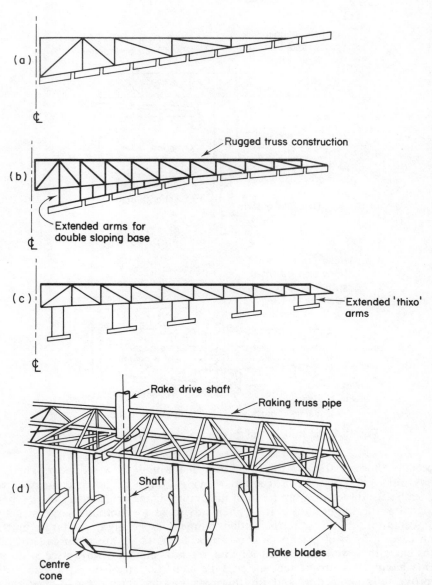

Fig. 10.7 Alternative types of thickener rake-arm design in common use: (a) conventional, centre supported; (b) large-diameter, sloped-base thickener; (c) large-diameter thickener, thixotropic material; (d) Dorr–Oliver pipe construction truss.

is hauled around the column by means of cables connected to a torque arm. This torque arm is rigidly connected to the centre column below the overflow levels. This type of rake is designed to lift automatically when the torque, developed as the rake moves through the sediment, counterbalances the mass of the rake. As

SOLID–LIQUID SEPARATION

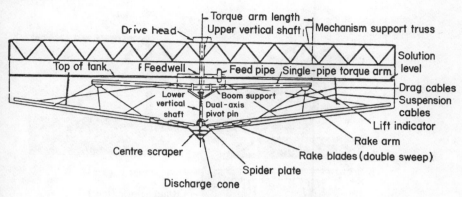

Fig. 10.8 Cable support-type thickener.

torque demand is reduced, the rake moves downwards while at the same time continuing to transport sediment to the centre. This type of thickener is commonly used for thixotropic or pseudoplastic sediments and is ideally suited to large thickeners handling coal-flotation tailings rich in bentonitic clays.

Another alternative is the traction thickener. This type comprises a centre column which is only partly employed to support the rakes. Movement of the rakes is provided by a drive mechanism or trolley, which is connected to a single long arm extending from the centre of the thickener to the tank periphery (see Fig. 10.9). The drive trolley runs along a peripheral rail providing the necessary traction for the raking system which usually comprises two or more shorter rake arms in addition to the main driving arm. Traction thickeners are available from 50 to 150 m diameter and are easily identifiable by the peripheral rail. They suffer the disadvantage of being unable to respond (i.e. rakes cannot be raised) to the sudden surges in solid loading commonly encountered in coal plants and probably for this reason they are rarely used in coal treatment.

Most conventional thickeners are installed with some type of lifting device for the raking arms. The fine clayey sediment involved tends to retard flow to the withdrawal point causing the tendency for ring or 'donut' formation to occur, especially with thixotropic sediments. In order to overcome any difficulties of this type or to counteract surges in solids loading, it is essential that some form of lifting device be provided. Manual or motorized lifting devices with automatic or

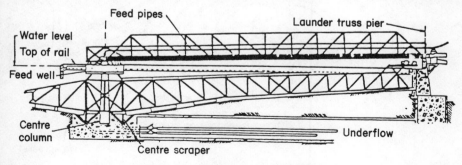

Fig. 10.9 Traction-type thickener.

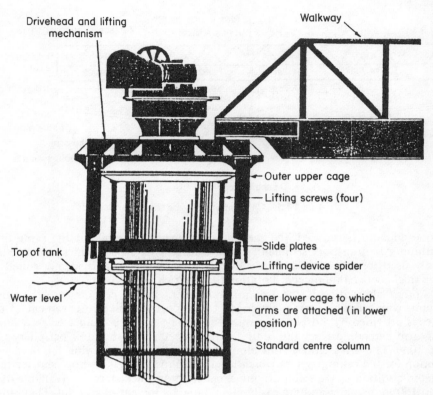

Fig. 10.10 Platform-type lifting device as used on centre-column-supported thickeners (courtesy Eimco Equipment).

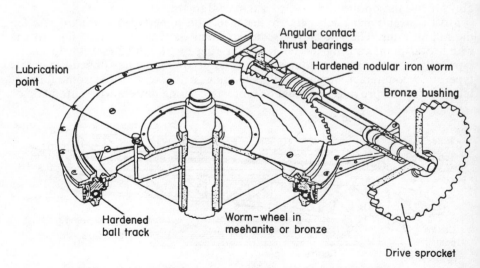

Fig. 10.11 Single-worm–spur-gear drive (courtesy of Dorr–Oliver).

SOLID–LIQUID SEPARATION

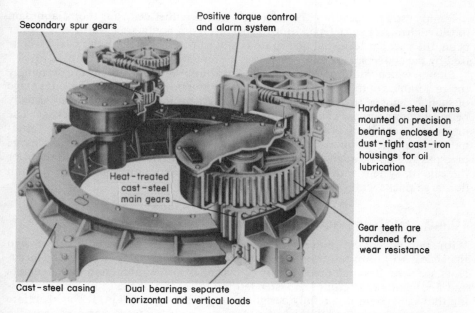

Fig. 10.12 Double-worm and spur-gear arrangement (courtesy of Eimco Equipment).

semi-automatic controls are available which raise or lower the rakes without affecting the normal rotation of the rake drive. Figure 10.10 shows a typical system.

Bridge type thickeners are usually driven by a single-worm–spur-gear combination as shown in Fig. 10.11. Larger, centre-column units up to about 50 m in diameter usually have drive heads with a double-worm and spur-gear arrangement, sometimes manufactured with secondary gearing as shown in Fig. 10.12. Central pier thickeners are normally fitted with drives which are so designed as to provide drivage via a worm gear set to a large-diameter ring gear through a pinion wheel. The ring gear provides a locus to which the drive is connected. As thickener diameters become larger, and especially for caisson types of centre column, this gearing becomes more complex and the load-carrying capability and rotational stability, combined with the very large torque loads involved, become limiting factors in design. The most powerful drives are those fitted with hydrostatic bearing systems which ensure the necessary protection against metal wear and fatigue. These bearing systems have been installed with diameters of up to 8 m. They comprise two main parts: the runner and the thrust pads. An oil film capable of supporting the load is maintained between the runner and each pad. Oil is continuously replenished under pressure by a controlled hydrostatic pumping system, thus ensuring very high thrust and radial load-carrying capacity.

Commonly used accessories for conventional thickeners include overload-protecting devices, torque-indication devices of various types, level-sensing devices and flocculant- and other chemical agent-mixing, and dosing- and performance-monitoring systems.

Pumps used to withdraw the underflow can be either centrifugal or diaphragm (positive-displacement) types. Both types are commonly used, with the former being the cheaper to install and more costly to run whilst the latter is characterized by the converse.

Pumps which withdraw the underflow through a tunnel system are seemingly the preferable route but also the most costly to construct. The major advantage of such systems is that, especially when diaphragm pumps are used, they usually permit high underflow concentration to be handled without fear of blockage. The main disadvantage is obviously the cost of constructing the tunnel. Other alternatives are: the submersible pump systems comprising a removable discharge pipe passing up through the centre column, connected to a centrifugal pump; or the caisson system with the centrifugal pumps located within the centre column (for large thickeners).

10.4.2 High-capacity Thickeners

With the advent of polymer flocculants came the opportunity to design thickeners to utilize fully the higher settling rates obtainable when such flocculants are used. The combined ability of flocculants or combination of coagulants and flocculants has resulted in faster settling rates, and at the same time, providing that the system is carefully designed, very clear supernatant. It was therefore inevitable that traditional thickener designs would undergo some review, the outcome of which has been that several new high-rate or high-capacity thickeners have been introduced to coal treatment. Varying degrees of success have been reported, but generally the factors limiting effectiveness have been the following:[9]

- low residence time of solids passing through the thickener;
- characteristic inconsistency in feed-solids concentration resulting in surges which seriously affect efficiency;
- high flocculant consumption and cost;
- poor mixing and dispersion of flocculant;
- chemical changes in the composition of the feed water, causing severe fluctuation in the extent of flocculation achieved;
- ineffectiveness of thickener-control equipment to monitor conditions and respond appropriately quickly—exemplified by level-control devices.

Nevertheless, there are now numerous effective installations of several different types in the coal industries of the world.

High-rate thickeners are characterized as being modified conventional thickeners which usually have larger and deeper feed wells. Flocculated feed solids are therefore introduced almost directly into the thickening pulp, and the peripheral volume between feed well and overflow launder becomes almost a reservoir for clarified water. Such designs provide for solids-handling capacities lying between conventional and the so-called high-capacity types, thereby deriving some of the advantages and disadvantages of both. High-rate thickeners are becoming increasingly common in coal-preparation usage.[10]

High-capacity thickeners are primarily designed to utilize fully the obvious benefits of high-molecular-weight polymer flocculants. Those resembling the familiar thickener design all require the incoming, highly flocculated slurry to be directed beneath a fluidized sludge blanket. This blanket is maintained in dynamic suspension by the counterbalanced downflow of flocculated solids and upflow of displaced water. This phenomenon results in final entrainment of loose flocs or fine solids and ensures the peak effectiveness of the total thickener performance. Because of the delicate balance involved in maintaining this very essential condition it becomes easy to understand how the foregoing factors can limit thickener performance.

High-capacity thickeners of this type are the deep cone type, the so-called Enviro-Clear thickener and the Eimco Hi-Capacity thickener. Only one high-capacity thickener/clarifier can be regarded as a totally different concept, this being the lamella type.

The Deep Cone Thickener

The use of thickening tanks with a deep, steep-sided conical profile is by no means a new approach. The first designs, which have led to the currently used versions, first appeared in the 1960s when synthetic polymer flocculants were becoming widely used. The current most commonly used deep cone is probably that developed by British Coal,[11] which is principally used to treat froth-flotation tailings for disposal without the necessity for further dewatering. The shell of this type of cone comprises an upper cylindrical launder section located above the

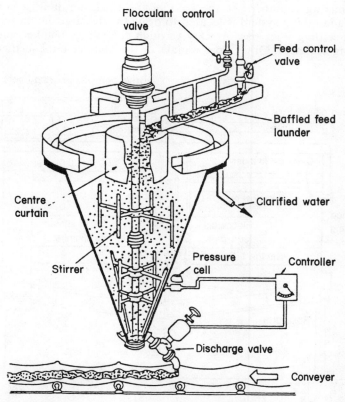

Fig. 10.13 British Coal Corporation standard four-metre-deep cone thickener.

main conical section (see Fig. 10.13). The stirrer mechanism is composed of three parts: the lower section of the shaft to which are attached the paddles; the mid-section main-drive shaft; and the coupling to the gear box which provides a stirring speed of 2 r.p.m. The discharge mechanism consists of a pneumatically operated 150-mm Saunders valve, which automatically receives a signal to open, from a pressure transducer (load cell) located in the wall of the cone apex. Tailings are either directly fed to the cone from a flocculant-mixing tank, or via a feed head-box to the cone from a flocculant-mixing tank, or via a feed head-box to the cone feed-launder.

Cones of this design range from 3 to 3.7 m in diameter, are 4.3 m deep and are capable of handling 70 m^3/h of feed at approximately 5% solids by weight. A final underflow discharge of 65–75% solids by weight is usually attainable.

An additional stage in the predisposal treatment has been the introduction of powdered cement mixture to solidify further the underflow. This has been shown to produce a more stabilized product for subsequent waste-heap disposal.

The Eimco Hi-Capacity Thickener[10]

This unit is shown in Fig. 10.14. The feed enters through a hollow drive-shaft into which further addition of flocculant solution can, if required, be introduced. The feed then passes into a formed sludge-blanket located adjacent to the discharge from the mixing chamber. At the base of the thickener tank a conventional inclined-blade raking system is located to draw the dense sediment to the centre cone. Fast-settling flocs pass quickly through the sludge-blanket and into the compressing sediment. Other flocs remain in the blanket until further cohesion

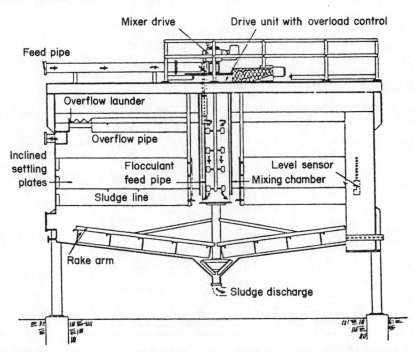

Fig. 10.14 The Eimco Hi-capacity thickener (courtesy Eimco Equipment Company).

SOLID–LIQUID SEPARATION

occurs. Clarified water accumulates above the sludge-blanket in moderately quiescent, streamline-flow conditions, before passing to the overflow launder. The thickness of the sludge-blanket is automatically controlled by means of a level sensor. This type of thickener is regarded as being specially suitable for inclusion in existing plants where space is restricted. Its manufacturers produce the entire range of commonly used coal-treatment thickeners and, therefore, their recommendations concerning selection of thickener type is of interest in view of this. The Eimco Hi-Capacity thickener has been installed in sizes ranging from 5 m to

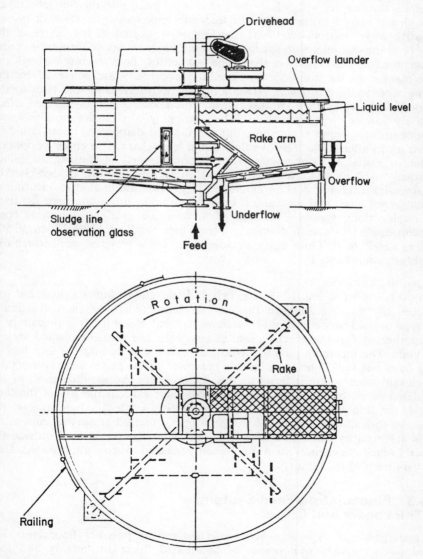

Fig. 10.15 The Enviro-Clear clarifier/thickener (courtesy Enviro-Clear Corporation).

15 m, and current coal-plant applications cover all types of slurry. Use of a de-aeration tank prior to the thickener is recommended to eliminate air bubbles which can disturb the sludge-blanket formation.

The Enviro-Clear Thickener

One version of this unit is shown in Fig. 10.15. Several slightly different configurations do, however, exist.[13] The feed slurry is introduced into the central inverted feed cone, which forces the incoming feed into an initial horizontal direction. The gap between the baffle plate and the end of the feed pipe therefore determines the velocity with which the feed is introduced into the sludge-blanket. Although not shown in the figure, the feed inlet pipe can enter the unit in two alternative ways: bottom inlet with the feed pipe located in the centre of the sludge boot; or side inlet with the feed pipe entering from above through a centre well surrounding the rake-drive shaft. The fast-settling flocs migrate through the sludge-blanket to the compression zone and eventually are conveyed by the rotating rakes to the centre well surrounding the rake-drive shaft and housing the mud boot from which they are discharged. The discharge can be effected either by gravity flow or by pump suction. Where pumping is selected a variable-speed driven centrifugal pump is common, but a two-speed diaphragm pump could be selected as an alternative. The overflow liquid is discharged to either peripheral launder or radial overflow weirs. The cylindrical shell of the tank contains a vertical sight-glass to facilitate visual inspection of the sludge-blanket. Ultrasonic-level sensing is often employed to provide automated monitoring of the sludge-bed level, and mechanical or hydraulic rake-drives are used for fixed or variable raking speeds. This type of thickener is available for most coal-treatment applications with diameters from 3.8 m handling 170 m^3/h to 36.5 m handling 15 000 m^3/h. Once again a de-aeration tank is recommended to remove air bubbles from the feed.

The Lamella Thickener

This unit is shown in Fig. 10.16. The originally developed lamella thickener was produced by the Axel Johnson Institute for Industrial Research in Sweden.[14] This type of thickener comprises a series of inclined plates in close proximity to one another so that the effective area becomes the horizontal projected area of each plate. The incoming pulp is introduced either directly into the feed box or into a flash-mix tank. The feed slurry is presented to the plates via a rectangular feed curtain from where it flows to the plates, where the sedimentation of the flocculated solids occurs. The rising clarified water exits at the top of the tank through the flow distribution orifices. The settling solids gravitate against the plates and slide downwards to the sludge hopper, in which additional concentration is achieved by the action of a low-amplitude vibrator located outside the hopper. Lamella thickeners are not commonly selected for coal applications. Unit sizes vary from 45 to 200 m^2.

10.4.3 Flocculation Feed Systems for Thickeners and Cones

The upsurgence of expensive high-molecular-weight polymer flocculants has caused a concurrent development of automated flocculant mixing and feed equipment. These types of flocculants require a definite period of time to go into

SOLID–LIQUID SEPARATION

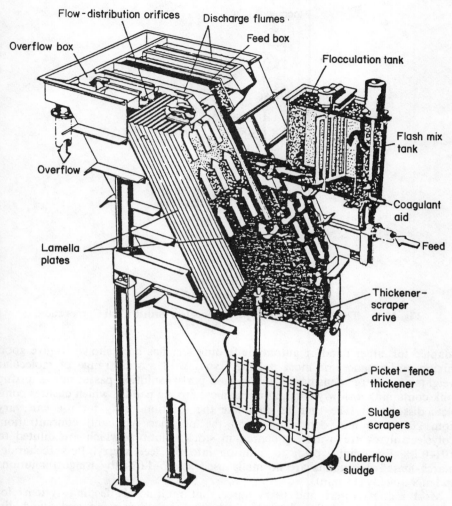

Fig. 10.16 The Axel Johnson 'lamella' thickener.

solution, because the water must hydrate the long molecules before they can uncoil to form a uniform gel-like solution. The time required for dissolution can be appreciably reduced if the dry polymer granules are properly dispersed in the water solute. This is achieved by means of inductors or some form of cyclonic device which both combine and agitate the solids and liquids continuously. The old systems involving gravity-feeding powder or granules into a flowing stream are now regarded as wasteful and extremely vulnerable to the formation of glutinous gels. British Coal's Mining Research and Development Establishment (MRDE) pioneered the development of the system which now forms the basis of many used throughout the international coal industry.[15] The Bretby Autex unit, shown in Fig. 10.17, is a cyclonic flocculant granule mixer. This dispenser can be

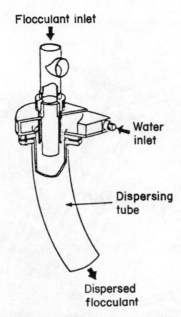

Fig. 10.17 The Bretby Autex disperser (courtesy British Coal Corporation).

adapted for either hand or automatic feeding and has been shown to give good mixing performance for most polymer types with a wide range of molecular weights and charge densities. The dispersed partial solution passes into a mixing tank containing a slowly rotating mechanical mixing paddle, which ensures complete dissolution. The time to complete the solution ready for use can vary from 20 to 60 min depending upon the polymer type and concentration. Polyelectrolytes are usually prepared in stock solutions which are diluted to 0.01–0.1% (by weight) prior to addition into the feed slurry. Polysaccharide-(starch-)based flocculants are normally used as 0.5–1.0% (by weight) solutions and mix quickly (15 min).

Most manufacturers and users agree that multinozzle feeding systems for adding flocculant solution into the feed give good dispersion and avoid disruption of the slurry likely to cause floc breakdown. Nevertheless, most treatment plants still employ inadequate methods for both mixing and adding flocculants.[16] A surprising number still employ hand mixing of powder into a flowing stream of water to create the polymer solution, and then single-point addition of the solution into the slurry. The first problem has been overcome to some extent by the manufacture of concentrated polymer solutions which only require to be diluted for use.

Fully automatic mixing and feeding systems have been developed and their cost-effectiveness can usually be clearly demonstrated. The danger with such systems is that some operators will ignore the ongoing attention and adjustment which is essential in order for the cost-effectiveness to be fully realized. One such unit is shown in Fig. 10.18. Like the Autex unit that it incorporates, the Bretby automatic flocculant-mixer was developed by the MRDE. This mixer is in turn

SOLID–LIQUID SEPARATION

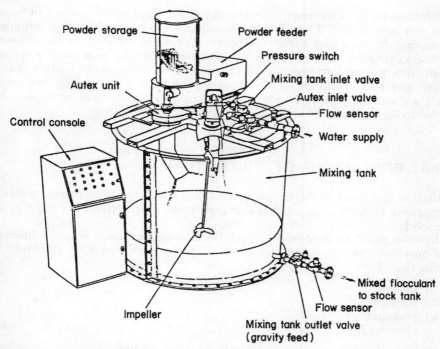

Fig. 10.18 The Bretby Mark II automatic flocculant mixer (courtesy British Coal Corporation).

incorporated into a complete flocculation-control system, as shown in Fig. 10.19. The main controlling feature of this system is the clarometer which receives a sample of the thickener feed and automatically tests the settling rate. The value

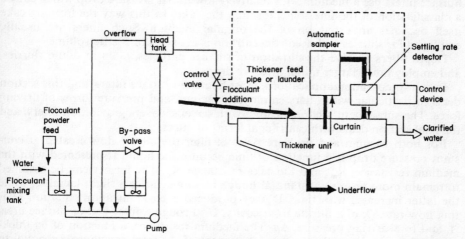

Fig. 10.19 MRDE flocculation control system incorporating the clarometer and Bretby Mark II automatic flocculant mixer (courtesy British Coal Corporation).

determined by the clarometer is compared to a set-point which governs the reaction of a controller. If flocculant dosage requires adjustment following this measurement, the controller causes the flocculant-control valve to be repositioned accordingly. Other similar systems are available employing some form of water-clarity determining instrument to effect thickener operating control. Most can be set to give good performance if maintained properly and regularly inspected. They are especially valuable when more than a single chemical agent is to be added to the thickener feed.

10.5 PRINCIPLES OF FILTRATION

Filtration is defined as the separation of solids from liquids by passing a slurry comprising both through a permeable membrane which retains the majority of the solids.

In order to obtain fluid flow through the filtering medium, a pressure differential must exist and there are four types of driving force which are employed to create this effect:[17]

(a) gravity;
(b) vacuum;
(c) pressure;
(d) centrifugal.

There are basically two types of filtration used in practice:

1. *surface filters*—commonly called cake filters, in which the solids are deposited in the form of a cake on the upstream side of a relatively thin filter medium;
2. *depth filters*—in which particle deposition takes place inside the medium.

Surface filters use a medium with relatively low initial pressure drop which causes a classification of the fine solids forming the cake. In this way the forming cake itself becomes an extension of the original medium. Such filters are usually employed for slurries with feed concentrations in excess of 1% by volume.

Depth filters involve the filtration of smaller particles, often in dilute slurries, and employ much larger initial pressure drops.

In coal-preparation applications, most filters are surface filters, and this section deals in particular with filters employing vacuum and pressure forms of driving force. The subsequent section, dealing with screens, covers gravitation filters, and the final section deals with centrifugal forms of filters.

For both vacuum and pressure types of filter the filtrate flow created at constant pressure drop is a function of time because the liquid is presented with the medium resistance, R_m, and the cake resistance, R_c. The former can be assumed to remain more or less constant (although blinding causes a slight increase) while the latter increases with time. D'Arcy produced a basic filtration equation relating flow rate, Q, of a filtrate to viscosity, U, through a filter bed of surface area, A, and to a driving pressure, Δp. The medium resistance, a function of its thickness and construction, becomes a component of the total resistance, i.e. equal to $(R_m + R_c)$.

SOLID–LIQUID SEPARATION

D'Arcy's equation for cake filtration on a filter cloth therefore becomes:

$$Q = \frac{A\Delta p}{U(R_m + R_c)}$$

In practice, medium-resistance blinding may be catered for by including an allowance in R_m. Also, as the resistance of the cake, if it is incompressible, may be assumed to be directly proportional to the amount of cake deposited, then

$$R_c = \alpha W$$

where W is the mass of cake deposited per unit area (in kg/m^2) and α is known as the specific cake resistance (in m/kg). This is often regarded as being fairly characteristic of a certain type of material of known size-distribution.

The mass of cake deposited is a function of time in batch-filtration systems and can therefore be related to the accumulated filtrate volume, V, obtained during time, t.

$$WA = cV$$

where c is the concentration of solids in the suspension (in kg/m^3). The general filtration equation, taking all this into account, therefore becomes:

$$Q = \frac{A\Delta p}{\alpha U c \left(\dfrac{V}{A}\right) + U R_m}$$

and this equation forms the basis of most practical filtration equations used to determine filtering conditions.

Since

$$Q = \frac{dV}{dt}$$

$$\frac{dt}{dV} = \alpha U c \frac{V}{A^2 \Delta p} + \frac{U R_m}{A \Delta p}$$

Development of this formula for compressible cakes and for either constant-pressure or constant-rate filtration permits it to be used for more or less all continuous, semicontinuous or batch-type filters. For constant pressure, integration of the above equation produces a useful relationship:

$$\frac{t}{V} = a\left(\frac{\alpha U c V}{2 A^2 \Delta p}\right) + b\left(\frac{U R_m}{A \Delta p}\right)$$

which gives a straight line if t/V is plotted graphically against V (see Fig. 10.20(a)). Slope a and intercept b, used together with experimentally determined values of α and R_m, permit this equation to be used for a constant-pressure system.

For constant-rate filtration, i.e. Q is constant and Δp is varied, $V = Qt$ and

$$\Delta p = \alpha U c \frac{Q^2}{A^2} \times t = U R_m \frac{Q}{A}$$

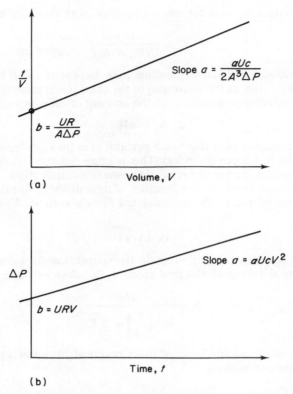

Fig. 10.20 (a) Plot of $t/V = f(V)$ for constant pressure and incompressible filter cakes. (b) Plot of $\Delta p = f(t)$ for constant rate and incompressible filter cakes.

A plot of Δp against t results in a graph similar to that obtained previously (see Fig. 10.20(b)).

10.6 VACUUM FILTRATION

10.6.1 General

There are numerous different types of vacuum filter but all can be categorized into two groups:

- batch;
- continuous.

Batch vacuum filters are rarely employed in coal applications and will not be discussed. Continuous vacuum filters are widely employed and three main forms are encountered:[18-20]

- rotary drum;
- rotary disc;
- horizontal belt or disc.

10.6.2 Rotary Vacuum Drum Filter

The drum filters used in coal applications are of the external type. They consist of a drum with connections from the periphery to an automatic filter valve. The drum rotates, partially submerged in the slurry and cake is deposited on the drum surface and filtrate is drawn into a network of piping within the drum towards a central valve located in the drum trunnion (see Fig. 10.21). The manufacturers employ different methods of providing the drainage system and securing the filter cloth. Generally, drum filters of current design are of multipanel type with cloth secured by caulking or wire winding.

The drum deck is formed by replaceable drainage grids held in position by division strips. These grids support the filter cloth and are of such a design as to provide maximum free area and therefore maximum filtering capacity. They are often made from polypropylene or similar strong, light plastic material to facilitate easy removal and, if required, replacement, while at the same time providing lower potential to become coated with scale or to corrode. The method of discharging the cake is of great importance in maintaining a clean cloth. Recent developments have included a belt-type of drum filter which involves the use of a continuous filter-cloth which, by a system of pulleys, leaves the drum surface at the discharge point carrying the filter cake with it to a discharge roll of small diameter which causes the cake to be dislodged. Facilities for washing the filter-cloth belt on both sides are usually provided following the discharge roll. Figure 10.22(a) shows a typical system. For coal applications the belt discharge filter has become widely used, especially for slurries containing clayey material which caused serious cloth-blinding problems with other types of filter.

For cakes which do not blind or where good filtrate clarity is required, a layer

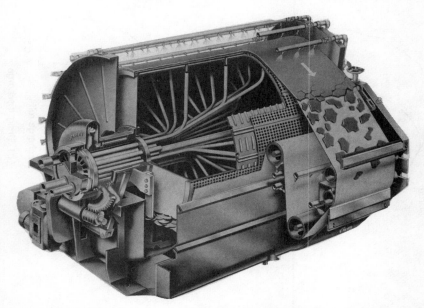

Fig. 10.21 Rotary-vacuum drum filter (courtesy Filtres Vernay).

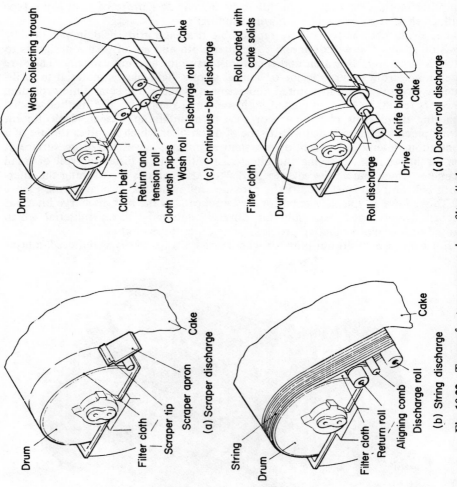

Fig. 10.22 Types of rotary-vacuum drum filter discharge arrangements.

SOLID–LIQUID SEPARATION

of cake can be built up on the drum to the limits of the position of a fixed knife, which scrapes off the cake. The scraper form of discharge is shown in Fig. 10.22(b). This discharge mechanism, usually consisting of a metallic frame fitted with hard plastic tips, can be used for almost all applications. It is usually accompanied by means of a blow-back valve which, by positive pressure applied from behind the filter cloth, causes the cake to be dislodged, reducing the wearing tendency by the blade contact.

Another technique, used when the cake has some strength (i.e. resilient, clayey cakes) is to discharge it by using endless strings, in much the same arrangement as the belt. These strings lift the cake away from the drum surface (see Fig. 10.22(c)). Because of the tendency for the strings to tear the cloth, creating poor discharge, string discharge types have been more or less superseded in coal plants by the belt types.

The performance of drum filters is controlled by adjustments in three main variables:

- drum rotational speed;
- applied vacuum;
- submergence.

A change in any one of these affects cake formation, drying, throughput and the degree of dewatering achieved. Most drum filters are supplied with the means for adjusting all three variables. Common sizes and capacities of rotary vacuum drum filters for coal application range from approximately 7.5 m^2 area × 1.875 m diameter to 115 m^2 area × 4.3 m diameter. Installed masses of these filter units range from approximately 5.5 t to 30 t.

The major advantages and disadvantages of drum filters compared to the other types can be summarized as follows:

Advantages
1. continuous and automatic in operation with ensuing low labour and operating costs;
2. design usually caters for reasonably wide operation variations;
3. clean in operation and generally easily maintained;
4. effective washing and dewatering.

Disadvantages
1. high capital cost;
2. larger space requirements;
3. unsuitable for fast-settling slurries;
4. low efficiency with ultrafine material tending to blind the filter cloth.

10.6.3 Rotary Vacuum Disc Filters

The fundamental design of almost all vacuum disc filters is the same. They consist of a number of flat filter elements, mounted on a central shaft and connected to a normal rotary vacuum filter valve. Each disc, and there may be as many as 15, is composed of up to 12 such elements which accommodate filter cloth on either side, as can be seen from Fig. 10.23. As the unit rotates, the discs become submerged in slurry contained in the slurry bowl and usually agitated by an impeller mechanism. Gradually a filter cake is formed and then dewatered as

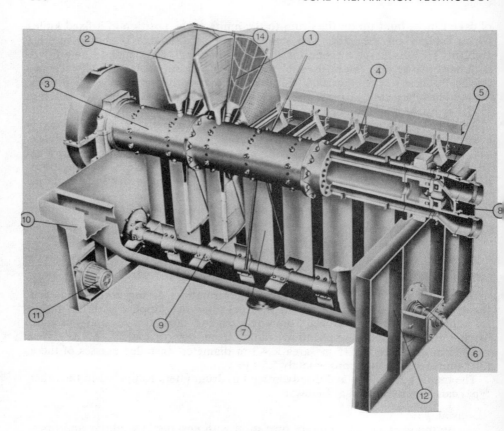

Fig. 10.23 Rotary-vacuum disc filter (courtesy Eimco Equipment). 1. Choice of moulded plastic, plastic grid insert, wood or metallic *sectors* tailored to specific applications. 2. *Filter media* in a wide selection of fabrics are available to provide the most effective liquid removal and cake formation. 3. *Centre barrels* are available in cast iron, fabricated trapezoidal port, or fabricated removable pipe for maximum hydraulic/pneumatic efficiency. 4. *Guided scrapers* protect filter media for maximum life (alternative: adjustable deflector blade). 5. *Eimco Hy-Flow*™ *valve* accommodates liquid and pneumatic flows without restriction (alternative: fabricated design). 6. *Floating-type stuffing box* prevents leakage and requires less maintenance. 7. *Wide discharge chutes* prevent cake clogging. 8. *Pressurized air* loosens cake from the filter media—choice of sudden blow, continuous blow, or snap blow with optional stagger blow centrebarrel orientation available for various applications. 9. *Agitation system* maintains slurry in suspension with solids at specific gravities as high as 7.5. 10. Overflow box, running the full length of the tank, prevents spillage and maintains uniform slurry level for optimum performance. 11. *Agitator* constant-speed chain-and-sprocket drive (alternative: variable-speed drive). 12. *Heavy-duty tank* supported by structural steel, external frame (alternative: complete rubber lining for corrosion resistance, or a four-foot-wide layer of rubber under agitator for abrasion resistance.) 13. *Feed* to the filter is from a feed-distribution manifold discharging into tank between alternate discs on the rising side of the centrebarrel. This method is advantageous for filtering flotation products (alternative: full-length feed trough on discharge side of tank). 14. *Interlocking sector clamps* keep discs in alignment.

Fig. 10.24 Krauss–Maffei large-diameter disc filter installation.

the unit rotates out of submergence in the bowl. The filter cake is usually removed by a combination of scraper blades and a blow-back system. A further development of this system is a form of electropneumatic pulser valve which works in conjunction with a multilayering technique. This permits resubmergence of undischarged cake, creating in some cases substantially improved cake yields for clean-coal slurries in particular. Rotary vacuum disc filters available for use in coal preparation normally vary from $2 \text{ m}^2 \times 1.2 \text{ m}$ diameter to $310 \text{ m}^2 \times 3.8 \text{ m}$ diameter, utilizing up to 15 discs in the larger sizes.

Figure 10.24 shows an exceptionally large disc filter manufactured by Krauss–Maffei which has a single disc-filtering area of $37.5 \text{ m}^2 \times 5.3 \text{ m}$ diameter. Each disc is composed of 30 segments compared to the 8–12 of the more conventional range of disc filters. The advantages and disadvantages associated with disc filters in general, when compared to other alternatives, are as follows.[21]

Advantages
1. low capital cost per unit area;
2. large filter areas and minimal floor area requirement;
3. rapid change of filter cloth;
4. two different slurries can be handled by one machine;
5. dilute slurries can be effectively filtered.

Disadvantages
1. washing of cake of filter media is not efficient;
2. cakes have higher moisture than from the drum type;

3. cake discharge not as effective than for various drum alternatives—higher moisture may be caused by blow-back;
4. media wear or damage more rapid with disc sectors.

10.6.4 Rotary Horizontal Filters

Many slurries filter best when assisted by gravity, and horizontal filters attempt to exploit this potential advantage.

Several varieties exist, all of which have been used in coal preparation at some time or another, but generally most coal applications have tended to be supplied with either a drum or disc type and only occasionally with the horizontal-belt type.

Continuous horizontal filters can therefore be subdivided into two groups:

(a) belt-type which include continuous rubber-belt and tray-belt types;
(b) horizontal disc filters which include single-disc units with either scroll or paddle cake discharge and the tilting-pan filter.

Only type (a) will be discussed, because they are occasionally used in coal treatment.

Basically, there is a slotted or perforated endless belt supporting the filter cloth which is transported over the vacuum- or suction-box (see Fig. 10.25). The slurry is pumped into the filter at one end and filtrate is drawn through to the vacuum-box leaving behind the filter cake which is then dewatered before it reaches the discharge point at the other end of the units. Washing of the cake *en route* to discharge is possible but is rarely practised in coal treatment. However, it has been attempted, with varying degrees of success, to steam-dry the cake by suspending a steam hood over the dewatering section. The filter belt can be washed thoroughly on the return journey to the feed end.

Horizontal filters are available in sizes up to 30 m^2 in area and have been successfully used for froth flotation concentrates and coal-fines dewatering.

10.6.5 Rotary Vacuum Filter Systems

Figure 10.26 shows a more or less standard form of filter installation. The filter valve-head forms the connection between the individual filter cells and the filtrate receiver. This is subdivided into various zones for each individual filtering step, i.e. forming the cake, drying, blow-back, etc. The lower connection from the filter valve transmits filtrate, via a pipe, to the receiver. Similarly, the upper connection recovers filtrate and air from the dewatering operation. A third valve, usually present on vacuum disc and drum filters, transmits low-pressure air to assist with the discharge of the cake. The filtrates and air enter the side connection of the filtrate receiver where the filtrate gravitates. A filtrate pump is then used to discharge the accumulated filtrate and to overcome the suction head. The air is pulled through the top connection of the receiver to a moisture trap or condenser. The required pressure differential is usually achieved by using a water-ring type of pump. Fresh water is used for seal water and a discharge silencer is located on the air discharge to reduce the noise.

Drying-zone filtrate containing air can, if required, be collected in a separate filtrate receiver by using a barometric leg. This may be beneficial, if practical, because a simpler pumping system can be used. However, using a barometric leg

SOLID–LIQUID SEPARATION

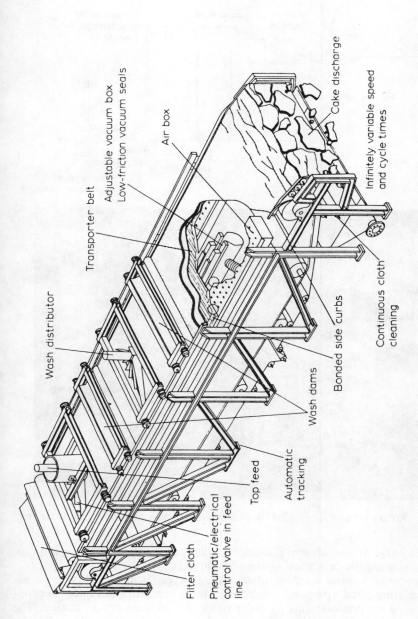

Fig. 10.25 Horizontal belt filter (courtesy Eimco Equipment Company).

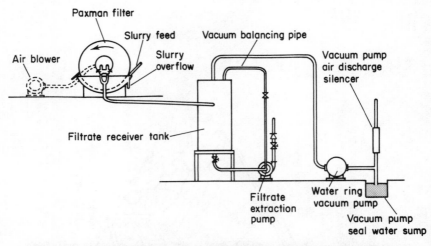

Fig. 10.26 Rotary vacuum drum filter: (top) diagrammatically; (bottom) from valve end.

type of discharge is not always possible because of plant-elevation limitations. The leg usually requires a vertical column height of more than 9 m.

Consistency in feeding vacuum filters of any type is an absolute criterion towards ensuring good operating conditions. Slurry concentration must be properly controlled by prethickening in order to ensure uniform concentration. The feed-point location will vary depending upon the type of filter, but side feeding by means of an overflow box to the filter tank is most common for rotary filters. The introduction of various controlling and safety systems to filter installations has

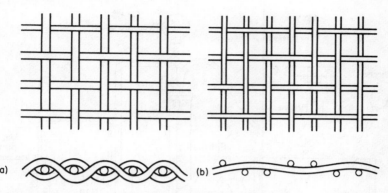

Fig. 10.27 Basic weave types: (a) plain square weave; (b) plain twilled weave.

created much improvement in performance. Among these are: density gauges for the feed slurry; water-sensing electrodes in air lines from the filtrate receivers; level devices in the filter tank; pump- and drive-failure electronic interlocking devices; and some novel variable-speed drive arrangements.

Vacuum filters in coal treatment usually employ some form of woven-wire or fabric-filtering media.

Woven wire lends itself easily to fabrication of disc-filter bags covering each segment. It is also easily applied to filter drums. It is resistant to blinding and can be easily cleaned, does not shrink or stretch but costs much more than fabric. It then becomes a question of whether it is achieving the required technical performance, while at the same time lasting for a sufficiently longer time than a fabric filter medium to offset the increased cost. Weave types available include varieties of the basic plain and twilled patterns as the most commonly used (see Fig. 10.27).

Woven fabrics are made with natural and synthetic fibres. For vacuum filters, synthetics are most commonly used and monofilament and multifilament fabrics are both fairly common. Filter fabrics are mainly woven in four weaves: plain, twill, plain reverse Dutch and satin. The first three are overlaps from wire-cloth weaves. The last two weaves are popular because they tend to provide the best filtering characteristics. Filter fabrics are commonly woven from nylon, polyester, polypropylene and polyethylene, but polyester and polypropylene are probably the most common vacuum-filter fabrics of all.

10.6.6 Vacuum Filter Performance

Much research work has been dedicated to finding out the ways and means of determining the performance of vacuum filters.[23] The filtration equations given earlier in Section 10.5 have generally formed the basis for developing empirical formulae for design purposes and for assessing filter performance.[24]

The normal approach adopted is to carry out a series of laboratory tests with the slurry to be filtered in order to establish the filtration characteristics under a variety of test conditions.[25] The simple testing apparatus usually comprises a filter leaf which is submerged in a filter bowl containing the slurry, agitated by a mixing impeller. A small vacuum pump connected to the leaf via filtrate-receiving

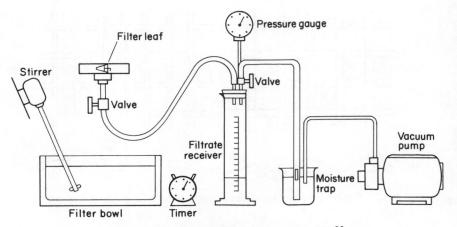

Fig. 10.28 Simple filter-leaf test assembly.[22]

jars permits filtrate rates to be measured and pressure to be controlled. This apparatus is shown in Fig. 10.28. By this form of test, filtering time, cake-formation rate, filtrate-flow rate, cake moisture, filtrate clarity, etc., can be obtained for various feed-solids concentrations, different flocculant conditions, varied applied pressures, etc. From these data the filtration characteristics can be defined. It is then usually advisable to proceed with testing using pilot-scale equipment such as the drum filter shown in Fig. 10.29, and utilizing operational data and experience obtained from previous installations with similar equipment. The graph shown in Fig. 10.30 gives an idea of the potential correlation which can be obtained by the three sources of data.

Fig. 10.29 Paxman 1 m² transportable pilot-test filter installation units.

SOLID–LIQUID SEPARATION

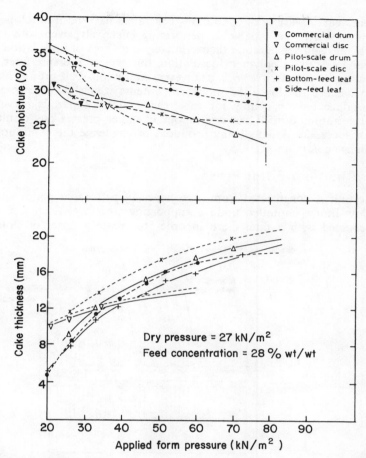

Fig. 10.30 Cake thickness and moisture against form pressure for coal concentrate slurry.[25]

10.7 PRESSURE FILTRATION

10.7.1 General

Two distinct types of pressure filter are utilized in coal treatment:

- batch, chamber filter presses;
- continuous, belt filter presses.

Usually, whichever type is selected, their application is a common one—to treat coal-flotation tailings or occasionally raw-coal slurries, i.e. refuse material.

The use of the chamber filter press has long been the traditional approach adopted in Europe, especially Britain, but it has always suffered the disadvantage of being a batch process necessitating high capital cost for no productive return.[26]

In recent years a variety of continuous-belt filters have been developed which have been shown to be capable of performing effectively, even with the most difficult tailings slurries. Despite their relatively high cost of installation and the great dependence upon polymer flocculation, this type of pressure filter appears to be growing in popularity. So far, however, continuous-belt presses have been unable to achieve the low cake moisture contents which chamber press types readily produce, and probably for this reason alone a parallel development towards the automation of semicontinuous chamber presses has resulted in a number of successful designs being produced to challenge the continuous belts. This challenge continues.

10.7.2 Chamber Filter Press

The basic unit is constructed from a sequence of perforated plates alternating with hollow frames mounted upon a supporting structure. Each face of every plate is covered with a filter cloth in order to create a series of cloth-walled

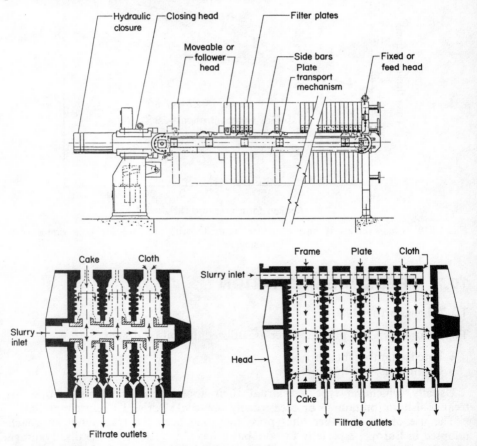

Fig. 10.31 Chamber filter-press assemblies: (a) assembly of the press; (b) recessed-plate filter assembly; (c) flush-plate and frame-filter assembly.

SOLID–LIQUID SEPARATION

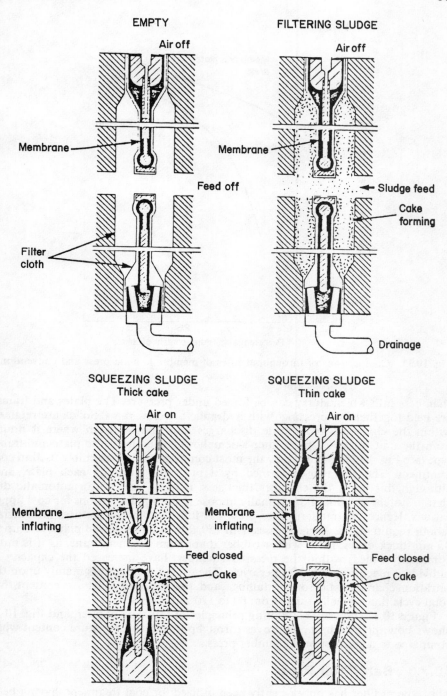

Fig. 10.32 Operation of a membrane plate.

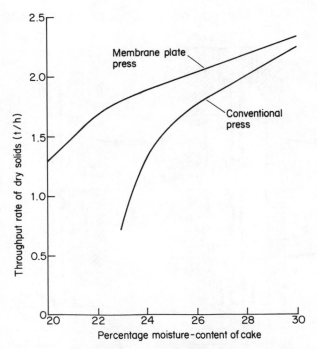

Fig. 10.33 Comparisons of throughput rates of membrane plate press and conventional press.

chambers into which slurry can be forced under pressure. The plates and frames are held together by 'pressing' with hydraulic or screw rams. Solids are retained within the chambers and filtrate discharges into hollows from where it drains from the unit. Recessed-plate filter assemblies evolved from the plate-and-frame type (see Fig. 10.31) and are now the most common form of assembly used in coal treatment. The chamber is formed by a protruding rim on each plate, and, although this limits the chamber thickness, it lends itself to an automatic discharge system and does not normally impose operating limitations for coal applications. Hence, the modern filter press has fully automated features minimizing labour requirements. Large presses with 150, 2 m × 2 m recessed plates and up to 35-mm-thick cakes, fitted with a rubber diaphragm (or membrane, as it is more commonly called) within the recessed chamber, have increased the capacity of individual machines to 8–10 t per cycle. The cycle time depends mainly upon the filtration characteristics of the tailings and the required moisture content, but total cycle times usually range from 80 to 120 min.

Figure 10.32 shows the operating principles of this type of filter, and Fig. 10.33 shows how the membrane improves throughput rate and moisture content when compared to a conventional chamber press.

10.7.3 Belt Filter Press

This type of filter has only recently been utilized for coal treatment, having been developed and tested in sewage treatment and other industries over a period of almost 15 years. Although designs from various different manufacturers vary in

SOLID–LIQUID SEPARATION

construction, the belt filter press technique involves continuously producing a highly flocculated and thickened feed which is then distributed across a wide, porous-belt conveyor. This continuous-belt conveyor transports the slurry through a tortuous route of several pulley rolls, culminating in compression, which produces a cake from the solids and removes the filtrate. The resultant filter cake can contain 70–80% solids and is usually readily handled and, if required, can be mixed with other, coarser solids for disposal.[27]

The general claim made by most manufacturers is that the principal advantages of this type of filter compared to the alternatives (in addition to continuous-operation ones) are as follows:

- lower energy consumption;
- ease of operation and maintenance;
- consistency of cake combined with high solids recovery (in most cases);
- more compact installation (in most cases).

Because of the high dependence which this type of filter has on successful and consistent flocculation, conditions from mine to mine can be widely varied. It has become almost recognized practice for potential users to conduct on-site tests whenever possible or else carry out comprehensive pilot-scale tests on slurry samples. Consequently, most manufacturers have small-scale machines in order to carry out such work. Except for flocculation testing, only a very limited amount of useful data can be derived from bench-scale test-work for compression filters of this type. Figure 10.34 shows an installation of Aries–Andritz 3.5 m units which dewater 145 t/h of clay-rich coal refuse at a coal-preparation plant in the United States of America.[28] Similar types of filter, from a growing number of manufacturers, have now been installed in over 50 coal-preparation plants in different parts of the world. Despite the fact that flocculant costs are great and the resultant filter-cake moisture and solids in the filtrate are usually higher than can be obtained with chamber presses, the press filters are generally capable of providing adequate performance. When this is combined with the advantages mentioned earlier, the belt press becomes a strong competitor to the alternatives.

10.7.4 Continuous-pressure Filter

The continuous-pressure filter combines the benefits of conventional vacuum filtration with those of pressure filtration. Vacuum filtration has the advantage of continuous operation; the filtration process does not need to be stopped for cake removal. Pressure filtration has the advantage of a high capacity for a given filter area and a high efficiency in liquid removal from the filter cake. The combination of the useful properties of both filter systems is incorporated in a relatively recent continuous-pressure filter, called KDF (Fig. 10.35).[29] This filter has the following advantages:

(a) energy saving by lower energy consumption per kilogram of filtered product in subsequent cake treatment;
(b) the higher capacity for a given filter area means that a KDF takes up less space than a conventional filtration system;
(c) time saving; the production process does not need to be stopped for cake removal.

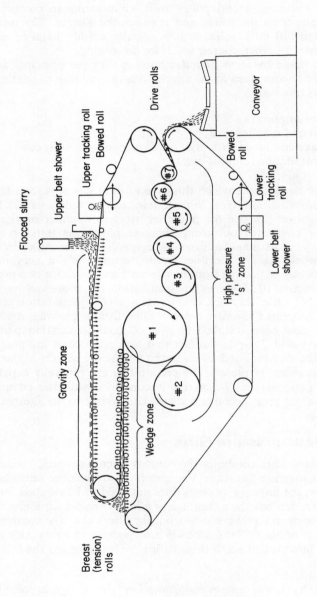

Fig. 10.34 Belt filter press (courtesy Aries–Andritz).

SOLID–LIQUID SEPARATION

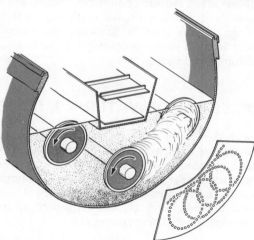

Fig. 10.35 (Top) Amafilter KDF continuous-pressure filter installation in Zolder, Belgium. (Below) The filter principle (courtesy Amafilter).

In the filter are six filter axles, each with a particular number of elements attached. The filter elements consist of three layers of wire mesh which supports the outer layers and also allows free flow of the filtrate. The two outer layers are made of finer mesh, on which the filter cake forms. The elements rotate on their axes, which move along a circular trajectory. Periodically the elements submerge into and emerge out of the slurry. A constant air pressure is maintained above the suppension, thus providing the pressure gradient necessary for filtration. The filtrate flows through the elements into the hollow axle. The pressurized air and the water from the dewatering process flow out of the filter through the same outlet as the filtrate.

10.7.5 Pressure-filter Systems

All types of pressure filter require a consistent and highly concentrated feed-slurry and this requirement is usually fulfilled by thickeners of one type or another. The thickened underflow is generally pumped into a filter buffer tank, where additional flocculation can be carried out as required. The buffer tank serves primarily to facilitate feed distribution to the various filters. Capacities of the individual filter units rarely match the thickener underflow rate, and occasionally, as many as nine separate units are required. This occurs for both chamber and belt press types.

Chamber presses normally utilize monofilament filter cloth made from polyester or other similarly strong fabric capable of withstanding the high pressures and frequent flexing involved. Current practice favours monofilament cloths which have been calendered or polished on one side to produce a very smooth surface finish, thereby ensuring good unassisted cake discharge. For many belt press filters the texture of the belt is also most important in so far as it influences sludge adhesion, and its physical strength is of paramount importance. Special-purpose polyester monofilament woven belts are specifically manufactured for each type of filter. Some filters require endless belts which are costly to manufacture. Heavy and continuous tension on the cloth demands special qualities of material, because if an endless cloth is damaged, the entire belt may require replacing. The present trend is towards developing clip-fastening devices for the belts in order to facilitate rapid changeover.

10.8 PRINCIPLES OF DEWATERING SCREENING

This section deals with the use of a specific group of screening devices, more or less gravitational filters, for dewatering fine-grained solids from slurries. Above 0.5 mm, the dewatering process is a simple one involving straightforward drainage on a screening surface of finer aperture size than the majority of the solids.

Because the prime force involved in this form of dewatering is gravity, much depends upon careful consideration of several important factors, in order to obtain a successful application.[30] These factors are:

1. fragment shape characteristics;
2. size distribution and range;
3. nature of screening surface and aperture;
4. amount of open area;
5. relative densities of the solid components;
6. solids–liquid concentration (and range).

From these factors the specific screening data are obtained as follows:

7. drainage area required;
8. screening surface acceleration (for moving screens);
9. operating frequency, angle of throw and screen slope;
10. width-to-length relationship of the screen.

Most of these screen design features have already been addressed in Chapter 4, but certain specialized aspects relevant to dewatering screens require some expansion.

In dewatering, the effectiveness of a screen depends greatly on the formation of a filter bed created by coarser fragments, larger than the aperture size, or by bridging of smaller fragments to arrest percolation during subsequent drainage of water. For successful operation, approximately 40% of the size-consist of the screen feed should exceed the screen aperture, and the feed slurry solids concentration should always be above 50% on a volumetric basis.

The screen deck itself is usually either stainless-steel wedge-wire or woven-wire, with the former arranged in a cross-flow manner being the most widely employed deck. In a cross-flow arrangement the individual fragments are exposed to the aperture for only a relatively short duration while the water is free to drain almost continuously. Wedge-wire decks can be turned through 180° and interchanged periodically to ensure even wear and maintain a sharp leading edge to the flow.

The open area provided by the screen deck determines the drainage capacity of the screen. The relative density of the fragments becomes a significant factor if light materials (e.g. low-ash coal fines from froth flotation) are likely to be overagitated by the mechanical action of the screen. If the filter bed is disrupted or is unable to form properly, the efficiency of the screen will be greatly impaired. In extreme cases a fixed screen or sieve bend is usually preferable.

The solid–liquid concentration of the feed has a great influence on the efficiency. Ideally, a feed with a sufficiently low fluid volume to permit drainage largely by capillary action is desirable. For fine solids, mainly less than 500 μm in size, about 30–40% by weight should be the objective. In order to achieve this it is often necessary to densify the feed by means of hydrocyclones or mechanical classifiers. The earlier Fig. 4.4 shows a commonly employed hydrocyclone–dewatering-screen combination.

The drainage-area requirement for a screen is a function of the screen type and the size distribution of the feed. Very fine suspensions yield solids possessing large surface area. Drainage calculations are therefore often based on empirical data obtained from commercial installations.[31]

Acceleration components in moving screens are almost diametrically opposite at the feed and discharge ends of the screen. A low acceleration should occur at the feed end in order to avoid counteracting the capillary action, which would result in solids loss through the deck. At the discharge end, where most of the water has drained away, higher acceleration of the well-formed cake is essential to maximize the dewatering achieved. Such a combination, despite being highly desirable, is rarely achieved in practice, although some high-frequency drives come impressively close. The significance of frequency and angle of throw in mechanical and electromechanical screens is to determine the rate of travel and the amount of life which the screen deck can contribute to the formation and subsequent dewatering of the solids cake. High-speed vibrator screens driven by an out-of-balance mechanism are operated at relatively low frequencies of about 1500–1800 cycles/s. Some of the electromagnetic types are operated at much higher levels, i.e. 5000–7000 cycles/s, but with stroke settings of less than 2 mm in some cases.

Experience has shown that for high-speed, mechanical vibrator-driven units a 40–45% angle of throw gives good results combined with transport speeds of 0.2–0.3 m/s. The rate of travel is also influenced by the inclination of the screen.

One of the major limitations of dewatering screens is that, mainly because of the high frequencies and varied acceleration requirements discussed earlier, they are characteristically limited in capacity. Few units are available which can effectively dewater more than 30–40 t/h of minus 0.5-mm fine solids. It therefore becomes necessary to employ multiples of units arranged together in a parallel system. For most dewatering screens, the length is more or less fixed by the design, and the width determines the capacity. It is therefore usually the width that is limited by the mechanical design.

10.9 DEWATERING SCREENS

10.9.1 General

Two basic groups of screens are employed for dewatering:

1. fixed screens, including inclined flat screens and sieve bends;
2. moving screens, including electromagnetic and high-frequency vibrating screens.

In fact, two specific types of screens are the most prevalent of all. These are the sieve bend and the high-frequency vibrating screen. There are numerous varieties of each but the basic operating conditions differ little.

10.9.2 Sieve Bend

This screen is shown previously in Fig. 4.17 and is almost equally widely employed for sizing and for dewatering applications in coal treatment. It usually comprises a wedge-wire, cross-flow deck curved to a radius of either 45° or 60°. The 45/60° units can handle 8–10 m^3/h per metre of width of minus 0.5-mm coal-fines slurry, with a feed concentration of 30–45% solids by weight.

By prethickening using hydrocyclones, the correct feed conditions can be ensured on an ongoing basis. This presumes that facilities are also provided to enable the subsequent thickening of the dilute overflow from the hydrocyclone.

In some cases two, or even three, stages of sieve bend arranged in series may be employed, but this arrangement is more commonly employed in sizing applications. The addition of a high-speed electromechanical or electropneumatic form of rapping device, or preferably a high-frequency vibrator, can enhance drainage. Of the two types of device, the vibrator is probably less damaging to the screen structure, but both tend to cause irregularities in drainage and can hinder cake formation, causing solids loss. Such devices are therefore probably better applied to sizing rather than to dewatering units. The severe limitation imposed on any static screen in only being capable of utilizing gravitational force, generally means that such types of device are usually employed to prepare fine solids for subsequent centrifuge dewatering.

SOLID–LIQUID SEPARATION

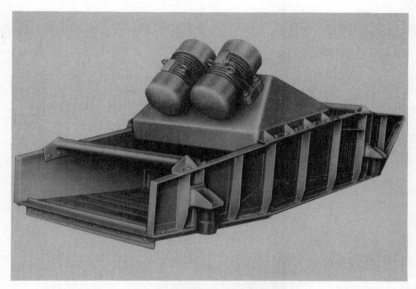

Fig. 10.36 Fordertechnik high-frequency dewatering screen (courtesy Tema Siebtechnik).

10.9.3 Vibrated Screens

An example of this type of screen is shown in Fig. 10.36. The screen deck is usually wedge-wire but other, similar screens can utilize woven-wire decks to enable finer solids to be dewatered. In the featured screen the wedge-wire, cross-flow panels usually increase in aperture from feed to discharge, and side panels located at the feed end assist primary drainage. It is common to employ either a hydrocyclone or a mechanical classifier to prethicken the feed to this type of screen.

In dewatering minus 0.5 mm coal fines, this screen can produce cake with a moisture of 20–30% by weight, and the largest unit would handle up to 25 t/h. Normally, the screen is arranged with a 5° negative slope which can be augmented by the addition of an adjustable discharge weir. By adjustment a bed-depth of between 200 and 300 mm is normal.

Instead of utilizing a wedge-wire or woven-wire screen deck it is also possible to use a flexible rubber slotted deck, as is employed by the screen shown in Fig. 10.37. This positively inclined screen is vibrated by means of a high-frequency, 3600 cycles/s, vibrator motor. An intense patting motion, causing the vertical slots to oscillate between open and closed, assists in drawing the water rapidly through the sedimenting solids and the rubber deck. The motion of the rapidly forming cake down the surface of the screen is unaffected by this flexing action. The dewatering capability of this screen is further enhanced by as much as an additional 20% by means of a vacuum-assisted drain-pan located close to the discharge point. The rubber deck is about 12.5 mm in thickness and the screen is more or less of standard 3 m long × 1.2 m wide dimensions. It will handle 25–30 t/h of minus 0.5-mm coal-fines slurry, ideally after prethickening using a hydrocyclone, as shown previously in Fig. 4.4.

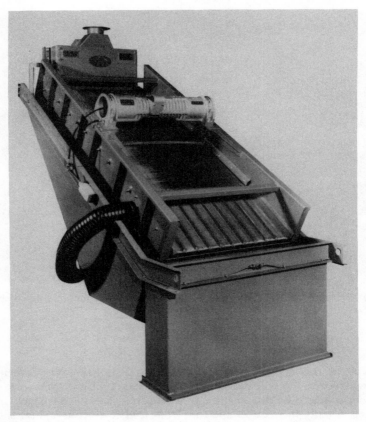

Fig. 10.37 Derrick high-frequency vacuum-assisted screen incorporating a 2.25–3.75 kW industrial blower. (Courtesy, Derrick Equipment Company.)

10.10 PRINCIPLES OF CENTRIFUGATION

Centrifugal dewatering machines commonly used for coal treatment function by a combined centrifugal sedimentation and filtration mechanism which is, to a large extent, based upon relative-density differences between the various contained solids and water. The laws governing both gravitational sedimentation and filtration are thus applicable, once suitable modifications are applied to accommodate the substantially enhanced prime force-field supplied by the centrifugal action of the machine.

The solid fragments are subjected to centrifugal forces which make them move radially outwards through the rotating slurry to form a compact shell of sediment on the inner periphery of the rotating bowl of the centrifuge. Hence, a centrifugal dewatering machine comprises an imperforate or solid bowl into which a slurry is continuously fed, to be rotated at high speed with the bowl. The mechanism by which the solid cake and the water, or centrate, are individually removed varies with machine type.[32] Of the wide range of such machines current-

SOLID–LIQUID SEPARATION

ly available the coal industry make use of only a relatively small proportion. These various relevant types will be discussed later, but for the time being they will be classified into two modes:

(a) vertical axis of rotation or planetary centrifuge;
(b) horizontal axis of rotation or side-feed centrifuge.

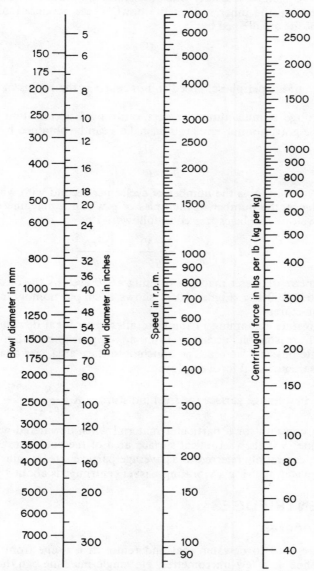

Fig. 10.38 Chart for determining centrifugal force. To determine force, draw a straight line from bowl diameter through the speed employed. Continuation to the column head 'centrifugal force' gives the × gravity force developed.

The mechanics involved in defining the behaviour of each of these two modes is complex and its inclusion here would serve little practical purpose. It is, however, useful to understand the effect which the rotational speed of the machine can have on the solids in terms of centrifugal force. In the case of the planetary mode, the rotating bowl exerts a centripetal force on the incoming slurry as soon as contact is made. Once the slurry is rotating it in turn exerts a centrifugal force on the inner wall of the bowl. These accelerations being now equal and opposite are defined by:

$$F = \frac{V^2}{r}$$

where V is the linear peripheral speed in m/s and r is the radius of curvature in metres.

Most centrifuge manufacturers express centrifugal acceleration in terms of multiples of the gravitational acceleration, g. This can be obtained by combining:

$$\frac{F}{g} = \frac{V^2}{gr}$$

and $V = 2\pi rN$, where N is the number of cycles per second with which the bowl is rotating. Hence, the number of multiples of g which a machine rotating at a speed of N develops, can be expressed as follows:

$$F_1 = \frac{V^2}{gr} = \frac{(2\pi rN)^2}{gr} = \frac{4\pi^2 rN^2}{g}$$

For a one-metre diameter machine rotating at 10 cycles/s the centrifugal force is approximately $200 \times g$. Figure 10.38 shows a simple nomogram commonly used by manufacturers.

The free moisture remaining in the cake after centrifugal dewatering is often predicted with reasonable accuracy from an empirical equation taking into account definite physical factors. These include surface area, size analysis, surface tension and the centrifugal force employed:

$$\text{Percentage surface (or free) moisture} = K_c \left(\frac{S}{dF_1}\right)^{0.25}$$

where K_c is a constant for a particular material when dewatered with a specific type of machine; S is the theoretical surface area of the coal in m^2/kg; and d is the size factor, commonly referred to as average particle size, in mm. The value of K_c for coal dewatering using a vibrating-basket centrifuge is about 18.5.[33]

10.11 CENTRIFUGES

10.11.1 General

Centrifuges are used to dewater coal and refuse in a range from as coarse as 50 mm to as fine as a few micrometres. No single machine can, however, treat this entire range satisfactorily. They therefore depend for their effectiveness on being fed a relatively close size-range of solids under very uniform conditions of feed rate and solids concentration.

SOLID–LIQUID SEPARATION

In general, centrifuges may be classified into two groups:

(a) perforate basket;
(b) solid or screen bowl.

There are many types of machines in each group but those commonly employed in coal treatment can be further subdivided as follows:

- *perforate-basket type* (horizontal or vertical axis of rotation)
    —without transporting device
    —with positive discharge system
    —vibrating basket

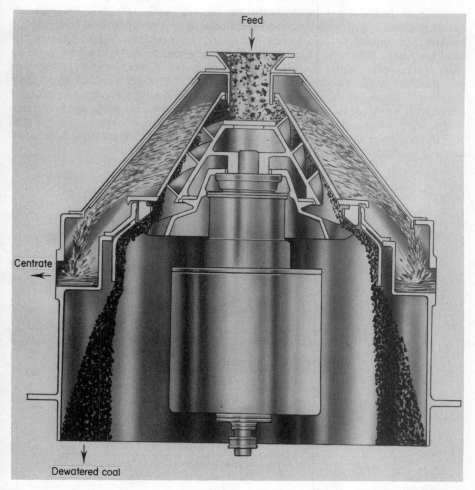

Fig. 10.39 Perforate-basket centrifuge with transport device (vertical axis) (courtesy CMI).

- *bowl type* (horizontal axis of rotation)
  —cocurrent solid bowl
  —countercurrent solid bowl
  —screen bowl.

10.11.2 Perforate-basket Centrifuges

Perforate-basket machines without transport devices comprise a simple, truncated, vertical conical basket with an upwardly pointed apex into which is fed a partially dewatered coal. The coal solids are accelerated by the rotating basket and the water drains through the coal and the basket perforations. These types of machines are commonly used to dewater coal from approximately 10 mm to 1.0 mm size-range. Capacities of up to 50 t/h can be handled to produce a surface moisture of 6–8% by weight. Perforate baskets with transport devices are common. One variety, with a vertical rotational axis, is comprised of two rotating elements, an outside conical screen frame and an inside solid cone that carries spiral hindrance flights (see Fig. 10.39). Gears powered by a single motor produce a differential speed in the two elements. The wet coal enters the apex of the cone and the cake is discharged through the base of the machine. Coal sized from 10 mm to 0.5 mm can be handled at up to 80 t/h in a single machine to produce a cake with a surface moisture of 6–7% by weight. Fine-coal feeds are dewatered in specially developed machines, and a recently designed horizontally rotated machine has been shown to be capable of handling 35 t/h of minus 0.5 mm coal slurry to produce a cake with about 10–12% moisture content.

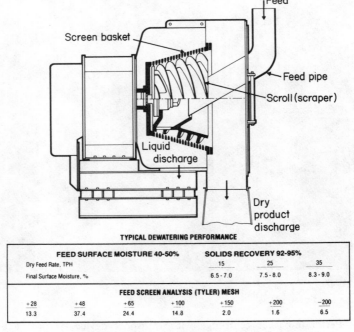

TYPICAL DEWATERING PERFORMANCE

FEED SURFACE MOISTURE 40-50%			SOLIDS RECOVERY 92-95%		
Dry Feed Rate, TPH			15	25	35
Final Surface Moisture, %			6.5 - 7.0	7.5 - 8.0	8.3 - 9.0

FEED SCREEN ANALYSIS (TYLER) MESH						
+28	+48	+65	+100	+150	+200	−200
13.3	37.4	24.4	14.8	2.0	1.6	6.5

Fig. 10.40 Perforate-basket centrifuge with transport device (horizontal axis) (courtesy Wemco).

SOLID–LIQUID SEPARATION

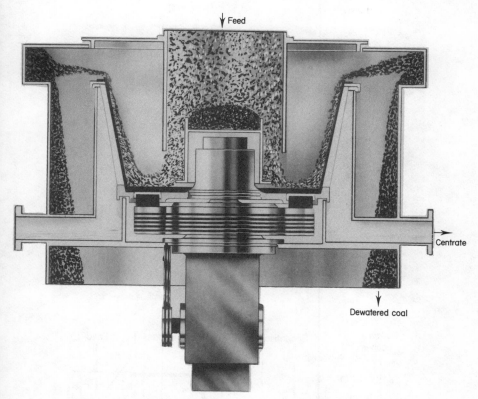

Fig. 10.41 Vibrating-basket centrifuge (vertical axis) (courtesy CMI)

This centrifuge is shown in Fig. 10.40 and employs a scroll scraper arrangement to transport the cake from the apex to the discharge point. Solids recovery with this type of machine has been stated as being 92–95%. The vertical spindle counterpart can dewater a similar feed with 40% moisture content to obtain a cake with about 15% surface moisture.

Vibrating baskets are very commonly employed almost equally as often in vertical and horizontal rotational modes. The vibration creates a loosening effect on the dewatering solids as they travel through the basket. These types of machines are not operated at such high speeds as those with transport devices, and two motors are generally required—one for rotating and one for vibrating the basket.

In the vertical-axis type the coal enters the machine from the top through a hopper which assists distribution (see Fig. 10.41). The solids pass to the inner periphery of the base of the basket and are conveyed up the side by the vibrating action. Feed size-ranges vary from as coarse as 75×6.7 mm, which can be handled at up to 200 t/h to produce a surface moisture of 2%, to fine feeds of 6.7–0.5 mm, which can be handled at up to 110 t/h to produce a surface moisture of less than 7.5%. An upper moisture-content limit of 30% is usually set.

The horizontal-axis machines are typified by Fig. 10.42. Coal enters the apex end of the basket via a feed pipe and the acceleration causes the cake to deposit

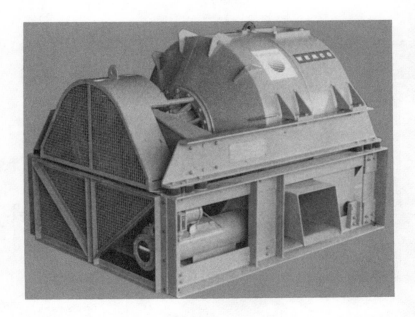

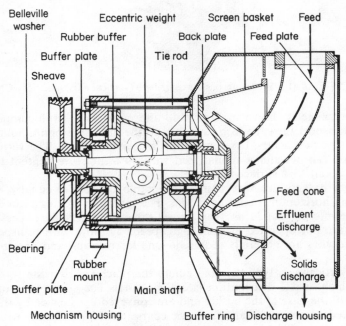

CENTRIFUGE MECHANISM

Fig. 10.42 Vibrating-basket centrifuge (horizontal axis) (courtesy Wemco).

SOLID–LIQUID SEPARATION

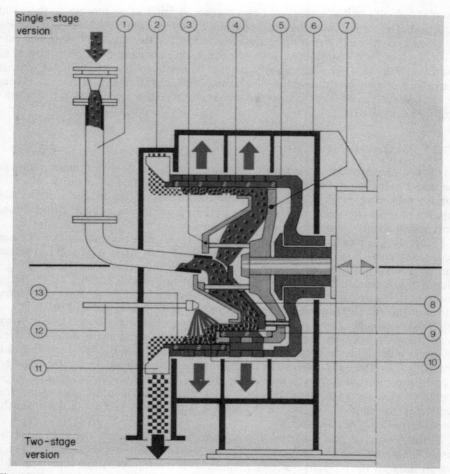

Fig. 10.43 Pusher centrifuge (single-stage and two-stage version): 1, feed pipe; 2, solids collector; 3, feed cone; 4, slotted screen; 5, basket; 6, filtrate collector; 7, pusher plate; 8, support ring; 9, first stage; 10, push ring; 11, scraper; 12, washpipe; 13, second stage.

against the screen through which the centrate passes. The vibratory action of the basket distributes and transports the cake to the discharge housing. This type of vibrating-basket centrifuge has similar capabilities to the vertical type. Both varieties are well tried and tested in coal applications.

A special category of horizontally rotated basket centrifuge is the pusher centrifuge, which has found some application in dewatering partly drained 0.5 × 0.1 mm size-fraction material from dewatering screens. The feed enters the machine in the axial direction and flows over a distribution cone past a more or less vertical disc and then onto a cylindrical screen basket formed from a trapezoidal cross-sectional wedge-wire surface. The cake formed during one cycle is subsequently pushed by the disc to a point on the inner periphery of the cone closer to the discharge point (see Fig. 10.43). The cone and disc rotate at the same angular velocity as the basket and carry out a reciprocating motion in the axial direction. The stroke frequency ranges from 1.5 to 3.5 cycles/s. The basket has no lip edge, and in order to maintain the cohesiveness of the cake an upper feed rate must be rigorously maintained. Instead of featuring a simple cylindrical basket as shown in Fig. 10.43 pusher centrifuges can be made to have multistage screens consisting of two or more steps of successively large diameter. These machines are at present rarely used in coal treatment, despite the fact that good applications probably exist.

10.11.3 Bowl Centrifuges

Bowl-type decanter centrifuges were first introduced into coal-treatment plants in the mid-1960s with the solid-bowl unit being utilized in the treatment of refuse or tailings following prethickening by gravity thickeners. This machine is usually capable of achieving good centrate clarity (1000 ppm maximum) and can produce a cake with 30–40% moisture content.

Figure 10.44 gives a cutaway view of a solid-bowl centrifuge. It comprises three major elements: bowl, conveyor and drive unit. The bowl is supported at either end by a bowl head with an integral trunnion. The conveyor is constructed from one or more helical flights attached to an axial hub which rotates on the same horizontal centre line as the bowl. Both conveyor and bowl have the same contour and the clearance between the tip of the conveyor flight and the inside of the bowl wall is held to a practical minimum. Bush bearings located in the bowl-head trunnion support the conveyor and allow it to rotate at a different speed from the bowl. This differential speed is a set percentage of the bowl speed and is controlled by the gear-unit ratio. A stationary, cantilevered feed-pipe transfers feed to the rotating conveyor hub in the case of the countercurrent feed arrangement, or else enters directly the feed inlet located adjacent to the cylindrical section in the case of a cocurrent machine. Figures 10.44 and 10.45 illustrate the two alternative types of machine.[34]

In the cocurrent-flow concept whereby the feed slurry enters the pool near the larger-diameter end of the bowl, both the water and the solids are conveyed cocurrently to their relative discharge points. The long detention time available can produce high cake-solids content from compressible material which continues to concentrate under moderate gravitational force, e.g. flocculated clay tailings. Cocurrent machines are usually operated with the pool level near the solids discharge-port elevation and at much lower speeds than countercurrent machines. For this reason they have proved to be attractive for coal-refuse applications where a slightly higher (35–45%) cake moisture can be tolerated. High solids-recoveries are obtained—96 to 99%—and consequently, very clear centrates.

Other benefits are longer conveyor, feed and discharge-port life (three to six times that of the countercurrent machine) especially if ceramic lining is utilized (see Fig. 10.46).

In the countercurrent-flow concept the feed slurry enters the machine near the junction of the cylindrical–conical section. The solids which settle are therefore conveyed to the discharge-port countercurrent to the centrate which overflows the large end of the bowl. The pool depth in the centrifuge bowl is adjustable by means of plate dams located at the centrate overflow. The higher power con-

SOLID–LIQUID SEPARATION

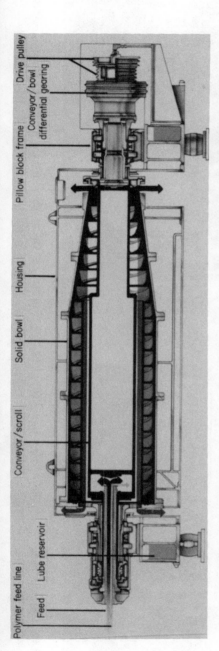

Fig. 10.44 Solid-bowl decanter centrifuge (cocurrent machine) (courtesy KHD Humboldt–Wedag).

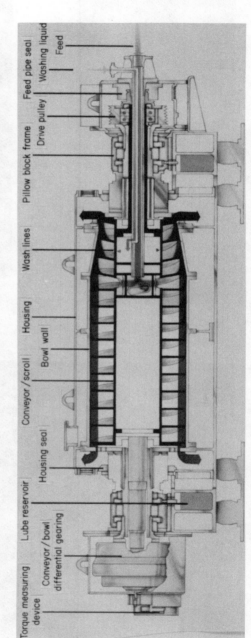

Fig. 10.45 Countercurrent solid-bowl centrifuge (courtesy KHD Humboldt–Wedag).

Fig. 10.46 Ceramic lining of solid-bowl/screen-bowl centrifuges: (a) bowl—screen section showing ceramic screen; (b) bowl—screen section showing longitudinal slots; (c) conveyor assembly; (d) bowl—solid section showing ceramic tiling. (Courtesy Broadbent.)

SOLID–LIQUID SEPARATION

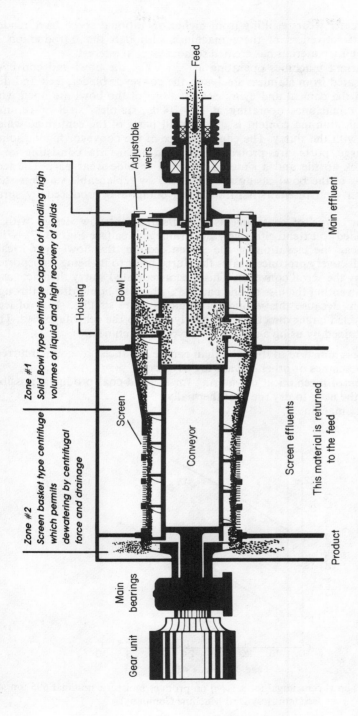

Fig. 10.47 Screen-bowl centrifuge. Typical cross-section showing major components.

sumption and greater wear resulting from higher operational speed have tended to reduce the attractiveness of these machines, although the introduction of highly resistant lining materials has created some revival in interest.

For countercurrent machines operating at 250–900g, the bowl and conveyor are usually fabricated from stainless steel, and the conveyor blades, feed and discharge ports and the conical and cylindrical sections of the bowl are lined with ceramic tiles. For machines operating at 900–2000g, stainless steel or ceramic material with high alumina content is used for all parts of the centrifuge which come in contact with the slurry. The edge and face of the conveyor flights, solids feed and discharge ports are protected by weld-on inserts consisting of a stainless-steel back-up tile and a sintered tungsten carbide wear part. The feed zone and sections of the bowl casing are protected by replaceable stainless-steel liners coated with wear-resistant hard surfacing compound produced by spray welding.

The screen-bowl type of centrifuge is a countercurrent-flow design with a screen section added to extend effectively the drainage area (see Fig. 10.47). With the pool level near the elevation of the screen section, the bowl cylindrical–conical section effectively preconcentrates the slurry prior to its being transported onto the screen by the scroll conveyor. This preconcentrated slurry then acts as a filter medium, preventing the loss of fine particles, while the liquid drains through the screen section. Because this volume of liquid is quite small (5–10% of feed volume), it is recycled either directly or indirectly back to the centrifuge feed. The benefits to be obtained by using a screen-bowl type of machine are:

- up to 60% less moisture in the cake than can be produced by countercurrent solid-bowl machines or other dewatering devices;
- maximization of mechanical dewatering for cleaned-coal products, possibly eliminating the need to dry the coal thermally;
- high-capacity machine.

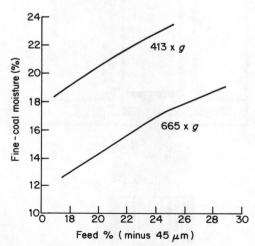

Fig. 10.48 Screen-bowl centrifugal force ($\times g$) on product moisture, nominal 595 μm × 0 feed (courtesy Bird Machine Company).

SOLID–LIQUID SEPARATION

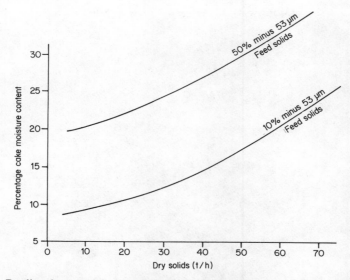

Fig. 10.49 Predicted outputs from 1200 × 2500 mm screen-bowl centrifuge for 0.5 mm–0 coal fines.

In an attempt to improve recovery and yield the driest cake possible, today's machines are designed to operate at speeds that generate over $1000g$. For applications dewatering extremely fine coal (i.e. minus 44 µm), high centrifugal-force levels are desirable in order to ensure that a well-consolidated cake is formed prior to passage into the screen section of the bowl. Consequently, as the size-consist becomes finer, the selection of the correct centrifugal-force level becomes more and more important, as can be seen from Figs 10.48 and 10.49. For coarser size-consists (i.e. those containing appreciable amounts of plus 600 µm), lower

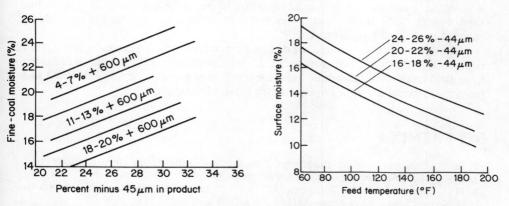

Fig. 10.50 (left) Screen-bowl centrifuge influence of size-consist on product moisture (courtesy Bird Machine Company).

Fig. 10.51 (right) Screen-bowl centrifuge tests on slurry pipeline (courtesy Bird Machine Company).

Fig. 10.52 Pilot-scale solid-bowl centrifuge for field trials (courtesy Broadbent).

gravitational force-fields are acceptable; for example the vibrating-basket machines referred to in Section 10.11.2 typically operate at 50–100g. Figure 10.50 shows the effects of particle size-consist on the moisture content of the centrifuge cake.

Another means for reducing cake moisture still further is to preheat the feed slurry. Test programmes have shown that inlet temperature can sometimes have a dramatic effect on the final cake moisture of free-draining material such as fine coal. Figure 10.51 shows that a 6–8% reduction in cake moisture is possible by increasing the feed temperature to 95 °C.

Although test data valuable for design purposes can be obtained by laboratory bench-scale test-work, the most valuable data are, without doubt, those obtained from field trials with pilot-scale machines such as the one shown in Fig. 10.52 and from commercial machines operating on similar feed material.

REFERENCES

1. D. G. Osborne. Gravity thickening, Chapter 5 in *Solid–Liquid Separation*, (Ed.) L. Svarovsky, Butterworth, Guildford, 1981, pp. 120–161.
2. H. S. Coe and G. H. Clevenger. Methods for determining the capacities of slime-thickening tanks, *Transactions AIME* **55**, 356–384, 1916.
3. E. B. Fitch. Current theory and thickener design, *Filtration and Separation*, **12**, 355–359, **13**, 480–488 and **14**, 636–638, 1975.
4. K. J. Scott. Experimental study of continuous thickening of a flocculant silica slurry, *Industrial Engineering Chemistry* **7**, 4, 582–588, 1968.

5. E. B. Fitch. Gravity separation equipment—clarification and thickening, *Solid–Liquid Separation Equipment Scale-up*, (Ed.) B. D. Purchas, Uplands Press, Croydon, 1977.
6. G. J. Kynch. A theory of sedimentation, *Transactions Faraday Society* **48**, 166–172.
7. W. P. Talmage and E. B. Fitch. Determining thickener unit areas, *Industrial Engineering Chemistry* **47**, 1, 38–41, 1955.
8. L. A. Dale and D. A. Dahlstrom. Design and operation of thickening equipment for closed-water circuits in coal preparation, *Transactions AIME* 1965 also publ. *Coal Preparation*, fourth edn. Chapter 12 (Ed.) J. W. Leonard, American Institute of Mining Engineers, New York, pp. 12.33–12.54, 1979.
9. R. C. Emmett and R. P. Klepper. Technology and performance of the Hi-Capacity thickener, *Mining Engineering* **32**, 1264–1269, 1980.
10. P. L. King. Thickeners, Chapter 27, *Mineral Processing Plant Design*, (Ed.) A. L. Mular and R. B. Bhappu, Society Mining Engineers AIME, New York, 1980, 541–577.
11. J. Abbott, C. C. Dell, B. Denison, D. Knott and N. W. Hill. Coal preparation plant effluent disposal by means of deep cone thickeners, 6th International Coal Preparation Congress, Paris, Paper 20C, 1972.
12. J. M. Keane. Sedimentation: theory, equipment and methods, *World Mining*, Nov. 1979.
13. N. P. Chironis. New clarifier thickener boosts output of older coal preparation plant, *Coal Age*, **80**, Jan. 1976.
14. R. L. Cook and J. J. Childress. Performance of lamella thickening in coal preparation plants, *SME. AIME.* 1980.
15. J. Lightfoot. Practical aspects of flocculation, *Minerals Engineering Handbook, Mine and Quarry*, **81**, 51–53, 1981.
16. H. A. Hamza. Least-cost flocculation of clay minerals by polyelectrolytes, *Transactions IMM*, Section C, **87**, C212.
17. L. Svarovsky. Filtration fundamentals, Chapter 9 in *Solid–Liquid Separation*, (Ed.) L. Svarovsky, Butterworth, Guildford, 1981, pp. 242–264.
18. D. G. Osborne. Vacuum filters, Chapter 13 in *Solid–Liquid Separation*, (Ed.) L. Svarovsky, Butterworth, Guildford, 1981, pp. 321-357.
19. S. M. Moos and R. E. Dugger. Vacuum filtration: available equipment and recent innovations, *Mining Engineering*, **31**, 1473–1486, 1979.
20. D. A. Dahlstrom. How to select and size filters, Chapter 28 in *Mineral Processing Plant Design*, (Eds.) A. L. Mular and R. B. Bhappu, American Institute of Mining Engineers, New York, 1980, pp. 578–600.
21. B. Wetzel. Disc filter performance improved by equipment redesign, *Filtration and Separation*, **11**, 270–274, 1974.
22. D. G. Osborne. Rotary vacuum filtration of coal flotation concentrates, *International Journal Mineral Processing* **3**, 175–191, 1976.
23. F. M. Tiller. What the filter man should know about theory, *Filtration and Separation*, **12**, 386, 1975.
24. D. A. Dahlstrom. Practical use of applied theory of continuous filtration, 2nd Pacific Chem. Eng. Congress. Congress Proceedings published by American Institute of Chemical Engineers, Denver 1977.
25. D. G. Osborne. Scale-up of rotary vacuum filter capacity, *Transactions Institute Mining and Metallurgy (London)* **84** (826), C158–C166, 1975.
26. R. J. Wakeman. Pressure filters, Chapter 12 in *Solid–Liquid Separation*, (Ed.) L. Svarovsky, Butterworth, Guildford, 1981, pp. 302–320.
27. R. H. Mason. Belt filter presses squeeze into preparation plants, *Coal Mining and Processing*, Jan. 1982, **19**, 50–56.
28. T. M. Fraser. The continuous pressure filter—a transfer of fines dewatering technology to the coal industry, 1st International Society of Mining Engineers of American Institute of Mining Engineers (SME-AIME) Fall Meeting, Honolulu. Sept. 1982.

29. M. L. oudil. Aspects of continuous pressure filtration engineering, *Aufbereitungs-Technik*, **27** (7), 387–395, 1986.
30. D. G. Osborne. Screening, Chapter 8 in *Solid–Liquid Separation*, (Ed.) L. Svarovsky, Butterworth, Guildford, 1981, pp. 213–241.
31. E. Ohl. New dimensions for dewatering screens, *Coal Mining and Processing*, Sept. 1979, **16**, 80–82.
32. L. Svarovsky. *Solid–Liquid Separation Processes and Technology*, Handbooks of Powder Technology, Vol. 5, Elsevier, New York, 1985, pp. 81–93. (Centrifugal Sedimentation).
33. Anon. Prediction of moisture of centrifugally dewatered solids, *Wemco Data Sheet*, April 1964. Wemco Handbook.
34. N. D. Policow and J. S. Orphanos. Development of the screen bowl centrifuge for dewatering coal fines, *Mining Engineering* **35** (4), 333–336, 1983.

Chapter 11

THERMAL DRYING

11.1 INTRODUCTION

Thermally drying coal is the most expensive unit operation performed on the coal during its preparation. Usually, therefore, it is only carried out when proven to be absolutely necessary.

Since the surface area of granular solids increases proportionally with fineness of size, the need to dewater the coal and its ultimate cost are closely related to the proportion of fines which it contains.

A good example of this is coal which is mined within the Rocky Mountains area in the western provinces of Alberta and British Columbia in Canada. In this area, run-of-mine coal is especially fine in size-consistency and consequently, mining operators in this area have become major users of thermal drying of clean-coal production. The coal from this area can often contain as much as 50% of the total mass in the minus 500 µm size-range. With mechanical dewatering, a surface moisture of below 15% is usually difficult and costly to achieve, whereas thermal methods permit moisture contents of 6–8% to be readily obtained. In the case of western Canada, thermal drying is justified by the extremely cold winter conditions and the shipping distance to export markets.

Removal of surface moisture by drying is done for one or more of the following reasons:

(a) to avoid freezing difficulties and to facilitate handling during shipment, storage and transfer to the market;
(b) to maintain high coal-pulverizer capacity in power-plant applications;
(c) to reduce heat loss due to evaporation of surface moisture during combustion;
(d) to beneficiate the coal for coke making and briquetting;
(e) to decrease transportation cost;
(f) to facilitate dry beneficiation.

A wide variety of dryers are available but only a relatively small number are used in coal preparation. The types can be grouped into direct or indirect heat-exchanger types, as shown in Table 11.1[1] and selection is influenced mainly by size distribution and the degree of dewatering required.

The cost benefits to coal transportation and utilization, especially for thermal coals, are the two reasons most commonly cited to justify thermal drying. Savings in transportation costs are directly proportional to the degree of beneficiation

TABLE 11.1 Thermal-dryer Option Analysis[1]

No.	Description	Coal size handled	Type of heating[a]	Heat-transfer media	Air handling emission consideration	Relative capacity/ number of units	Operating temperature	Comments
1.	Disc	−0.5 mm	i	steam or hot oil	yes/moderate	low/several 70 t/h max. solids	315°C max.	abrasion sensitive
2.	Fluidized bed	max. 45% −0.5 mm	d	air/ combustion gases	yes/high	high/single 50–600 t/h solids	480°C max.	large dust-recovery facility needed
3.	Holo-Flite	−0.5 mm	i	hot oil or oil base fluid	no	low/several 20–30 t/h solids	315°C	Joy Manufacturing Company patent
4.	Mech. dewatering	−0.5 mm	d	steam or exhaust gases	no	low/several 20–30 t/h solids	180°C	reduced fluid viscosity giving better drainage
5.	Radiant	−0.5 mm	i	electricity	no	very low/several	315°C max.	abrasion sensitive
6.	Flash	−10 mm	d	air/steam	yes/moderate	medium/several	650°C inlet	turbulent flow essential
7.	Rotary shell	max. 60% −0.5 mm	d	air	yes/moderate	medium/several	340°C	
8.	Multilouvre	max. 30% −0.5 mm 10 mm max.	d	brine/oil	no	low/several	315°C	
9.	Rotary-louvre (drum)	max. 60% −0.5 mm	d	air	yes/moderate	low/several	315°C	also an effective cooler
10.	Spray	−0.5 mm	d	air	yes/high	medium/several	315°C	coal slurry required
11.	Steam tube	−0.5 mm max. 50%	i	steam	yes/low	low/several	150–430°C	

[a] Type is direct, d, or indirect, i.

achieved with the coal, and therefore removal of water is almost as significant as removing non-coal material, especially for coals containing large amounts of fine material. As an example: for each 100-tonne rail-car load of coal, a rail charge of say £5/t may be charged to move the coal to a port 200 km from the mine. If the coal sales contract calls for a total moisture of 10% by weight, each rail car will contain up to 10 t of water, costing £50 to transport. Any improvement in this amount of water may represent a net saving if the cost of dewatering is lower than the rail cost. Usually, modest improvements can be very cost effective because a bonus may also be obtained for improvement in coal quality, e.g. higher heat content. Hence, benefits to utilization may also be realized. At a power plant, a reduction in moisture of about 1% will result in a slight increase in thermal efficiency in addition to the higher heating value obtained, due mainly to reduced heat-wastage in evaporating water.

11.2 PRINCIPLES OF THERMAL DRYING

The majority of industrial coal dryers are of the direct heat-exchange type that employ convection as the principle of heat transfer. In such a system, hot gases and wet coal are brought into intimate contact on a continuous basis. These gases contain nitrogen, oxygen, water vapour, carbon dioxide and carbon monoxide, together with small quantities of other gases, such as hydrogen and oxides of sulphur and nitrogen. The behaviour in the dryer is determined by the design of the drying system and the temperatures, pressures and composition of the gas mixture employed. If the mixture obeyed the Ideal Gas law, it would occupy a volume of exactly 22.4 litres at a pressure of one atmosphere (1 bar or 1.03 kg/cm^2) and a temperature of 0 °C. The weight of the gas would then equal its molecular weight (1 gram mole), e.g. 32 g oxygen (molecular weight 32) and the gas would occupy 22.4 litres. Only rarely do gases occur under standard conditions and other laws must also be considered. When gases are heated and temperatures therefore increased, expansion occurs (Charles' law) and when pressure is lowered, gases expand (Boyle's law). Conversely, when temperature is reduced or pressure increases, gases contract in volume. Further discussion of these laws lies outside the scope of this book, but knowledge of them is very useful in understanding gas-convection drying in particular, and other references should perhaps be sought.[2,3]

Calculation of the adjusted gas volume for varying pressure is achieved by multiplying the standard volume (22.4 litres) by 1.03 and dividing by the new absolute pressure. Adjusted gas volume for temperature variation is obtained by dividing the standard volume by the absolute temperature (273 K) and multiplying by 273 plus the new temperature.

Theoretical interpretation on the mechanism of drying and other general details are available in numerous chemical engineering texts.[2,4,5] The heat required for drying is supplied either through conduction, or radiation or convection. Despite the fact that most commercial coal dryers are of the convection type, all three forms of heat transfer occur, although heat is primarily convective by virtue of the design of the dryer. Drying with heated air involves both heat and mass transfer. When hot, unsaturated gas (predominantly air) flows through or across wet coal, vaporization of moisture will occur and water vapour is transferred to the gas in response to the humidity gradient across the gas–solid interface. Moisture concentration at the surface of the coal decreases and further moisture diffuses from the interior of the flowing coal or from within the coal itself by capillary action. As long as moisture is transferred to the surface at a rate similar to the rate of evaporation, the quantity of water evaporated will remain constant providing that the gas velocity, temperature and humidity also remain unchanged.

During this so-called *constant-rate* period where steady-state conditions prevail, the rate of drying is really controlled by the rate of vapour diffusion and therefore totally dependent upon the humidity differential.

Hence, in gas convection-dryer design, gas-flow calculations must cater for the required degree of moisture reduction by providing an evaporative capacity based on calculation of this differential converted to volumetric flow rate. At this point, it is perhaps useful to remember that it requires 610 kcal to evaporate 1 kg water.

When the diffusion rate begins to decrease, i.e. moisture migration rate to the surface falls below rate of evaporation at the surface (gas–solid interface), the drying rate drops rapidly, despite further moisture reduction, until an equilibrium condition is reached.[6]

A typical drying curve is shown in Fig. 11.1,[9] which relates surface moisture of the coal to the rate at which water is evaporated from the coal by the gas. Drying is usually preceded by an induction period of unsteady state during which the wet coal feed and the dryer gases become mixed or come into contact with each other. This period is usually very short and is often ignored in the analysis of time required for drying. Following this, constant-rate drying occurs, during

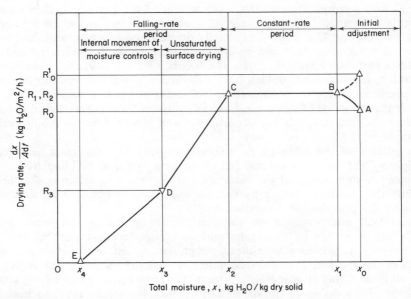

Fig. 11.1 Typical rate-of-drying curve on the basis of total moisture content: x_2 = critical moisture content; x_4 = relative equilibrium moisture content.[9]

THERMAL DRYING

which the most effective drying takes place. The final period is usually determined by the dryer type and design characteristics but is more noticeable in the case when low-rank coals are thermally dried and moisture reduction to below the normal equilibrium of the coal is required. For these types of coal, diffusion by capillary action from within the coal structure is essential and drying of surface coal forms only a small part of the total dewatering requirement.

Adjustments to total drying-time requirement will usually be possible with most gas-convection types of thermal dryer, and parameters to be considered are feed-size consistency, gas velocity and temperature, temperature differential between gas and feed coal and the physical dimensions of the feed-bed of the drying vessel. Controlling the dryer to achieve the best performance possible involves optimum application of the hot convection gases to the coal as well as the mechanical handling of coal in the dryer. It is important to remember that as the drying approaches the minimum attainable moisture level, the temperature of the coal will begin to increase, approaching the temperature of the drying gases. Great care must be exercised in preventing too rapid a temperature rise in the coal, in order to avoid ignition or the initiation of autogenous heating of those coals known to be prone to spontaneous combustion.

11.3 THERMAL DRYER TYPES

Coal thermal dryers can be divided into two basic groups, i.e. *direct* or *indirect* heat exchange, as shown in Fig. 11.2.

As previously mentioned, the most common dryer in use today for coal preparation is the direct-heated, fluid-bed type for which the fuel source can be any one of numerous alternatives, i.e. pulverized or stoker coal, gas or fuel oil; and in some cases gas is used as a back-up to low-volatile coal where the dryer uses part of the coal product it is drying as a fuel source. Direct heat-exchange dryers may

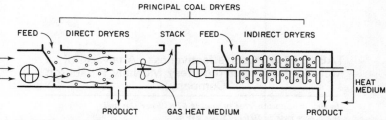

Heat transfer is by direct contact between damp coal and hot gases. Resultant vapour is carried away by the hot gases (heat medium); also called CONVECTION DRYERS.

examples:
(a) Pneumatic conveyor dryers (flash or suspension types)
(b) Fluidized-bed dryers
(c) Rotary dryers (single-shell or rotary louvre types)
(d) Steam filtration or centrifugation

Heat transfer is via metal casing or shell. Vapour is removed independently of heat medium. Rate of drying is dependent upon contact and rate of heat transfer through the wall; also called CONDUCTION DRYERS.

examples:
(a) Screw conveyor type using hot oil, steam or hot air (Holo-Flite® and Torus Disc)
(b) Rotary dryers using steam or hot air

Fig. 11.2 Categories of thermal dryers for coal.

TABLE 11.2 Thermal Drying of Bituminous and Lignitic Coals, in North America in 1975[7]

Dryer type	Number units	Approximate tonnage treated (10^3 t)	Average[a] per unit (t/h)
Fluidized bed	56	22 400	100
Multilouvre	12	2 735	57
Rotary	3	635	53
Screen	5	1 785	89
Suspension (flash)	30	5 250	44
Cascade	1	50	12.5
Total	107	32 855	76.75

[a] Average hours operated per annum taken as 4000.

suffer the disadvantages of high potential loss of fine particles, requiring expensive dust-extraction equipment in order to meet stack gas-emission environmental regulations. In extreme cases, alternative types of dryers may be sought. Also, coal types which are friable or decompose rapidly during heating, such as low-rank coals, exhibit large potential quantities of fine particulate content. This depreciated product may be avoided by selection of other forms of dryers, in particular by indirect heat-exchange types which, despite being more expensive and limited in unit capacity, may not result in significant product degradation or the additional equipment requirement for particulate-emission control.

Fluidized-bed dryers, some with capacities in excess of 1000 t/h, are currently in operation in coal-preparation plants and this form of dryer represents the largest unit capacity of any dryer currently in use for coal-drying applications, as can be seen from Table 11.2. Some indication of the cost-effectiveness of fluidized-bed thermal dryer is given in Table 11.3.

TABLE 11.3 Cost-effectiveness Analysis of a McNally Flowdryer (courtesy McNally Pittsburgh Company)

In this installation, cost of coal is $2.839 per ton and the power rate is $0.01175 per kWh.

Operations	Cents per ton
Operating labour	0.955
Fuel (start-up oil and coal)	3.475
Power	2.140
Miscellaneous (operating supplies)	0.105
Total operating costs	6.675
Maintenance	
Maintenance labour	1.366
Maintenance supplies	0.332
Total maintenance costs	1.698
Total cost per ton of dried coal	8.373

THERMAL DRYING

11.4 CONVECTION DRYERS

Hot gases generated in a combustion chamber are used as the medium to transfer heat to the coal-drying zone. This must be done carefully so that ignition or overheating of the coal does not occur, causing devolatilization or partial oxidation. This latter effect applies to coking coals and in particular those having special coke-blending properties, e.g. high fluidity and dilatation. In many cases, however, laboratory examinations into the effects of heating on coking-coal properties have shown that in the temperature range encountered during gas-convection drying, the extent of oxidation caused by heating has little practical consequence.[8] Nevertheless, excessive exposure caused by dryer operation at well below rated capacity can prove detrimental, and this is carefully avoided by providing automated feed-rate control systems. Exhaust gases transporting moisture away from the coal must remain sufficiently hot to prevent condensation. They must also accommodate gaseous and solid combustion products in some cases, and particulate removal by both wet and dry methods is commonly an integral part of the total dryer plant in the case of certain types of fluid-bed and rotary-louvre dryers.

The two most common direct, convection types of dryer are:

(a) rotary: single-shell or multilouvre dryer designs;
(b) fluidized-bed dryer designs.

11.4.1 Directly Heated Rotary Dryers

Single-shell Rotary Dryer

In its simplest form, the single-shell rotary dryer comprises a rotating cylindrical shell, set with its axis inclined at a slight angle and mounted on rollers to facilitate rotation (see Fig. 11.3). The coal is introduced at the highest end and agitat-

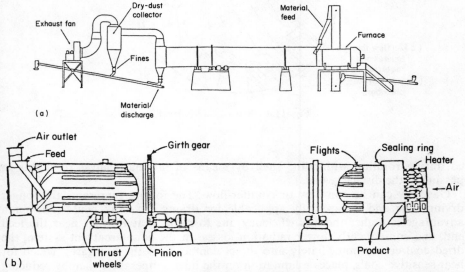

Fig. 11.3 Single-shell rotary dryers: (a) parallel flow; (b) countercurrent flow.

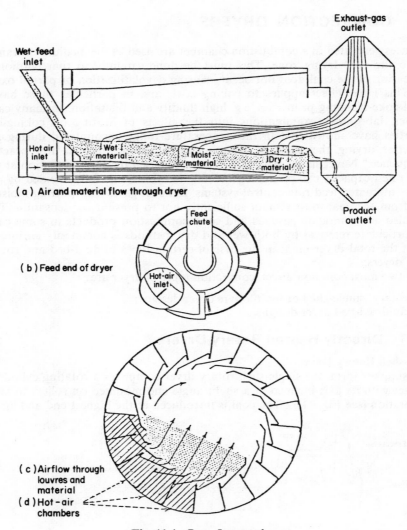

Fig. 11.4 Roto-Louvre dryer.

ed during rotation within the shell by means of internal shelves or flights in passing to the discharge end.

Gas flow can be con-flow or counter-flow. The former represents the longer drying time and is especially applicable for drying low-rank coal.[9] The main advantage is higher thermal efficiency due to rapid heat transfer near the feed end, and consequently, lower radiation losses. In the countercurrent system, the dried coal comes immediately into direct contact with the hot gases, which for heat-sensitive coals, places a limitation on the inlet temperature thereby reducing the attainable thermal efficiency. In some cases, this causes higher gas-flow rate and consequent increased dust-handling capacity.

Rotary Louvre Dryer

This type of dryer consists of a solid outer cylindrical and an inner shell composed of stainless-steel louvres arranged and shaped as shown in Fig. 11.4. The inner shell of overlapping louvres supports the bed of coal and usually increases in diameter in the direction of flow. The damp coal is gently conveyed by the louvres to the discharge end as the drum slowly revolves. During the drying operation, hot gases (usually air plus combustion gases), which are introduced through the louvre openings, permeate the coal bed and thereby come into intimate contact with every component of the bed.

These machines are built in sizes from 0.75 to 4.0 m in diameter and the diameter-to-length ratio is usually about 1 : 4. The thermal efficiency of rotary dryers is a function of the temperature levels and usually ranges from 30% to 60% for coal-drying applications. Rotary-louvre designs of dryer can have up to twice the evaporative capacity of the single-shell type, ranging from about 0.50 to 1.00 kg/m^3/h.

11.4.2 Design Considerations for Rotary Dryers

A major factor in the design of a rotary dryer for fine-coal drying is the selection of the gas velocity at just below the critical level at which coal dust is carried along by the gas flow. In rotary-louvre designs, gas velocity is only 30% of that through a similarly sized single-shell unit, because of the louvre system which creates a reduction in gas disturbance during normal operation.

Heat transmitted by the gas stream passes to the coal during the cascading action and also to the hot walls of the shell. The heat-transfer equation for this system is as follows:

$$Q = kaV\Delta T$$

where Q is the rate of heat transfer, k is the overall heat-transfer coefficient, a is the area of contact between the coal grains and the gas per unit volume, V is the volume of the dryer, and ΔT is the mean value of the temperature difference between the gases and the coal.

Many empirically derived values are used by dryer manufacturers in determining gas rates and corresponding heat-transfer rates, and caution should be exercised in using such data for commercial applications. Unless similar applications of comparable size exist, the design of new systems is usually preceded by bench-scale and pilot-scale test programmes in order to obtain reliable design data.

11.4.3 Indirectly Heated Rotary Dryers

One form of indirectly heated dryer, the Bearce dryer, is shown in Fig. 11.5.[10] A rotating drum 13 m long and 2 m in diameter is partially enclosed within a stationary shell 11 m long and 2.7 m in diameter. The rotating drum is jacketed by the stationary tube and the hot combustion gases are circulated between the two. Attached to the exterior of the rotating drum, are metal fins that run parallel to its axis. These fins help to distribute and transfer heat from the hot gases to the surface of the drum. The fuel source can be coal, oil or gas, and the hot products of combustion are introduced through openings on one side of the outer tube. An exhaust fan removes the gases via a stationary tube at the coal-discharge or lower end of the dryer. Moisture removed from the coal is carried through a stack at the discharge end.

Fig. 11.5 Photograph of the Bearce dryer showing the five burners (courtesy Indiana Steel and Fabricating Company).

Five burners spaced evenly at 1.5 m intervals along the feed portion of the drum are capable of 1.5 MW. At full operating capacity, this type of dryer will achieve a temperature of 150–430 °C depending upon the moisture content of the feed.

For very damp coal feeds, the use of chains (as shown in Fig. 11.6) will assist in coal transportation and also prevent the coal from sticking to the interior of the drum.

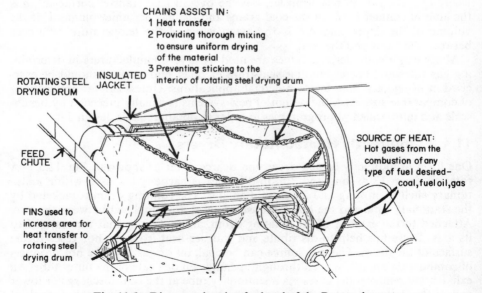

Fig. 11.6 Diagram showing feed end of the Bearce dryer.

THERMAL DRYING

11.5 FLUIDIZED-BED DRYERS

11.5.1 General

The use of the fluidized-solids technique emerged with applications in the petroleum and chemicals industries for processes where it was required to bring about intimate contact between small solid grains and a gas stream. One major application in coal treatment has emerged, this being the combined drying and combustion of coal. The phenomenon of fluidization, which has been discussed in detail in several publications,[12] is an operation whereby coal grains are transformed into a fluid-like state by passage of heated flowing gases. At low flow-rates, the gas percolates through the void spaces between the grains without causing any physical disturbance. Increased gas-flow causes the grains to dilate and move relative to one another. Eventually, a gas flow-rate is achieved when the grains become suspended in the upwards-flowing gas. This condition is called incipient fluidization. Further slight increase in flow rate creates more disturbance of the solids, and a thicker but more dilute bed is formed. Such a bed is called aggregated or bubbling fluidization due to the fact that the grains move vigorously. Any further increase will cause entrainment of solids, and some solids will be carried out of the bed with the gas stream. Figure 11.7 demonstrates these phases.

11.5.2 Fluidized-bed Systems

Fluidized-bed dryers, such as the McNally Flowdryer, Link-Belt Fluid Flo, ENI Coal-Flo and Heyl and Patterson Fluid Bed units, are widely used in North America for drying large flow rates, and individual units capable of evaporating over 100 t/h of water from a 1500 t/h feed are currently in operation.

Typical fluid-bed dryer capacities are shown by the graph in Fig. 11.8, with bed-plate areas in commercial units ranging from 4 to about 20 m². Figure 11.9 shows the major components of a fluid-bed dryer and indicates the typical temperature variations that occur during normal operation. Fuel source can be stoker (lump) or pulverized coal, natural gas or oil and all of these alternatives are frequently employed.

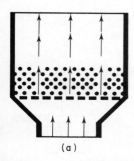 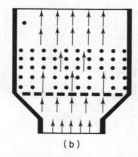 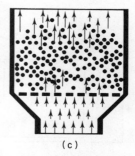

Fig. 11.7 Fluidization phenomena. (a) Gas velocity < fluidizing velocity; bed undisturbed. (b) Gas velocity > minimum fluidizing velocity; bed suspended. (c) Gas velocity approximately three times minimum fluid velocity; bed turbulent, behaves like a boiling liquid.

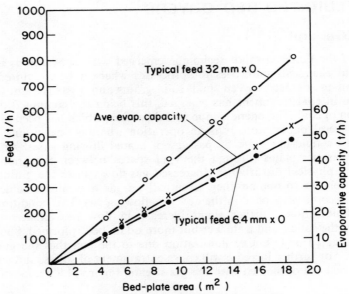

Fig. 11.8 Typical fluid-bed dryer capacity.

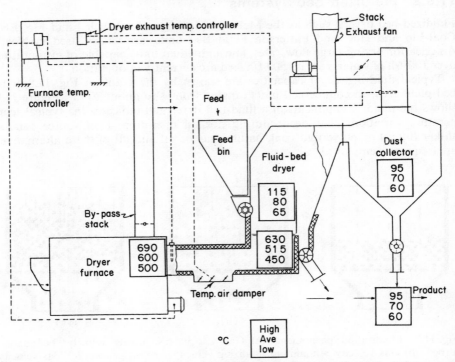

Fig. 11.9 Major components of fluid-bed dryer design and high, average and low temperatures during normal operation.

THERMAL DRYING

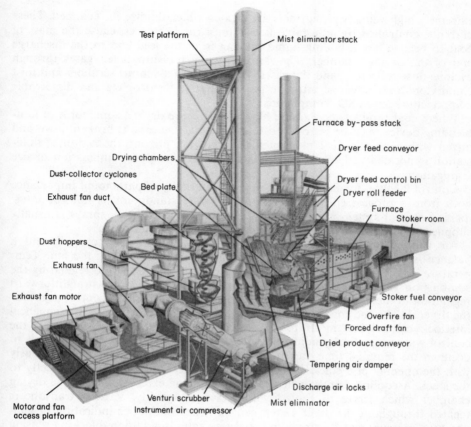

Fig. 11.10 Typical fluidized-bed thermal coal dryer (courtesy ENI).

Most dryers were originally designed to handle coal of up to 32 mm top size and ideally, the amount of gas-transportable fines in the product coal should not exceed 10%. There are, however, many dryer installations which dry coals of up to 50 mm top size and products with over 30% fines. In order to cater for the latter problem, large dust-handling capacities are required.

A typical dryer installation is shown in Fig. 11.10. In the drawing shown, the furnace is fired by either a spread-stoker or a pulverized-coal system. With the stoker, a portion of the coarse dryer product would normally be used for the fuel source. When the pulverizer is used, a complete pulverizer–burner system is provided. It is also possible to use a portion of the fines recovered from the dry dust-collection cyclones as furnace fuel to reduce pulverizing costs.

Coal to be dried is conveyed to a feed-control hopper fitted with some form of feeding system which can be readily and accurately controlled. Screw feeders or a vibrating-bin discharge mechanism are common methods for introducing a uniform flow of damp coal onto the constriction deck of the bed plate. The heated air from the furnace is pulled through the constriction plate under negative pressure by an induced-draft fan. As it passes through the constriction plate, extremely high-velocity air currents are created which fluidize the coal bed. These currents, controlled by adjustable flow-rate of the hot gases, cause the mass of coal to behave like a boiling liquid moving from the feed end to the discharge end of the drying chamber. The fan draws the moisture-laden gases through cyclone dust collectors, and the gases discharge into a venturi scrubber and mist eliminator for removal of ultrafine particulates, water droplets and dissolvable components such as SO_x compounds.

Where coal products containing filter cake require drying, some form of feed-blending device must be introduced to ensure that the cake is broken down and mixed with coarser coal. Unless such materials are present, the action of fluidization is normally adequate for achieving good mixing and elimination of size segregation.

Control facilities for fluidized-bed dryers are usually of paramount importance both from a process efficiency and from a safety standpoint. A fail-safe interlocking control system with automatic and manual emergency sprays is usually supplied and any malfunction will normally cause total shut-down.

Air volume is controlled by sensing the induced-draft fan current, and a balance is maintained by opening and closing the damper for the fan. Temperature is controlled by sensing the exhaust-gas temperature and regulating the combustion rate accordingly. This is usually accomplished by straightforward regulation of fuel supply. The exhaust temperature is the main control point for the dryer. High- and low-level set-points are established as part of the sequence-interlock control system. If the inlet temperature exceeds the set-point, the control will automatically adjust the fuel supply. If the temperature drop fails to occur in the required time, the drying system will be shut-down simultaneously with the opening of a stack damper to by-pass the hot furnace gases directly to the stack. A cooling damper will also open, allowing external air into the drying chamber which serves to cool down the system quickly. Various transducers located throughout the dryer system will provide continuous indication of temperature, pressure, gas-flow rate and coal-feed rate. In addition, blocking sensing devices will indicate coal-flow problems. With such facilities, supported by microprocessor centralized control, the plant operator is usually able to ascertain the cause of problems very quickly.

11.5.3 Design Consideration for Fluidized-bed Dryers

The most important parameter in terms of dryer design is temperature. For peak efficiency to be attained, the highest safe temperature should be created in the drying chamber in order to minimize sensible heat losses to the exhaust gases. Lower temperatures often mean that below-optimum levels are obtained in fuel consumption and power requirements and consequently, in thermal efficiency, as well as increase in dust carry-over. In order to determine the key design data, i.e. fuel consumption, heat output, inlet and exhaust temperature and gas volume requirements, a check-list of operating data must be completed. Table 11.4 gives an example. In some cases, particularly for new applications, such as in drying low-rank coal, bench- and pilot-scale drying tests are advisable in order to obtain the necessary design data. In fluid-bed dryers, the bed-volume requirements are of less importance than the drying area, which should be selected to conform to the carefully designed area and the enclosed constriction plate. The selected

THERMAL DRYING

TABLE 11.4 Design Data Sheet for New Thermal-dryer Installation

GENERAL INFORMATION
COMPANY... DATE.........
ADDRESS................. TELEPHONE.............. TELEX.........
CONTACT PERSON.................. TITLE............................
PLANT LOCATION................... PLANT ELEVATION...................
EQUIPMENT FUNCTION..
MATERIAL TO BE DRIED..

DESIGN INFORMATION
CAPACITY REQUIRED......... tonne/h
CHEMICAL REACTION (if any).........
HEAT OF REACTION.................. @ 25°C
REQUIRED REACTION TEMPERATURE......... °C
SPECIFIC HEAT OF SOLIDS..........................

FEED MATERIAL
FEED TEMPERATURE......... °C LOOSE BULK WEIGHT......... tonne/m^3
FEED MOISTURE OR SOLVENT CONTENT........................... kg/kg FEED
SOLVENT......... IF BOUND MOISTURE PRESENT WHAT % IS BOUND?....%

DISCHARGE MATERIAL
DESIRED TEMPERATURE......... °C LOOSE BULK WEIGHT......... tonne/m^3
MAXIMUM TEMPERATURE......... °C RESULT OF OVERHEATING.................
DISCHARGE MOISTURE OR SOLVENT CONTENT.................. kg/kg FEED
SIZE DISTRIBUTION ANALYSIS.....................................

DUST COLLECTION
( ) CYCLONE   ( ) WET SCRUBBER   ( ) BAG FILTER   ( ) OTHER
MAXIMUM ACCEPTABLE DUST LOADING OF EXHAUST GASES.............. p.p. million
IS MATERIAL CONTAMINATED BY CONTACT WITH COMBUSTION GASES?.........
IS MATERIAL AVAILABLE FOR PILOT PLANT TESTS? (MINIMUM 500 kg)...............
OPERATING TIME......... HOURS PER DAY......... DAYS PER WEEK.........
WHAT IS PROJECT PROGRAMME?......... START-UP DATE.........

UTILITIES
STEAM PRESSURE.................. N/m^2 TEMPERATURE.................. °C
NATURAL GAS......... N/m^2 CALORIFIC VALUE.... mJ/m^3
COAL SOURCE......... PULVERIZED OR STOKER......... CALORIFIC VALUE.... mJ/kg
OIL TYPE.........   CALORIFIC VALUE.... mJ/kg
COAL FUEL ANALYSIS: PROXIMATE:–ASH.... MOIST.... VOLS.............. %
(OIL FUEL ANALYSIS): ULTIMATE: C.... H.... N.... O.... S.............. %
COOLING WATER TEMPERATURE......... °C
ELECTRICAL POWER SUPPLY......... W   .... kVA
INSTRUMENT AIR PRESSURE......... N/m^2

OTHER PERTINENT INFORMATION
ENVIRONMENTAL IMPACT.....................................

drying chamber area should be vertically telescoped for an elevation preferably more than twice the effective length (in the feed direction) of this plate. Another limitation in designing fluidized-bed dryers is the difficulty encountered in accu-

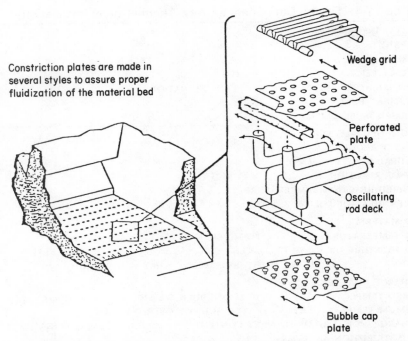

Fig. 11.11 Fluidized-bed construction deck detail (courtesy FMC).

rately predicting bed pressure drops, bed expansion height and the height above the bed above which no further redeposition of gas-borne solids occurs.[13]

Much of the information necessary to determine values for each of these dimensions is obtained from two sources:

(a) other operational drying installations;
(b) bench- and pilot-plant testing with a particular coal.

Usually, the basis for design is the constriction plate.[14] The constriction plate contains uniformly spaced orifices or perforations as shown in Fig. 11.11. The total opening usually represents between 7 and 15% of the total area and creates a high-pressure drop between the approach and fluidizing sides of the plate. This differential is crucial, ensuring a high and uniformly distributed superficial gas velocity from each perforation, and the combined effect will then be good fluidization.

11.6 OTHER FORMS OF DIRECT DRYER

Several different types of directly heated drying units are usually found in coal preparation in addition to the two main types—rotary and fluidized-bed—so far discussed. Most employ steam to improve filtration drainage rather than direct-evaporation methods. Such 'dryers' therefore include:

THERMAL DRYING

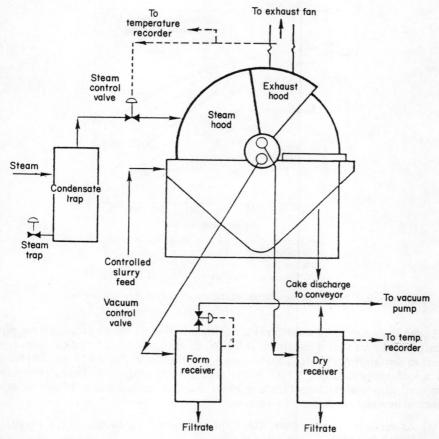

Fig. 11.12 Schematic of typical control method—steam drying with the Agidisc filter (courtesy Eimco Equipment Company).

(a) continuous-rotary or belt vacuum filtration; and
(b) solid- or screen-bowl centrifugal dewatering.

Two types of vacuum filter have been shown to benefit from the introduction of steam. These are the rotary vacuum drum and horizontal-belt types. The steam is applied via a steam hood located above the drying zone of the unit. For the horizontal-belt unit, the hood may be divided into several compartments in order to ensure good distribution. Figure 11.12 shows a typical arrangement for a drum filter. Steam at a maximum temperature of about 175 °C and atmospheric pressure is used and a moisture reduction occurs as a result of reduced viscosity from one centipoise (1 cP) at standard temperature and pressure (STP), to as low as 0.3 cP. Further airflow, occurring immediately prior to discharge or later during the transportation of the filter cake, results in an additional loss in moisture by adiabatic cooling from humidification and the release of sensible heat.[15]

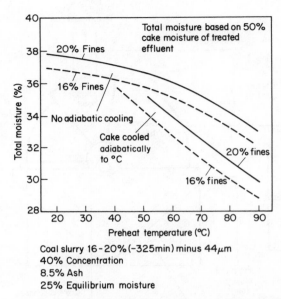

Fig. 11.13 Screen-bowl centrifuge tests coal slurry pipeline.

Large-scale testing of solid-bowl centrifuges with coal slurry produced for pipelining has shown that if the slurry is heated to above 80 °C, a significant improvement in dewatering occurs. This is shown graphically in Fig. 10.51 in Chapter 10. At a feed temperature of 93 °C, 6–8% reduction in surface moisture above that achieved at ambient temperature is obtained. The heated slurry results in more effective dewatering due to two reasons:

(a) At elevated temperature, the fluid viscosity is reduced, which results in better drainage.
(b) Adiabatic cooling then occurs and the latent heat associated with the hot cake is effectively used to produce further dewatering.

This second application is shown in Fig. 11.13 whereby a centrifuge is steam assisted in dewatering coal-fines slurry.

The economics of steam filtration or centrifugation will greatly depend upon whether or not steam generation costs can be recouped by premium earned from producing a lower-moisture filter cake. It is often very difficult to quantify this benefit, and installations are usually found in coal-preparation plants where a low-cost source of steam is readily available. A dedicated steam-generating unit is rarely justifiable.

11.7 CONDUCTION OR RADIANT-HEAT DRYERS

These indirect types of dryers use heated surfaces in intimate contact with coal to achieve drying. They require a minimum of air and therefore present a low environmental pollution potential.

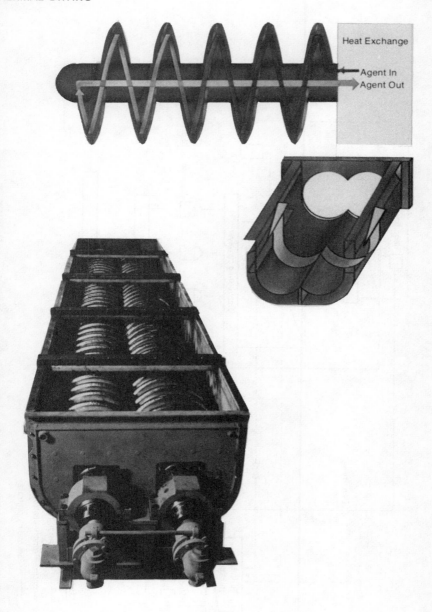

Fig. 11.14 Holo-Flite® thermal dryer (courtesy Joy–Denver). (a) A stainless-steel D2420-6 Holo-Flite® Processor without cover. (b) Holo-Flite® screw. The heat exchange agent enters the first flight on the rotary joint end, travels the length of the Holo-Flite® Processor screw, and then exits the unit from the same end. (c) Trough jacket. Trough walls provide additional heat-transfer surfaces as shown. The heat-exchange agent moves down and back along the length of the jacketed trough.

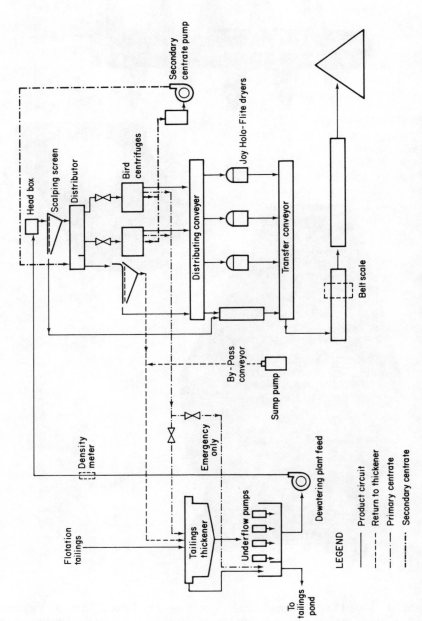

Fig. 11.15 Flowsheet of tailings dewatering plant, incorporating Holo-Flite® dryers.

Numerous units are in use for drying coal, although none can be regarded as being commonly applied. Most indirect heat-exchange type coal dryers have low material-flow capacities, making their application an expensive multi-unit system, and they are therefore most often used for finer-sized coal.

The Bearce dryer described earlier comprises a rotary double-shell cylinder slightly inclined, into which wet coal is introduced. External heater units supply the hot gases which pass between the two shells. This dries the fine-coal feed by direct contact with the inner shell.

The Holo-Flite® dryer is an auger-type continuous heat exchanger comprising a heated chest containing a hollow-flight helicoid screw conveyor. Both the chest walls and the hollow flights are filled with circulating hot oil which dries the wet coal. This unit is shown in Fig. 11.14. There are now several installations of this dryer in North America including one in Canada where three Holo-Flite® units are used in parallel to dewater froth-flotation tailings which are then used as a thermal fuel for a power-generating station.[17] Figure 11.15 shows the circuit that is employed. The evaporative capacity of each unit is 4.5 t/h from 30 t/h wet cake feed, containing 25–30% moisture by weight. The dried product contains about 8% moisture. The three Holo-Flite® units, each containing intermeshing hollow flights 610 mm in diameter and 7.23 m long, employ a heat-transfer fluid called Therminal 66. This fluid is heated to 315 °C and pumped at a rate of 60 litre/s. The heater is rated at 15 MW. The rate at which the wet solids pass through the dryer is variable because the dryer unit is driven by a 93-kW hydraulic drive which can be varied from 0 to 12 r.p.m.

A similar type of dryer is the Torus Disc heat exchanger[13] which consists of a stationary horizontal vessel with a tubular rotor on which are mounted hollow discs which provide about 85% of the heating surface of the unit. The remainder is provided by the rotor shaft and the inner wall of the vessel. Figure 11.16 is a diagram of this unit.

In order to obtain a high heat-transfer coefficient, it is essential that new surfaces be generated as material passes through the unit. This is achieved by stationary agitation ploughs located between each pair of discs. The material is moved along the axial direction by adjustable conveying vanes fixed to the outer rim of each disc, which stir the material through the annular space between the discs and the vessel.

In contrast to the Holo-Flite® dryer in which rotor speed determines product-flow rate, it is claimed that the Torus Disc rotor can be operated at the speed that gives optimum heat transfer. Residence time is independent of rotor speed and is instead adjusted by means of an overflow weir at the discharge end. This ensures an average material hold-up of up to 90% of the total available volume within the drying chamber. Surface temperature is about 300–320 °C. The usual heat source is steam or hot oil which is required at a ratio of 1–1.5 kg per kg water to be evaporated. Units with working volumes of up to 25 m^3 are available with body dimensions of 12.4 m long × 2.6 m wide × 4 m high, but applications in coal drying are rare and smaller units have been used. Disc diameters of up to 2.4 m are available and one unit handles 64 t/h of 0.6 mm × 0 mm fine coal, reducing the moisture from 25 to 10%. The rotor unit rotates at between 5 and 10 r.p.m. but is not variable.

The three examples of indirect thermal dryers so far quoted are intended to demonstrate alternatives which have found coal application. There are numerous

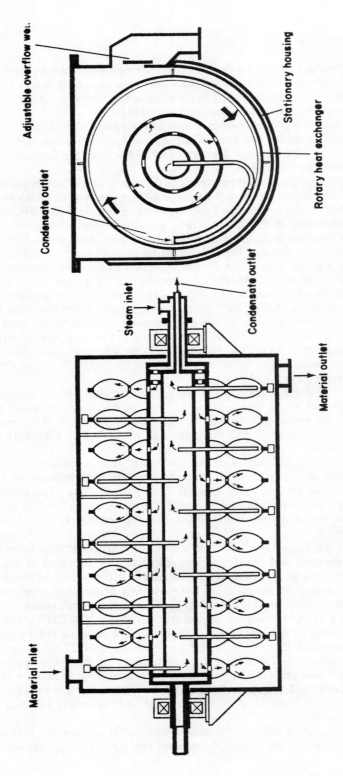

Fig. 11.16 The Torus disc thermal dryer (courtesy Bepex Corporation).

THERMAL DRYING

others which could be used for fine-coal drying, and a growing awareness of their advantage over direct dryers for coal applications is evident, especially in the treatment of fine coal. Indirect dryers claim several distinctive advantages of which the principal ones are:

(a) The product does not contact the heat-transfer medium, and risk of contamination is avoided.
(b) The finest particles contained in the feed material are protected from dusting because of the small volume of vapour involved.
(c) The relatively low transport velocities reduce product degradation.
(d) The explosion hazard is much lower and more readily controlled.
(e) The potential thermal efficiency is higher in many cases, due in part to agitation and also to the close proximity of the material to the heating surface.

Typically, the direct dryer is used for smaller- or medium-sized production, i.e. 10–30 t/h per unit. In most cases, mechanical dewatering by filtration, centrifugation or screening should precede the drying step because less energy is required and the dewatering cost can be optimized. Probably the most important phase in determining the application of this type of dryer is pilot testing, and most manufacturers of indirect thermal dryers which are applicable to fine-coal drying have pilot-scale units available for such tests.

11.8 NOVEL DRYING SYSTEMS

With the growing awareness of the need to address the problem of fine-coal beneficiation, an increasing amount of attention is being devoted to thermal dewatering. In addition to the several indirect thermal-dryer installations described in the previous section, there have also emerged a number of novel methods which have included the use of sonic energy, microwave energy and liquified carbon dioxide and nitrogen. Technically effective drying systems have been developed for drying inert granular solids, but their application to coal is still under development.

The MCS Biomass Converter[19] which is shown in the diagram in Fig. 11.17 was developed in Canada for drying fish meal. It is a dehydration system which employs a principle similar to pulse jet engines used in the early days of rocket propulsion. Fuel is 'detonated' in the combustion chamber generating shock waves that atomize the moisture which is then evaporated by hot, exhaust, gaseous products of the combustion. Drying efficiency is high with most units requiring only 700–800 kcal/kg of water evaporated compared to more than 1000 kcal/kg of water for direct rotary dryers. This improvement is credited to the sound waves which generate rapidly fluctuating pressures creating some drying at subatmospheric pressure.

The dryer plant shown in Fig. 11.17 is 21.4 m long, 10.7 m wide and 6.7 m high. It is fuelled by diesel and propane gas and has an exhaust gas emission of over 1000 m^3/h at atmospheric pressure.

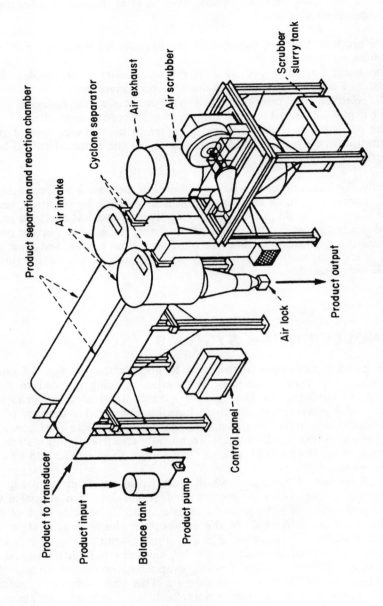

Fig. 11.17 The MCS biomass converter.

11.9 DRYER SELECTION

The preliminary selection of a dryer usually falls into two steps:[2]
  (a) listing those dryers that can handle the material requiring drying; and
  (b) estimating the total annual cost (including capital charges and operational cost) of each potential dryer and eliminating the most costly alternatives.

The graph in Fig. 11.18[21] indicates the scope of the drying equipment according to the method by which the material is heated and transported. Most coal dryers will accept free-flowing granular solids but damp coals demand more specialized methods of either blending or transportation. The number of alternatives will therefore be quickly reduced by considering the mode of operation, mode of heating, whether to employ a direct or indirect system and the specific peculiarities of the material being considered.

When the alternatives have been reduced to a minimum, the basic design criteria must be obtained, an indication of which was presented previously in Table 11.3. From these data, and bench-scale testing to obtain a typical drying curve (similar to that shown in Fig. 11.1), it should be possible to make a selection of a particular dryer type. A crude plot of the drying curve can be obtained from an infrared analysis or by using a balance and a drying oven to dry a few kilograms of sample material. Other factors, such as the type of contacting surface with which the drying chamber is lined, or the stickiness of a filter cake, must also be assessed by simple testing because in indirect-dryer selection these could be significant factors in selecting a specific type of unit.

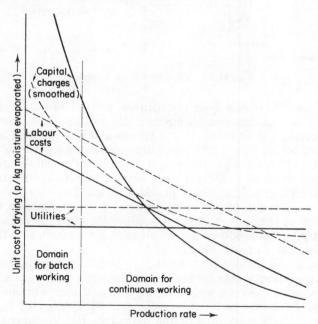

Fig. 11.18 Variation of unit costs of drying with production rate --- batch, —— continuous working.

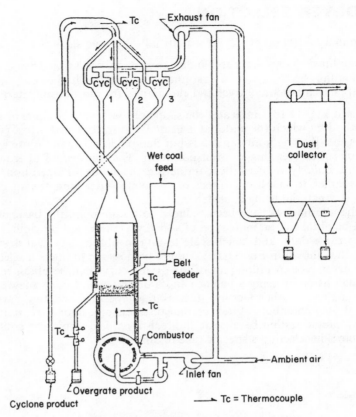

Fig. 11.19 Pilot flowdryer system.

Eventually, for 85% of new dryer installations, pilot-plant testing is required to ascertain selection and determine equipment size and ancillaries. Several tonnes of 'typical' material will be required as a sample and it is common for such tests to be conducted on the mine site in order to ensure that a representative wet-coal sample is treated. According to several researchers[22] pilot-plant test data obtained with a fluidized bed dryer have verified everything the drying curves have predicted and, in addition, have provided insight into other aspects such as the degree of degradation which occurs with drying, dust emission to the stack, drying to below equilibrium moisture conditions for low-rank coals, etc. Figure 11.19 shows a pilot fluidized-bed dryer unit used for research into the drying of sub-bituminous coals from the western USA.

For direct-heat dryers, as a rule of thumb, a simple rotary counter-flow dryer with average gas velocity may be five to six diameters long in order to achieve the required product moisture with a 50 mm × 0 coal feed. For con-flow drying the length would increase to seven or eight diameters. Alternatively, the rotary louvre type might only need three to four diameters. For the same result, a fluid bed of rectangular configuration could be as small as 0.6 diameters square, requiring much less floor area, but its height would be significantly greater.[20] It is

THERMAL DRYING

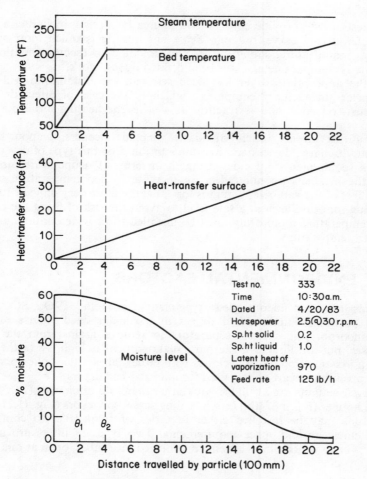

Fig. 11.20 Continuous-dryer data.[18]

usually the case for coal drying that the selection procedures when applied to these alternatives will point to a clear choice.

The simplest method for assessing all indirect dryers is to collect data that will enable a comparison to be made of feed rate per unit area versus final moisture. The object is to develop a curve like that shown in Fig. 11.20 which involves the following data:[18]

(a) bed samples for moisture analysis;
(b) bed temperatures at each sample point;
(c) heating-medium temperature;
(d) agitator speed;
(e) power reading;
(f) operating pressure.

Batch testing will provide conservative figures for design, including a reliable heat-transfer drying curve, but pilot-scale unit testing is much more valuable. Each dryer manufacturer should be able to provide clear recommendations once the specific system parameters have been defined by test work.

Once the appropriate dryer has been selected, the installation will require various pieces of ancillary equipment, which in the case of direct dryers will include dust-collection and -extraction as well as the heating furnace and fuel-handling system.

For indirect dryers, such auxiliaries include feeders, vapour-handling equipment, discharge devices and instrumentation. For this type of dryer, instrumentation requirements are most stringent in terms of temperature monitoring because the products do not reflect an increase in bed temperature until the cooling effect of the vaporized liquid is lost. It is therefore very difficult to control the product moisture level at a safe level to avoid the risk of firing the coal, and reliable temperature monitoring must be installed to provide continuous indication of bed temperature.

11.10 ENVIRONMENTAL FACTORS

Stack emissions from uncontrolled dryer plants can include NO_x, SO_x and CO and fine particles. Fluidized-bed dryers in particular need to use extraction systems incorporating cyclone separators to remove grains from the exhaust dryer gases, but are limited if the particulate content is high. Nevertheless, dry extraction systems employing cyclones with either fabric filter collectors or electrostatic precipitators are widely used for fluidized-bed dryers.

Multicyclone units are available with either manifold discharges (Fig. 11.21(a)), common hopper (Fig. 11.21(b)) or self-sealing screw conveyors (Fig. 11.21(c)). Collecting units may be installed either in series or in parallel to comply with required efficiency and space requirements. Two types of filters are used: the compact type in which filter elements are formed into tubes, open at one end and

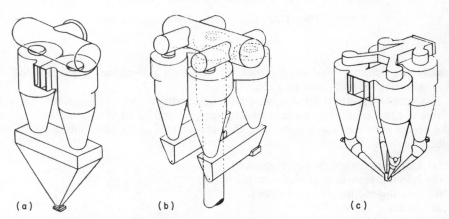

Fig. 11.21 Multicyclone dust collectors: (a) manifold; (b) trough hopper; (c) screw conveyor.

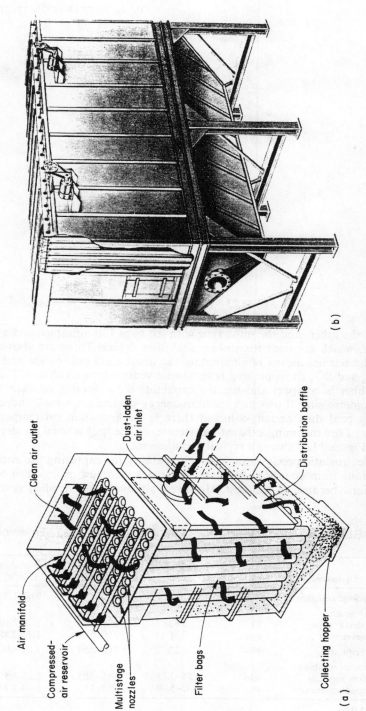

Fig. 11.22 Dust filter units: (a) contact type; (b) pulse type.

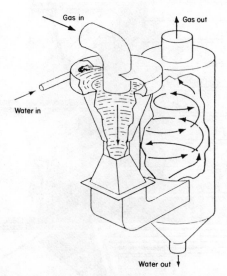

Fig. 11.23 Kinetic venturi scrubber.

closed at the other; and the pulse type with felt bags 150 mm thick and up to 4 m in length, which are used for stoker, coal-fired dryers. These are shown in Fig. 11.22. The normal means of eliminating the finest-sized particulate and controlling NO_x and SO_x emission level is by kinetic venturi-type scrubbers. This type of wet scrubber is compact and has the capability for collecting particles down to submicrometre sizes. Low- and medium-energy devices are usually adequate for capturing coal dust and fly-ash and there is no limitation on temperature or humidity of the incoming exhaust-gas stream, so the unit is ideal for dryer applications. Figure 11.23 shows a typical arrangement.

The electrostatic precipitator has the capability for capturing the entire range of emissions from coal-fired furnaces but is not common in fluidized-bed dryer installations because of the cost. However, although the installed cost is high,

TABLE 11.5 Comparing Average Values of Key Emission-control Devices

Equipment	Efficiency (%)	Pressure drop (kPa)	Gas velocity (m/s)	Installed cost ($/m³h)
Cyclone collector	85–95	0.85–1.25	—	0.60–1.80
Fabric filter				
shaker type	99	0.62	0.01–0.03	1.27–2.10
reverse jet	99+	0.85	0.09–0.14	1.70–2.55
Precipitator	99+	0.22	1.02–2.54	1.70–6.00
Venturi scrubber				
low medium	95+	1.25–3.75	30.5–3.05	1.27–2.55
medium energy	98+	3.75–5.00	45.7–3.05	1.70–3.40

[a] At throat
[b] Through cyclone separator.

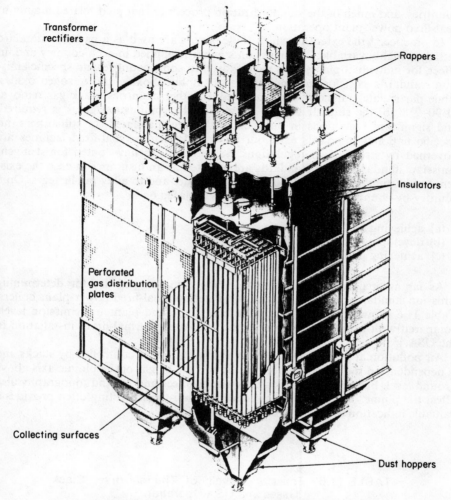

Fig. 11.24 Electrostatic precipitator (source *Power*, April 1974).

operating cost is low because the gas stream need not be accelerated. Figure 11.24 shows a typical system. Ironically, if low-sulphur-content coals are burnt in a system with an electrostatic precipitator, the efficiency in recovering fly-ash may drop due to the low resistivity of the ash constituents, caused principally by the lower iron content.

Table 11.5 gives average values[23] of key emission-control devices used in coal-combustion systems of the type employed with fluidized-bed thermal dryers.

Air-quality standards can, either separately or in combination, relate to ambient air quality, stack emission or full quality specification, and standards are generally set in terms of average values or maximum values on an hourly, daily, monthly or yearly basis. There is fairly wide variation between the coal-user countries, and much of the standardization procedures adopted related mainly to coal-fired power-plant operation.

In practice,[24] the extent to which these factors are used as a basis for standard setting is highly variable. The emission standards applied in each country and, in effect, for individual plants, also vary widely. In some countries, no specific emission standards exist for such pollutants as sulphur dioxide and nitrogen oxides, while particulate emissions are controlled. In other countries, stack gas removal of 90–95% of the sulphur dioxide present in the combustion gases is required, and significant reductions in nitrogen oxides are mandated. In still others, no specific national standards exist but air emissions from coal-fired facilities are governed by strong local regulations. Some countries have opted for stringent emission standards: such standards are adopted as an insurance in case the existence of harmful effects is established in the future and to allay public fears. On a country-by-country basis different emphasis is placed on:

(a) achieving a balance between costs and benefits;
(b) developing a cost-effective strategy; and
(c) achieving public acceptance.

As far as coal-dryer installations go, the common practice for determining emission standards is to apply those stipulated for coal-fired power-plant boilers. Table 11.6 shows a comparison of Federal, State and plant test-emission levels compared with original design data for a coal-fired thermal-dryer installation in the USA.[22]

Air pollution at ground level can, to a large extent, be controlled by stacks and is dependent on the concentration source and the height of the plumic axis above ground level. Further, meteorological circumstances and ground topography also affect the plume. The use of scale models in wind-tunnel testing often provides a valuable indication of practical emission behaviour.

TABLE 11.6 Emission Levels of Thermal-dryer Stack Gases in a USA Operation

Particulates	Design	Plant tests	Federal EPA requirement	State requirement
gr/d.s.c.f.[a]	0.031	0.011	0.031	0.10
lb/h	12.680	5.950	12.680	40.80
mg/d.s.c.m.[b]	70.000	24.800	70.000	225.80
SO_2				
ppm (volume)	60.000	0.760	60.000	60.00
Venturi scrubber				
max. pressure	25.000	16.000	25/40	—
Scrubber limestone				
lb/h	43.750	0	0	43.75

[a] grains per dry standard cubic foot.
[b] milligrams per dry standard cubic metre.

REFERENCES

1. D. G. Osborne. Coal fines dewatering, *Transactions Society Mining Engineers of AIME* **276**, 1843–1849, 1985.
2. C. E. Sloan. Drying systems and equipment, *Chemical Engineering*, June 19, 167–175, 1967.
3. J. M. Coulson and J. F. Richardson. *Chemical Engineering*, Vol. 2, third edn (revised by J. H. Harker and D. A. Allen), Chapter 16 (Drying), Pergamon, Oxford, 1982, pp. 710–762.
4. P. Levin. Thermodynamics of fine coal drying, *Mining Congress Journal*, April 1960, 79–81.
5. W. C. Lapple and W. E. Clark. Drying methods and equipment, *Chemical Engineering*, October 1955, pp. 191–210 and November 1955, pp. 177–183.
6. W. R. Marshall and S. J. Friedman. Section 13, Drying, in *Chemical Engineering Handbook*, (Ed. R. H. Perry), 5th edn, McGraw-Hill, New York, 1973.
7. L. Westerstrom. Mineral Industry Surveys, USBM Report No. 1745, 1975.
8. D. E. Pearson. The quality of Western Canadian coking coal, *CIM Bulletin* **73** (813), 1–15, 1980.
9. M. A. Rashid and R. J. Germain. Drying low-rank coal, Paper No. 7, CIM Second Technical Conference on Western Canadian Coals, May 1982. Proceedings published by Canadian Institute of Mining and Metallurgy.
10. H. B. Charmbury. The Bearce dryer goes commercial, *Coal Mining and Processing*, October 1977.
11. F. J. Coward. Development of a fluid-bed thermal dryer. *Minerals Processing*, June 1967, pp. 22–25.
12. J. M. Coulson and J. F. Richardson. *Chemical Engineering*, Vol. 2, third edn (revised by J. H. Harker and D. A. Allen), Chapter 6 (Fluidization), Pergamon, Oxford, 1982, pp. 270–278.
13. J. W. Leonard and T. S. Spicer. Thermal dewatering, Chapter 13 in *Coal Preparation*, fourth edn (Ed.) J. Leonard, AIME, New York, 1979, pp. 13–3 to 13–57.
14. J. C. Agarwal, W. L. Davis and D. T. King. Fluidized-bed coal dryer, *Chemical Engineering Progress* **58** (11), 85–90, 1962.
15. C. E. Silverblatt and D. A. Dahlstrom. Improved dewatering and coal by steam filtration, SME-AIME Annual Meeting, New York, 1964.
16. N. D. Policow and J. S. Orphanos. Development of the screen bowl centrifuge for the dewatering of coal fines, First International SME-AIME Fall Meeting, Honolulu, Hawaii, September 1982.
17. J. J. M. Van den Broek. From metallurgical coal tailings to thermal feed, *Mining Engineering*, January 1982, 49–52.
18. W. L. Root. Indirect drying of solids, *Chemical Engineering*, May 1983, pp. 52–64.
19. J. Chowdhury. Pulse combustion lower drying costs, *Chemical Engineering* December 10, 1984.
20. V. A. Cheney. How to select a dryer, *Chemical Engineering Progress*, April 1979, **85**, 40–43.
21. R. B. Keey. *Drying: Principles and Practice*, Vol. 13, International Series of Monographs in Chemical Engineering, Pergamon, Oxford, 1972.
22. P. T. Luckie and E. A. Draeger. The very special considerations involved in thermal-drying of western region coals, SME-AIME Fall Meeting, Salt Lake City, Utah, September 1975.
23. Anon. Combustion pollution controls, Part III of *Power from Coal*, a special report by *Power*, February 1974, pp. 557–564.
24. R. E. Bailey. Coal use and the environment, IEA. OECD, Report by the Coal Industry Advisory Board, Vol. 1, March 1983.

Chapter 12

CHEMICALS IN COAL PREPARATION

12.1 INTRODUCTION

The use of chemicals in coal preparation has grown significantly during the past 20 years, and the range and diversification in their current applications now warrants treatment in a separate chapter in any book dealing with coal treatment.

Most large plants would be unable to operate effectively without assistance from chemicals, and in terms of environmental control their use is also of paramount importance.

For simplicity it is convenient to divide reagent usage into application groups:

(a) materials handling;
(b) coal beneficiation;
(c) solid–liquid separation;
(d) environmental control.

Throughout the entire coal chain from mining to the customer, chemical reagents will be employed for one or more of the above functions. However, this chapter will confine discussion to the coal-preparation and handling activities at the mine site.

Little or no reagent usage occurs in the raw-coal handling and stockpiling except for especially friable coals where surfactants may be used to reduce dust problems.

For usage of any form of agent to be technically effective and/or financially justifiable, it is most important that some form of test work be carried out using representative samples to establish the following:

(a) selection of the most effective type of agent;
(b) selection of the most effective dosage;
(c) benefits of combination of several chemicals, e.g. selective flotation—modifier and depressant for pH regulation and collector plus frother, etc., in order to separate coal and pyrite.

In other words, there are often groups of chemicals associated with one specific application and test work must determine an optimized set of dosage combinations. Testing is thus a very important facet in the selection and efficient utilization of chemical agents in coal applications.

CHEMICALS IN COAL PREPARATION

Materials handling and transportation:	— dust suppression — freeze control — pipelining of coal — spontaneous combustion prevention
Beneficiation:	— flotation — agglomeration — dense-medium stabilization — chemical cleaning
Solid–liquid separation:	— coagulation — flocculation — filtration aids — property modifiers — acid neutralization
Environmental control:	— water quality — dissolved and suspended solids — erosion — autogenous heating — corrosion resistance

Fig. 12.1 Areas for reagent use.

The major areas for the application of chemical substances in aiding the preparation of coal are shown in Fig. 12.1. In the area of mechanical handling and transportation of coal, chemicals can help overcome environmental problems of dust suppression from stockpiles and the prevention of freezing of coal in rail cars, bins and conveyors. Movement of coal through pipelines could also be aided considerably by the use of chemicals.

In most cases, the beneficiation of coal in the fine- and ultrafine-coal size-ranges, can only be readily and economically achieved through the modification

Materials handling:	— freezing-point depressants — organic surfactants — binders and latex emulsions — low-viscosity hydrocarbons — coating reagents
Beneficiation:	— frothers — conditioning agents — diesel oils as collectors — light oils as agglomerants — bentonite and viscosity adjusters
Solid–liquid separation:	— anionic, cationic, polyelectrolytes, non-ionic polymer flocculants — inorganic salts — starches, gums, alginates — surfactants
Environmental control:	— pH controllers — coagulants and flocculants — stabilizers — organic surfactants — corrosion-reducing agents

Fig. 12.2 Common chemicals used.

of the surface chemical properties. Techniques currently applied include froth flotation and oil agglomeration. Chemicals may also be used in the larger size-ranges in order to stabilize dense-medium suspensions, adjust pH, and aid the removal or control of slimes, which might otherwise impair beneficiation. But their main area of application is in the treatment of fine coal.

The most common area of chemical use is that of coagulation/flocculation of suspended solids. Due to the considerable amount of ultrafine coal and clays generated by modern mechanized mining, chemical removal of such suspended particles is often the only economical method. In order to meet contractual moisture contents and reduce freight costs, the dewatering of thickened sludges and froth-flotation concentrates may similarly be effectively, and thus economically, achieved by the use of chemicals.

The present range of chemicals used in the industry is summarized in Fig. 12.2. This range is very broad and includes simple inorganic salts, naturally occurring starches and gums, hydrocarbons, oil, polymers of varying molecular weights and types, and complex mixtures of organic and inorganic compounds which form the basis of many proprietary brands.

When handling product coal, simple salt solutions, glycols and organic surfactants may be used to prevent freezing or reduce the strength of ice bonds. Low-viscosity hydrocarbons and other coating reagents can prevent freezing of coal in railway wagons, road trucks or belt conveyors. Dust suppression and the control of wind erosion may be achieved by the use of surfactants, oils, latex emulsions or polymeric binding agents.

In the beneficiation of coal, polymers and clays may be used to give stability and greater uniformity to dense-medium suspensions. Froth-flotation techniques require coal-conditioning agents, hydrocarbons as collectors, and surfactants as frothers.

Considerable work has been undertaken to produce an effective range of filtration aids, flocculants and coagulants. These vary from the simple and traditional natural compounds, such as inorganic salts and starches, gums or alginates, to complex formulated polymers that may have very high molecular weights and may contain a combination of anionic, cationic and neutral groups within the one polymer. Such formulations of polymers, although essentially similar to each other, have sufficient differences to produce varying degrees of flocculation. Extensive testing may be required to determine the optimum formulation and dosage. Filtration aids, in the form of surfactants, may reduce general moisture levels and decrease heat requirements in thermal drying.

Chemical usage in environmental control is becoming an increasingly significant applications area despite the fact that economic advantages are rarely apparent at present. Reagents are required to perform more or less the same functions as previously described in materials handling and solid–liquid separation, but the methods of application may be quite different. Surfactants for wind erosion and weathering, inorganic salt solutions for pH control, coagulants for clarification, and chemicals to reduce or inhibit corrosion and provide metal protection are probably the most common reagents used in this field.

There is little doubt that in today's international coal market-place the increased use of chemicals can now be regarded as being a critical factor in the profitability of mining and preparing coal.

Dust control:	— water — surfactants — oils
Wind erosion:	— binding agents — resin emulsions — latex emulsions — chemical binders
Freeze control:	— calcium chloride — glycols — oils — multicomponent compounds

Fig. 12.3 Materials-handling aids.

12.2 MATERIALS HANDLING

The mechanical handling of coal usually requires the suppression of dust, reduction of coal loss through wind erosion, and prevention, and control of ice in wet coal (Fig. 12.3). Such measures must take into account the mean grain-size of the coal, the extent of its probable hydrophobic nature and the impact upon any subsequent treatment of the coal.

12.2.1 Dust Control

For localized temporary dust control, water is the simplest and least expensive control agent. Figure 12.4 shows a coal stockpile serviced by a dust-prevention tube which, together with water spraying, is often adequate to contain all but the most dusty coal feeds. However, when dry hydrophobic dust is encountered, the addition of surfactants to the water sprays will usually overcome severe dust problems. Commercial surfactants are formulated from various low-molecular-weight organic compounds, such as primary and secondary alcohols, and are

Fig. 12.4 Coal stockpile with dust-tube accompanied by local water-spray suppression.

Fig. 12.5 Spraying binder on loaded coal (courtesy Dow Chemicals).

readily available and economical to use. They act by reducing the surface tensions of the water–coal interface, thus allowing a film of water to cover the entire coal particle. Commercial dust-control surfactants may also incorporate other organic compounds that produce a sticky or resinous film on evaporation of the water. Such methods of dust control vary in effectiveness usually in direct proportion with the amount of organic compounds contained in the formulation.

For the control of dust from stockpiles, or the prevention of erosion of coal in railway wagons, a surface crust may be formed by application of binding agents (Fig. 12.5) which are sprayed onto the surface of the coal. Figure 12.5 shows the spray system and Fig. 12.6 shows the crusted surface. Such agents may take the form of resin or latex emulsions, organic polymers, oils sprays and foaming reagents. Resin particles in aqueous suspensions, together with a surfactant, coat the coal and subsequently solidify to form a crust. Latex emulsions may be similarly prepared and subsequently take 3–5 h to cure. Resin emulsions are normally diluted to between 2 and 3% of the original solution strength and latex emulsions to 4–6% of original concentration. Application rates are given as 1–3 litre/m^2 for resin sprays, and latex sprays as 100–200 litre/railway wagon, i.e.

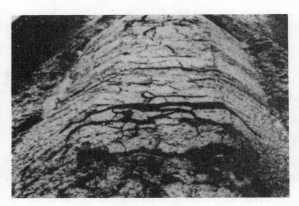

Fig. 12.6 Dried, crusted surface of chemically bound coal (courtesy Dow Chemicals).

100 t. A cost of £2.50–5.00/railway wagon is considered as normal for both these reagents. Dust losses under fairly severe weather conditions have been estimated as being as much as £50 a railway wagon, and the potential benefit of using such agents is therefore clearly evident.

Polymer-based applications may dry to a flexible plastic film which can compensate for any shrinkage that occurs. Oil sprays may use any available heavy oil and can often be proved inexpensive depending on the source of the oil. However, objections to oil sprays may be raised—notably, environmental pollution, odour and application problems.

12.2.2 Prevention of Ice

Prevention of freezing of broken coal can be achieved through the use of freezing-point depressants or by the use of conditioning agents which considerably reduce the strength of the ice mass.[1,2] Figure 12.7 shows ice fragments with and without treatment by freeze-reduction agents. The treated ice is clearly structurally weaker. Calcium chloride solutions can depress the freezing point of water to $-40\,°C$ by the application of a 35% solution of the salt. There are generally only slight corrosion problems, which the introduction of lime or sodium chromate will usually eliminate. Application of 4–8 litres of calcium chloride solution per tonne of coal, at a cost of £0.50–1.00/t of coal, is normal.

Aqueous solutions of glycols, i.e. ethylene glycol, propylene glycol and diethylene glycol, are the normal constituents of 'antifreeze' type preparations. Commercial preparations may also include acetate salts, rust inhibitors and surfactants for increased performance. Such products act in the dual manner of depressing the freezing point of water and disrupting the ice crystals. Product use is at a level of 0.4–1.7 litre/t of coal at a cost of £0.15–0.60/t.

Oil-based products may be applied either directly or by means of an emulsion. Such products may range from a simple light oil spray to more complex formulations. However, all act by coating the coal particles with a thin oil film, which effectively prevents ice bonding to the coal or railway wagon sides. Any remaining oil emulsion interferes with ice-crystal formation.

Multicomponent organic systems are formulated to combine as many features as possible and will include surfactants, glycols and oils. They act by retarding all

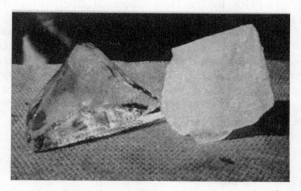

Fig. 12.7 Ice with (right) and without (left) treatment by freeze-conditioning agent (courtesy Dow Chemicals).

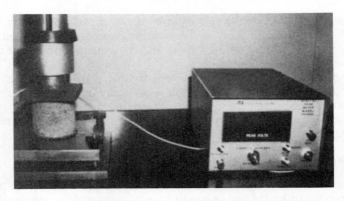

Fig. 12.8 The Dow freeze-reduction agent shatter-test apparatus: (a) set up for shatter test; (b) typical results after shatter testing.

features of ice bonding and adhesion, so producing a very weak ice mass. These systems are not corrosive, are easy to apply, and have the low dosage rates of 0.5–2 litre/t coal at a cost of £0.25–1.00/t.

Freeze-conditioning agents have been increasingly used since their introduction in the mid-1970s by the Dow Chemical Company of the USA. This company has pioneered their usage and development. A technique for testing the performance of agents has been developed by Dow. It does not test efficiency, but instead examines the ability of frozen lumps to shatter under impact.

A test specimen of fine coal is cast into a short rod using a measured dosage of agent with a normal amount of surface moisture. The frozen specimen is then broken by dropping a standard (free-fall) weight onto the specimen. An impact-force transducer is then used to measure the failure point of the specimen and the shattering force is calculated from the equation:

$$\text{shatter force (dN)} = \frac{\text{shatter strength (kg)} \times \text{acceleration}}{10}$$

where acceleration = 9.8 m/s^2

Figure 12.8 shows the apparatus, and the graph in Fig. 12.9 gives some typical comparative-test results.

CHEMICALS IN COAL PREPARATION

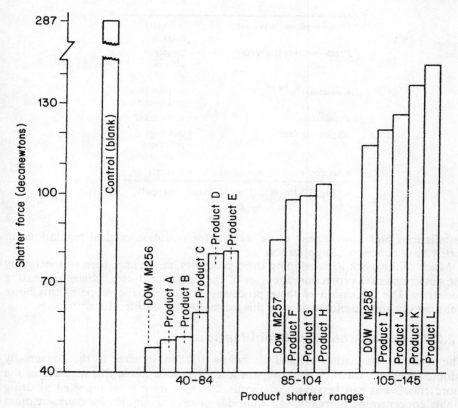

Fig. 12.9 Typical comparative results of shatter-testing-treated fine-coal samples.

12.3 REAGENTS IN COAL BENEFICIATION

Coal cleaning can generally be split into two size-ranges: the $+0.5$ mm fraction for which gravimetric techniques predominate; and the -0.5 mm fraction for which surface-chemistry-based separation techniques are becoming increasingly used, together with the more traditional gravity methods that have predominated in the past (Fig. 12.10).

For the larger coal, the main uses of chemicals are in: pH control, removal of slimes and clays, and the stabilization of dense-medium suspensions. The pH control, and slimes and clay control are achieved by well-established techniques. Dense-medium stabilizing agents include polymers and refined clays, which give lower media-settling rates, thus allowing the use of coarser media or reducing the need for tight control in media-quality specifications.

Chemical coal-cleaning methods (i.e. excluding froth flotation and oil agglomeration) have become increasingly motivated by environmental pollution constraints such as emission levels of sulphur oxides and nitrogen oxides. Numerous chemical cleaning processes have been developed for commercial applications; however, most have failed to emerge as economically viable commercial

Dense-medium suspensions	— polymers
	— clays
Froth-flotation frothers:	— cresols
	— pine oils
	— alcohols
Collectors:	— sulphonates
	— hydrocarbons
Promoters:	— surfactants
Agglomeration:	— oils
	— polymers
	— flocculants
Chemical-cleaning processes:	— solvents

Fig. 12.10 Coal beneficiation reagents.

installations and now appear to be waiting for pollution-control regulations to justify their commercial application.

Much of the work in developing these processes has, in fact, been supported by the environmental protection agencies of the major coal-user, power-generating countries. The discussion of these processes lies outside the scope of this book but there are some useful reviews of this technology in the literature.[12,13]

12.3.1 Dense-medium Stabilization

The mineral, magnetite, ground to below 100 μm in size, is the universally adopted medium used for coal preparation. The more stable this medium is, the more effective it can become. Numerous attempts have been reported at using highly concentrated inorganic salt solutions or organic liquids for dense-medium separation of coal because of the relatively low range in operating density. Despite the obvious potential benefit of high separation efficiency, commercial applications have generally failed, due either to high corrosion, serious evaporation loss of the separation fluid, or failure to comply with safety regulations necessary because of the highly toxic nature of most of these fluids. Other, non-toxic fluids have been tried, including liquefied carbon dioxide, polymer-stabilized water, etc., but no commercial applications exist for any of these alternatives. The most recent and most promising of these has been the Otisca process[20] which utilizes non-polar organic fluids of very low viscosity in which the coal is virtually insoluble. A 20 t/h pilot-plant unit has been developed for testing this process. The exact chemical composition of the fluids used with this process is at present being maintained a commercial secret.

Natural organic polymers such a guar gums, alginates and dextrin[3] may be used for medium stabilization. Synthetic polymers such as polyethylene oxide and polyacrylamides may also be used. A polymer found to be of great use in recent studies is the biopolymer xanthan, which has the advantage of a low absorption rate on to coal, and the ability to achieve considerable cross-linking through the use of chromic salts. With solution strengths of 0.1%, much higher relative densities than usual may be achieved.

Very fine bentonitic clays may also be used as stabilizing agents for magnetic suspensions. However, as they are of natural origin and form a suspension, their reliability and ease of control is poorer than that of synthetic polymers. Optimum clay contents of 3–4% are usually recommended.

Although unit costs of clays are lower than those of polymers, the dosage rates for polymers are 20–50 times less than those for clays, resulting in a lower overall cost. Using very finely ground magnetite as the solid component of the medium, together with a polymer solution, it is reported that extremely stable suspensions with little or no settling over a prolonged period of time can be achieved.

Organic liquids such as benzene, perchlorethylene, carbon tetrachloride, bromoform, acetylene tetrabromide and pentachloroethane are all widely used for laboratory float–sink testing, together with concentrated inorganic salt solutions of calcium chloride and zinc chloride. The use of these chemicals in laboratory applications is generally acceptable providing that good ventilation and other safety precautions prevail.

12.3.2 Froth-flotation Reagents

Flotation of coal is aided by the fact that coal is normally naturally hydrophobic. However, problems may occur when oxidized coal or large amounts of hydrophobic impurities are present in the coal. Current coal-flotation practice[4] involves the bulk flotation of coal, rather than selective flotation of individual macrolithotypes, and therefore the reagents used are generally fairly unsophisticated.[5]

Frothers, collectors and promoters or conditioning chemicals are all readily available and relatively inexpensive. In many cases the ease of collection still obviates any need for detailed economic appraisal.

Frothers are generally lightweight organic molecules with polar groups attached. Aliphatic alcohols and carboxylic acids are good frothers; however, similar molecules containing more than eight atoms produce significantly poorer results. Simple phenolic compounds, in particular cresylic acid and zylenols, are

TABLE 12.1 Alcohols and Ethers which may be Suitable for use as Frothers

Alcohol	Carbon-chain length
Butyl	4
Amyl	5
Hexyl	6
Heptyl	7
Octyl	8
Methyl-isobutyl-carbinol (MIBC)	6
2-Ethyl hexanol (or 2-octanol)	8
Phenol	6
d-Cresol	6
x-Naphthol	10
b-Naphthol	10
x-Terpineol	10
Cyclomexanol	6
Diacetone alcohol (4-hydroxyl, 4-methyl, 2-pentone)	6
Ether	Number available
Polyoxypropylene glycols	at least 4
Polyoxethylene glycols	at least 3

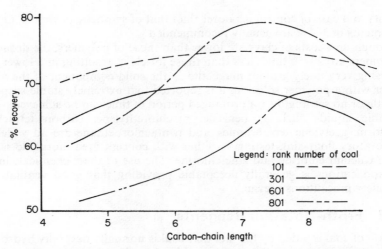

Fig. 12.11 Graph showing effect of carbon-chain lengths of frothers with coal tank (also see Table 12.1).

also very effective frothers. Some frothing reagents may also act as surface-active agents and help by dispersing clay slimes. Care must be exercised in selection because the use of certain types of reagents may cause reduced hydrophobicity of coal and/or enhanced coating of coal particles leading to excessive consumption by the collector. Cresols, alcohols and industrial by-products containing these molecules have all been used successfully. Excessively high reagent consumption can usually be avoided by multistage conditioning during flotation, i.e. adding some of the reagent to the cells.

Table 12.1 lists the alcohol- and ether-based compounds which may be suitable for use as frothers.

The use of aliphatic alcohols and polyglycol ethers as frothers has become increasingly widespread during recent years.[10] In many cases these have superseded the use of phenolic-type agents in their effectiveness and for environmental reasons. Work with these compounds has led to the conclusion that optimum recovery can be determined by examination of the carbon chain length of the alcohol or polyglycol used with relation to the type or rank of coal being treated. Figure 12.11 shows this phenomenon clearly. However, in order to determine an optimum reagent, i.e. to tailor the compound to the coal requiring treatment, researchers have shown that it is necessary to examine combinations of up to four compounds, i.e. three alcohols and a polyglycol in varying proportions with each other. This approach now appears to be widely practised, especially in the treatment of weathered coal or in desulphurization applications.[15]

The most popular types of coal collectors are emulsions of oils such as diesel, kerosene or paraffin oils. Optimum dispersion of oils results in droplet diameters of approximately 5 μm. Collectors function by coating the particle with a thin film of oil, which increases the hydrophobicity of the particle. The efficiency of the oil-film coating may be increased by the use of promoters or conditioners.[5,6,15] Clays and shales may be dispersed by the use of sodium carbonate or sodium silicate, although strict control of pH is probably more useful.

CHEMICALS IN COAL PREPARATION

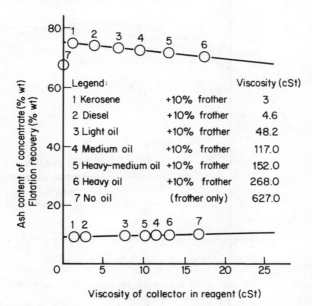

Fig. 12.12 Graph showing viscosity of collector in reagent recovery.

One way of relating various oils used as collectors is to compare their relative viscosities. The graph shown in Fig. 12.12 tends to suggest linearity between reagent viscosity with recovery and ash content of the concentrate, and this may further indicate the influence which reagent viscosity has upon reagent dispersion in the feed slurry. Various researchers have emphasized the need to emulsify properly oily collectors which are water insoluble as supplied.[15] Dispersion is clearly a critical requirement in achieving high flotation efficiencies for any application. For this reason, coal flotation is often called emulsion-flotation of coal, in order to differentiate from conventional (mineral) flotation using water-soluble reagents. The process is based on selective wetting by ultrafine oil droplets and is only applicable for solids such as coal that are naturally hydrophobic to some extent.

Recent improvements in coal-flotation efficiency have mostly been attributed to the development of collector reagents which incorporate emulsifying agents. Both non-ionic and anionic additives have been examined, and their effects are shown in Fig. 12.13 which is a graph comparing various reagents in terms of combustible recovery, while Fig. 12.14 shows similar effects on clean-coal ash content. Both graphs compare conventional hydrocarbon collector/frother-containing emulsifying additives with cresylic acid in the treatment of bituminous coal.

In coal flotation, it is a practical proposition to depress either the coal or the non-coal constituents, which are usually predominantly silicious or calcareous. It is also environmentally desirable to reduce iron sulphide content, and the depression of pyrite has become of particular interest in recent research. For coal depression, starches, especially dextrin and the natural hydrocolloids have proved effective. Also, some inorganic salts such as stannous chloride, hydrophosphoric

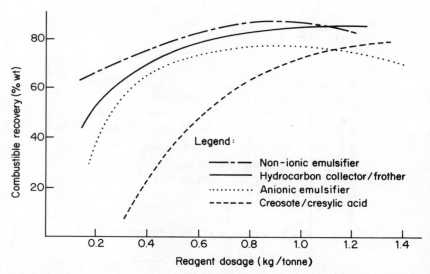

Fig. 12.13 Graph showing effects of additions of emulsifying agents into reagents in terms of recovery.

acid and potassium permanganate depress certain coal types effectively. In fact, any reagent which causes surface oxidation of coal will behave as a depressant, and even indiscriminate pH regulation can have a significant effect. Pyrite depression is achieved by increasing pH by adding lime; sodium cyanide and oxidation products of pyrite, ferrous and ferric sulphate act as depressants.

For slime coating of either the coal-laden froth or the coal surface itself, a slime-depressing agent such as sodium silicate may be effective. Normally, high-ash slurries contain clay and silicate minerals, especially quartz, and in addition to sodium silicate, phosphates such as hexametaphosphate, lignin and other polymeric sulphonates, quebracho and tannins may be useful, especially for cal-

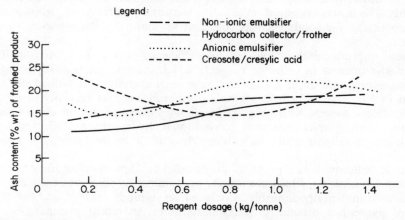

Fig. 12.14 Graph showing effects of additions of emulsifying agents into reagents on ash content of frothed product.

careous slimes.[4,5] Oxidized coal surfaces are very unreactive to coal flotation, especially when alcohol/oil–reagent combinations are used. For surface oxidation caused by weathering after mining, it is possible to restore hydrophobicity to an effective extent by attrition and/or by the addition of a froth-conditioning agent.[16] Such agents are reportedly able to eliminate surface moisture and this oxide film by chemical action, thus restoring the water-repellent coal surface for the oil to coat.[17] In investigation of the use and effectiveness of these agents, it was observed that comparison of atomic ratios of hydrogen to oxygen (from ultimate analysis) provides a useful indication of coal floatability. When the H:O ratio is high (above 20) the coal normally floats well, and when it is low the coal becomes hard to float. Coals which have become weathered *in situ* are often not readily treatable unless this ratio is improved by chemical treatment. Attrition alone is rarely successful, unless the coal has a naturally low oxygen content. For oxidized coal, a typical reagent blend has been suggested as being 9.5 parts diesel oil plus 4 parts glycol ether and 1.5 parts conditioning agent, but only laboratory and in-plant testing will ascertain the appropriate dosage combination.[18] In order to obtain high flotation recoveries, it is important not only to use a hydrocarbon oil collector but also to ensure that the preconditioning involves emulsification of the collector–frother as well as thorough dispersion of the conditioning agent. Tests have shown that in this way, surface-active components of this agent most effectively assist in the abstraction of the oil collector onto the surface of the coal. In this way, their effectiveness is optimized and the dosage of the conditioner is maintained at or below 10% of the amount of the oil collector.

12.3.3 Oil Agglomeration Agents

A wide variety of bridging liquids, all of which are predominantly hydrocarbon in content, have been tried as agglomerating agents with varying degrees of success.[7,8] These have included No. 2 fuel oil, paraffinic crude oil, kerosene, mixtures of crude oil and tar, Varsol, and raffinates.

In most oil-agglomeration processes, none of the bridging liquid is recovered for re-use and, therefore, its cost becomes a predominant factor in determining the viability of the individual process. Only those processes which are found to be most economical in their use of oil, relative to the value of the product coal, stand any real chance of being applied. Metallurgical coking coals have therefore proved to be the most attractive targets. However, with the last oil crisis in the late 1970s, the attractiveness of agglomeration processing of coal diminished, and consequently the concept still remains unused. Recent developments[9,10] in recovering the oils for re-use promise to revive interest and, with more attractive oil costs, there is a strong possibility that this process may still be commercially adopted. Such a process, incorporating both spherical oil agglomeration of the coal, and one of a number of potential thermal methods for recovery of the solvent or light oil fraction from the coal pellets, can significantly lower the cost of cleaning the coal. For the spherical agglomeration process incorporating solvent recovery, the cost saving has been determined to be in excess of 30%[9] and low moisture content in the final product further enhances both the coking and the thermal properties of the product coal.

The capability of an oil to wet the coal surface and the degree of selectivity attainable are the major means for assessing the effectiveness of the oil as an agglomerating agent. With bituminous coal, most hydrocarbons are successful agglomerants capable of recovering in excess of 90% of liberated combustible matter from a coal slurry. In terms of selectivity, however, the lighter or more refined oils prove more successful. In spherical agglomeration processes, as a rule of thumb,[7] hydrocarbons of relative density less than 1 g/cm^3 provide the best selectivity. This is mainly attributed to their lower viscosity and the resultant ability to coat the coal particles better during mixing. Denser oils are therefore much more effective in the subsequent pelletizing step as binding agents because of their ability to draw together further the fine particulate containing agglomerates. This relationship between the relative density and the oil and the the ability to agglomerate coal particles is shown in Fig. 9.43. This graph compares oil density with a dimensionless ash ratio obtained by comparing the coal agglomerate ash contents obtained using various oils with that obtained using Stoddard solvent.

Oil agglomeration is far less prone to the effects of oxidation or weathering of the coal surface and even coals of low rank can produce good agglomerates with high recoveries. Improvements may, however, be obtained in many cases by the use of chemical additives. The most common improvement is found to occur as a result of pH adjustment by the addition of alkali such as sodium carbonate, but improved selectivity has been shown to occur in some cases with addition of fatty acid. Pyrite rejection can be produced by the addition of potassium ferrocyanide and/or sodium carbonate, but pyrite has wetting properties closely resembling those of coal and there is always the risk of reducing coal recovery in eliminating pyrite content.

The subsequent treatment of the tailings resulting from the agglomeration process is by conventional solid–liquid separation equipment, i.e. thickeners, vacuum or pressure filters, or centrifuges. Polymer flocculants, together with slurry-modifying chemicals such as lime, are likely to be employed as settling or filtering aids.

12.3.4 Selective Flocculants

Selective flocculation is a comparatively recent mineral-beneficiation technique that is currently being employed for iron ore and potash separations from clay minerals.[19] A number of attempted coal-cleaning approaches have been reported.[11] However, no commercial application is believed to be in operation at the present time.

Various physicochemical relationships, such as the nature and magnitude of the surface charge and pH, are of great importance in determining the exact conditions for selectivity. In this regard, the process is considerably more vulnerable to changes in the chemical conditions of the feed than its flotation counterpart.

Because of the higher density of the non-coal constituents and their smaller proportion in the feed to the process, it has been suggested that selective flocculation would work best under conditions where the coal is dispersed and the non-coal components are flocculated.[20] This can in some cases be readily achieved in laboratory- and bench-scale test work, despite the variety of non-coal components present in some raw coals. However, in a commercial circuit, where further variables occur and can fluctuate widely, e.g. ionic components and concentration in the process water, application of this concept seems impractical.

For coal flocculation, polyethylene oxide, a non-ionic polymer has been recommended with adjustment of the pH of the slurry to sensitize the separation following addition of gangue-depressing agents.

12.3.5 Chemical Cleaning Processes

Chemical coal-cleaning[13] has been under active development for the past decade, but has not yet reached the commercial-plant stage. The main impetus for chemical cleaning has come from a serious need to control the emission of sulphur oxides from coal-fired power-generation plants. Much of the coal in the United States, for example, contains excessive sulphur, and environmental pollution control regulations currently imposed by state and federal governments severely limit the emission of sulphur oxides.

There are several potential methods of controlling or avoiding this form of pollution. These include flue gas desulphurization, physical coal cleaning, fluidized-bed combustion with sulphur capture, and conversion of coal into clean gaseous or liquid fuels. None of these methods is, however, viewed as ideal and fully capable of contending with the special circumstances of every coal user. All of these methods have their shortcomings. Flue gas desulphurization, for example, is impractical for small plants and uneconomical for plants that operate intermittently. Physical coal-cleaning cannot remove organically bound sulphur and is usually not cost effective for removing finely disseminated pyrites in any significant quantity.

Chemically cleaned coal will be of uniform quality and low in sulphur. It will also be low in ash and, in most cases, it will be combined with coal obtained from conventional cleaning in order to optimize total coal-cleaning costs. The chemically cleaned product itself may be supplied as a powder or slurry or in the form of uniformly sized pellets, granules or briquettes. Some of these forms will be more amenable than others for ensuring good transportation and storage characteristics. Chemically cleaned coal will always prove to be more expensive to produce than conventionally washed coal, but less expensive than gaseous or liquid fuels made from coal when the cost is determined on the basis of net heat-content received.

Chemically cleaned coal may also find application in coal–oil or coal–water mixtures which require finely sized coal of uniform quality. These mixtures can be burned in boilers designed for oil firing with only slight modification to the burners.

A number of different chemical cleaning processes are under development at present. While all of these processes are designed to remove most of the inorganic sulphur from coal, not all will reduce the organic sulphur content or ash content to the same extent. The processes also differ in how they affect other properties of the coal in the nature of the chemical reactions and conditions employed. Most involve the digestion of the coal by a solvent or by a chemical comminution liberating the coal from non-coal constituents. The resulting cleaned coal obtained from all such processes is usually finely divided and with a distinctly lower level of impurities than would be achieved by physical coal-cleaning.

Leaching processes involve the use of strongly acidic or alkaline solutions aimed at reducing the sulphur content of the coal following pulverization and subsequent treatment by one other group of chemical comminution processes. Usually, the raw coal requires grinding to facilitate maximum utilization of the leachate, and hence crushing and grinding become major components of the processing cost. Another group employs a hot caustic solution to dissolve and extract the organic matter from suspension. The extract is then filtered to remove mineral matter and the filtrate obtained is treated with acid to precipitate the cleaned coal. This group also involves pulverization of the raw coal prior to processing.

Other possible chemical-treatment methods which have been attempted for either comminution or beneficiation or a combination of both include those employing corrosive gases. By bubbling a concentrated gas capable of degrading the coal structure into a mixing tank containing a high-speed impeller (1200–1600 r.p.m.), rapid comminution of coal can be achieved. Chlorine and ozone gases have both proved effective comminution agents in this type of process but economics and safety considerations have so far prevented commercial application.

12.4 SOLID–LIQUID SEPARATION

The types of chemicals most commonly encountered in solid–liquid separation applications are shown in Fig. 12.15. Generally, the amount and type of chemicals used varies with the volumetric ratio of water and solids. Very low particulate concentrations require the formation of large floccules, while higher concentrations of solids require compact, granular floccules. Very high concentrations of solids require the modification of the physical properties of the interstitial water, i.e. the viscosity or surface tension must be reduced.

Inorganic salts, such as aluminium and ferric sulphates, act by neutralizing the charge on particles, thus allowing coagulation to take place. Strict dosage and pH control, often created by the addition of lime or soda ash, is essential for optimum performance. Such salts also neutralize the charge on clays, thereby preventing adhesion to coal particles (slime coating) that would later hinder selectivity in flotation. Figure 12.16 gives a tabulation of some of the more commonly used chemical coagulants.

Natural polymers are generally short-chain neutral organic compounds—starches, gums, alginates and polysaccharides. The polysaccharides are most

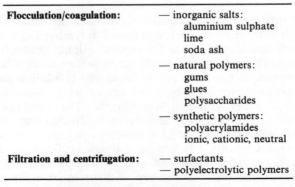

Fig. 12.15 Solid–liquid separation chemicals

Substance	Formula	Mol. wt.	Most common form	Solubility in water[a]	Reaction of solution in water
Aluminium sulphate	$Al_2(SO_4)_3 \cdot 18H_2O$	666	lumps, powder	36.3	acid
Sodium aluminate	$NaAlO_2$	82	powder		alkaline
Ferric chloride[b]	$FeCl_3 \cdot 6H_2O$	270	lumps	91.9	acid
Ferric sulphate	$Fe_2(SO_4)_3 \cdot 9H_2O$	562	small crystals		acid
Ferrous sulphate	$FeSO_4 \cdot 7H_2O$	278	small crystals	26.6	acid
Lime (quick)	CaO	56	lumps powder		alkaline
Lime (hydrated)	$Ca(OH)_2$	74	powder	0.128	alkaline

[a] Solubility in g water-free substance per 100 g water at 20 °C
[b] Ferric chloride is very hygroscopic

Fig. 12.16 Commonly used coagulant chemicals.

effective in neutral to alkaline conditions, whilst the gums, alginates and glues are often most effective in acid media. Due to the short rigid-chain structure and low bonding strengths, the flocculants have very low shear strengths, and this tends to encourage the use of excessive amounts of these flocculants.

Synthetic flocculants can have extremely large polymer chains that may contain ionic, cationic or neutral groups. Most synthetic flocculants are based on polyacrylamide or one of its derivatives. The ionic groups cause the polymer to uncoil and bond to the surface of the coal or clay minerals. As a result, some selectivity may occur, either intentionally or otherwise.

Extremely long-chain polymers have the greatest chance of contacting all the particles, producing large, open floccules having a high sedimentation rate. Such large floccules also have a high residual moisture content and low shear strength. Shorter, more highly charged polymer chains will tend to produce compact granular flocculants having much improved filtration characteristics.[14] Figure 12.17 shows the chemical structure of starch- and polyacrylic polymer-based flocculants.

All synthetic-polymer flocculants function in two stages when the solution is dispersed in a slurry: i.e. ion or charge neutralization: and bridging. The exact mechanism and the extent to which adsorption occurs are still not fully understood but it is postulated that initial adsorption occurs at one or more segments, resulting in the formation of strong bonds between the polymer and the solid. After one end of the long molecule has been adsorbed onto a particle, the remainder moves freely in the suspension to bridge and adsorbs onto other particles. With many molecules in solution, this action results in quick and irreversible flocculation and agglomeration of the formed flocs into a fast-settling mass of solids.

Solutions of polyacrylic and polyethylene oxide flocculant types exhibit non-Newtonian (pseudoplastic) properties, i.e. viscosities decrease with increasing shear rate. At low-shear conditions, effects on the flocculant are minimal but at

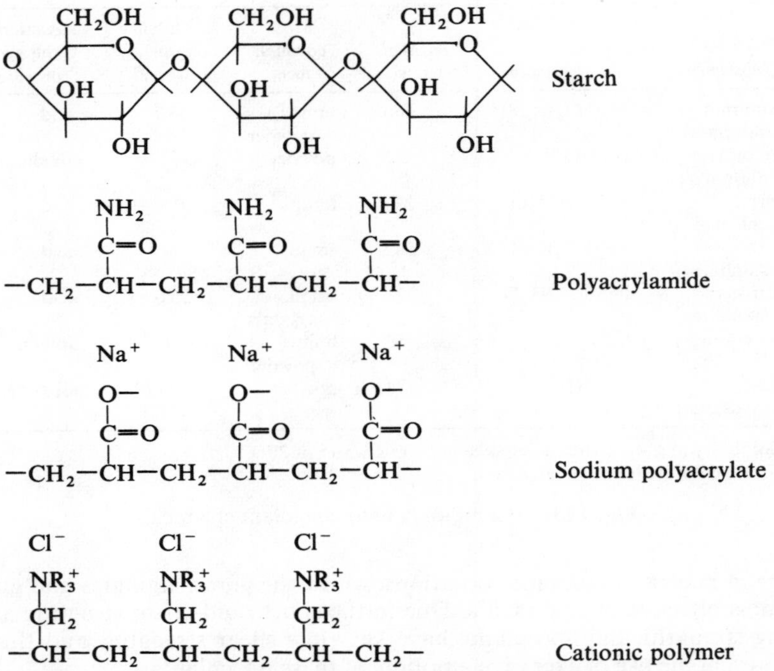

Fig. 12.17 Chemical structure of polymer flocculants.

the very high shear rates encountered during violent mixing or in pumping the solution, degradation of the polymers, resulting from rupture of the polymer chain, will occur. Tests have shown that up to 50% viscosity loss results from shear encountered during eight minutes of stirring in a high-speed blender. If, however, the correct type of pumps and mixing equipment are employed such degradation will not occur.

In coal and tailings thickening and clarification applications polymer solution strengths of between 0.1 and 1.0% are prepared from the dry, granular solid form using an eductor. Generally, the natural polymers and low-molecular-weight types are prepared in the more concentrated form while high-molecular-weight types, being more viscous, are prepared in diluted solutions. Many cationic polymers are now supplied in liquid form and can either be used undiluted for filter aid application or diluted up to 100 : 1 for thickening and clarification.

When solutions have been prepared, they should be continuously stirred for at least one hour after mixing by a slowly rotating agitator, and then pumped by positive-displacement pumps or else gravitated to the stock tank or to the feed system. Stock solutions usually remain stable for at least two weeks after mixing with clean water. Further dilution of mixed solutions is usually necessary in up to 100 : 1 ratio using clean water.

Dosage ranges for starch-based and low-molecular-weight synthetic polymers will usually be 0.025–0.5 kg/t. High-molecular-weight polymer dosages will be similar to this for conventional thickeners, lower for clarification, and 0.25–1.0 kg/t for high-capacity thickener applications.[23]

Dewatering of slurries may be aided by lowering the viscosity or surface tension of the coal–water bonds. This is achieved either by the use of surfactants or by the application of heat. Steam or hot air is usually only economical if there is a readily available source of waste heat. Surfactants, consisting of organic solvents and other low-molecular-weight organic compounds, are able to reduce effectively both the surface tension and the viscosity of water. Dosage rates of 0.25–2.5 kg/t of dry suspended solids can readily lead to a moisture reduction of 2–4% in centrifuge or filter-press cakes.[21]

For more dilute slurries (those containing less than 30% solids by weight), short-chain, highly charged polymers, often used in conjunction with inorganic electrolyes, can produce very compact, uniform floccules. Such floccules can have excellent vacuum filtration characteristics. Dosage rates of 0.1–0.25 kg/t of dry solids have resulted in moisture reductions of 3–5% and increased cake yields of 40–200%.[24]

The cost of synthetic polymers ranges from £0.75 to £1.50 per tonne for solid polymers, and is perhaps a little less for liquid polymers.

12.5 ENVIRONMENTAL CONTROL

Perhaps the most significant contribution to environmental improvement has been the improved techniques of solid–liquid separation combined with the use of polymeric flocculants (Fig. 12.18). Less than 20 years ago, the world-wide use of settling ponds and lagoons, similar to the one shown in Fig. 12.19 was a common occurrence in coal-preparation plants. Now, with the advent of efficient mechanical dewatering equipment typified by solid-bowl centrifuges, automated filter presses and continuous-belt filters, most new coal-preparation plant water circuits have been closed. All of these dewatering machines, however, require pretreatment of the slurry using a combination of inorganic salts and organic polymer flocculants in order to perform at the required efficiency.

The use of dust control and erosion control surfactants is becoming more widespread. Many of these substances contain both an effective surfactant, which renders the surface more hydrophilic, and a water-holding agent, which represses evaporation. This combination has been found very effective in three different areas of application:

Flocculation/coagulation:	— inorganic salts: ferric sulphate aluminium sulphate lime soda ash
	— natural polymers — synthetic polymers
Erosion/dust control:	— organic surfactants
Autogenous heating/ spontaneous combustion:	— heavy oil/tar — organic surfactants

Fig. 12.18 Chemicals for environmental control.

Fig. 12.19 Coal tailings settling pond with dredge to remove settled sludge.

(a) erosion control from coal stockpiles and slurry impoundment areas;
(b) erosion control for transport systems;
(c) repression of autogenous heating and the prevention of spontaneous combustion.

For erosion control from stockpiles, etc., the compound is diluted with water and sprayed on to the settled surface. Testing in a wind-test tunnel apparatus may be carried out to ascertain the appropriate compound and its dosage. For erosion control in handling coal, the compound must usually be sprayed in diluted form into a flowing stream in order to ensure adequate coverage. Special diluted forms of the compound, which facilitate economical usage, are available for this application.

The compound used to prevent spontaneous combustion usually contains a larger binding-agent component to encourage the rapid formation of a surface crust on the pile. This then becomes a suitably impervious air barrier, thereby sealing off the air flow necessary for autogenous heating to occur.

Solution strengths with these types of compounds range from 0.2–0.5% for (a) and (b) at 2–3 litre/m^2 coverage, to 2–5% for (c), also at 2–3 litre/m^2 coverage. The cost range is Can$0.10–0.30/t treated.

12.6 CASE STUDY

In order to demonstrate the impact of reagents on specific aspects of coal preparation, the following case-study is included.[21,22,25]

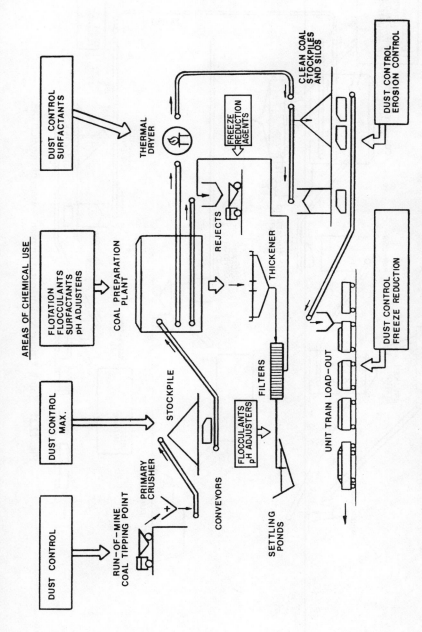

Fig. 12.20 Canadian coal-preparation plant—areas of chemical use.

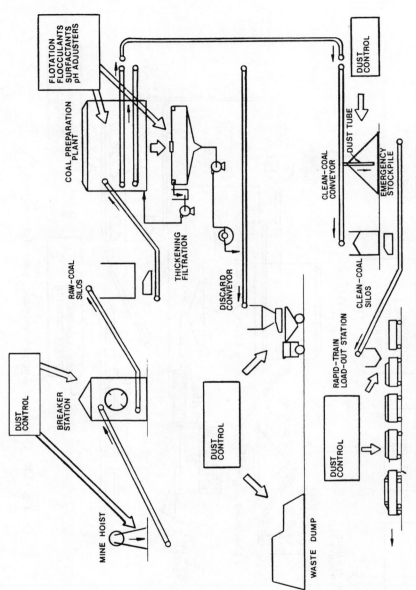

Fig. 12.21 South African coal-preparation plant—areas of chemical use.

TABLE 12.2 Reagent Costs for Canadian and South African Plants (Can¢/t Clean Coal)

	Canadian plant	South African plant
Dust control	15	35
Flotation	70	70
Flocculation	30	30
Acid neutralization	5	10
Surfactants	35	45
Freeze reduction	120	—
Erosion control	20	—
Others, including rust inhibitors, etc.	25	15
Total	320	205

Total Reagent Costs for a Western Canadian Operation and a South African Operation[26]

Figures 12.20 and 12.21 show coal-preparation plant flow-diagrams for a Canadian export metallurgical coking-coal producer and a South African export thermal-coal producer. Table 12.2 summarizes the chemical costs per tonne of clean coal, and the cost advantage for the South African plant is clearly the result of more favourable climatic conditions.

12.7 THE VALUE OF CHEMICALS TEST WORK

The use of chemicals in coal preparation is now of invaluable help and an economic necessity. In fact, without the use of chemicals it would be impossible to achieve today's environmental standards, process the increasing amounts of fine coal and still make a profit. Coal-preparation engineers should therefore be aware of all chemicals available to them, the potential areas of application and the methods of testing their effectiveness; and perhaps, most importantly, their economic benefits.

Many operators now use chemicals with almost addictive commitment, believing that chemical reagents are the answer to all problems.

Few select and test reagents objectively, comparing one with another and properly determining the benefits or disadvantages. Many instances occur where reagents are being used for one purpose that is counterproductive in other areas in the plant, e.g. flocculants used in thickening and filtering reduce froth-flotation performance when unused polymer solution is recycled to the froth cells. Indiscriminate use of chemicals should be avoided by careful evaluation, regular testing and re-evaluation, with other new alternatives being considered.

The chemicals bill for a large coal-mining operation can provide a strong argument in favour of a philosophy of spending money on recurrent re-evaluation of chemical usage rather than of 'saving' money by using more reagents.

REFERENCES

1. J. O. Glanville and L. H. Haley. Physical chemistry of frozen coal, *Mining Engineering* **34**, 182–186, 1982.
2. K. H. Nimerick, B. E. Scott and F. J. Beafore. Freeze conditioning frozen coal to ease handling and unloading problems, *Mining Engineering* **31**, 1380–1383, 1979.
3. L. Valentink and J. T. Patton. Rheological properties of heavy-media suspensions stabilized by polymers and betonites, *Transactions SME-AIME* **260**, 113–118, 1969.
4. F. F. Aplan, C. M. Bonner, W. C. Hiri and R. C. Rastogi. Recent advances in coal flotation, 2nd Annual Coal Preparation Conference, Lexington, Kentucky, April/May 1985, sponsored by Coal Mining and Processing, Inc.
5. D. J. Brown. Coal flotation, in *Froth Flotation*, 50th anniversary volume, AIME, New York, 1962, pp. 518–538.
6. W. Farley and M. Coebank. How chemically promoted flotation boosts fines recovery, *Coal Mining and Processing*, **15**, 48–51, 1979.
7. C. E. Capes, A. E. Smith and I. E. Puddington. Economic assessment of the application of oil agglomeration to coal preparation, *CIM Bulletin* (July 1984), 115–119.
8. I. E. Puddington and B. D. Sparks. Spherical agglomeration processes, *Mineral Science Engineering* **7** and **3**, pp. 282–287, 1975.
9. C. H. Cheh, R. W. Glass and R. Sehgal. Solvent recovery for the oil agglomeration coal-cleaning process, SME-AIME Annual Meeting, Dallas, Texas (1982).
10. G. R. Rigby, A. D. Thomas, C. U. Jones and D. E. Mainwaring. Slurry pipeline studies on the BPA-BHP 30 t/h demonstration plant, *Hydrotransport* **8** (BHRA), Johannesburg (August 25–27, 1982).
11. F. Townsend. Flotation reagent technology—the next decade, *Mine and Quarry*, **14**, 32–37, 1985.
12. D. G. Osborne. Flocculant behaviour with coal–shale slurries, *International Journal Mineral Processing* **2**, 243–260, 1974.
13. T. D. Wheelock. Status of chemical coal-cleaning processes, Fifth International Conference on Coal Research, Dusseldorf, West Germany (September 1980).
14. G. Y. Contos, I. F. Frankel and L. C. McCandless. Review of chemical coal-cleaning processes, EPA Report No. EPA-600/7-78-173a, Versar Inc. (1978).
15. H. A. Hamza. Flocculation of froth flotation boosts fines recovery, *CIM Bulletin* **72**, 48–51, 1979.
16. J. S. Laskowski and J. D. Miller. Reagents in the minerals industry, Conference IMM/CNRITM, Rome (September 1984).
17. K. H. Nimerick and B. E. Scott. New method of oxidized coal flotation, *Mining Congress Journal* **61**, 21–22 and 37, 1980.
18. K. H. Nimerick. Characterization of coals responding to froth conditioning, 1st International SME-AIME Fall Meeting. Honolulu, Hawaii, (September 1982).
19. M. J. Scanlon, P. V. Avotins, S. S. Wang and P. Strydon. Flotation promoters improve fine coal recovery, *World Coal* **9**, 54–56, 1983.
20. A. F. Paananen and W. A. Truscott. Factors influencing selective flocculation—desliming practice at the Tilden mine, *Mining Engineering* **32**, 1244–1247, 1980.
21. Z. Blaschke. Beneficiation of coal by selective flocculation, Proc. 7th International Coal Preparation Congress, Sydney, Australia, 1976, pp. 1–11.
22. D. G. Osborne. Coal fines dewatering, *Transactions Society Mining Engineers of AIME* **276**, 1843–1849, 1985.
23. D. G. Osborne and R. G. Smith. Reagents to save the day in coal preparation, 57th Colloid and Surface Science Symposium, Toronto (June 1983).
24. D. G. Osborne. Gravity thickening, Chapter 5 in *Solid–Liquid Separation*, (Ed.) L. Svarovsky, Butterworth, Guildford, 2nd edn, 1981, pp. 120–161.
25. D. G. Osborne. Vacuum filtration, Part 1, Chapter 13 in *Solid–Liquid Separation*, (Ed.) L. Svarovsky, Butterworth, Guildford, 2nd edn, 1981, pp. 321–357.
26. D. G. Osborne. Cost considerations in the use of reagents in coal preparation, *Process Economics International* **5**(2), 32–41, 1985.